一个“睡客”之路

住哪？4

区伟勤 著

中国建筑工业出版社

图书在版编目(CIP)数据

住哪？4/ 区伟勤 著.—北京：中国建筑工业出版社，2021.3
ISBN 978-7-112-25994-6

Ⅰ. ①住… Ⅱ. ①区… Ⅲ. ①室内装饰设计-中国-图集②散文集-中国-当代 Ⅳ. ①TU238-64②I267

中国版本图书馆CIP数据核字（2021）第046877号

责任编辑：杨 晓 唐 旭
责任校对：张 颖

住哪？4
区伟勤 著
*
中国建筑工业出版社出版、发行（北京海淀三里河路9号）
各地新华书店、建筑书店经销
恒美印务（广州）有限公司印刷
*
开本：889毫米×1194 毫米 1/20 印张：23 字数：770千字
2021年5月第一版 2021年5月第一次印刷
定价：138.00元
ISBN 978-7-112-25994-6
（37165）

本书

一个设计师住酒店的体验和手迹，用当时所住的每间客房内的信纸与笔记录平面及细节、心得及感受，简单地以全原稿、全手写的形式再现，借此倡导手画手写的“原始”思维方式和“笨拙”的学习方法，与读者一起寻找一条怎样住、住哪的线索！

About the Book

This book is the experience and handwritings of Grand O. He used the pens and paper in the hotels where he stayed to draw the planes and details of the hotels. The book presents all his manuscriptssimply in order to advocate the pure record by hand, the original and clumsy way of thinking and studying. Mr. O hopes to search for places where and how to live with readers.

出版说明

由于是即兴所作，手稿中难免存在字迹潦草、语句不通等不当之处。为保留作者原汁原味的推敲与记录，编辑仅对手稿部分进行了细微的调整，而在“文章注释”部分中，在原稿文字的基础上进行了加工整理。因此，本书文字内容以“文章注释”为准，所有照片除标注外，均来源自区伟勤先生在入住过程中的原创拍摄。特此说明。

For Your Reference

Because the manuscripts were improvised during Mr. O.'s travels, mistakes were, of course, inevitable. In order to maintain his original thoughts and records, the editor made only small adjustments to the handwriting, while processing them in the commentaries. Therefore, the texts in the book are subject to the commentaries. All the pictures were taken by Mr. O.during his stays in hotels, with the exception of those marked otherwise.

序

李仪

我与区伟勤先生只有一面之缘。

I had the honor to meet Mr. O Weiqin once.

虽只见过一面，但朋友圈常互动，常看到区先生的小诗，感叹区先生的热爱生活和灵感无涯；虽只见过一面，但有读过区先生已出版的3本《住哪？》，从设计师的眼光看酒店，有着丰富的想象与画面感，更有亲眼观察、亲身感受后的独到见解；

Nevertheless, we often communicate via Moments. From there, I read some poems from Mr.O. Thus, I appreciate his feelings for life and the infinite inspiration. I have finished reading 3 copies of his Where to Stay?. In his books, he shows us hotels with the eyes of a designer, with abundant imagination and picturesque feeling, with his unique opinion after the experience by himself.

虽只见过一面，但知道区先生要在三十年内出10本《住哪？》，贾岛说，“十年磨一剑，霜刃未曾试”。做一件事容易，用心做十年，做到极致却很难。而区先生，是三十年。

Mr. O plans to publish 10 "Where to Stay?" copies in 30 years. Jia Dao said, "Sharpening a sward for ten years, yet, I haven't tried its snowy edge". It is easy to do one thing. But to insist it for 10 years, and to do it to extreme, it must be difficult. Mr.O, he keeps doing something for 30 years.

长年在世界各地差旅途中的区先生，视野开阔、中西融合、格物致知。《住哪？4》区先生同样是以随笔形式描绘了最近旅程中的上百家酒店：从意式浪漫的西西里岛圣多梅尼科皇宫酒店到日式唯美的京都虹夕诺雅酒店，从美洲的阿文蒂诺凯悦拉霍亚酒店到澳洲的皇后镇俱乐部盛橡套房酒店，从中国第一家五星级酒店白天鹅宾馆到近年新的潮牌酒店主题酒店……再配上区先生手绘的房间平面图，相得益彰，别具风格。

Mr. O is always on the way to travel all over the world, all year around. His travels widen his view. His open creation ideas are combination of Chinese and western cultures, He studies so hard on something, from knowing it to its knowledge. In Where to Stay? 4, as usual, he described more than a hundred hotels during his recent trips, in the form of casual notes. These hotels vary from Italian romantic San Domenico Palace Hotel in Sicily to Japanese delicate Hoshinoya Hotel in Tokyo, from Hyatt Regency La Jolla at Aventine in America to Oaks Queenstown Club suites in Australia, from White Swan—the first Five-star hotel in China to some trendy or theme hotels which newly rose in recent years…They are illustrated with O.'s hand-drawn plans of the guestrooms, matched and unique.

这本书适合在学习时看，因为它是设计图册，可以给年轻设计师一些借鉴；这本书适合在旅行中看，因为它是酒店攻略，可以点亮我们的旅途；这本书适合在心情闲暇时看，因为又轻快又好读，随时拿起来都能够续得上，是没有负担的享受；这本书适合在心情烦闷时看，因为书中记录了作者的心得感悟，是一本有意思的私人日记，如同一位风趣的伙伴，带你走进一段美好时光，继而发现生活的小确幸。

You can read the book when studying, because it is a design diagram for the young designers; you can study it during travelling, as it is the guide book for hotels. It also lights up your way. You can learn it in your leisure time, for it is delightful and easy-going. You can pick it up or put it down anytime, anywhere. You can enjoy it with no burden; you can enjoy it when you are blue as well. The writer recorded his feelings in it. It is an interesting private diary like a funny companion who leads you to a wonderful time. And accordingly you may encounter a little happiness in your life.

凡心所向，素履所往。
一纸一笔，记录下了黑白分明；一部手机，留存下了色彩斑斓。
因为热爱，永不止步的前行者区先生又要踏上新的酒店设计之旅了，永远在创新，始终在路上。
Follow your heart, your steps will lead you to where you really want to go.
With his pen and paper, Mr .O. marks the black and white signs of age;
With his mobile phone, he keeps the colorful moments.
With his passion, he steps forward on his new, non-stop trip of hotel design.
Forever creative,
Always on the way.

祝福区先生，坚持本身就是一件很酷的事。
最后，也希望读者们有所收获，有所共鸣。
Give my best wishes to Mr. O. His insistence is really cool.
Finally, I hope the book will always have something for everyone, and can always touch the readers' hearts.

李仪 2020.10.13

李仪女士，“酒店高参”副总裁，豪华酒店的品论人，拥有多年的模特经验，曾为多家酒店、地产、珠宝、时尚杂志拍摄广告及宣传片，同时担任自媒体“玫瑰住在这里”的主理人，致力于酒店行业的发展。

Ms. LiYi, Vice President of HOTELN, is a commentator of luxury hotels. She has many years of experience as a model. She has shot advertisements and promotional videos for many hotels, real estate, jewelry and fashion magazines. She is also the manager of We Media "Rose Lives Here". Ms. Li always dedicates herself to the development of hotel industry.

彭长歆

区伟勤先生是我的大学同班同学，他的姓氏曾经让我大感新奇，我们都叫他阿勤。阿勤也是我的舍友，我们在编号为307的宿舍里共同生活了四年。我们都睡下铺，我的床与他的床隔着一条窄窄的过道，我相信他对酒店客房的兴趣从那个时候已经开始。阿勤是我们班最早进入室内设计领域的建筑师。我记得四年级的时候，他已经在外面或兼职或实习，回宿舍聊天时不时冒出一些室内装修的词汇。那时正值改革开放大潮席卷广东，广州作为商业之都有大量的装修业务，阿勤以其广州仔的精明和睿智义无反顾地投身其间，最终成为知名的室内设计师。他1997年发起创办韦格斯杨，其中“韦”即代表他。

Grand O. is my classmate in university. His family name made me surprised first. We all called him Qin. He was also my roommate and we had stayed in Room 307 for four years. Both of us stayed on the lower berth and separated by a narrow path. I believe he became interested in hotels at that time. Qin is one of the earliest designers who entered the field of interior design in our class. He began to take a part-time job or internship when he was a graduate. He often said some technical terms about interior design when we were chatting. At that time, the tide of Reform and Opening Policy swept through Guangdong Province. Guangzhou is the commercial capital. There were lots of decoration businesses.Qin, a son of the city, smart and wise, threw himself into the wave and eventually became a famous interior designer. He founded GrandGhostCanyon Designers Associates Ltd. Grand stands for Qin.

阿勤是一位极富才情的室内设计师。他的作品空间感强，有温度，反映了一位建筑学人对空间的执着，以及空间操作的人性化与在地化。他和他的团队拥有大量的建成作品，让我这位老同学、老舍友无比羡慕。他在朋友圈里还不时晒出他的“区体”诗——一种结合了作品感悟与产品推广、怡然自得的感情抒发。他的诗情意真切、毫不虚伪，在“网红”极度膨胀赚快钱的当下，阿勤三十年矢志不渝地热爱并坚持室内设计事业，堪称清流！

Qin is a brilliant interior designer. His works are full of strong spaciousness and consideration,they show his dedication to space as an architect, as well as the humanity and localization of the space operations. He and his team have finished so many works that I,his classmate and roommate, admire him so much. From time to time, Qin shows his O.style poems in Moments, which are his feelings about his work, product promotion and also his free emotion. His poems are true and unsophisticated. In these days, when Internet celebrity is exploding and making fast money, Qin loves and insists on the interior design for thirty years, he is truly a pure stream.

回到阿勤嘱我写序这件事。我很早就获赠并阅读过阿勤的《住哪？》系列图书。我知道，阿勤经常参加徒步穿越这类苦逼的运动，他的公司里不时会举办赛图会，即通过内部快题竞赛、点评提升员工的设计素养，所以他是用一种水滴石穿的精神磨炼自己并影响自己的员工。他每到一处酒店即测绘酒店房间的做法坚实了我的这一看法。我能想到的就是他对这个行业的热爱，他愿意在繁重的工作之后，拖着疲惫的身躯做着我们建筑学一、二年级学生做的尺度及空间体验训练，还把自己的感受书写出来，这不是热爱是什么？！《住哪？》既有空间，又有温度，每每阅读都感到代入感很强，可以想象作者如何在房间里面观察和思考，实在妙不可言！

Let's get back to my foreward for Where to Stay?. I have these series from Qin as a gift and read them with honor. From these books, I learnt that Qin often took part in some hard activities like hiking through the desert. Design events are also held regularly in his company in order to improve the design quality via competitions and comments among the staff.Therefore, Qin keeps tempering himself and also effects his workmates with the spirit of dropping of water outwear the stone. I am convinced by his persistent practice which he measures and draws the guestrooms when and wherever he stays.I can feel his love for design. Though he is tired after hard work, he keeps the scale and spatial experience training which is often taken by freshmen and juniors. He also keeps writing a series of books to record his feelings. What is it if it isn't love?!Where to Stay? has not only imagination but also humanization. I often have the sense of immersion while reading. I can imagine how the writer observes and thinks in the guestrooms. How amazing!

最后，祝愿阿勤的《住哪?》系列一直写下去！

Finally, I hope Qin will keep writing the series of Where to Stay?.

彭长歆.
2020.9.30

彭长歆博士，华南理工大学建筑学院副院长、知名教授，致力于研究中国近现代建筑、地域性建筑设计与理论、建筑遗产保护、历史环境保护与再生设计等，是岭南学派的重要实践者，以及遗产保护与乡建权威，代表著作有《现代性・地方性——岭南城市与建筑的近代转型》等。

Dr.Peng Changxin, Vice President of the South China University of Technology ,School of Architecture, a well-known professor. Dr. Peng dedicates to the study of modern buildings, regional architecture design and theory, architectural heritage protection, historical environment protection and regeneration design and so on. He is an important practitioner of the Lingnan School, as well as the heritage protection and the authority of township construction. His representative works include: Modernity. Regionality—— the Modern Transition of City and Construction in South China.

吕邵苍

一千个人读红楼，会有一千个红楼故事。

Honglou Meng gives its readers a thousand stories.

(P.S. Honglou Meng is one of the four famous ancient novels in China, written in Qing Dynasty by Cao Xueqin.)

某天，忽接到广州区大设计师来电，邀我写个序什么的。我这种“不正经”中老年设计师能写什么序啊。没办法，都是设计师，设计也是江湖啊，先答应着呗。最好区帅哥能来院子，品茶、喝酒、聊设计、聊生活、聊院子，这场景应该不错的……

One day, I received an unexpected call from the design master of GZ. He asked me to write something like a foreword. I wondered how I could write it, for I am a mid-to-old designer. I had no choice but to say yes. Why? We are both designers, and our area is also Jianghu.

It would be amazing if brother O. comes to my little courtyard, to have some tea, or a drink. We will talk about design, about life and about my court.

没曾想，不几天，《住哪？4》样本就寄到，场景如梦幻泡影，正（已）消散……

I had not expected that I could receive the sample of Where to Stay? 4 within just a few days. My dream disappeared like a bubble…

四本《住哪？》，不是一序能写清楚的吧？一如现在……

我想一序亦表达不了，读《住哪？》的各种心思与情绪……

There will never be one single foreword for the whole four editions of Where to Stay?, nor to express my feelings and thoughts when reading them.

其中的究竟不就是人与人，人与物，人与事的万千写照吗？

Where to Stay? is a documentary diverse from people to people, someone to something, a man to a matter.

如不经历，能乐乎，能品乎，能谈乎，能悟乎，能读乎，能思否，能否停留片刻？

If you have never experienced, can you be happy with it? Enjoy it? Talk about it? Understand it? Read it?Think about it or even stay for a moment for it?

粗看漫不经心地涂涂画画，潦草之极。

心里道来：这也能成为书？亦可以出书？

殊不知，这世道能做很多可为之事，等你下手做成，那是隔着十万八千里呢。

At first glance, the book is just like something with casual paintings and drawings. How illegible!
I thought to myself, how could something like this become a book and be published?
We are even unaware that there are a lot of things that can be done. However, there is a long way between your action and your goal.

设计师的工作看似美好，确是美好。无数个美好的背后是无数个这样的草图与经验、经历、甲方、同行、世道，与自己博弈，最终才能成其设，呈其好。如此书，浸透着时间、阅历、经历、陪伴、笔头、功夫……台上一分钟，台下十年功，大概就是如此吧！

The designers' work looks great, and it really is. While a great deal of amazing work is made up of these sketches, experience, Party A, companions, the world and the fights with oneself. Finally, it made the great design and showed the wonderful work. A book like this is full of time, observation, experience, companion, drawing and practice. The old saying "a minute on stage deserves ten years of practice" is probably that.

感谢区帅哥的邀序，看到了设计同行的另外一个天地，只是惭愧得很，从受邀至今，已二月有余。提笔、下笔、停下、再提笔、下笔又停下，至今时，终成。

Thanks for Mr. O.'s invitation. I could see another view of design. I just feel ashamed that it has been more than two months since he invited me. I picked up the pen, put it down and stop to think, again and again till now, I made it.

期区帅哥什么时候不只是拆遍世间酒店，亦可住上一回院子，拆哪个院子，院子又可再生……

I hope Mr. O. will not only stay at hotels around the world, but also try to stay in my "Courts". Renovating courts can give them new energy and make them more attractive.

时间虽不语，却回答了所有问题。

Time answers all the questions without a word.

吕邵苍
于2020.10.22

吕邵苍，一个爱折腾的非著名设计师，一个中老年跨界的二次创业者，北大光华管理学院EMBA、意大利米兰理工室内硕士，担任多家学院客座教授及硕士生专业指导老师，是"中国院子·自在生活方式"的发起人和倡导者，创建吕邵苍设计事务所，为"云隐东方·院"HOTEL的创始人和产品总设。作品多次荣获国内国际顶尖赛事大奖，代表作有：吕邵苍本人，"云隐东方·院"系列作品。

Mr. Lv Shaocang,an unrestrained and infamous designer, and a senior crossover double entrepreneur.He has got EMBA from Guanghua School of Management of Peking University and as well as a master of Interior Design from Polytechnic University of Milan, Italy.Mr. Lv is a visiting professor and master supervisor in many colleges. He is the founder and pioneer of Chinese Courts and Free life style. He founded Lv Shaocang Architectural Design Firm, the Oriental Courtyard in the Clouds and holds a post of the Chief Product Designer. His works won the top domestic and foreign awards for many times. His representative works include: Lv Shaocang himself, and the Oriental Courtyard in the clouds .

何津

区先生是一名非常优秀的设计师，也是一位很棒的合作伙伴。

Mr. O. is a very excellent designer as well as a perfect partner.

专业、用心、别裁的酒店设计往往可以给予一间酒店无限的想象和赋能。作为一名酒店业主，也是一名资深酒店住客，我对酒店业一直都有着独特的感情，我会要求我们酒店的设计方必须重视住客体验，注重细节。与区先生的合作亦是如此！

Atechnical,attentive and unique design always gives a hotel infinite imagination,also make it attractive and dynamic. As a hotel owner and a senior guest, I have a deep and special feeling for hotels. I require the designers must put the emphasis on the guests' experience and details. So does the cooperation with Mr. O.

回到区先生的《住哪？》一书，区先生以流水日记方式述说及分享其在世界各地的住店经历，原始的手稿、简单的文字带着读者进入一名资深设计师视角的“睡客”之旅，很棒！

Let's come back to Where to Stay?, Mr. O. describes his travelling and share his experiences of staying in hotels from all over the world in the form of diaries. His original manuscripts and simple words lead the readers to a trip of sleeper, which is in a senior designer's view. Terrific!

期待区先生更优秀的作品，不论设计、文学或者人生！

I am looking forward to more excellent works by Mr. O., whether designs, literature, life or anything.

于2020年10月16日顺德乐从

From Lecong Shunde , October 16, 2020

何津先生，广东佛奥集团董事长，威珀斯酒店创始人，广州、三亚、佛山资深商业运营商。

Mr. He Jin is the chairman of Guangdong FOAO Group, the founder of VAPERSE Hotel and also the senior commercial operator in Guangzhou, Sanya and Foshan.

学酒的人：

[illegible]《[illegible]?》[illegible]“学酒的人”[illegible]，[illegible]，[illegible]。[illegible]，[illegible]。

Lily [illegible]，《[illegible]》[illegible]，[illegible]，[illegible]《[illegible]》[illegible]；[illegible]博士，[illegible]大学同窗[illegible]，[illegible]人士，[illegible]；[illegible]，[illegible]——[illegible]》；[illegible]，[illegible]，[illegible]，[illegible]（[illegible]酒店）

[illegible]，[illegible]；[illegible]不同，[illegible]，[illegible]；[illegible]，[illegible]、[illegible]、[illegible]，[illegible]，《[illegible]?》[illegible]

[illegible]学[illegible]！[illegible]

[illegible]
[illegible]

Those Who Write the Foreword

写序的人

浓缩的都是精华。每一本的《住哪？》邀请“写序的人”也是费煞思量，有设计师朋友、酒店业主或酒店管理者、媒体人，而这一次也尝试请网红达人来捧捧场。

Every concentration is the essence. Inviting the people who write the foreword for every Where to Stay?was serious and thoughtful, including designer friends, hotel owners or manager, media people. This time, I try to invite Internet celebrity.

Lily李仪小姐，“酒店高参”的高层，网红，自媒体“玫瑰住在这里”的主角，自信；彭长歆博士，我的大学室友，现在是华南理工大学建筑学院的技术权威人士，乡建的先驱者之一；吕邵苍设计大师，拥有自己的酒店品牌——“云隐东方・院”；何津先生，地产业界老前辈，商业屡创奇迹，酒店出手更是不凡，正在合作的老朋友（威珀斯酒店）。

Miss Lily Liyi, the senior manager of HOTELN, as well as an Internet celebrity. She is the protagonist of We Media “Rose Lives Here,”a pretty lady with confidence; Dr. Peng Changxin, my college roommate,is a technical authority in School of Architecture, the South China University of Technology. Now he is one of the pioneers of rural construction; LvShaocang, a design master, has his own hotel brand—the Oriental Courtyard in the Clouds; Mr. He Jin is the senior of the real estates. He has made many miracles in business and is outstanding in running hotels. We are friends who are working together (Wipers Hotel).

大行业一致，有共知共识，细分不同，有碰撞、互补、互动，也许酒店业就是靠大家共同努力，共建，共生，共赢。

The big industry is consistent with common knowledge and consensus but different segments. We have collision, complementation and interaction. Perhaps the hotel industry relies on our joint efforts, joint construction,symbiosis and win-win cooperation.

小小积累，《住哪？》有你们，荣光！

It is my little accumulation, also my glory to be with you, and with Where to Stay?.

遐思录（二）

28/6，2020

有机缘，这回"录"又有专辑了。第一，也许会使得了感觉同类的酒店也都类同；不同类型的酒店（餐饮、度假、商务或城市）也相互借鉴，能让你眼前一亮的感觉越来越少了。那，是什么原因[illegible]去呢？还不用说汇集成书了！

针对断"录"，会回看出来了很久；闲时慢慢给了一些"闲书"，特别是[illegible]的老[illegible]，特别有意思（"[illegible]酒不醉"）。我们的去入30年，十本，80字，80句（左右）之[illegible]，那不是也很有意思吗？就这样吧：十本，30年，这十（相信）可以让我不再退，也可以有更加好的[illegible]情和故事，[illegible]（不过不是我也会[illegible]了）。

一些[illegible]，[illegible]一个人的[illegible]、[illegible]、[illegible]、[illegible]的日子，用眼耳心去[illegible]，[illegible]酒店的一日房[illegible]，[illegible]的心情[illegible]，[illegible]那就[illegible]吧，[illegible]！（[illegible]）

Indecision (Part Two)
退堂鼓（二）

有几个点，让这个“鼓”又打起来了。第一，也许是住多了，感觉同类的酒店趋于类同；不同类型的酒店（商务、度假、亲子式等）也相互借鉴，能让你“眼前一亮”的越来越少了。那，怎么才能写下去啊，更不用说汇集成书了！

For some reasons, I play the "drum" again. First of all, maybe I have stayed in too many hotels, I have the same feeling about the similar ones. Different kinds of hotels can be learnt from each other.Fewer and fewer hotels will surprise me. So how to keep writing, needless to publish books with my manuscripts!

针对这个“鼓”，是退是进，想了很久。闲时喜欢看看一些“闲书”，特别是主张七字真言的老者蔡澜先生的杂文（“抽烟，喝酒，不运动”），特别有趣。突然想，我可以：30年，十本，80岁，800间（左右）的五星级酒店，那不是也很是有趣吗？就这样吧：十本，30年。这个小目标可以让“鼓”不再退，也可以有更加好的激情和持续。当然，我会有“好东西”呈现给你们，我身边的朋友、同行和年轻人。（不知不觉我也是行家了。）

For this "drum", I wonder whether to go ahead or give up for long. I like some light reading in my spare time, especially the books written by Cai Lan who claims his seven-word mantra "smoking, drinking , no exercise". How interesting! Suddenly I think I can finish ten books in thirty years till my eighty years old, stay in around 800 five-star hotels, will it be interesting?In that case, ten books, thirty years. This small goal makes me go ahead and also have more passion to stick on. Of course, I will show some wonderful things to my friends, my colleagues and the youth. (I suddenly find that I become a professor.)

也许，传统是要有人去传承的，包括这样简单地记录一个人的吃吃、喝喝、走走、看看、玩玩、睡睡的日子，用眼、耳、口、鼻、手去耗费时光，包括出书。酒店的一间房最简单，让复杂的心情停留在一间酒店的房问里面。住哪？那就住那吧！“鼓”，动着！

我还是喜欢手捧香的书。

Maybe, traditions should be passed on by someone, including recording someone's days of eating, drinking, traveling, watching, playing and sleeping simply, spending some time with eyes, ears, mouth, nose and hands, as well as publishing books. A guestroom in the hotel is the most simple. So leave the complicatied feeling in it. Where to stay? Just stay there! My drum is playing now!

I prefer books in hands.

生命时刻

从第一本《陪你哪？》开始就感恩出书人的不离不弃，仍以"入职数码时代"的一路走来出版，到第二、第三本都是在本书的"生命时刻"来分段式完成；

但在这序言完成之前我关注的仍得给你们最深的内容，在行程上完成书的文字或插图的心得，旅游路上完成每天的游记，或记录"写生"，回国后出版时快速地把它完成"日记"，让大面积插图本身，也就变得容易了（心情也轻松）许多）

别人练剑的时候，我用一点点时间；别人游泳的时候，我用一点点时间；别人发呆的时候，我用一点点时间……这本手稿也就在我多年不同时候、笔，也不断地积累——还有很多从旅途带回的字稿和涂鸦——这也就出书了啊；第一次出版我们全人家关于手稿的第一句话，第一本还有心"手稿"的开了一页，后来觉得"真实"，那就上手稿吧。有路上生命时刻完成的手稿，让《陪你哪？》变成了一个"有故事"在路上的记录，"生命时刻"也就有了意义，当然也就会了；感谢的，你们我的，我本的"幸福"了，生命就在，我们会在……多么美好，管它呢！随手，随笔，随心！

—2020

Spare Time
业余时间

从第一本《住哪？》开始就感叹出书人的不容易，佩服。我也慢慢地从“积重难返”式的一口气的出版方式到第二、第三本尽量在平时的“业余时间”来分散式完成。

From the first Where to stay?,I admired that writers were not easy. At the beginning, I finished the first book without any stop, I felt difficult and pressed, and then the second and the third books were separately finished in my spare time.

住店，退房前完成平面及关注或值得手绘记录下来的内容，在行程中完成本店的文字或相关的心得；旅游路上完成每天的游记，或记录“写点”，回国后就可以快捷准确地完成“回忆”。让大面团搓成小条条，写作也就变得容易了。（心情也轻松了许多。）

I usually finished drawing the plane and some special hand-drawn works before checking out when I stayed in hotels, wrote down my views about hotels on the way; I finished my everyday journal or marked the "writing points" while traveling. When I returned home, I could remember all my experiences quickly and correctly. It seemed easier to write my books in this way and also made me relaxed.

别人看剧的时候，我用一点点时间；别人打游戏的时候，我用一点点时间；别人发呆的时候，我用一点点时间……于是，手稿也就各式各样，不同的纸、笔和抖动程度，更有修改、涂鸦的痕迹——这怎么出书啊？这是中国建筑工业出版社的编辑第一次拿到手稿时的第一句话。第一本还“专心专意”地抄正一次，后来觉得“真实最重要”，那就上手稿吧。那些路上，业余时间完成的手稿，让《住哪？》真正成为一个专业的“睡客”在路上的记录。“业余时间”也就相当充实，当然也失去了看剧的、打游戏的、发呆的“幸福”了。业余就好，没有压力，管它呢！

I spent a little time in recording my experiences and feelings, while others were watching TV, playing games and having daydreams… So my manuscripts were all in different sizes, different paper, different pens and shaking writings with corrections and graffiti. How could these manuscripts be published? It was the first sentence when the editor in China Architecture & Building Press saw them for the first time. I copied them carefully for my first book. Later, I thought truth was the most important. So I decided to get my manuscripts published.The manuscripts were finished in my spare time and on my journey. They made Where to stay? become a journal by a professional sleeper. My spare time was full and rich; of course I lost the "happiness"of watching TV, playing games and having daydreams. Spare time was so good that I didn't have any pressure. Let it be!

随手、随笔、随心！

I recorded it with my hand, my pen and my heart!

Contents

目　录

Contents

目 录

Contents

目　录

意大利西西里游记（2017年4月7~20日）

The Trip to Sicily, Italy

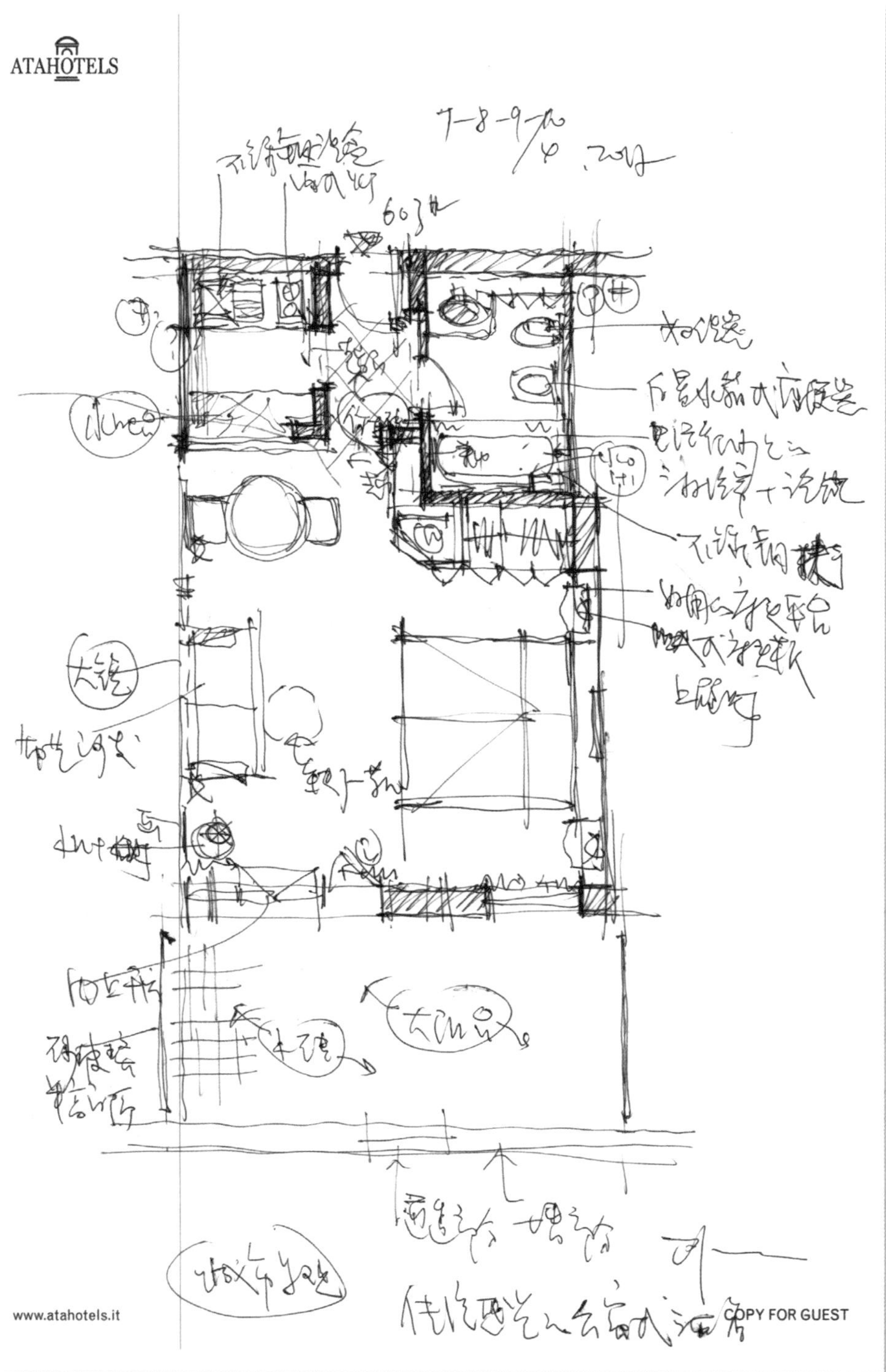

ATAHOTELS

阿塔酒店

★★★★

ATAHOTELS
MILAN ITALY

Address : Sede Legale Via Gioacchino Murat,23,20159 Milan,Italy
Telephone : +39 02 895261
Http : //www.atahotels.it

ATAHOTELS

9/4 晚
2017 9:10

到米兰的第一晚。

[illegible]

“[illegible]”[illegible]！

[illegible]

Tamerlane [illegible] 酒店 Atahotels

www.atahotels.it

ATAHOTELS

[illegible] 2016 [illegible]

[illegible]

[illegible] 米兰 [illegible]

[illegible]

① [illegible]

[illegible]

www.atahotels.it

ATAHOTELS

的，专心于一些新的创意，[illegible]

[illegible]

[illegible]创意。

② Flos 的总经理，[illegible]

[illegible]

[illegible]

[illegible]

[illegible]

Smart [illegible]

[illegible]

[illegible]！

③ 如果[illegible]

[illegible]

[illegible]。

www.atahotels.it

ATAHOTELS

④ 不一样的

晚上[illegible] Fataly 一家

[illegible]

[illegible]！[illegible]

[illegible]十二点。

[illegible]

[illegible]

[illegible]

[illegible]。

吃了一顿，[illegible]

[illegible]

[illegible]

[illegible]

今天[illegible]！

[illegible]

www.atahotels.it

At 9a.m., April 9, 2017 On the Third Morning in Milan

2017年4月9日上午9:00 到米兰的第三天上午

阿塔酒店
Atahotels, Milan, Italy

4月7日凌晨从广州白云机场起飞，很久没有让自己出远门，有点小紧张。中文的白云机场还是简简单单就可以了，可能大家“出国”已经是习以为常了，谢谢我们的同事协助这次的旅行，当然还有我们的组织方：朗道公司，牛！

We took off from Guangzhou Baiyun Airport on the early morning of April 7. I felt a little nervous because I hadn't taken a long journey for long. It was convenient to check in at Guangzhou BaiyunAirport. Maybe people get used to going abroad. Thanks our colleagues for helping with this trip as well as our organizer. Langdao Company, so great!

“红酒之旅”开始了！

Our "red wine trip" began!

航班是在卡塔尔的多哈经停，9小时+经停2小时+7小时，共18小时的单程飞行时间。途中，在飞机上还不错，看电影，看书，看自带的报纸，晕睡，乱吃，也就不太难熬地到达了米兰。

We transferred at Doha, the capital of Qatar, 9 hours+2 hours+ 7 hours, OMG! We spent 18 hours on a one-way plane trip. We watched movies, read books and newspapers, slept and ate all the way, so it wasn't a tough time. Finally, we arrived at Milan.

乘坐出租车回到公寓式的阿塔酒店，楼顶层的早餐更是让人倾心，我们可以在此欣赏米兰城市的风景。还是挺亲切的，我们2016创客公寓应当像这样就好了——亲和、温馨、轻松、自然。

We took a taxi to the apartment-style hotel——Atahotel. We had breakfast on the top of the building, there we could enjoy the view of Milan. We felt excited. I thought our Creator Apartment 2016 should be like it—heartily, warm, relaxing and comfortable.

4月8日一早去看家具展，收获不错。谢谢同学老菜（蔡）的指引，看一些重点的馆，收获不少。其中：

1. 大牌就是大牌，Poliform，占地大，内容多，抓眼球，设计新产品，有引领全场的领导性，随便你怎么拍，反正都是高成本、高技术的产品，你不能轻易地模仿。牛！有自信。系列式的设计也是值得我们去学习的，专心于一套“新创意”，可以深化衍生出一系列的产品，事半功倍——前提是有它的“主”创意。

We went to visit the Milan Furniture Fair on the morning of April 8. Thanks to my classmate, Mr. Cai's guidance, we could see some important pavilions and learnt a lot from them including the followings:

1. Famous brands were really famous, Poliform occupied a large area with rich content, attracted our eyes, new designed products showed the leading of all. You could take photos freely as you liked because you could never copy these high-cost and high-tech products easily. They were so cool with confidence! A series of designs were worth studying. They concentrated on a set of new ideas to derive a series of products. If you got the main idea first, you would get twice the result with half the effort.

2. FLOS，灯光公司，绝对将同行抛离“几条街”，产品多，名师合作多，场面宏大，科技投入多（表面的和内在的都相当震撼），其中可控滑动式的灯具（轨道灯）有趣，Smart，会思考的灯光，大自然式的展示，将户外灯光表达得淋漓尽致！

2. FLOS, a lighting company, absolutely left the colleagues far behind. It had a great many products, and cooperated with many famous designers. It was also a grand occasion with high technology (We were all shocked by surface and internal) .The controllable sliding light with track was funny and smart. The thoughtful light and the natural show made the best use of outdoor lights.

3. 办公家具也流行立式与移动式，看到了一家不错的制造商。

3. vertical and removable office furniture was popular, and we saw a good manufacture.

4. 不一一尽达。

4. I couldn't tell all in detail.

晚上与老菜私聚Eataly，一家超市与餐厅结合的时尚店，让我们一睹什么是人气！据说排队一直到晚上十二点！

I had an appointment with Mr. Cai in Eataly in the evening. It was a new style shop which was mixed a supermarket and a restaurant. We saw the popularity with our own eyes. It was said that people lined up till midnight. So cool!

模式最重要，有好的模式就不愁没有生意做，这个在什么行业都是行得通的，佩服，要关注，更要学习。

Pattern is the most important. A good pattern is the guarantee of business. It does work in every industry. I admiredt his pattern. I would focus on it and also learned from it.

吃了一顿，然后到了一层吃了一个大雪糕，很甜，糖分肯定严重超标呢。

After the dinner, I had a big ice cream on the first floor, too sweet. I was sure the sugar was over the limit badly.

就一次，在米兰真好！

Only this time, it was good in Milan.

今天继续去看看家具展！

I continued to visit the furniture fair today.

用心投入，创意领先，产品超群，吸引了全世界的目光，有奢华古典的，更有时尚如Kartell的透明树脂产品，亦有科技与自然相结合的Flos灯具。大牌云集，如Baxter，永远不会让你失望。

All the thoughtful input, the leading creation and the extraordinary products attract the eyes all over the world. There are something classic, fashionable acrylic products like Kartell, and Flos lighting, with the combination of technology and nature. Ever more, brands flock, such as Baxter. It will never let you down.

O.'s File

（区生词典）

米兰国际家具展：米兰国际家具展自1961年开始举办，展会包括米兰国际家具展、米兰国际灯具展、米兰国际家具半成品及配件展、卫星沙龙展等系列展览。

米兰国际家具展被誉为世界家具设计及展示的“奥斯卡”与“奥林匹克”盛会，并扩展至建筑、家纺、灯具等家居领域，每年举办一次，是全世界家居设计者与产业相关者聚集、交流的圣地。（百度百科）

Milan International Furniture Fair has been held since 1961. It includes Milan International Furniture Exhibition, Milan International Lightning Exhibition, Milan International Furniture Semi-Finished Products and Accessories Exhibition, Satellite Salon Exhibition and other series of exhibitions.

Milan International Furniture Fair is known as"Oscar" and "Olympic"of furniture design and exhibition in the world. It has also been extended into the furniture field like architecture, home textiles and lightings. It is held once a year. Milan International Furniture Fair becomes the sacred place for gathering and communication of furniture designers and industrial stakeholders all over the world. (From Baidu Encyclopedia)

2017.04.09米兰家具展
Milan International Furniture Fair

ATAHOTELS

www.atahotels.it

ATAHOTELS

www.atahotels.it

At11p.m.,April 9,2017
2017年4月9日晚上11:00

继续看家具展Ferrari Millano。轻轻松松的早餐：面包、水果、两个鸡蛋和三杯美式咖啡，大太阳，相当的惬意！

I continued to visit Ferrari Millano. First I had a leisurely breakfast: bread, fruit, two eggs and three cups of Americano, the very big sun. How relaxed!

十一点多才到达展会，没有像昨天那样被“敲诈”90欧元，心情特别不一样，还是主要看看16号、20号馆的“大牌行动”。

I got to the fair at around 11 a.m. I felt pretty good because I was not “blackmailed” 90 Euros like yesterday. I wanted to see No.16 and No.20 pavilions: Famous Brands.

非常震撼和吸引眼球，一系列的设计目不暇接，大品牌之间的合作。Ferrari跑车的内饰设计成就这个品牌的再上一层楼，这一切的领先感觉是金钱起了很大作用，让其他品牌的产品难以超越。当然其他的也偶有不错的，其中对一个做户外泳池边家具的印象颇深，大面积的镜，巧妙地用半圆倒影成圆，半吊的植物有趣而有创意，牛！

A series of designs were so shock and attractive that I could hardly take them all in with my eyes, especially the cooperation among the famous brand companies. The interior designs of Ferrari Sports Car made this brand go further to another level. I think money played such an essential role that other brands with common things could't over take it.Of course, other brands had good designs occasionally. I was deeply impressed on a company which designed the poolside furniture. A very big mirror was made up of two semi circles. Half-hanged plants were interesting and creative. Phenomenal!

反观国内的某一些设计师参与的设计场内、场外展——势单力薄，拍马也追不上，起码有50年的差距——努力，自勉！

Looked backward at some exhibitions including indoors and outdoors, which some of Chinese designers were involved in them, were too weak to catch up , at least 50 years behind—work hard, encourage ourselves!

明天一早飞西西里，红酒之游开始了。

The next morning we would fly to Sicily and began our Red Wine Trip!

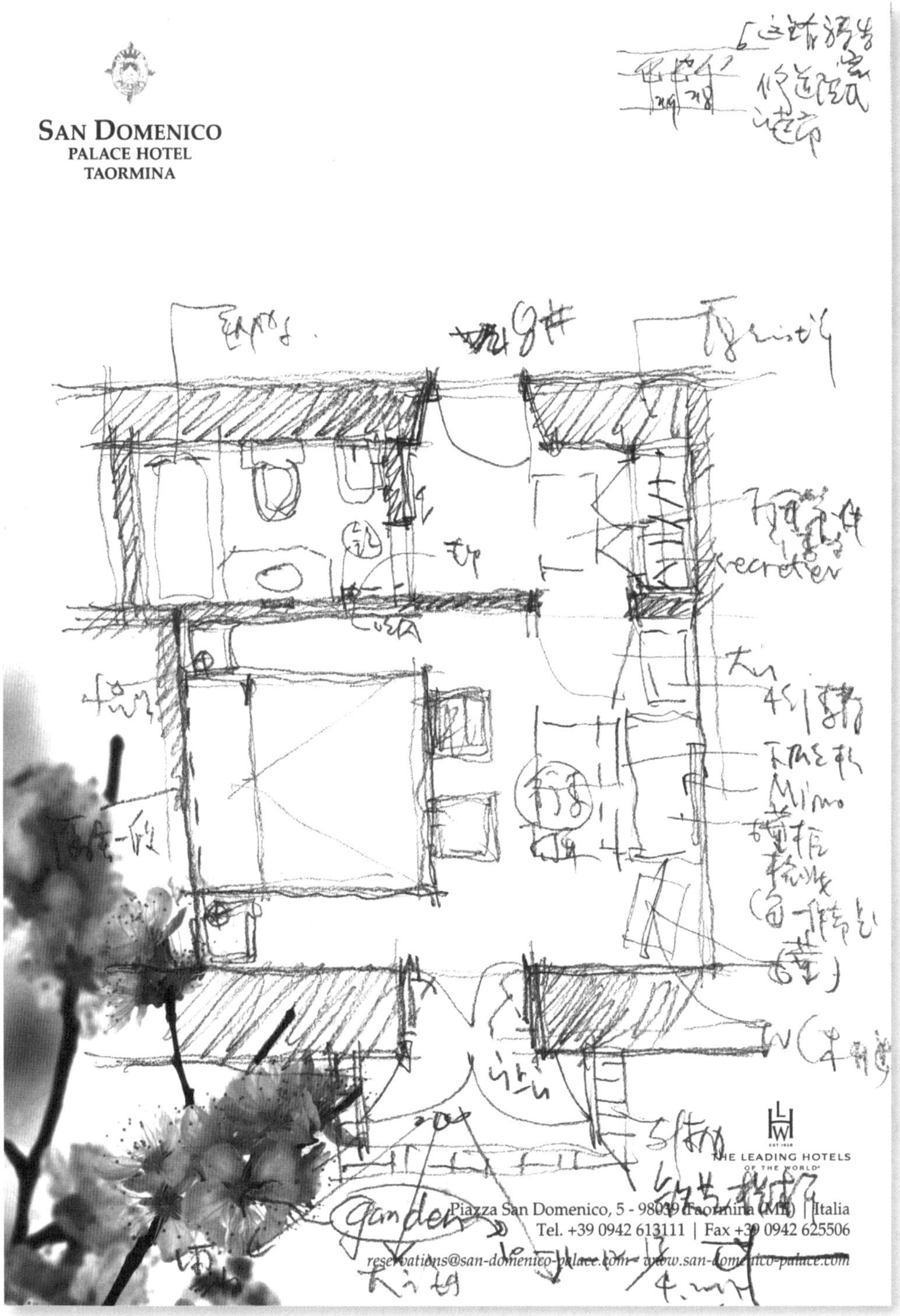

SAN DOMENICO
PALACE HOTEL
TAORMINA

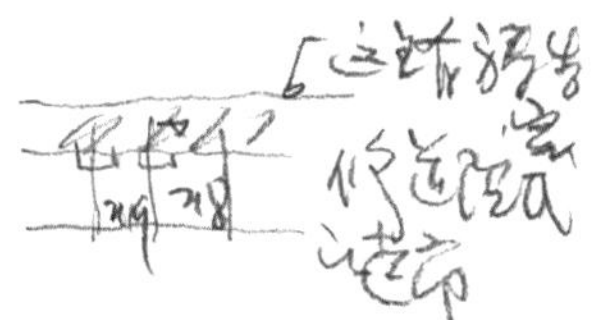

THE LEADING HOTELS OF THE WORLD

Piazza San Domenico, 5 - 98039 Taormina (ME) | Italia
Tel. +39 0942 613111 | Fax +39 0942 625506
reservations@san-domenico-palace.com - www.san-domenico-palace.com

SAN DOMENICO
PALACE HOTEL
TAORMINA

圣多梅尼科皇宫酒店

2

★★★★

SAN DOMENICO PALACE HOTEL TAORMINA ITALY

Address : Piazza San Domenico, 5-98039 Taormina(ME),Italy
Telephone : +39 0942 613111
Tel : +39 0942 625506
Http : //www.san-domenico-palace.com
E-mail : reservations@san-domenico-palace.com

圣多梅尼科皇宫酒店
San Domenico Palace Hotel, Taormina, Italy

圣多梅尼科皇宫酒店

San Domenico Palace Hotel, Taormina, Italy

SAN DOMENICO
PALACE HOTEL
TAORMINA

THE LEADING HOTELS OF THE WORLD

Piazza San Domenico, 5 - 98039 Taormina (ME) | Italia
Tel. +39 0942 613111 | Fax +39 0942 625506
reservations@san-domenico-palace.com - www.san-domenico-palace.com

SAN DOMENICO
PALACE HOTEL
TAORMINA

THE LEADING HOTELS OF THE WORLD

Piazza San Domenico, 5 - 98039 Taormina (ME) | Italia
Tel. +39 0942 613111 | Fax +39 0942 625506
reservations@san-domenico-palace.com - www.san-domenico-palace.com

2017年4月11日

April 11, 2017

圣多梅尼科皇宫酒店

San Domenico Palace Hotel, Taormina

米兰，一早去机场，两个小时。

Milan, It took us two hours to get to the airport in the early morning.

4月10日，我们终于上岛了。感谢我们的队友帅哥提前7个月订了这一家这么“离谱”的，在西西里岛是最高的酒店，有像“悬崖上的金鱼公主”一样的历史。建于1896年的Grand Hotel翼楼的一层是15世纪时期的修道院，牛！

Finally, we landed on the island on April 10. Thanks to our teammate—a handsome guy, he made the hotel reservation seven months in advance. It was too "ridiculous"! The hotel was the tallest one in Sicily just like the history of the Goldfish Pricess on Cliff. I just checked. Grand Hotel was built in 1896, the first floor of the Garden Wing was once a monastery in the fifteenth century. How amazing!

圣多梅尼科皇宫酒店

San Domenico Palace Hotel, Taormina, Italy

什么是人生最珍贵的？时间啊！历史是一种记录，也是一种价值的简单体现。入口、室内、藏品、内庭园、修道院的客房，海景、火山、日落、日出；这里是让人们不得不发懒的地方。早餐、倚海、晒背，可以慢慢地享受到中午，写写画画，还是难得的工作好环境！

What is the most precious in our life? Time! History is a kind of record as well as a simple embodiment of value. Entrance, Interior, Collections, Inner yard, guestrooms in the monastery; seascape, volcano, sunset, sunrise. It was a relaxing place that made people lazy. We had breakfast by the sea, felt the sunshine on my back. We enjoyed the leisure time till the noon. I wrote and drew. What a wonderful environment for work!

昨天下午和晚上逛逛小城市，Bontiqe+大冰棒+海鲜大餐，吃了近三个小时，大家都撑死了，累死了！

We walked around the little city at a leisurely pace yesterday afternoon and evening. Bonitque ,Ice bars and a nice seafood dinner. We spent nearly three hours enjoying the big meal. We were so stuffed that we felt tired!

好生活，时间造，“造孽”啊！

Good life is made by time. It is a sin!

A Long-Time Big Meal
浪"慢"意"大"利餐

早餐后2017年4月12日
April 12, 2017, after breakfast
太阳下，大海边
By the sea, in the sun
有伞未为黑
Never get a suntan because of umbrellas

圣多梅尼科皇宫酒店
San Domenico Palace Hotel, Taormina, Italy

吃饭从来都是"旅行团"的话题和问题，这次也不例外，但我们这个是"土豪团"——不土又豪的美女帅哥团，吃饭，是另外一个问题，究竟是什么呢？

Having meals was always the topic and problem of tour groups. This time was no exception. But our trip was a rich one. There were lots of rich beautiful ladies and handsome gentlemen with good taste. Having meals was another problem. What is it on earth?

意大利晚餐从8点开始，到晚上10点多、11点，对于美女多多、帅哥少少的我们来说，难以坚持，失衡，如果都是男亲女爱的"撑台脚"，那恐怕就嫌时间太短了，哈哈！

The Italian dinner began at eight o'clock and lasted till ten or eleven p.m. It was difficult for us to insist for such a long time because of losing the balance —more ladies and fewer gentlemen. Maybe lovers would think the time was too short if they had dinner together sweetly.

头盘、餐前酒—主食+红、白酒—主菜—餐后甜品—coffee or Drink（烈酒），多来劲啊！

Appetizer, some aperitif—staple food ,red wine and white wine, main course—dessert—coffee or drink(liquor). What a big meal!

但"中国人"那是高速度，小肚量，压根撑"死"了。

Chinese were at a fast speed but had a small stomach, we really stuffed to "die."

于是，便要早一点吃，但餐厅没开门；要少一两道菜，那就体味不到真正Sicily的饮食文化了。也难为我们的槐帅哥了！这碗饭，吃得不容易的！

Therefore, we wanted to eat early but the restaurant hadn't open yet; If we ordered one or two dishes fewer, we couldn't experience the real food culture of Sicily. That was really hard on our handsome guide, Huai. It was not easy to do this job!

从名字开始就有意思：Eat+Italy，时尚的超市与餐饮的混合体，琳琅满目，人山人海。门口的大辣椒，好喜欢，倒是排队让我们有了更多的交流时间，吃得过瘾（也许是盒马鲜生的小榜样吧）。

Even the name was quite interesting: Eat+Italy. It was a mixture of a stylish supermarket and restaurant, with all sorts of goods and full of people. I liked the big chili best at the door. Lining up made us communicate more and tuck in for a big meal. (Maybe it was the role model for Freshippo.)

Des Ètrangers Hotel & Spa

陌生人温泉酒店

★★★★★

3

Des Etrangers Hotel & Spa Italy

Address : Passeggio Adorno,10/12-96100 Siracusa,Italy

Telephone : +39 0931 319100

Fax : +39 0931 319100

Http : //www.desetrangers.com

E-mail :fo.desetrangers@amthotels.it

14-15-16/4.2012

Des Ètrangers Hotel & Spa

★★★★★

Passeggio Adorno, 10/12 - 96100 Siracusa - Italy - Tel. +39 0931 319100 - Fax +39 0931 319000

www.amthotels.it - fo.desetrangers@amthotels.it

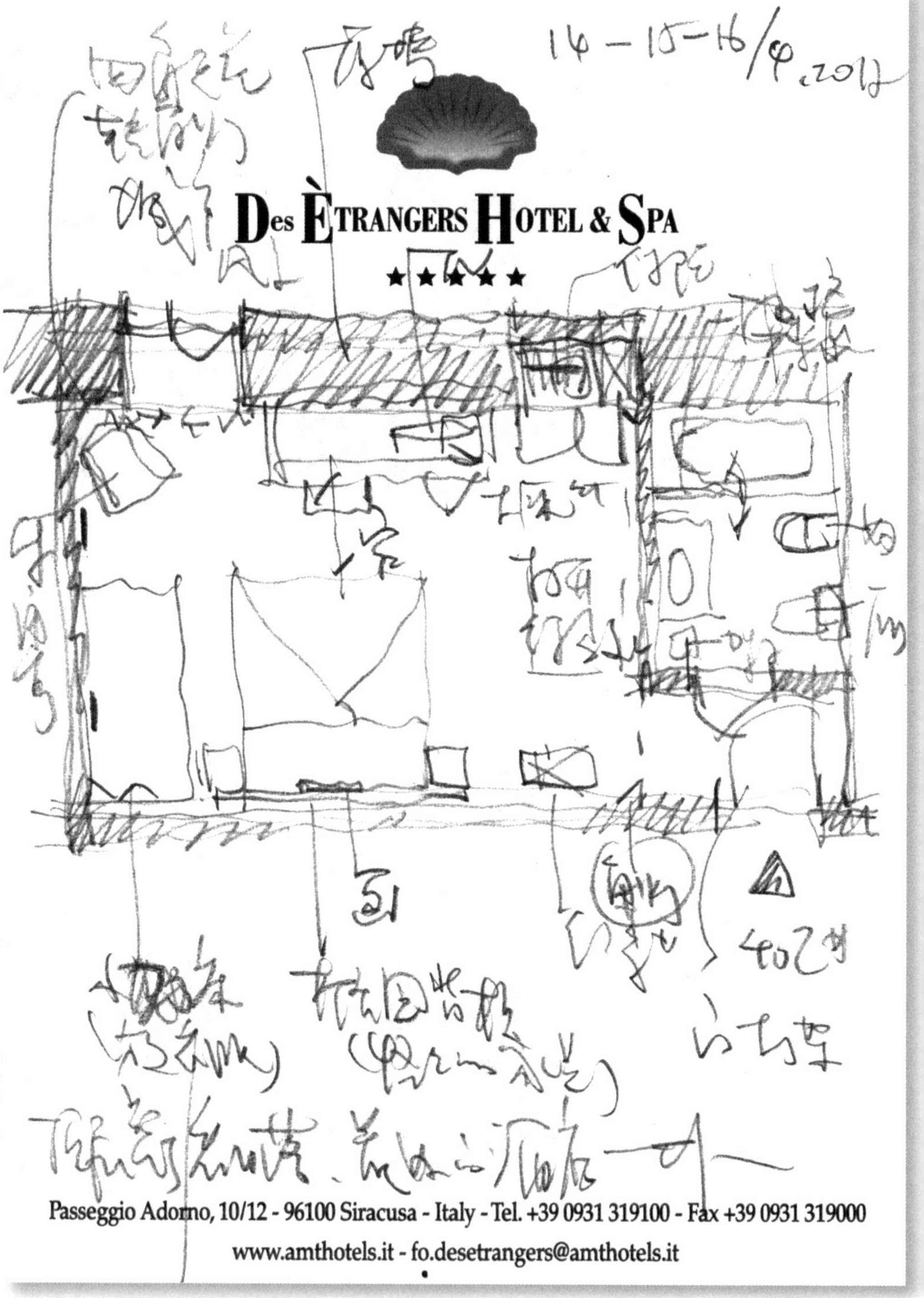

陌生人温泉酒店

Des Etrangers Hotel & Spa, Italy

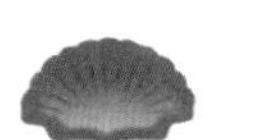

15/8, 2017

DES ÈTRANGERS HOTEL & SPA

★★★★★

Passeggio Adorno, 10/12 - 96100 Siracusa - Italy - Tel. +39 0931 319100 - Fax +39 0931 319000

www.amthotels.it - fo.desetrangers@amthotels.it

DES ÈTRANGERS HOTEL & SPA

★★★★★

Passeggio Adorno, 10/12 - 96100 Siracusa - Italy - Tel. +39 0931 319100 - Fax +39 0931 319000

www.amthotels.it - fo.desetrangers@amthotels.it

DES ÈTRANGERS HOTEL & SPA

★★★★★

Passeggio Adorno, 10/12 - 96100 Siracusa - Italy - Tel. +39 0931 319100 - Fax +39 0931 319000

www.amthotels.it - fo.desetrangers@amthotels.it

DES ÈTRANGERS HOTEL & SPA

★★★★★

Passeggio Adorno, 10/12 - 96100 Siracusa - Italy - Tel. +39 0931 319100 - Fax +39 0931 319000

www.amthotels.it - fo.desetrangers@amthotels.it

Happiness won't be Forgotten
幸福不会健忘

2017年 4月15日晚上11:00
At 11:00 p.m., April 15, 2017

陌生人温泉酒店
Des Etrangers Hotel & Spa, Italy

这几天都是“吃吃喝喝”，回到房间时都非常晚，当地时间11、12点了，累、困，倒头就睡，第三天早上也懒，匆匆忙忙地吃早餐，早已把“日记”抛到“九霄云外”了，要回想这两三天的事情，只能依赖手机相片和行程表，真的是记不清，那可以倒叙。

I did nothing but ate and drank these days. It was around eleven or twelve p.m. local time when I went back to the guestroom late. I was so lazy on the morning of the third day that I had breakfast in a hurry. And I threw my diary to the winds. I had to depend on my cell phone and the schedule to help me remember what I did these two or three days. If not sure, I tried to write back from the end.

今天4月15日周末，去酒店后面的广场，一家很“有范”的餐厅，衬衣+围巾，算是比较正装了！还好，有了这几天的“超分量”，大家都“学精”了，只有一道前菜，主菜，餐后酒，甜品和coffee，可谓9成饱。

Today was April 15, the weekend. We had dinner at a stylish restaurant on the square behind the hotel. I wore a shirt and a scarf as formal dress. So far so good, the meal here was so large that we got smarter to order only one appetizer, a main course, after- dinner drink, dessert and coffee. Not too full.

今天7:00日落，是在这家酒店五楼屋顶的餐厅看的，不用“舍近求远”，可谓近水楼台先得日，确实完美。

We saw the sunset at 7 p.m. on the fifth floor, the top of the hotel. We didn't need to go far away. It was perfect for us to enjoy the wonderful sunset conveniently.

白天多于15000步的“山城徒步”，感谢有小帅老头建筑师的“领导”，有吃、有喝、有解说，还让我们体味到了从近代到16、17世纪古城的时间演变。更重要的是让我们体验了大太阳下建筑的雕塑魅力，你喜欢吗?

I hiked in the mountain city for more than 15000 steps in the day. Thanks to the handsome old man's guide, we had something to eat and drink, a narration as well. We also experienced the change from modern time to the sixteenth or seventeenth century in the ancient city. More importantly, we experienced the charm of the architectural sculptures in the bright sun. Do you like it?

这次的安排还是我一贯的“只管交钱，像鸭子一样地跟着就可以了”，每天总有小惊喜，像昨天提供晚餐的酒庄，简直是太令人印象深刻了——古、简，果、酒，特别是餐前后花园的现摘枇杷，终生难忘，狂吃超过30粒，满足。

As usual, I just paid the money and followed the guide. So I felt a little surprised every day just like the winery where we had dinner yesterday. It gave me a deep impression—classical, simple, fruit, wine, especially we picked loquats in the back garden before dinner. I would remember this experience for a lifetime.I ate more than thirty loquats in total. Contented!

2017年4月15日于屋顶看落日

Watching the sunset on the roof

幸福其实很容易：一天做一件你未曾做过的事情；吃一样让你甜蜜的东西，看一件你心动的东西，一个酒庄能满足你三个愿望，夫复何求！

Happiness is easy: do a thing that you never experience, taste a kind of food that makes you feel sweet, see a thing that touches you. A winery can satisfy your three wishes. Is there anything happier than that?

况且还有这么多美丽的团友，真不愧是西西里的美丽传说！

What's more, there were so many beautiful group friends. It was really the Beautiful Legend of Sicily.

健忘，不会的！

Forgetful, I won't be.

全世界不同的日落。近水面看日出和在酒店屋顶餐厅观看确有不同，可以吹着空调，喝着鸡尾酒，吃着新炸的薯条，简单的奢侈。

Different sunsets from all over the world. Watching the sunset near the sea was quite different from that in the restaurant on the top of the house. We enjoyed the air conditioner, drank the cocktail and ate the fresh chips. What a simple luxury!

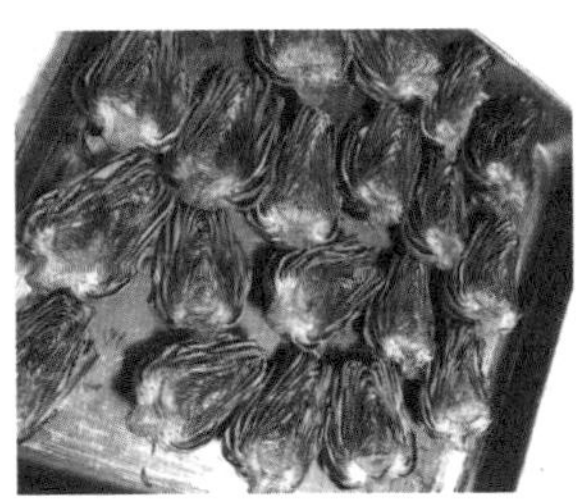

完美的艳遇：槐导游在复活节安排我们到他的同事——西西里美女Francisco家用餐，真正的乡村大餐，第一次体验到洋蓟的魅力，烤的，回味无穷！

A perfect romantic encounter, Huai, our tour guide arranged us to the Easter meal in his colleague's house, whose owner is a beautiful Sicily girl. What a really local big meal. It's my first time to taste the roast artichokes. How memorable!

4

VILLA ATHENA
ITALY

Address : Via Passeggiata Archeologica, 33. 92100 Agrigento,Italy
Telephone : +39 0922 596288
Tel : +39 0922 402180
Http : //http://www.hotelvillaathena.it
E-mail : info@hotelvillaathena.it

VILLA ATHENA

16-17-18

Music

Chair

HOTEL VILLA ATHENA s.r.l.
Via Passeggiata Archeologica, 33 • 92100 Agrigento (Italy)
Tel. (+39) 0922 596288 • Fax (+39) 0922 402180
www.hotelvillaathena.it • info@hotelvillaathena.it
P. Iva e C.F. 01930300841

slh.com
A MEMBER OF SMALL LUXURY HOTELS OF THE WORLD

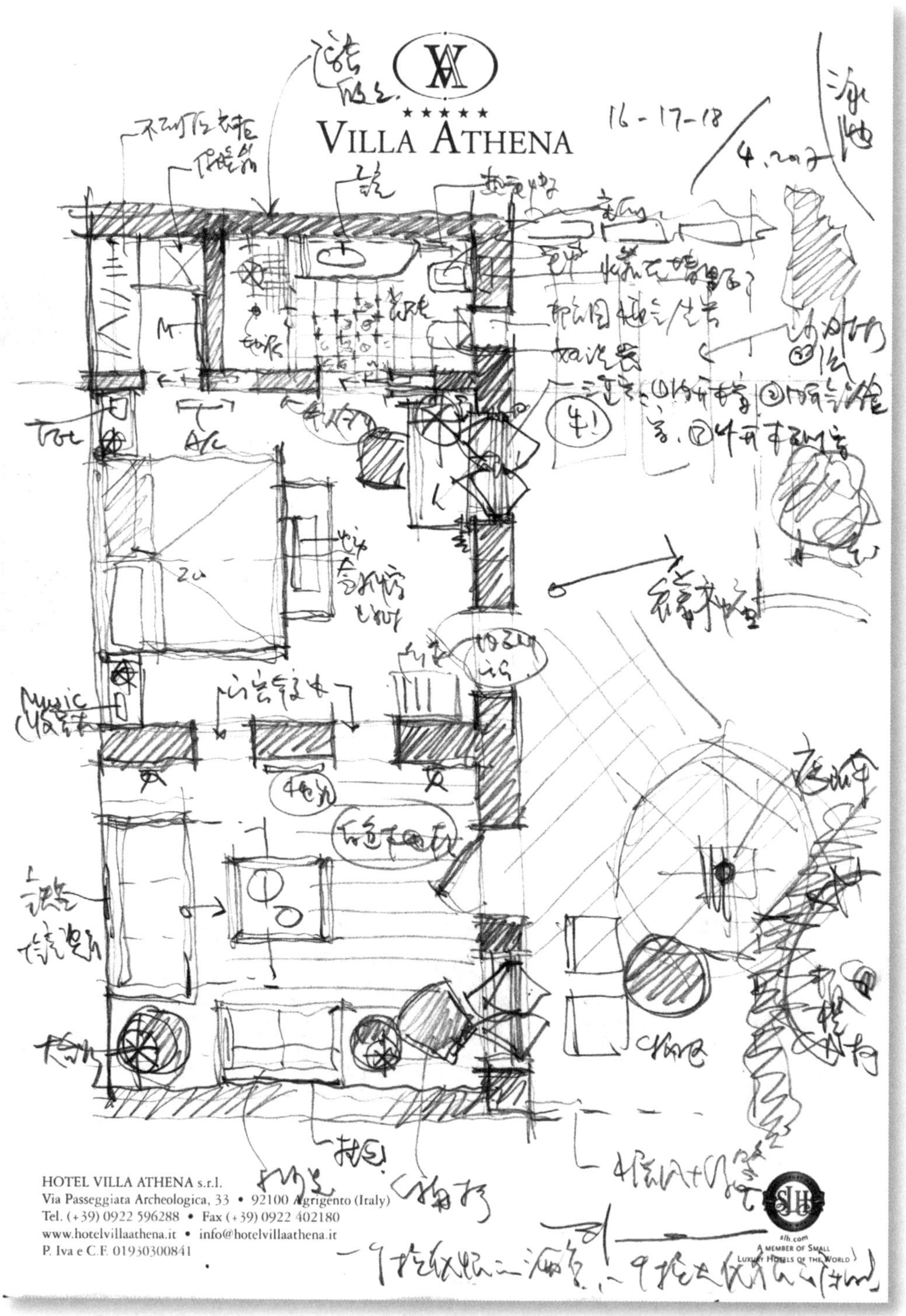

VILLA ATHENA

★★★★★

一间能看见"之神殿(庙)之酒店。

[illegible]

VILLA ATHENA，[illegible]

[illegible] Underwear [illegible] coffee [illegible]

[illegible]

HOTEL VILLA ATHENA s.r.l.
Via Passeggiata Archeologica, 33 • 92100 Agrigento (Italy)
Tel. (+39) 0922 596288 • Fax (+39) 0922 402180
www.hotelvillaathena.it • info@hotelvillaathena.it
P. Iva e C.F. 01930300841

SLH
slh.com
A MEMBER OF SMALL LUXURY HOTELS OF THE WORLD

The Most Enviable Wonderful Hotel
一家“拉仇恨”的神妙（庙）酒店

2017年 4月15日晚上11:00
At 11:00 p.m., April 15, 2017

从岛的一边用五个小时的车程去岛的另一边，其中一个睡觉的美女被车颠簸得直接摔下座位。

It took five hours from one side of the island to the other side by bus. A pretty girl fell down from the seat while sleeping on the bumpy road.

当看到以“神妙”的神庙为背景而近在咫尺的SLH酒店——Villa Athena，一间号称只有16间房的奢侈酒店，直接“醉”倒。

When seeing SLH hotel which looked like a wonderful temple—Villa Athena, all of us were fascinated by this deluxe hotel with only sixteen guestrooms.

很优厚的待遇，占了一个有客厅的套房：全白的调子，细节与灯光变为一体，白天与晚上有不同的气氛与感受，细细、慢慢地睡懒觉之后，一笔一笔地画着，感受着设计师的思考与“章法”，只是靠拍照难以有印象，客户行为和架构定义了客厅也可住人（一至两个小朋友），提供长居的可能性（这里太偏僻），可以有大行李打开的空间，方便衣服的摆放。洗手间的加热毛巾架（这里可是热带啊），让你可以每一天都洗洗内衣裤，让你可以每一天都有新的感觉，多么有细节啊！一“双”电视机，不用抢，大书台、大花园，到处都可以让你/你和情侣/你和家人发发呆，看看书，品品茶/咖啡, 那你还想走吗?

I got such a preferential treatment that I stayed in a suite with a living room. The suite was decorated in white color, details and lights were made into one, feelings and atmosphere were different in the daytime and at night. After a good rest, I began to draw the room slowly and carefully with sure strokes while thinking about the designer's ideas and methods. It was hard to make a deep impression only by taking photos. According to the consumer's behavior and the structure definition, one or two children could sleep in the living room. It was not possible for the guests to stay here for long (because it was really a far-off place). There was enough space to open the large suitcase and lay out clothes. There was a heated towel rack in the bathroom so that you could wash underwear every day. It made you fresh with the new hip. What considerate details! A pair of TV sets, you didn't have to fight for it. A big desk and a big garden allowed you or your lover or your family to relax everywhere, read books, and enjoy tea or coffee. Would you like to leave here?

抬头可以看看蓝天，低头有树影、鸟叫，关键是唾手可得的神庙就在眼前，希腊以外最完美的神妙（庙）。

Looked up to see the blue sky, looked down to see the shadow of the trees, listened to the singing of the birds, the key was that the wonderful temple was just in front of you. It was the most perfect temple outside Greece.

真可谓是一间“拉仇恨”的酒店，从里到外。

It was really the most enviable wonderful hotel, from inside to outside.

要住，请提前！

If you want to stay here, you must make a reservation in advance.

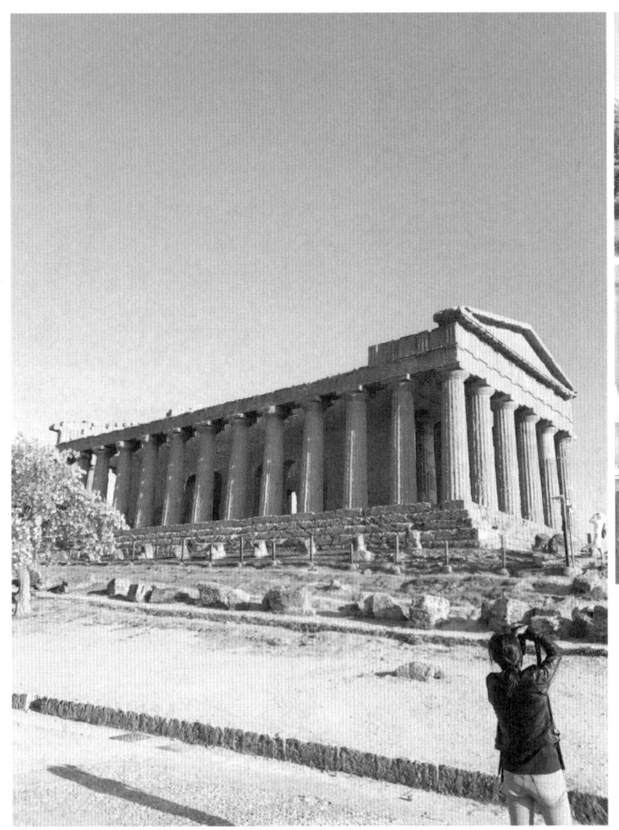

西西里岛神庙与酒店
The wonderful temple and Hotel Villa Athena in Sicily.

太近了，这个酒店的位置真的“无得弹”（棒极了），看着美景，晒着太阳，还有游泳池，“爽歪歪”。房间的纯色更是让酷暑瞬间降温。

It was too close. The location of the hotel was so perfect that we could watch the wonderful view, bask in the sunlight. The swimming pool was also cool! The pure color of the house cooled down the hot temperature immediately.

O.'s File

（区生词典）

现代化的Villa Athena是五星级典雅酒店，位于阿格里真托的联合国教科文组织世界遗产保护区寺庙谷内，距离公元前5世纪的完美杰作Temple of Concordi寺仅一步之遥。这是一座有300年历史的漂亮的花园别墅，内设有一家餐厅、康体中心、一个室外游泳池，还有令人神往的寺庙景色。（百度百科）

Hotel Villa Athena is a historic villa in Agrigento with wonderful views, which is within the Unesco World Heritage Site Valley of the Temples.It isalso a stone's throw away from the 5th-century B.C Temple of Concordia. The hotel is a handsome 300-year-old villa set in pretty landscaped gardens with a restaurant, a wellness center,an outside pool, and amazing views of the temple. (From Baidu Encyclopedia)

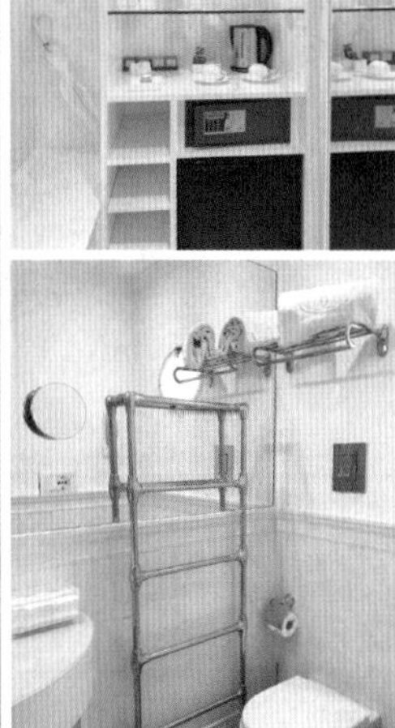

雅典娜别墅酒店
Villa Athena, Italia

5

伊吉亚别墅大酒店

★★★★★

GRAND HOTEL VILLA IGIEA PALERMO ITALY

Address : Salita Belmonte, 43 90142 Palermo,Italy
Tel : +39 091 631 2111
Fax : +39 091 547 654
Http : //www.villa-igiea.com
E-mail : H9836@accor.com

Grand Hotel Villa Igiea
Salita Belmonte, 43
90142 Palermo – Italia
Tel. +39 091 6312111 - Fax +39 091 547654
H9836@accor.com

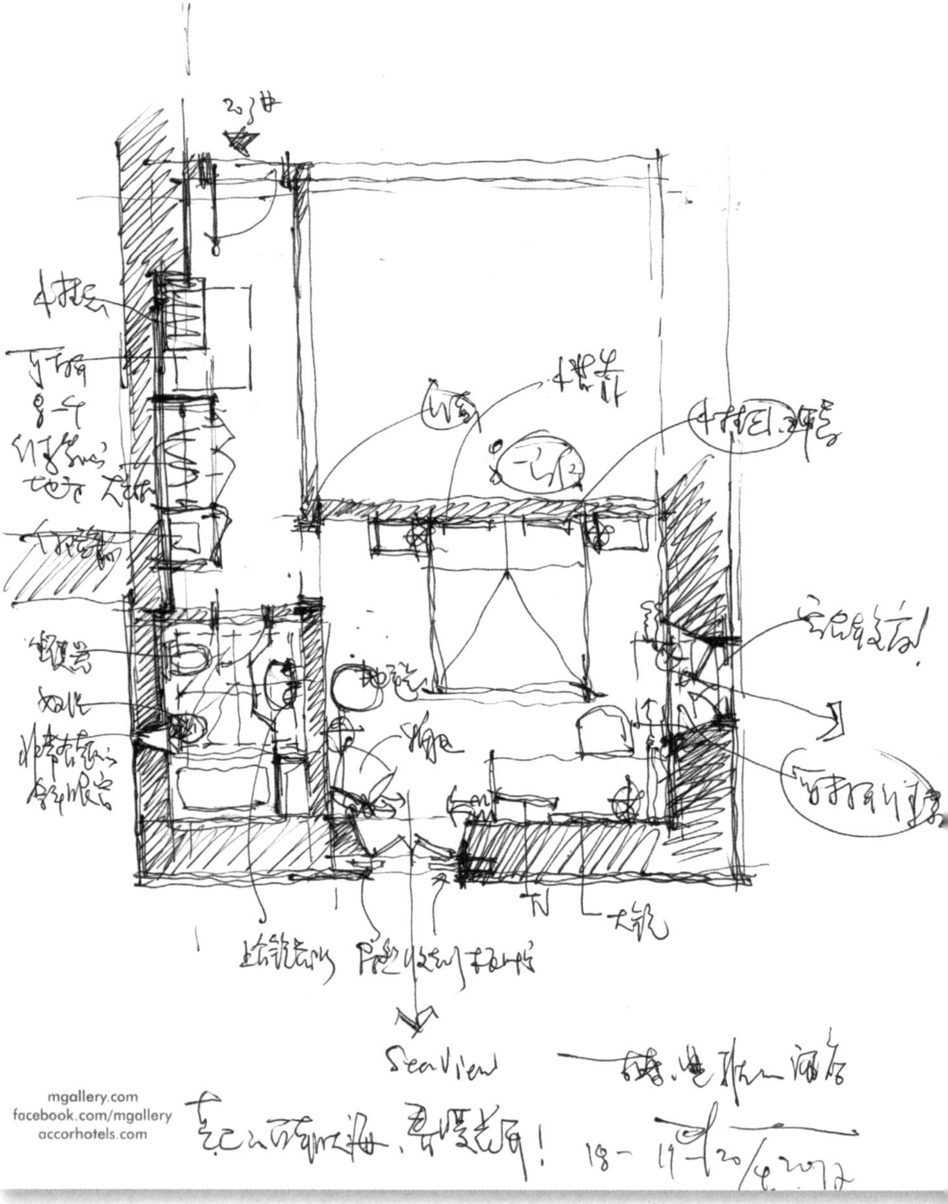

Sea View

Grand Hotel Villa Igiea
Salita Belmonte, 43
90142 Palermo – Italia
Tel. +39 091 6312111 - Fax +39 091 547654
H9836@accor.com
www.villa-igiea.com

[illegible]

最后两天入住大城市：西西里的Palermo这一家老酒店：Grand Hotel Villa Igiea（隶属Accor旗下的Sofitel系列）[illegible]

[illegible]

[illegible]

[illegible] 2017.[illegible]

岛上的历史定义了酒店的调子，每一处都是入画的级别。手机放不下来。早上的地中海真的很蓝，很蓝。

The style of the hotel is based on the history of the island. Every shooting is picturesque. You can hardly put down your mobile phone. In the morning, you find yourself in front of the blue, blue Mediterranean.

伊吉亚别墅大酒店
Grand Hotel Villa Igiea, Palermo, Italy

The Rooms are Pure Simple
简致房间

最后两天入住大城市：西西里的Palermo这一家老酒店Grand Hotel Villa Igiea（隶属Accor旗下的Sofitel系列），下午三点多入住，大太阳，很“土色”的外墙，艳阳下明的明，暗的特黑，雕塑感一流，任谁拍照效果都不错，古老的门头，大门，有序列感的大堂，拱形的长廊，大气，有历史的味道与琳琅满目的摆设、家具，柔美的灯具和古典的油画，当然也不时挂出一张“炫耀”历史的照片（虽然我不太懂照片里的人物和西西里岛的历史）。

I returned to the big city and stayed in a classical hotel—Grand Hotel Villa IGIEA in Sicily (it belonged to the Sofitel series of Accor). Checking in at three in the afternoon, the sun was shining brightly. The outer wall was dirt-colored. In the bright sun, the bright color was strong bright white, the dark was quite dark. It has a perfect sense of sculpture. everyone could take good photos of it. The ancient door header, gate and the lobby with an arched gallery were all distinguished. There were also a great variety of decorations with history, furniture, soft lamps and classical paintings. Of course sometimes a historic painting was hung to show off. (AlthoughI didn't know a lot about characters in the paintings or the history of Sicily island.)

再到达其他的空间，多个餐厅，特色和风格都有少许不同，特别是酒吧。古堡的儒雅，全石的天、地、墙，厚重，内敛，让人安稳其中，倍觉舒适；海边的露台，可坐可“葛优躺”，一望无际的大海，远处停满了各式长度、各种样子的小游艇，一副优哉游哉的生活状态。

And then I went to other places: so many restaurants with different features and styles, especially the bar. The castle was refined. The ceiling, ground and wall were made of stone. It looked heavy and introverted to make you feel peaceful and comfortable. You could either sit or "Geyou" lie on the terrace by the sea. All kinds of yachts in different sizes were parked far away on the endless sea. What a leisurely life style!

电梯只能坐三四个人，服务生让我们分批回房间，然后他们分批送达行李，迂回的走廊，到达房间，那就简单了，有前面的铺垫，不用再多说，不再用繁琐的设计，花墙纸，鲜黄色的地毯，古老的木家具，推窗，两侧均可面朝大海，真正的春暖花开。

The lift could only hold three or four people. The waiters told us to go to our rooms in batches. And then they took the luggage for us. We went through acircuitous corridor to our rooms. It was easy now. I didn't want to say any more. Fussy designs weren't used again.There was flower wallpaper, bright yellow carpet and antique furniture. Pushed the window open, you might face the sea on both sides, with real spring blossoms.

老酒店就是耐看，“耐人寻味”！

A classical hotel,the more you look at it, the more beautiful you will feel.

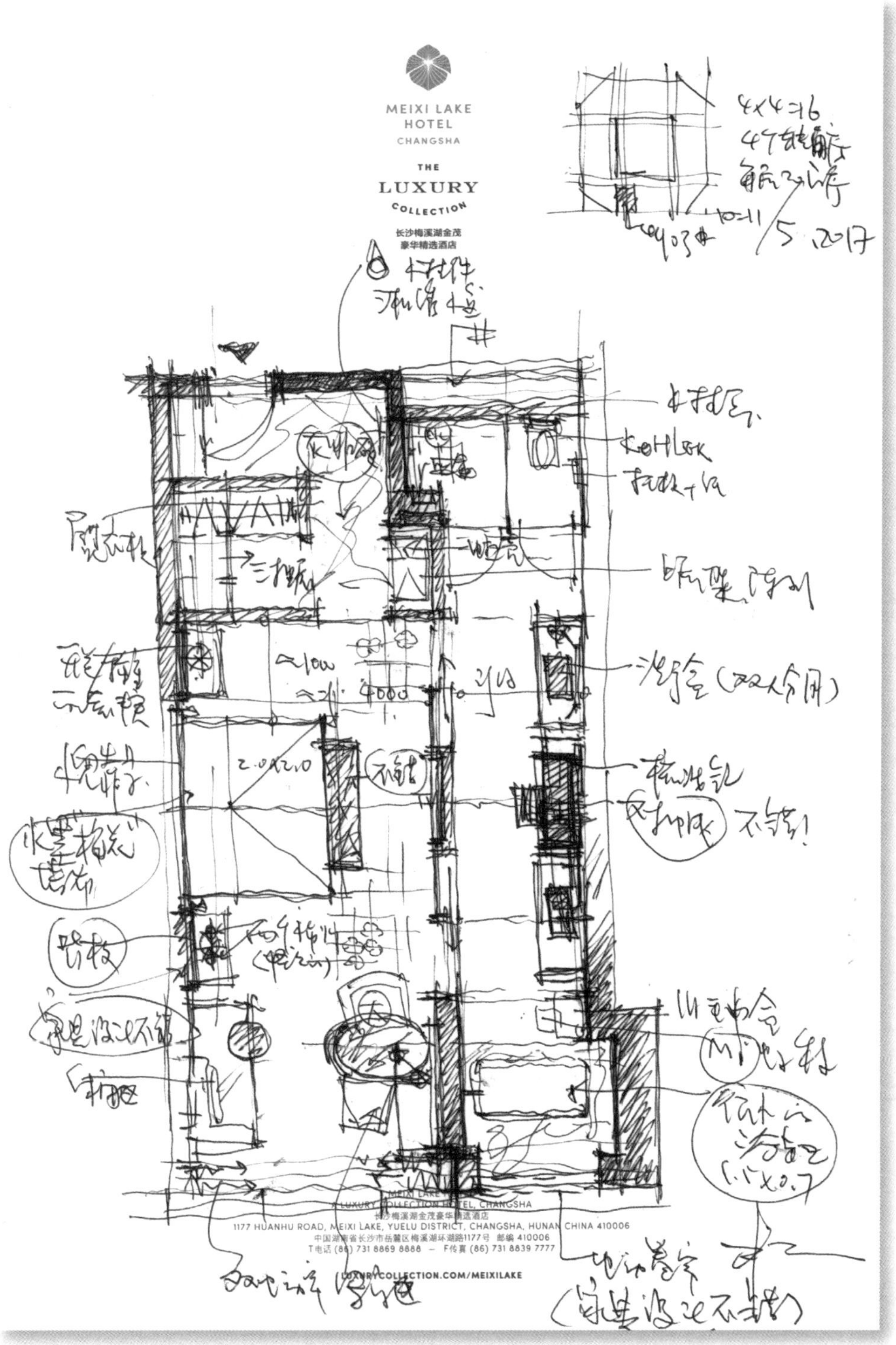

MEIXI LAKE
HOTEL
CHANGSHA

THE
LUXURY
COLLECTION

长沙梅溪湖金茂
豪华精选酒店

MEIXI LAKE HOTEL, A LUXURY COLLECTION HOTEL, CHANGSHA
长沙梅溪湖金茂豪华精选酒店
1177 HUANHU ROAD, MEIXI LAKE, YUELU DISTRICT, CHANGSHA, HUNAN CHINA 410006
中国湖南省长沙市岳麓区梅溪湖环湖路1177号 邮编 410006
T电话 (86) 731 8869 8888 — F传真 (86) 731 8839 7777
LUXURYCOLLECTION.COM/MEIXILAKE

6 长沙梅溪金茂豪华精选酒店

★★★★★

MEIXI LAKE HOTEL THE LUXURY COLLECTION CHANGSHA CHINA

Address : No.1177 Huanhu Road, Meixi Lake,Yuelu District, Changsha,Hunan,China
中国湖南省长沙市
岳麓区梅溪湖环湖路
1177号

Telephone : +86 731 8869 8888

Fax : +86 731 8839 7777

Http : //www.LUXURYCOLLECTION.com/MEIXILAKE

11/5.2017

有家美设计很不错的酒店

一个好的酒店首先，当然今天已是第一位的要素，是品位。现在是讯息平化时代，客户要有何种资讯不论难易，但要做到很好也是极需努力的。在这里我们看到设计公司的专业和设计经验了。但更重要的是他们的创意和精神，"外部看起来商，内部看起来道"。酒店的设计是一个不容易做到的"门道"。房间的灯光、配材、配色等等是第一重要的。而梅溪湖酒店在这些年里是越来越受到注重的。特别是室内设计的手法多变，比如更是有很好的设计细节，但也有一些花哨的"拼装"的，乱七八糟的设计方面简单大方的完美。

同时看了，到今天我们有接近200多家国内外又是很成功的酒店。在这里，在长沙豪华精选酒店一家是我们经典的，是极有意思的。木、金属、石材、玻璃来构成自然的设计材料，通俗大方的手法，引入很多的中国传统元素的细节，如斗拱、

MEIXI LAKE HOTEL
A LUXURY COLLECTION HOTEL, CHANGSHA
长沙梅溪湖金茂豪华精选酒店
1177 HUANHU ROAD, MEIXI LAKE, YUELU DISTRICT, CHANGSHA, HUNAN, CHINA 410006
LUXURYCOLLECTION.COM/MEIXILAKE

榫卯、金属地板家具都在这入住的空间有丝丝的情感。一个房间大大小小的设计不少于20件，其中还有很多让人觉得品味不错的，每件都是一个学问。室内设计要从大"空间"到小"细节"去考虑，设计和我都会与专业公司沟通好。不然一个房间就会因为细节的不够成"败笔"！

住上一家有意思的好设计的酒店，还是很幸福的。每一天都是让你幸福的之旅。一定会的。

MEIXI LAKE HOTEL
A LUXURY COLLECTION HOTEL, CHANGSHA
长沙梅溪湖金茂豪华精选酒店
1177 HUANHU ROAD, MEIXI LAKE, YUELU DISTRICT, CHANGSHA, HUNAN, CHINA 410006
中国湖南省长沙市岳麓区梅溪湖环湖路1177号 邮编 410006
T电话 (86) 731 8869 8888 — F传真 (86) 731 8839 7777
LUXURYCOLLECTION.COM/MEIXILAKE

Furniture Makes a Better Guestroom

家具成就更好的房间

长沙梅溪金茂豪华精选酒店

Meixi Lake Hotel Changsha the Luxury Collection, Changsha, China

一个好的酒店房间，当然布局是第一首要的。说实话，现在是资讯扁平化的时代，要突破、有惊喜确实不容易，但要做得很“烂”却是挺容易的，看你找到的设计公司的责任心和设计经验了，但更重要的是他们的创新精神。“外行看热闹，内行看门道”，酒店房间就是一个不容易做好的“门道”。房间的灯光、配材、配色等算是第一重要的，而家具的设计在这些年里越来越受到设计师，特别是室内设计师的参与与关注，当然是有成功的“跨界”设计师，偶有佳作，但也有一看就是“跨错界”的，乱七八糟的设计与贻笑大方的家具。

As a good hotel room, the most important thing is its layout. To be honest, nowadays information is flattening.It is not easy to make a breakthrough and surprise. But it is really easy to make it terrible. It depends on the responsibility and experience of the design company. Their spirit of innovation is more important. As the saying goes, "dilettante watch the scene of bustle, expert looks at the way."The hotel room is an uneasy "gateway" to make. The lighting, material and colors in the room are the most important. These years, designers, especially interior designers, care about and participate in furniture design. Of course, some are successful and make some perfect works. But some designs seem in the wrong job.

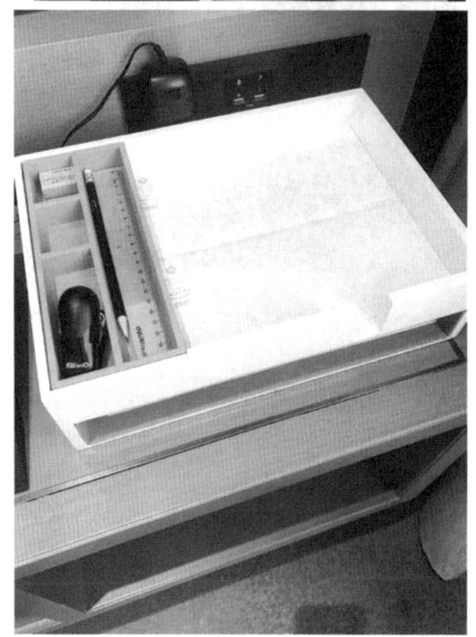

住多了酒店（到今天应该住了超过200家国内外五星级或相当级别的酒店），这次的长沙豪华精选酒店的家具有几件还是挺有趣的，木、金属、云石的搭配成为流行的主材，通彻大方的手法，引入细微的中国传统家具的细节，如折脚、小弧角、金属色边等等，都能让入住的宾客有丝丝的眷恋。一个房间中大大小小的家具不少于20件，其中三五件能引起你的共鸣和兴趣就是非常不错的了。家具设计是一门学问，室内设计师要从“大”空间到“小”家具都兼顾好，设计好或不好都要与家具专业公司沟通好，不然，一个房间就会因为家具不好成为“败笔”！

I have experienced too many hotels(In total over 200 five-star hotels at home and abroad so far). Some pieces of furniture in the Changsha Luxury Collection were interesting, the match of wood, metal and marble became the fashion main material. A decentstyle had some Chinese traditional furniture details, such as fold table feet, small arc angle, metal-colored edges and so on, the guests felt a little attached. There were at least twenty pieces of big or small furniture in the room, three or five of them could interest you and resonate. Furniture design is a kind of knowledge. And interior designer must care about furniture from large space to small one. However good or bad the design is, the designers should communicate with professional furniture company. If not, a guestroom will be a failure because of the unsuitable furniture.

住一间家具设计有意思的客房，还是挺幸福的。

It was really happy to stay in this interesting furniture design room.

每一天发现让你幸福的元素，一点点的！

Find something to make you happy every day, just a little!

长沙梅溪金茂豪华精选酒店

Meixi Lake Hotel Changsha the Luxury Collection, Changsha, China

Le MERIDIEN
艾美酒店及度假村

7

艾美酒店及度假村

★★★★★

LE MERIDIAN HOTEL ZHENGZHOU CHINA

Address : No.1188 Zhongzhou Avenue, Jinshui District,Zhengzhou, Henan, China
中国河南省郑州市
金水区中州大道1188号

Telephone : +86 0371 5599 8888

Http : //www.lemeridien-zhengzhou.com

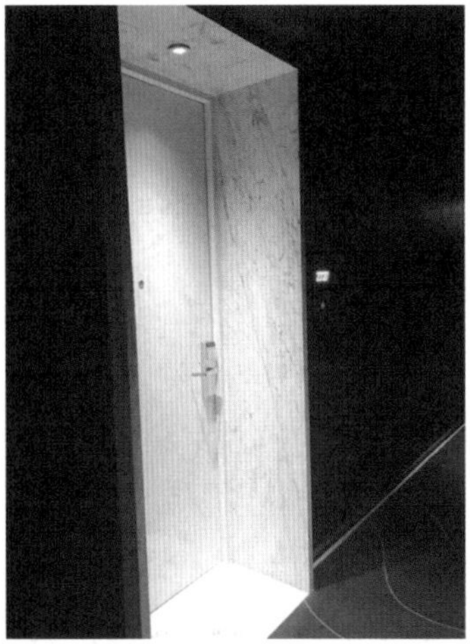

公共区域
The public area

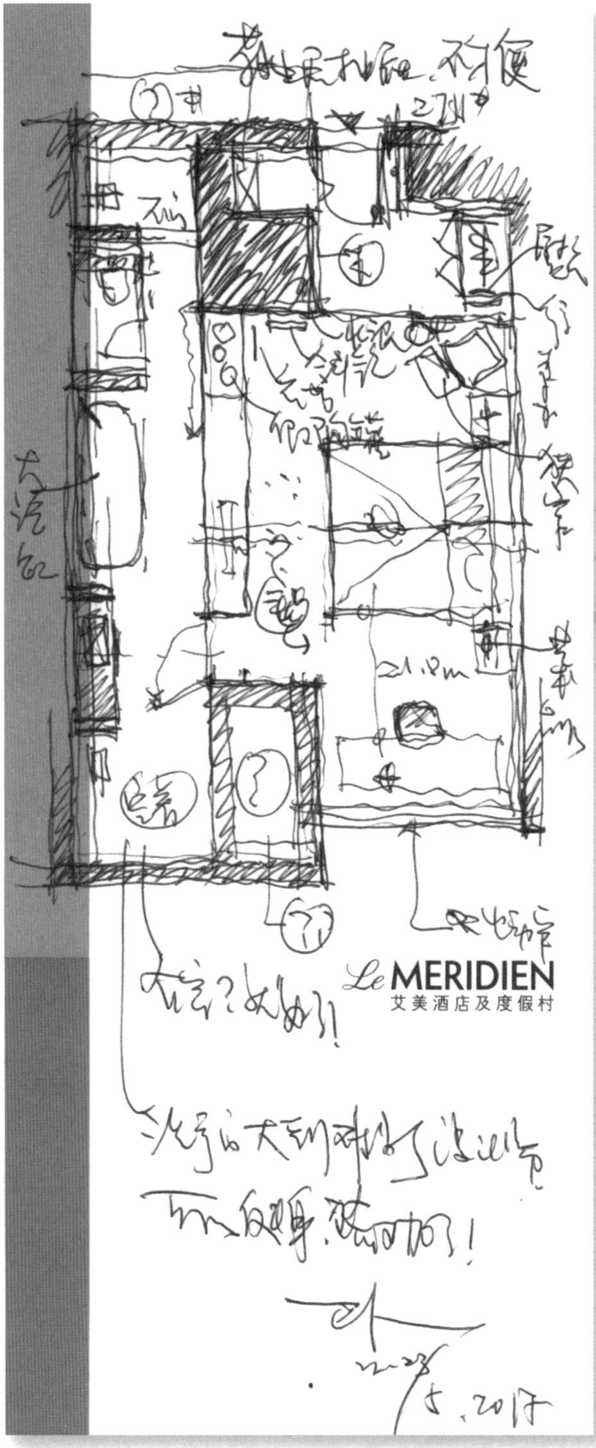

Le MERIDIEN
艾美酒店及度假村

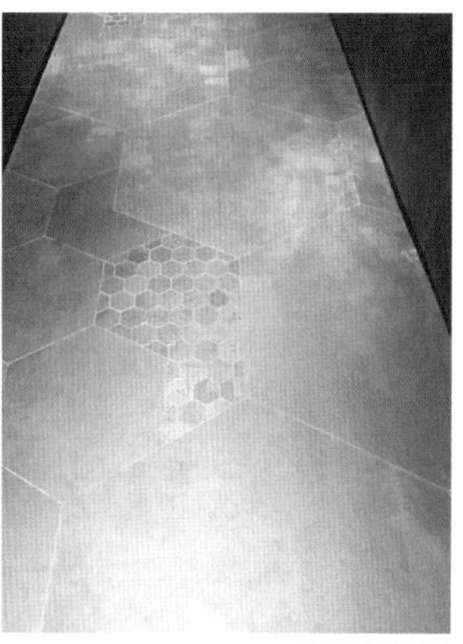

特别的地砖切割
The special Floor title cutting

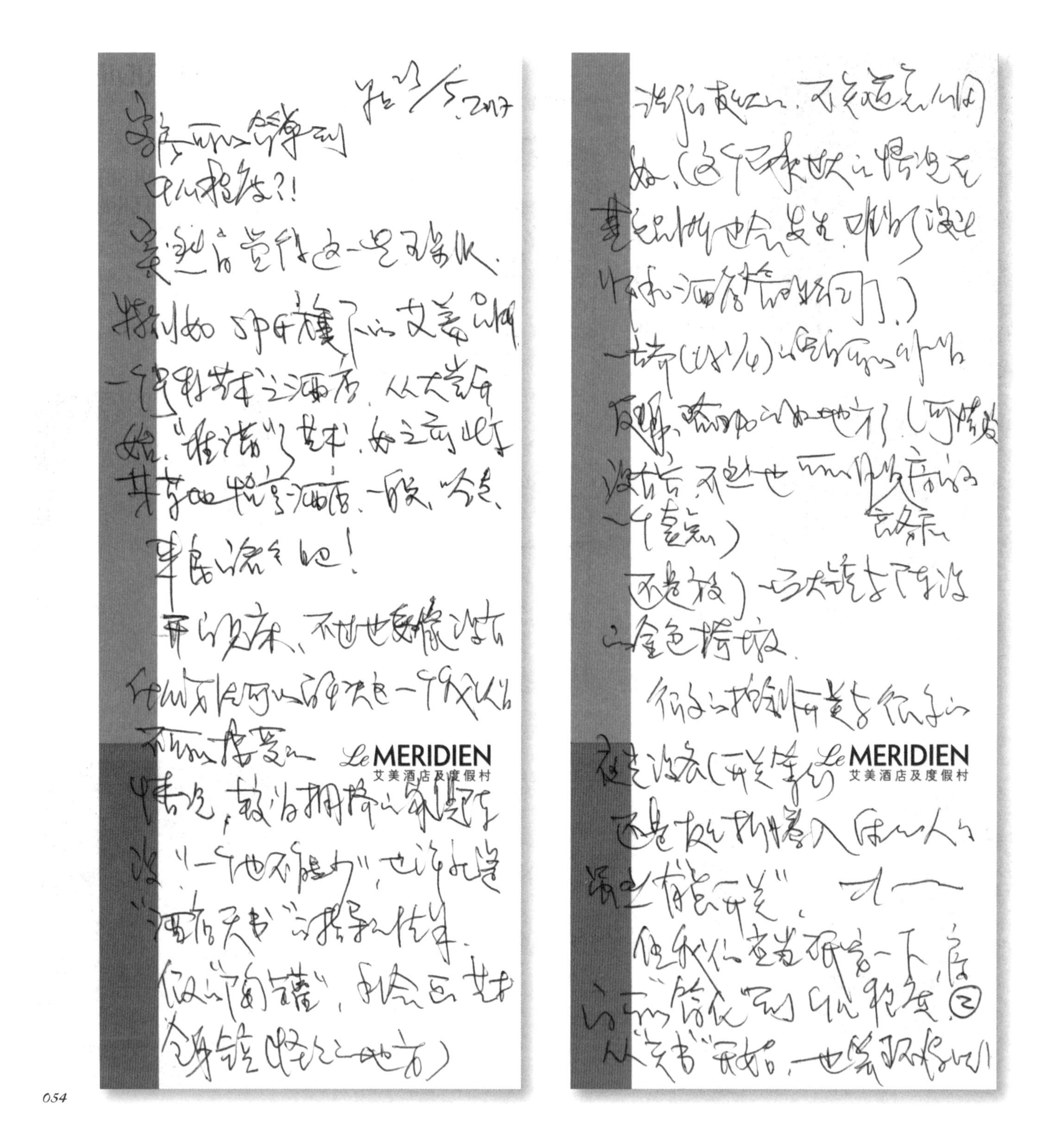

[illegible] 3/5, 2017

[illegible]

Le MERIDIEN
艾美酒店及度假村

[illegible]

[illegible]

Le MERIDIEN
艾美酒店及度假村

[illegible]

Simple is Beautiful

房以简为美

突然间觉得这些五星级，例如喜达屋旗下的艾美品牌，号称艺术酒店，都是从大堂开始便“堆满了”艺术品，就像之前北京芳草地怡亨酒店一样。哈哈，来房间看看吧。

Suddenly I found these five-star hotels, such as Le Meridien under SPG called Art hotel. They were full of works of art from the lobby, just as Hotel Eclat Beijing. Haha! Come to the guestroom and have a look.

开门见床，不过也好像没有什么办法可以解决这个我认为不可以接受的情况，较为拥挤的家具陈列，“一个也不能少”，也许就是“酒店天书”指导的结果。

Opening the door, I saw the bed. But maybe there wasn't any way to solve the problem which I couldn't accept. Too much furniture made the room crowded. "Nothing less" might be the result of the guidance of the Hotel Book.

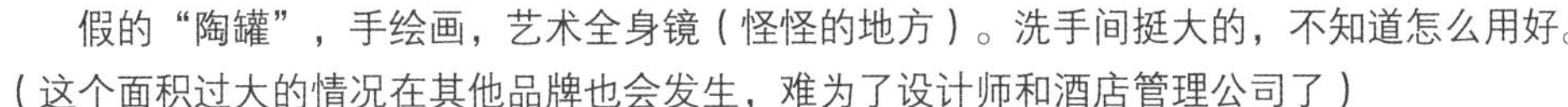

假的“陶罐”，手绘画，艺术全身镜（怪怪的地方）。洗手间挺大的，不知道怎么用好。（这个面积过大的情况在其他品牌也会发生，难为了设计师和酒店管理公司了）

There was fake ceramic, paintings and artistic full-length mirror which was a little strange. The bathroom was so big that the designer didn't have any ideas of using it. (Other brands also had this kind of situation. It was difficult for the designers and hotel Management Company.)

尽端（约1/4）的空间可以作为健身、瑜伽的好地方（可惜没有时间，不然也可以作为商务层房间的一个卖点）。

At the end of the room, around a quarter of it, it was a good place for fitness and Yoga. (It's regret that I didn't have any time to take exercise, or it could be a selling point as a business suite.)

还放了一面大镜，陈设了金色的椅墩。

There was also a big mirror and a golden low chair in the room.

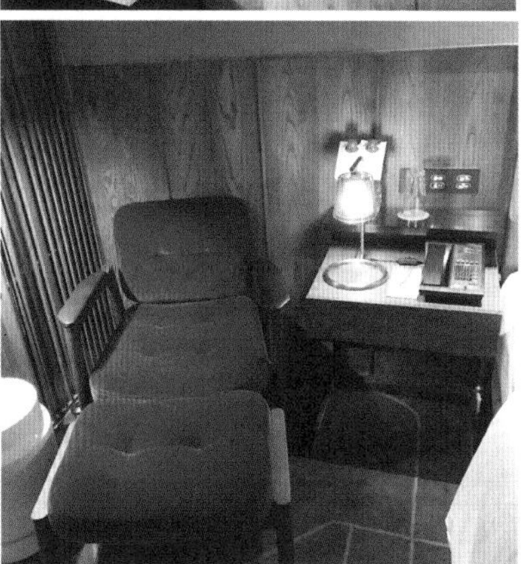

很多的控制开关与很多的设备（开关等），还是挺折腾入住的人的。

Too many control remotes and switches for appliances made the guests confused.

虽然有“总开关”，但我们应当研究一下，房间可以“简化到什么程度”，从“天书”开始，也算环保吧。

Although there was a master switch, we should think about the design. How simple can the guestroom be? Begin with the Book of Hotel, it's an environmental idea.

艾美酒店及度假村
Le Meridian Hotel,Zhengzhou,China

THE RITZ-CARLTON

HAIKOU

海口丽思卡尔顿酒店

8

★★★★★

THE RITZ-CARLTON HAIKOU CHINA

Address : No.39 Yangshan Boulevard, Longhua District, Haikou, Hainan, China
中国海南省海口市龙华区羊山大道39号

Telephone : +86 898 6683 6888

Http : //www.ritzcarlton.com

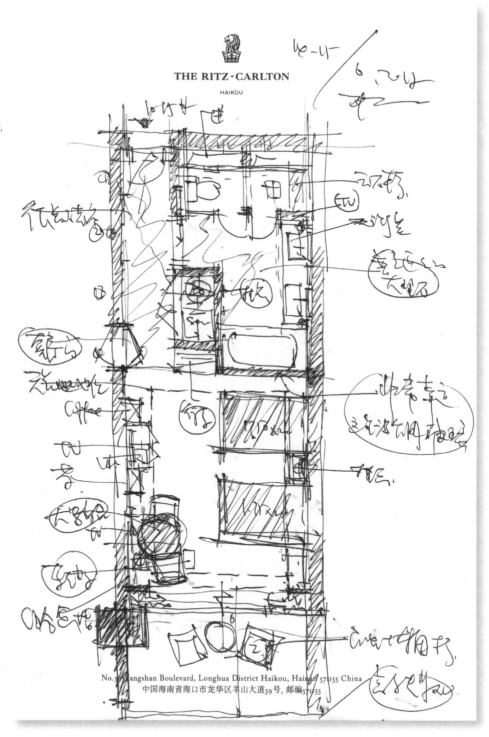

THE RITZ-CARLTON

HAIKOU

No.39 Yangshan Boulevard, Longhua District Haikou, Hainan 571155 China
中国海南省海口市龙华区羊山大道39号，邮编571155

15-16/6 2017

THE RITZ-CARLTON

HAIKOU

Ritz — Carlton

Luxury

THE RITZ-CARLTON

HAIKOU

THE RITZ-CARLTON

HAIKOU

No Feeling
没了感觉

两个人相处，最怕就是没了感觉。住的酒店多了，也怕这个。入住丽思卡尔顿，也突然有了这种感觉！可怕。

No feeling between lovers is the most horrible thing, just like staying in hotels too much. When I stayed in RITZ-CARLTON, I had this kind of feeling suddenly. How terrible!

打开房间的阳台门，可以看到一望无际的球场，这景色真奢华啊！

Opening the door of balcony, the endless golf course came into my eyes, the view was so luxury!

平心而论，这个品牌讲究的是实而不华的真切，经典而不失时尚的设计，踏踏实实的真金白银的投入与耐人寻味的设计，还是值得一住的，只是第一感觉平淡（也许是住得太多的结果）。

To be honest, this brand emphasized practicality but not luxury, classical and fashion design. They really spent a large amount of money investing and on intriguing design. The hotel was worth experiencing. But my first feeling was ordinary(Maybe I have stayed in too many hotels).

海口丽思卡尔顿酒店
The Ritz-Carlton, Haikou,China

怎样保持新鲜感与有感觉的确是值得我们设计师和做酒店管理的人士（专业）去思考的，既要经典又要情感打动，真不容易。

How to keep fresh and passionate is worth thinking about for designers and hotel executives. It is not easy to keep the hotel with feelings and touching, classical as well.

做酒店不易，要打动用户更加不易，坚持研究心理才是“一把年纪”的设计师要重点关注的。是时间，让感觉回来了，也是时间，让感觉没了。

It is not only difficult to build a hotel but also to move guests. Older designers should keep studying psychology. Time makes the feeling come back;and also the time makes the feeling disappear.

住下了，还是很牛的。

When I stayed here, I found it was still impressive.

让我们来感受和体验这样的品牌，相信还是有回味的价值的！

Let's enjoy and experience this brand. I believe it is memorable.

海口丽思卡尔顿酒店
The Ritz-Carlton, Haikou,China

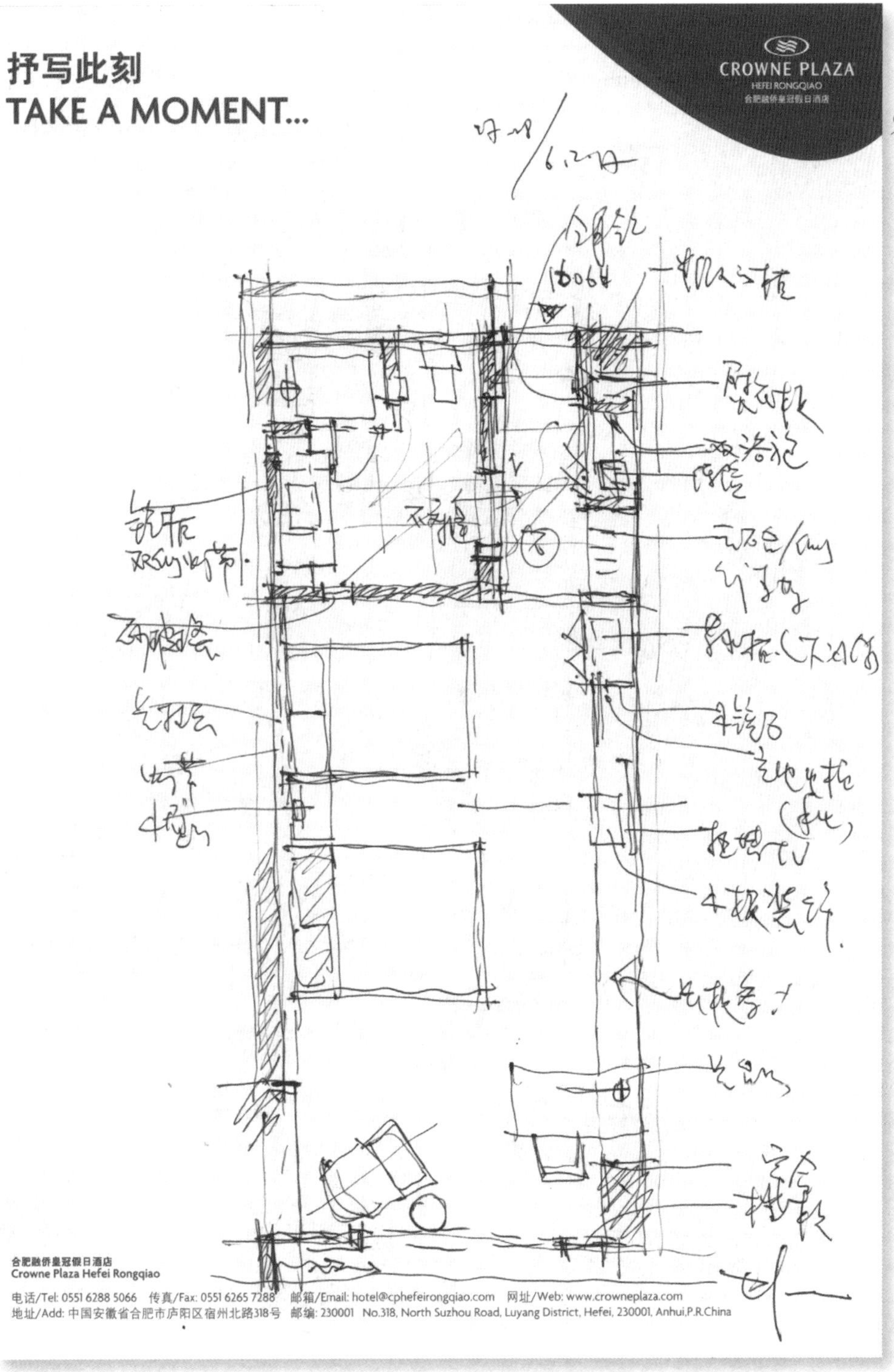

抒写此刻
TAKE A MOMENT...

CROWNE PLAZA
HEFEI RONGQIAO
合肥融侨皇冠假日酒店

合肥融侨皇冠假日酒店
Crowne Plaza Hefei Rongqiao
电话/Tel: 0551 6288 5066 传真/Fax: 0551 6265 7288 邮箱/Email: hotel@cphefeirongqiao.com 网址/Web: www.crowneplaza.com
地址/Add: 中国安徽省合肥市庐阳区宿州北路318号 邮编: 230001 No.318, North Suzhou Road, Luyang District, Hefei, 230001, Anhui, P.R.China

9 合肥融侨皇冠假日酒店

★★★★

CROWNE PLAZA HEFEI RONGQIAO CHINA

Address : No.318 North Suzhou Road, Luyang District, Hefei, Anhui, China
中国安徽省合肥市
庐阳区宿州北路318号
Telephone : +86 551 6288 5066
Fax : +86 551 6265 7288
Http : //www.crowneplaza.com

南昌格兰云天国际酒店

10

★★★★★

GRAND SKYLIGHT INTERNATIONAL NANCHANG CHINA

Address : No.1 Ganjiang North Avenue, Honggutan District, (AVIC PLAZA) Nanchang, Jiangxi, China
中国江西省南昌市
红谷滩赣江北大道1号
（中航广场）

Telephone : +86 791 8206 6666

Fax : +86 791 8207 7980

Http : //www.GSHMHOTELS.com

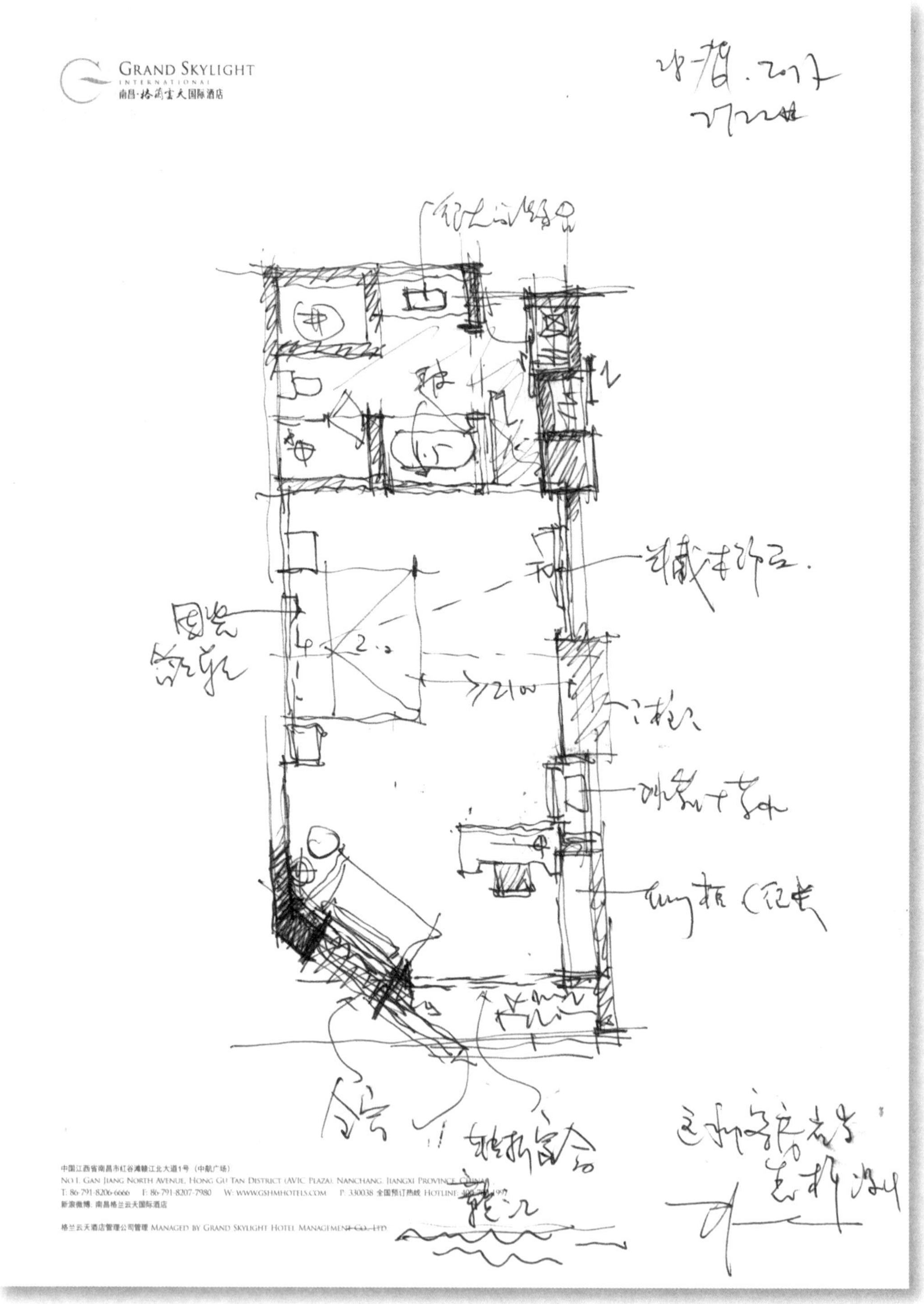

GRAND SKYLIGHT INTERNATIONAL
南昌·格兰云天国际酒店

中国江西省南昌市红谷滩赣江北大道1号（中航广场）
NO.1. GAN JIANG NORTH AVENUE, HONG GU TAN DISTRICT (AVIC PLAZA). NANCHANG. JIANGXI PROVINCE CHINA
T: 86-791-8206-6666 F: 86-791-8207-7980 W: WWW.GSHMHOTELS.COM P: 330038 全国预订热线 HOTLINE: [illegible]
新浪微博：南昌格兰云天国际酒店

格兰云天酒店管理公司管理 MANAGED BY GRAND SKYLIGHT HOTEL MANAGEMENT CO., LTD

长沙—广州，21/6，2017，高铁上

吴锦慧：

一个太大的房间，成本投入太大，应如何有效地利用空间面积或者做到，是否能够利用得更加合适呢。

入住这个位于江西省会城市南昌的红谷滩的格兰云天，感觉还是比较到位的，睡眠区更大，窗外的风景清，床还特别舒适的沙发，但给人更多尺度还是合适的心态上的良好的睡眠。

我们总的，这里有点，那个拉拉杆么，总觉得有的设计还是太多不足，特别是沿着床头的那种插座，会从床头上到地上太大太多插扣，太平时的用也是比较多余的设计，本来太满太空也是一个不少的问题，当然要考虑酒店公司，设计，布局以成本一同研究才能更好的把控定位的客房。大家做得很好的成为一种享受了。

—叶—

中国江西省南昌市红谷滩赣江北大道1号（中航广场）
NO.1 GAN JIANG NORTH AVENUE HONG GU TAN DISTRICT (AVIC PLAZA) NANCHANG JIANGXI PROVINCE CHINA
T: 86-791-8206-6666 F: 86-791-8207-7980 W: WWW.GSHMHOTELS.COM P: 330038 全国预订热线 HOTLINE: 400 700 1997
新浪微博：南昌格兰云天国际酒店

"Big" Wisdom

“大”智慧

一个太大的房间，成本投入有限，品牌有相对应的局限或指引，设计应当怎样去做更加合适呢?

Because of the too large guestrooms, limited cost, the brand restrictions and guidance, how to design it more reasonably?

入住这间位于江西省会城市南昌江边的塔式主楼的格兰云天酒店，感觉还是比较强烈的，睡眠区真大，有空空的冷清，床还特别矮，沙发、写字台的尺度其实还是合适的。心里想，问题出在哪里呢?

I had a strong feeling when staying in Grand Skylight Hotel which was in the main tower building by river in Nanchang, Jiangxi Province. The sleep area was really large, empty and lonely. The bed was especially low. The size of the sofa and table was fine. I wondered what was wrong with the design.

开灯，关灯，这里看看，那里拉拉柜子，发觉灯光的设计还是考虑不足，特别是浴缸与床之间的玻璃，会从视觉上让灯光大打折扣，太平均的固装投入也让整个房间没有了主次，木色太浅太亮也是一个不小的问题，当然要与业主、酒店管理公司、设计人员针对市场及成本一同研究才能做好中档定位的客房。大，有时候也能成为一种考验。

Turned on or off the lights, looked around,and opened the cabinet. I found that the design of the lights was not completely considered, especially the glass between the bathtub and the bed, and visually the lights were reduced their brightness. The cost of fixed installation was so average that the style was not highlighted. Wood color which was too shallow and too bright was also a problem. There must be a discussion among the owner, the hotel management company, the designers, the market and costs to decide how to make the mid-range guestrooms. Large, sometimes it is a test!

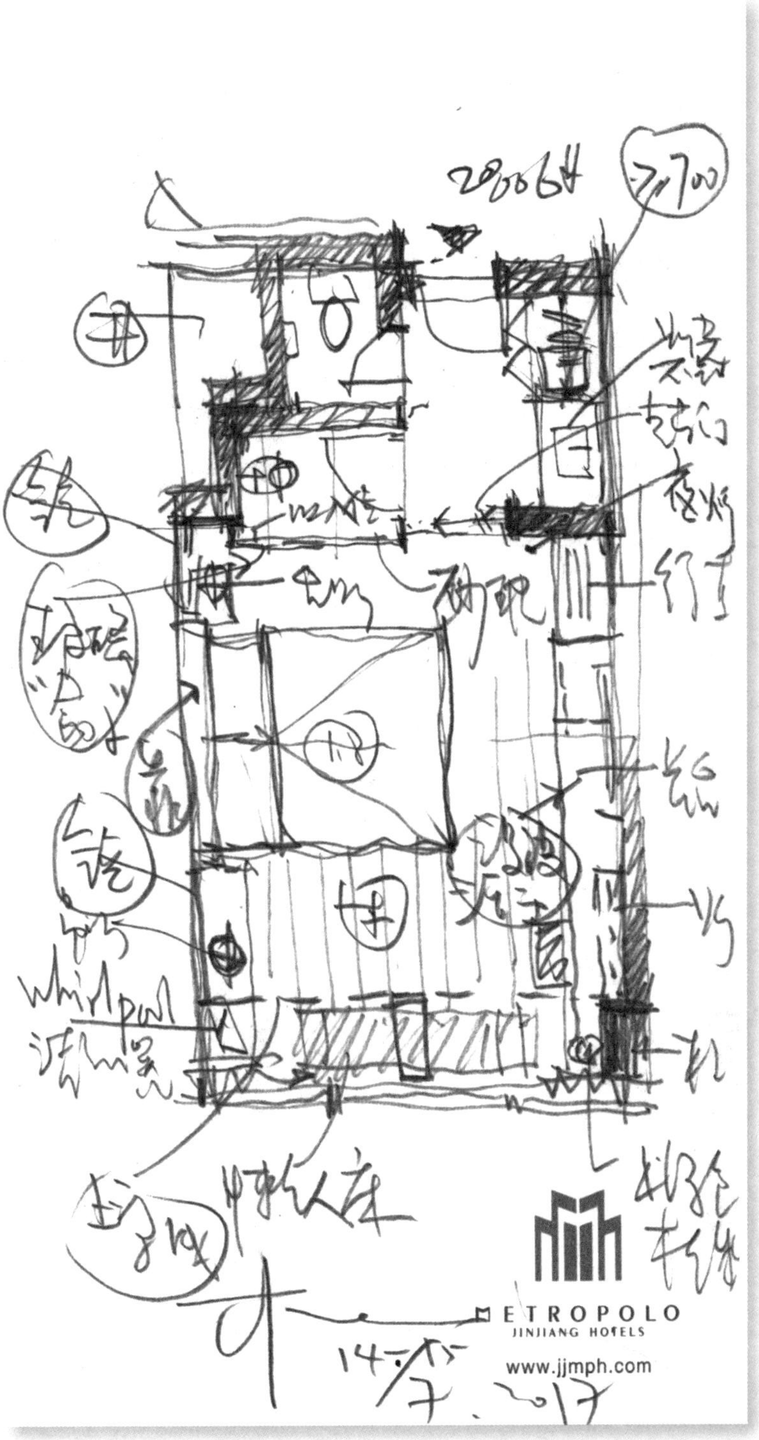

11

沈阳锦江都城

★★★★★

METROPOLO JINJIANG HOTELS SHENYANG CHINA

Address : No.1-1 Beiyi Middle Road, Tiexi District, Shenyang, Liaoning,China
中国辽宁省沈阳市
铁西区北一中路1-1号

Telephone : +86 24 8590 7777

Http : //www.jjmph.com

METROPOLO
JINJIANG HOTELS
www.jjmph.com

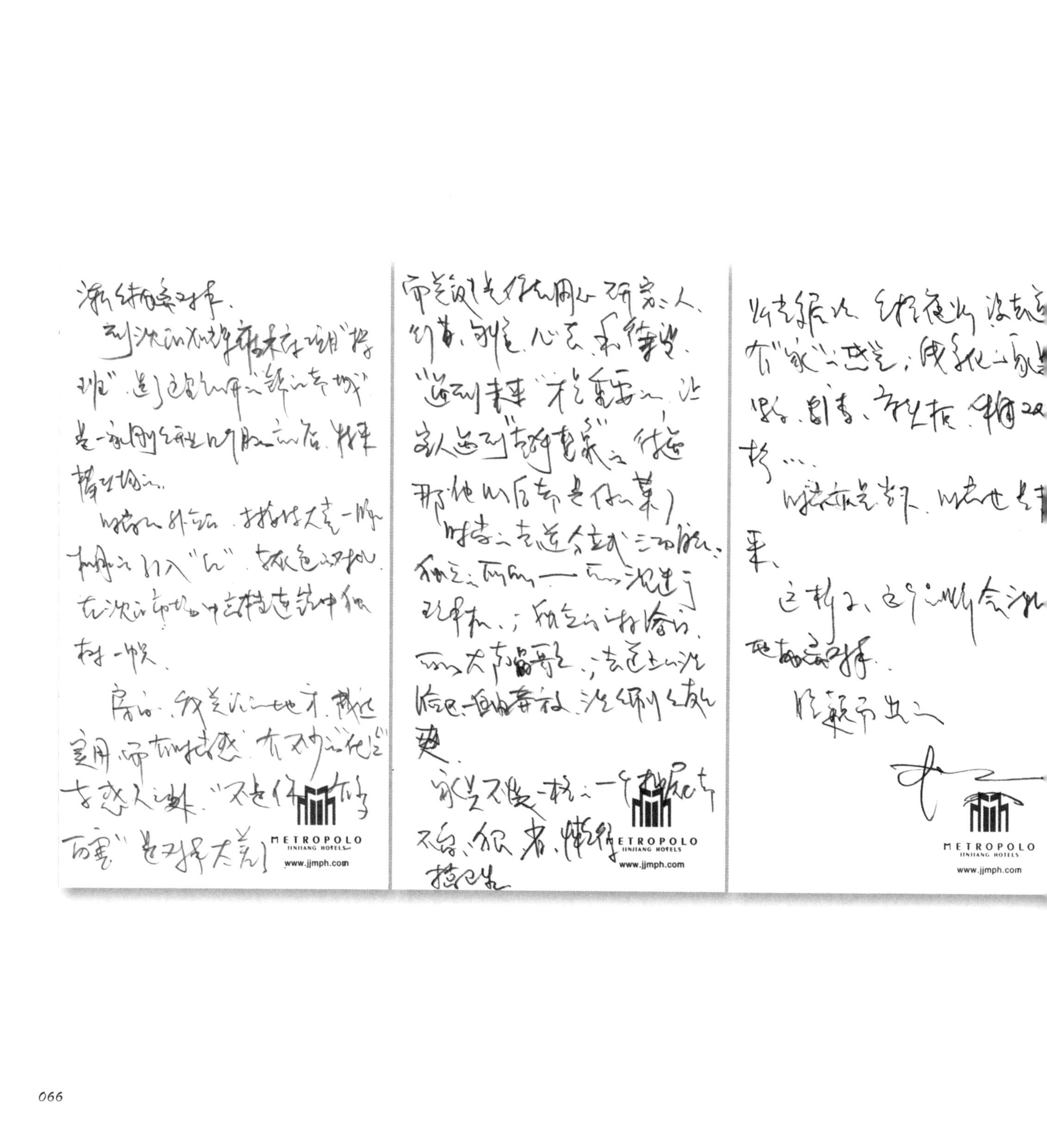

METROPOLO
JINJIANG HOTELS
www.jjmph.com

METROPOLO
JINJIANG HOTELS
www.jjmph.com

METROPOLO
JINJIANG HOTELS
www.jjmph.com

Leave Opponents behind Step by Step

渐渐抛离对手

到沈阳旭辉雍禾府项目“探班”，选了这里新开的“锦江都城”，是一家刚刚开业几个月的酒店，特来捧捧场。

I chose the new Jinjiang Metropolo when I came to visit Yonghefu Project in Shenyang. It was newly opened a few months ago. I should support it intentionally.

时尚的外立面，与接待大堂一脉相承地引入“红”，与灰色形成对比，在沈阳市场中高档连锁酒店中独树一帜。

The fashion façade and “Red”colour in the lobby came down in one continuous line. Compared with “Grey”colour, it was unique among the middle-high-grade hotels in Shenyang.

房间，我最关注的地方，成熟实用而有时尚感，有不少的“优点”与感人之处，“不是你有多厉害”，是对手太差了。

Guestrooms, which I focused on most, were mature and practical with fashion sense. There were lots of advantages and some touching details. The hotel wasn't quite perfect, just because opponents were too weak.

而关键是你用心研究：人、行为、习惯、心态和期望。

The key to success is that you have to study by heart: human beings, behavior, habits, attitude and hope.

“遇到未来”才是重要的，让客人遇到“超乎想象”的待遇，那他以后都是你的“菜”了！

Meeting the future is the most important. Let the guests enjoy unexpected treatment. And then he will be your hard fans.

沈阳锦江都城酒店
Metropolo Jinjiang Hotels, Shenyang, China

时尚的走道分立式三功能，独立的厕所——可以沉迷于玩手机；独立的淋浴间——可以大声唱歌；走道上的洗脸区——自由奔放，洗洗涮涮超级爽！

The fashion aisle was divided into three vertical parts: an independent toilet—you could indulge in playing cell phone; an independent bathroom—you could sing aloud in it; washing area on aisle—free and passionate, it was fantastic to shower !

家具不拘一格，一个抽屉都不留，狠，省，不用搞卫生。

The furnishing was diverse, without one single drawer, determined, saving. Not much cleaning was required.

灯光是多层次的，包括夜灯，没走廊，倒是有“家”的感觉。

The light is multi-level included night light. No corridor made us feel at home.

线条化的家具：写字台、行李架、床头柜、休闲双人椅……

The streamlined furniture: table, luggage rack, bedside table, comfortable love seat and so on.

时尚是当下，时尚也是未来。

Fashion is the present, also the future.

这样，这个品牌会渐渐地抛离对手。

In this way, this brand will leave opponents behind gradually.

脱颖而出。

And stand out!

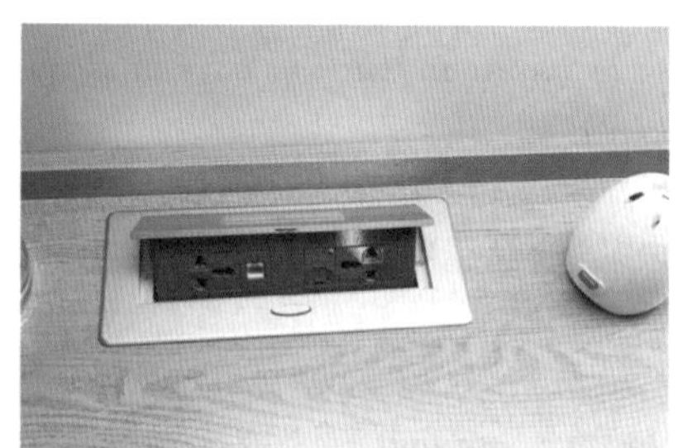

沈阳锦江都城酒店
Metropolo Jinjiang Hotels, Shenyang, China

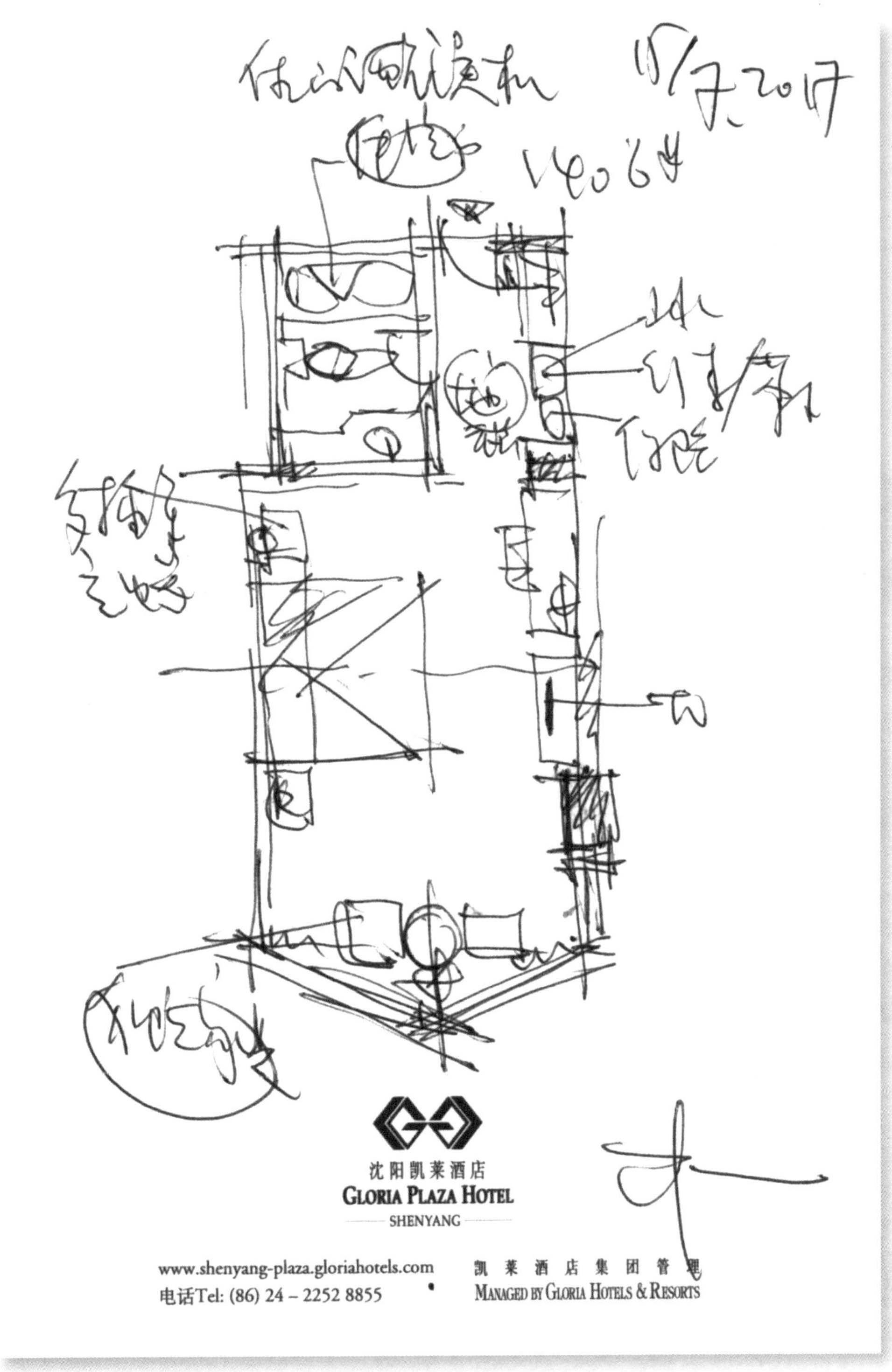

沈阳凯莱酒店

★★★★

12

GLORIA PLAZA HOTEL SHENYANG CHINA

Address : No.32, Yingbinstreet, Beizhan, Shenhe District, Shenyang, Liaoning, China
中国辽宁省沈阳市
北站迎宾街32号

Telephone : +86 24 2252 8855

Http : //www.Shenyang-plaza.gloriahotels.com

沈阳凯莱酒店
GLORIA PLAZA HOTEL
SHENYANG

www.shenyang-plaza.gloriahotels.com
电话Tel: (86) 24 – 2252 8855

凯 莱 酒 店 集 团 管 理
MANAGED BY GLORIA HOTELS & RESORTS

FOUR POINTS BY SHERATON

13 青岛城阳宝龙福朋喜来登酒店

★★★★★

FOUR POINTS BY SHERATON QINGDAO CHINA

Address : No.271 Wenyang Road, Chengyang District,Qingdao, Shandong, China
中国山东省青岛市城阳区文阳路271号

Telephone : +(86 532) 6696 8888

Fax : +(86 532) 6696 8806

Http : //www.FOURPOINTS.com/QINGDAO

青岛城阳宝龙福朋喜来登酒店

Four Points By Sheraton,Qingdao, China

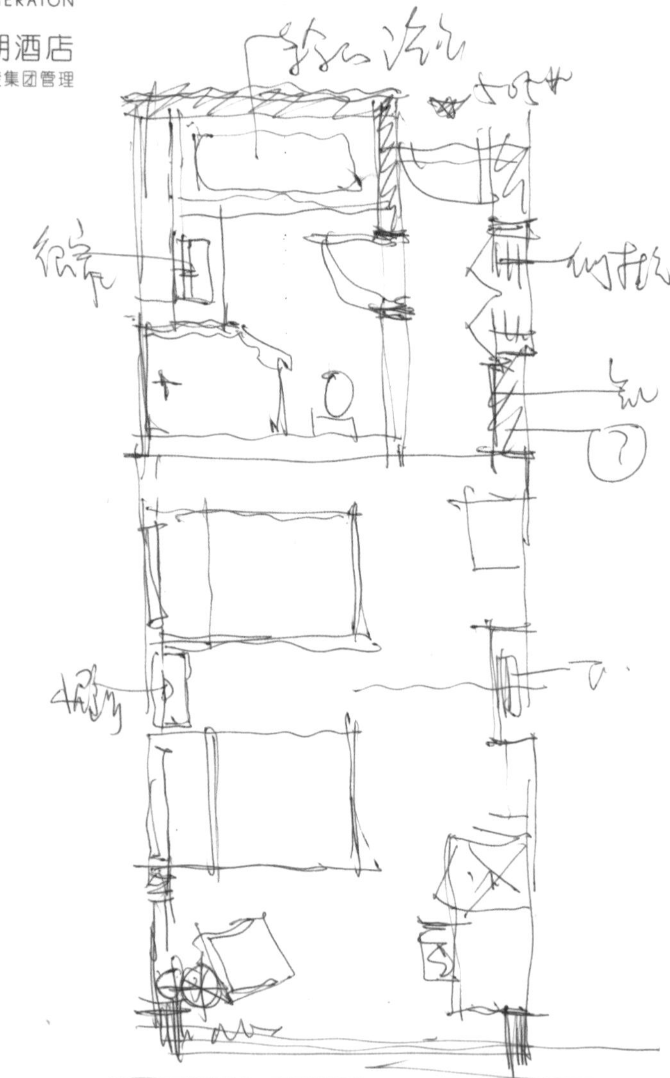

FOUR POINTS
BY SHERATON
福朋酒店
喜来登集团管理
青岛城阳宝龙福朋喜来登酒店
Four Points by Sheraton Qingdao, Chengyang
中国山东省青岛市城阳区文阳路271号，266109
No.271 Wenyang Road, Chengyang District,
Qingdao, Shandong 266109, China
电话 T 86 532 6696 8888 传真 F 86 532 6696 8806
FOURPOINTS.COM/QINGDAO

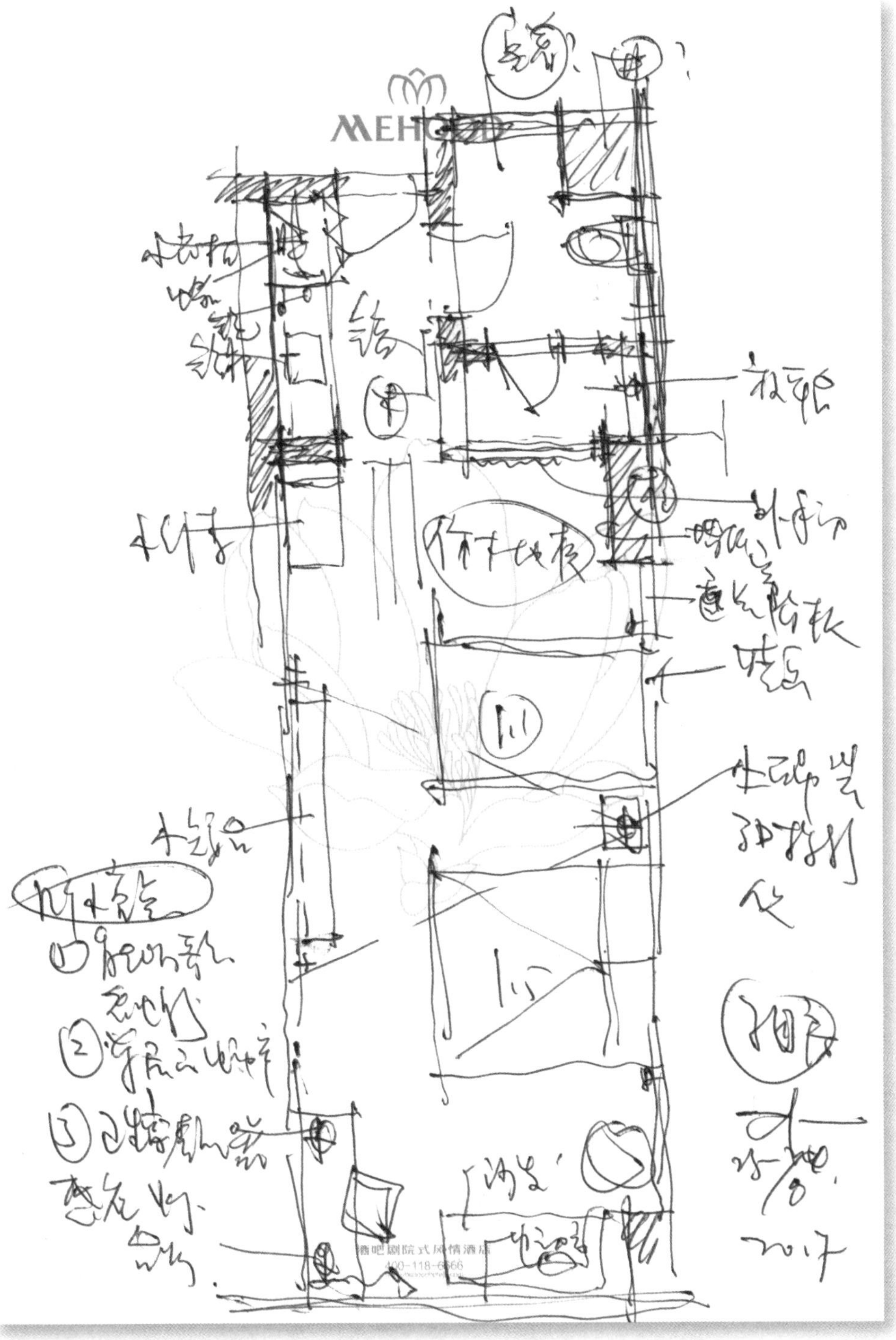

MEHOOD

美豪酒吧剧院式风情酒店

14

MEHOOD WEIFANG CHINA

Address : No.1 Building,Pingan Sijicheng, 3323 Beihai Road,Economic Development Zone,Hanting District, Weifang, Shangdong, China

中国山东省潍坊市
寒亭区经济开发区
北海路3323号
平安四季城一号楼

Telephone : +400 118 6666

Http : //www.mehoodhotels.com

ONE 96

15 香港ONE96酒店

★★★★

ONE 96HOTEL HONG KONG CHINA

Address : No. 196 Queen's Road Central, Hong Kong,China
中国香港皇后大道中196号
Telephone : +852 35196 196
Fax : +852 35196 100
Http : //www.96.com

香港ONE96酒店

One 96 Hotel, Hong Kong, China

T +852 35 196 196 F +852 35 196 100 196 Queen's Road Central, Hong Kong

www.one96.com

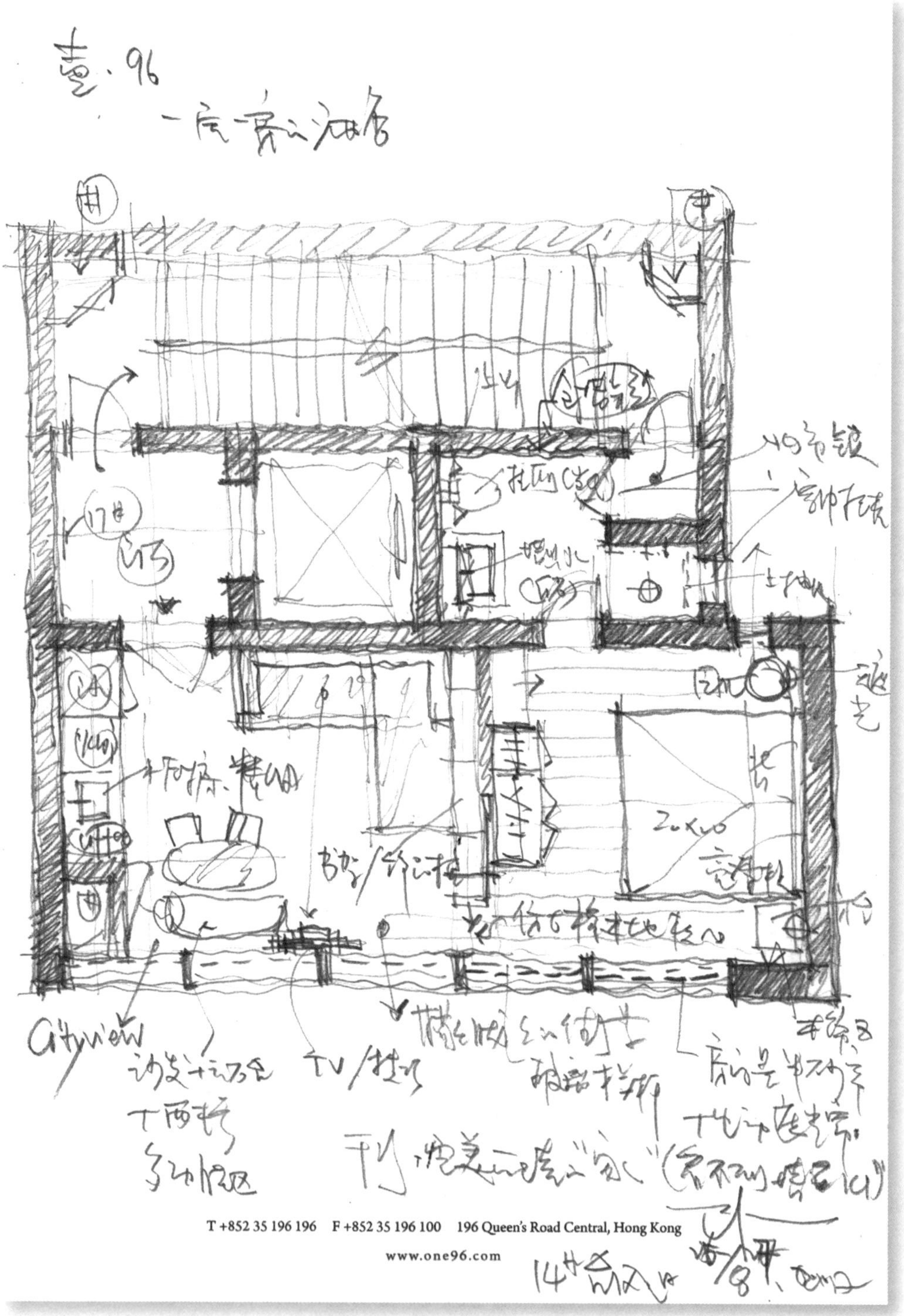

"壹"——一居一语之酒店　　[illegible] 8.2017

[illegible]

[illegible] 一居一语之酒店 [illegible]

[illegible] 新大道中196号之

"壹·96酒店" [illegible] 一层 [illegible] 六个房间 [illegible]

一层一户 [illegible]

[illegible]

一个服务员十一个客房 [illegible] 二楼（一层英式）一个20多平

[illegible] Lounge [illegible] 8层 [illegible] 30多个房间。

可能会给这个一篇幅吧！

一个酒店 [illegible]

"细节"每一处都是他们"亲手" [illegible]

[illegible]

[illegible]

一层一户 [illegible] 一层酒店，完美。

T +852 35 196 196　F +852 35 196 100　196 Queen's Road Central, Hong Kong

www.one96.com

[illegible]

"One"— the Hotel with Only One Guestroom on Every Floor

"壹"——一层一间房的酒店

香港寸土寸金，确是珍贵，连这种"竹笋楼"（就是标准层只有几十平方米且100多米高）都越来越被重新利用改造为精品酒店，这次就住到这一家网评不断，卖相相当极致的皇后大道中196号的"壹 96酒店"，以前以为它可以一层两三个房间。哇，这个是一层一户。太牛了！

Land is really precious in such a small island like Hong Kong. Even this kind of bamboo-shoot building, which its standard layout is only dozens of square meters and over 100 meters tall, is rebuilt to be a boutique hotel more often. I stayed in this hotel with lots of comments on the Internet. One 96 Hotel is at No 196, on Queen's Road Central looks extreme perfect. I thought there were two or three guestrooms on each floor. Wow! Every floor has only one guestroom. I stayed on the seventeenth floor, the same number as my birthday, so perfect!

香港ONE96酒店
One 96 Hotel, Hong Kong, China

配套嘛，没有，首层斜向退缩的入口成为遮雨篷，一个服务员+一个小前台，没了，二层为一个20多平方米的酒廊，再到38层就是30多个房间。

No supporting facilities. The entrance which was on the first floor became an awning. There was nothing else but a waiter and a small reception. The lounge on the second floor exceeded twenty square meters. There were more than thirty guestrooms in total from other floors to the thirty-eighth floor.

可能有61%的实用率吧！

Maybe the efficiency rate was about sixty-one percent.

一间房画起来还是非常有趣的，佩服设计师的"精打细算"，每一处都是这么的"寸土必争"，小厨房，多功能餐厅，工作小客厅，卧室+全功能的洗手间，完美!及至现代设备：浴室，有电动帘——总开关制，区域手机，而且有非常多的储物、挂衣空间。

It was funny to draw the guestroom. I admired the designer because he was so careful and wise that he could make full use of every inch of land. There was a mini kitchen, a multi-function dining hall, a little living room for work, a bedroom and a fully function bathroom. Perfect! There were also modern facilities: electrical curtain in bathroom, master switch, and regional mobile phones, a lot of space for storage and hanging clothes as well.

一房，可以让你长住的一间酒店，完美。

Only one guestroom, the hotel is so perfect that you will hope to stay for long.

華翔大酒店
HUAXIANG HOTEL

地址：中国·山西侯马市火车站南侧
电话：(0357) 4229688
传真：(0357) 4212299
邮编：043000

·H·U·A·X·I·A·N·G·H·O·T·E·L·

16

华翔大酒店

★★★★

HUAXIANG HOTEL HOUMA CHINA

Address : South of Houma Railway Station, Shanxi, China
中国山西省侯马市火车站南侧

Telephone : +86 357 422 9688

Fax : +86 357 421 2299

华翔大酒店
HUA XIANG HOTEL

同一个酒店、不同的心情

第二次来到侯马华翔，第一次[illegible]，而这一次是[illegible]

[illegible]

一个[illegible]城市的一个[illegible]酒店，十二年[illegible]

[illegible]

地址：中国·山西侯马市火车站南侧
电话：（0357）4219999 4219888
传真：（0357）4212299
邮编：043000

H·U·A·X·I·A·N·G·H·O·T·E·L

华翔大酒店
HUA XIANG HOTEL

感谢你们的信任，第一次[illegible]

[illegible]

[illegible]（本酒店、中档酒店）[illegible]

"每一步"[illegible]

[illegible]酒店[illegible]

地址：中国·山西侯马市火车站南侧
电话：（0357）4219999 4219888
传真：（0357）4212299
邮编：043000

H·U·A·X·I·A·N·G·H·O·T·E·L

华翔大酒店
HUA XIANG HOTEL

相信也是不错的选择，[illegible]

[illegible]

加油华翔，加油我们[illegible]

[illegible]

地址：中国·山西侯马市火车站南侧
电话：（0357）4219999 4219888
传真：（0357）4212299
邮编：043000

H·U·A·X·I·A·N·G·H·O·T·E·L

The Same Guestroom, but Different Feelings

同一个房间，不同的心情

第二次来到侯马的华翔，第一次来看看、聊聊，而这一次是来现场复核尺寸和汇报方案！

It was my second time staying in Huaxiang Hotel in Houma. At the first time, I just came to look around and have a talk. This time I came to the site and checked the size again and reported our plan.

一个不知道在几线城市的一个好地段的小酒店。十二年前的设计，显然过时了。我们的介入希望是以“合适的力度去改变这个酒店的时代定位”，像当年包头的“凯宾”（酒店）一样。

It was a small hotel which lies at a good location, but in a small city. The hotel was designed twelve years ago, so it was obviously out of fashion. We hoped our intervention would change the modern positioning of the hotel with our appropriate strength, just like Kempinski Hotel in Baotou in the past.

感谢甲方的信任，第一次非常简单的交流就定下了设计任务，也算是我们在酒店设计行业（小酒店，中档酒店）的再一次进步。

Thanks for Part A's trust, we agreed on the design plan at the first communication. We also made progress again on hotels (small hotels and mid-range hotels.)

“每一步都算数”，慢慢来，但要快，我们从这一类四星酒店深耕，相信也是不错的定位，相信像我们在地产的“江湖地位”一样，会三五年成为一面旗帜，和同事一起努力。

"Every step counts."It should be slow but must be quick. We begin with this kind of four-star hotels and do further study. I believe it is not a bad target. I also believe we will be a flag in three or five years, just like our position in real estate. I will work hard with my colleagues.

加油，侯马，加油，我们的一个小酒店，值得你的体验、期待！

Come on, Houma! Come on, our small hotel! It is worth your experience. I am looking forward to it.

改造、传统和勤俭的信仰。

也就那一段，这城市还遗存明清，更加不要说这个酒店之地——侯马，"新田"之地，并非没有底蕴！

酒店是十几年前的，当时仍是二十几年前的土建（建筑）设计，[illegible]……我们[illegible]，也许是双方[illegible]

其实是惦念的，我们两个人分别住在[illegible]

我们是住的两间房，是套房，[illegible]一些校友[illegible]，没有什么[illegible]，但在[illegible]和有机会[illegible]住到[illegible]

[illegible]

[illegible]

我们二人[illegible]信任[illegible]

[illegible]

[illegible]

[illegible]会儿！

[illegible]

地址：中国·山西侯马市火车站南侧
电话：(0357) 4229688
传真：(0357) 4212299
邮　　编：043000

Transformation, Deserves Your Expectation and Trust

改造，值得你的期待与信任

地理水平一般，运城不知道在哪里，更加不要说这个酒店的城市——侯马，与“猴年马月”不知道是否有关！

My geography was so poor that I didn't know where Yuncheng was, even the hotel was in it —Houma. Was it related to the Chinese saying "Horse month of the Monkey year"?

酒店是十几年前的室内设计，而且是二十多年前的土建（建筑）设计，所以有改造的必要性，找到我们，也许是双方的幸运。

The interior of the hotel was designed more than a decade ago. And it was built over twenty years ago. So its transformation had necessity and modernization. Our cooperation was lucky for both of us.

业主热情接待，我们两个人分别住尽端走廊两边的房间，我住的是这个有斜角的房间，类套房，虽然之前我们有了一些初步的方案，改动有大有小，但相信是可行的和有提高性的，但到现场一看，还是觉得设计真的要认认真真做，才能为业主带来好的商业形象和让租户租有所值。

The owner gave us warm welcom and we stayed in two guestrooms at the end of the corridor. I stayed in a room with a bevel, a little like a suite. Although we had some preliminary plans with some big or small alterations to the hotel, I believed it should be feasible and improving. When we got to the hotel, I found that we should make the design seriously so that it would bring a good business image for the owners and let all the tenants feel it well worth the investment.

我们的介入，多方面共同的努力，必然有好的结果，信任让期待顺利得以实现。半小时的汇报，董事长相当有见地地确认我们的设计。

I believed that our intervention needed extensive joint efforts and we would have a good result. Trust made the expectation come true successfully. After the half- an- hour report, the president confirmed our design insightfully.

一年后，地标式酒店将会面世。

In a year, a landmark building will be born!

期待。

Let's look forward to it.

HYATT

GRAND HYATT CHANGSHA

长沙君悦酒店

17

★★★★★

GRAND HYATT CHANGSHA CHINA

Address : No.36 Middle Xiangjiang Road, Tianxin District,Changsha, China
中国湖南省长沙市天心区
湘江中路36号

Telephone : +86 731 8823 1234

Http : //www.hyatt.com

E-mail : Changsha.grand@hyatt.com

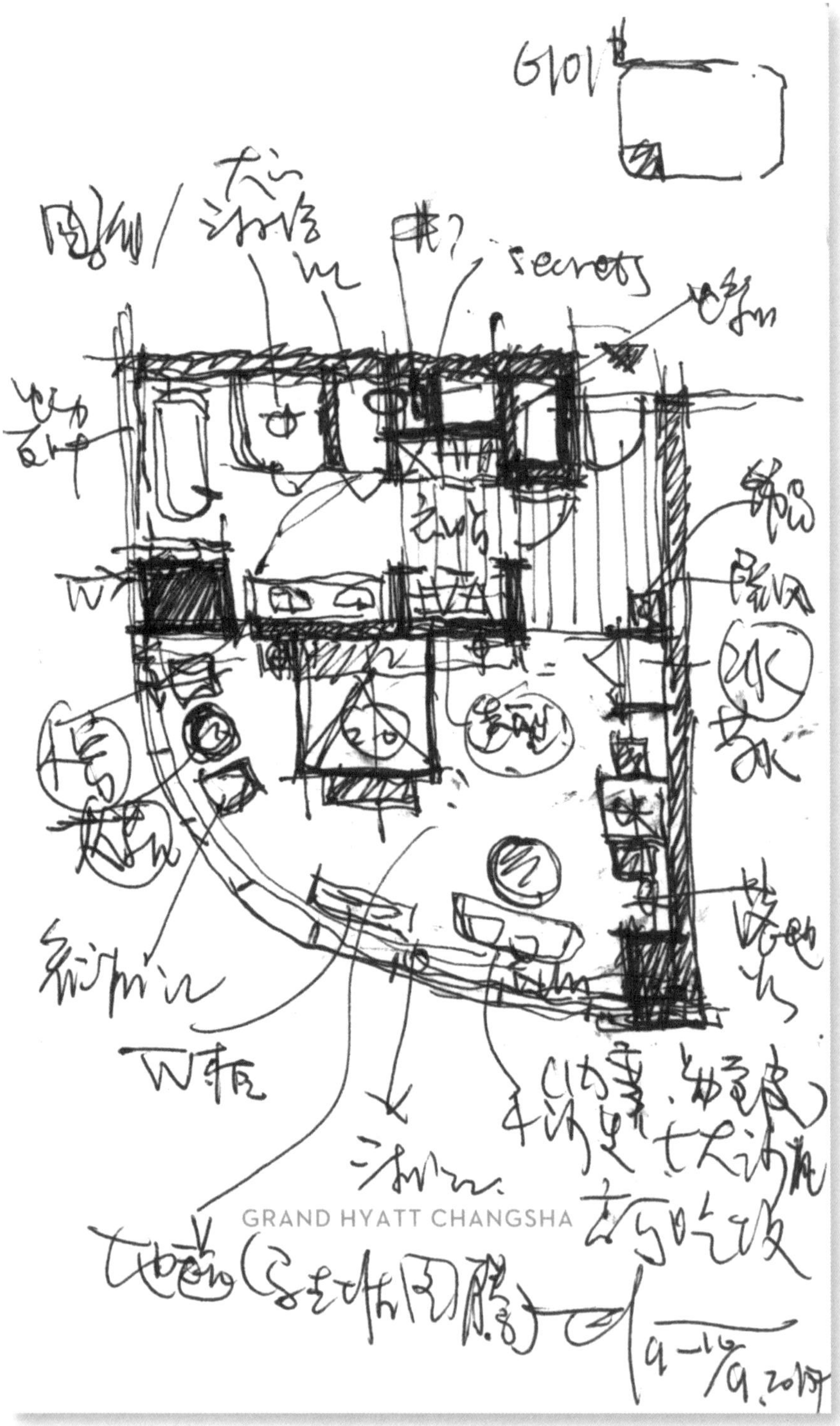

GRAND HYATT CHANGSHA

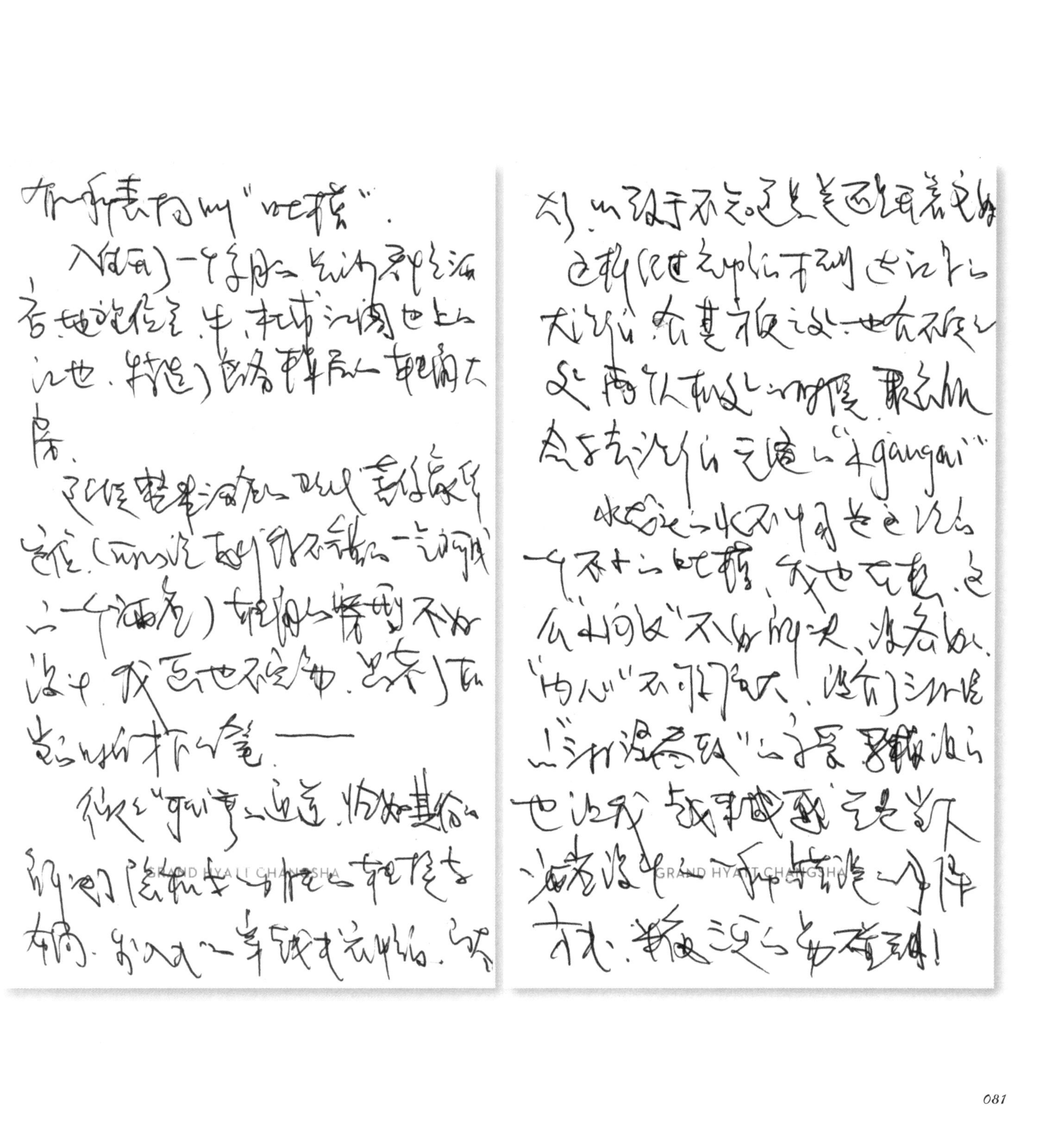

[illegible]"吃相"。

[illegible]

[illegible]

[illegible]"gangai"

[illegible]"不好"[illegible]

[illegible]

GRAND HYATT CHANGSHA

全的搭建出来的建筑[illegible]，结，立体！所有与时代并进的东西都不是今天的事物"。

太大了、[illegible]、[illegible]也[illegible]人[illegible]。[illegible]用餐[illegible]，[illegible]也[illegible]一下、[illegible]、[illegible]。

[illegible]

（[illegible]）

[illegible]一[illegible]。

[illegible]、[illegible]，[illegible]一种选择，没[illegible]，也可能不会失去一个地方。

[illegible]，不[illegible]，可以到此[illegible]，[illegible]一[illegible]"[illegible]"。

也许这是一种[illegible]，更加[illegible]。

[illegible]也[illegible]一。

GRAND HYATT CHANGSHA

长沙君悦酒店・老到的布局和家具
Grand Hyatt, Changsha, China

A Kind of Praise is a Complaint
有一种表扬叫“吐槽”

入住开了一个多月的长沙君悦酒店，地理位置优越，在杜甫江阁边上的江边，特选了商务楼层的转角房。

I stayed in Grand Hyatt Changsha which was open for more than one month. Its location was perfect—beside Du Fu River Pavillion, by Xiang River. I chose a business room at the corner intentionally.

延续整体酒店的现代奢侈豪华定位（可以说是设计得不错的一个一气呵成的酒店），转角的房型不好设计，我画也不容易，思考了相当长的时间才下的笔。

Continue the modern luxury positioning of the whole hotel (The hotel was pretty good and designed in one take) .It was not easy to design the corner room. It was also not easy to draw. It took me a lot of time to think and then to draw.

微微“拐弯的通道”，恰如其分地解决了隐私与功能的转换与布局，步入式的穿越式衣帽间，门大了，以至于不知道是否开着为好！这样经过衣帽间才到达可览江景的大洗手间，有其方便之处，也有不便之处，两个人相处的时候，取衣服会与去浴室产生“无缝对接”的小“尴尬”。

The transformation and layout of privacy and function was properly solved by a little bend tunnel. The door of a walk-in and pass-wise closet was so large that I wondered if it was good to open the door. I got to the big bathroom facing the river view through the closet. It would be convenient and inconvenient. It was a little embarrassed for two guests when one put on clothes while the other took a bath.

水龙头的水流不畅是这次一个不小的吐槽点，我也在想，这样的“小问题”不好解决，设备好，“内心”不够强大，没有了淋浴的“淋漓尽致”的享受，卫生间设门也让我越来越感受到这是当下酒店设计的一种错误的引导方式，半夜三更易碰到。

The poor water was a complaint. I thought this problem was too "small" to solve. The facility was good, but water supply was not strong enough to fully enjoy the shower. The toilet with door made me think that it misled on the design of hotels. It was easy to crash at midnight.

全江景落地窗的睡眠区，哇，无敌！确是长沙其他的酒店所不具备的“杀手锏”。太大了，弧形窗，尽揽湘江西北面的景色，有多种功能，设计师也颇尽心思，靠墙的书写、用餐区，由吊灯界定空间，休闲沙发座，内弯式扪布就难以避免的皱巴巴的，设计师也许还要细究一下。床居中，正规，霸气而自然而然。

Sleeping area had a large French window overlooking the river view. Cool! It was really the only hotel with such a perfect location while other hotels didn't have. The lunette was so large that the northwest of Xiang River could be seen. The designer tried his best to make the room with multiple functions. Dining and working area was along the wall, the space was divided by a chandelier. The curved cloth was hard to avoid being wrinkled. The designer should think it over. The bed was in the middle, normal and domineering.

西向布置了一组向江的休闲小椅（沙发），高的圆几，特别的大，颇有想法，可以在此看看书，晒晒午后的太阳，当然也可以顺便看看日落，发发呆。及至电动帘，圆弧形，噪声颇大，而且预位不够，窗纱开启会“磕磕碰碰”。这是不容易的一种选择，没有其他办法，也可谓不完美的一个地方。

In the west, there was a group of leisure sofa facing the river view. A high round tea table was quite big and thoughtful. Read books in the afternoon sun and enjoyed the sunset. Of course doing nothing in a daze was also a good idea. But the electric circular arc curtain was too noisy, the space was so narrow that the curtain was always crash when opening it. It was not easy to make a choice but no ways else. This might be regarded as an imperfect point.

小小吐槽，不碍首肯，可谓到此为止。长沙最值得一住的“新酒店”！

A little complaint, but it didn't interfere with my appreciation. Just stop here. It was the new hotel which was the most worth staying in Changsha.

也许正是这种挑刺才能让我更加佩服设计师将房间做到这么的有趣和有味道。

Maybe this kind of criticism could make me admire the designer. He made the guestroom so interesting and tasteful.

少有的湖湘文化气息，也让这里成为唯一。

The rare Huxiang culture made it become the only one.

长沙君悦酒店

Grand Hyatt, Changsha, China

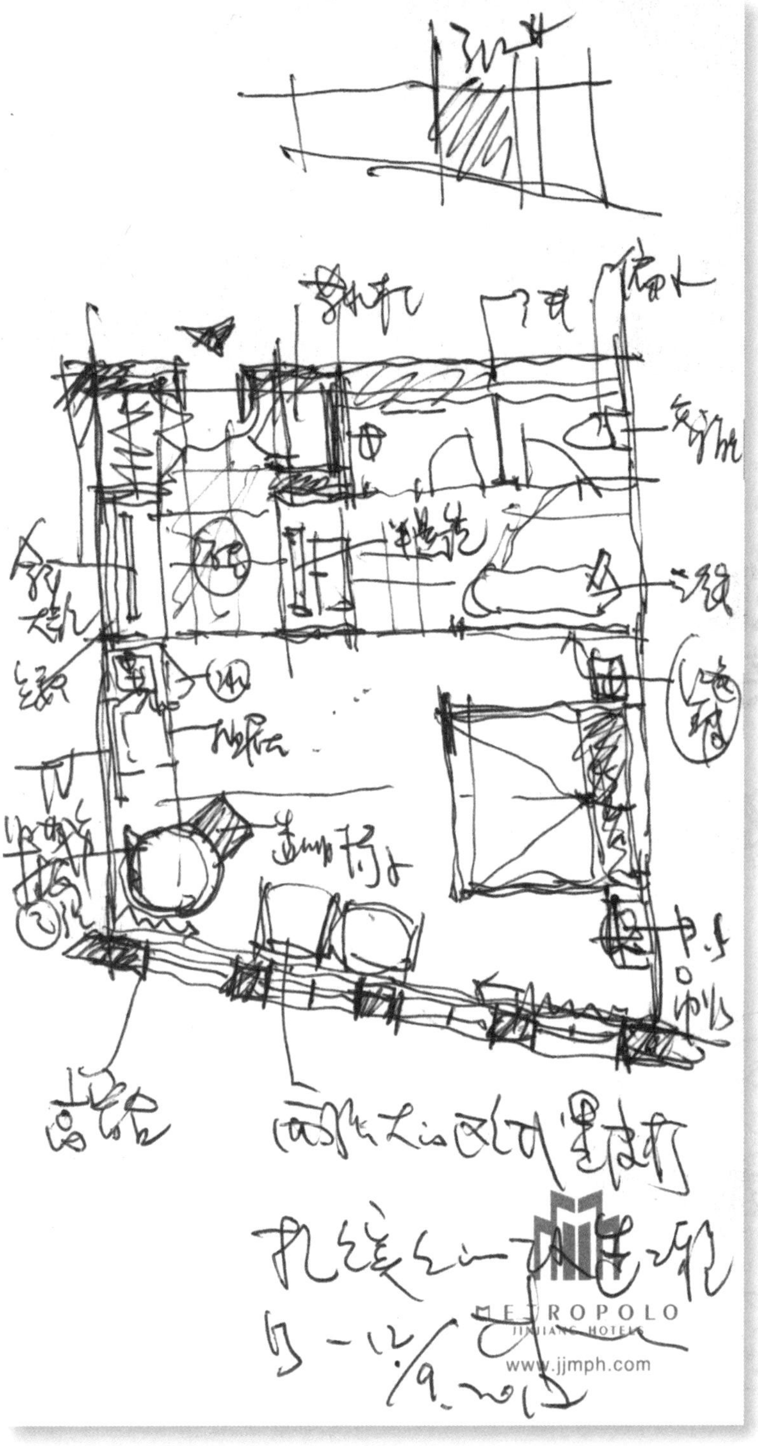

锦江都城经典南京饭店

18

★★★★★

METROPOLO CLASSIC NANJING HOTEL SHANGHAI CHINA

Address : No.200 Shanxi South Road,Huangpu District, Shanghai, China
中国上海市黄浦区
山西南路200号
（近南京路步行街）

Telephone : +86 21 6322 2888

Http : //www.metropolohotels.com

HOTELS
WORLDWIDE

上海W酒店

19

★★★★★★

W HOTEL SHANGHAI CHINA

Address : No.66 Lvshun Road, Shanghai, China
中国上海市旅顺路66号
Telephone : +86 21 5101 5566
Http : //www.wtwhotel.com
E-mail :info@wtwhotel.com

上海
W Hotel, Shanghai,

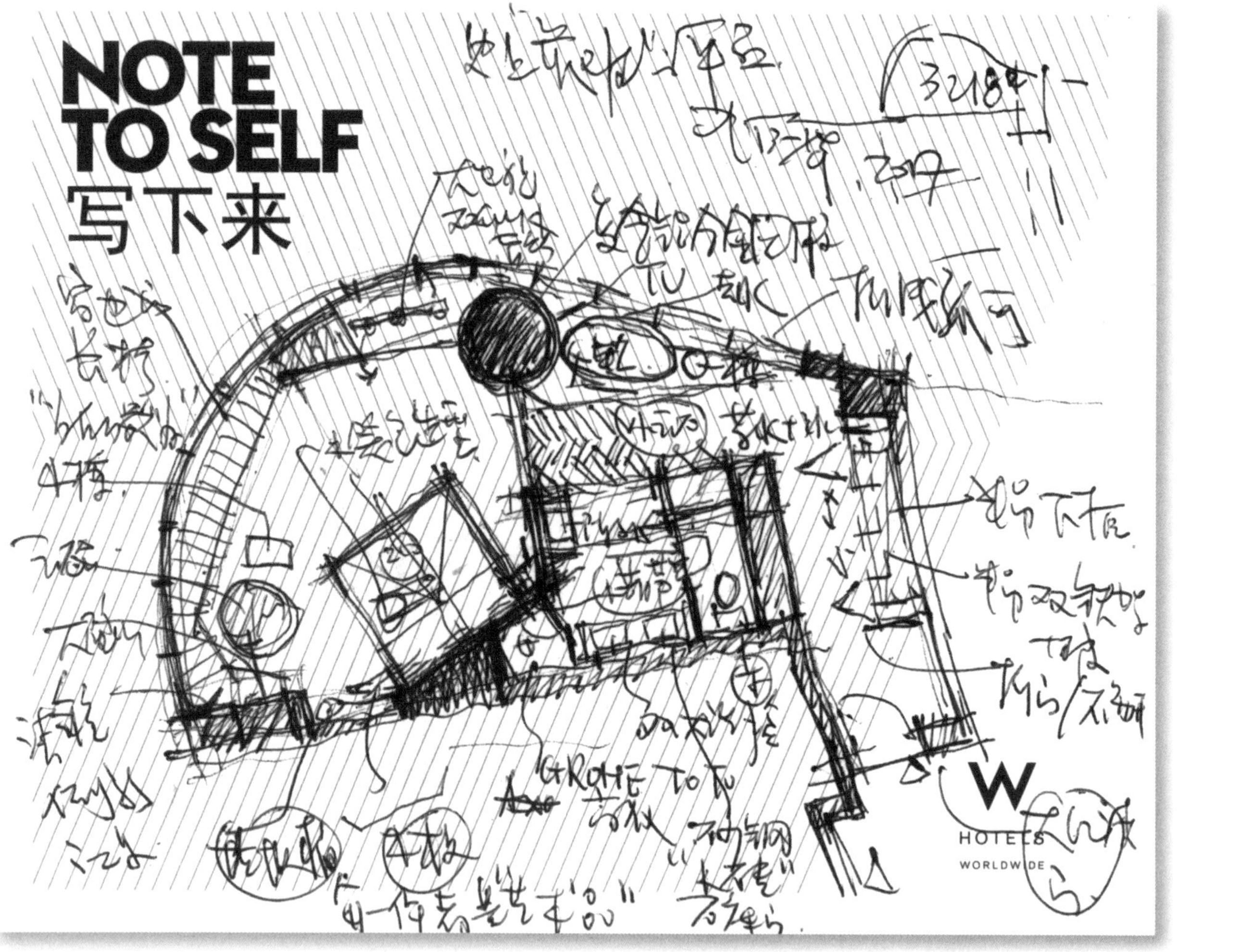

NOTE
TO SELF
写下来

W
HOTELS
WORLDWIDE

晓陆！ 11.4-9.2014

W+Shanghai 注定是一个深刻的地方。我们的关系，看着上海的设计居然来得这么近了！

走了一个城市的设计，看起来一个经典格局，就像中国的外方是大红的漆柱，里面的内墙做木搭的结构，不错的小砖

你、石库门的感觉（和我的样子的思路。）就像是年轻人的衣着和细节，小清新的也要好的东西派头。与外面的时尚和谐，不用放在心里不适的设计了（花费钱）就是娱乐的“小资”的

娱乐设计的关键是使人们在W的时候，我们这样的设计，是要像funny，走过也好，就是让你自己去了，所以这就是要向上，就是让娱乐了

晓陆说的，在，在这个……就是娱乐精神的一部分。我们一起会努力做得更好，加油……

晓陆加油！

Knowing A Little

略懂

上海W酒店注定是一个“潮”的地方，开业三个月，借着上海的家具展就来凑凑热闹了！选了一个转角的江景房间，有挑战的一个平面布局，挺练手的，外面是大红的漆门，里面以哑光橡木搭配哑光不锈钢小砖，仿“石库门”的感觉（类似我的小房子的想法），挺“娱乐”的是它的家具与配饰，小葫芦通电玻璃的床头漏窗与淋浴间呼应，热水蒸汽，不用通电就有不透的效果（浪费钱），最“娱乐”的当属床上的“小笼包”和“筷子”，娱乐设计师和我们住的人。

W Hotel Shanghai was certainly a fashion place. I came to join the fun by attending the Shanghai Furniture show. I chose a guestroom with river view at a corner. It was a challenge to draw plan layout, a little difficult. Outside was the bright red gate, while matt oak matched matt stainless steel small brick inside. All these imitated the feeling of "Shikumen" (just like my small house). The interesting things were its furniture and accessories. The windows behind the bed with a small electrical gourd and the bathroom had worked together, including hot water and steam. But it was opaque without electricity (waste money). The funniest was "Xiao Long Bao" and chopsticks on bed. They entertained designers and our guests.

“W”嘛，就是要有娱乐，想想也对，生活这么快节奏的当下，什么是最受欢迎的？当然是娱乐了。

W Hotel is just for fun. I think it is right. What is the most popular in this high speed era? Of course the entertainment.

上海W酒店
W Hotel, Shanghai, China

用色艳丽，灰、白、蓝、黄……挺有“娱乐精神”的一间，相信W会坚持娱乐精神，至于老外懂不懂，重要吗？

The color was gorgeous, grey, white,blue, yellow and so on. The hotel was full of fun. I believe W Hotel will stick to its spirit of entertainment. Is it important whether foreigners understand or not?

略懂就好了！

It is good to know a little.

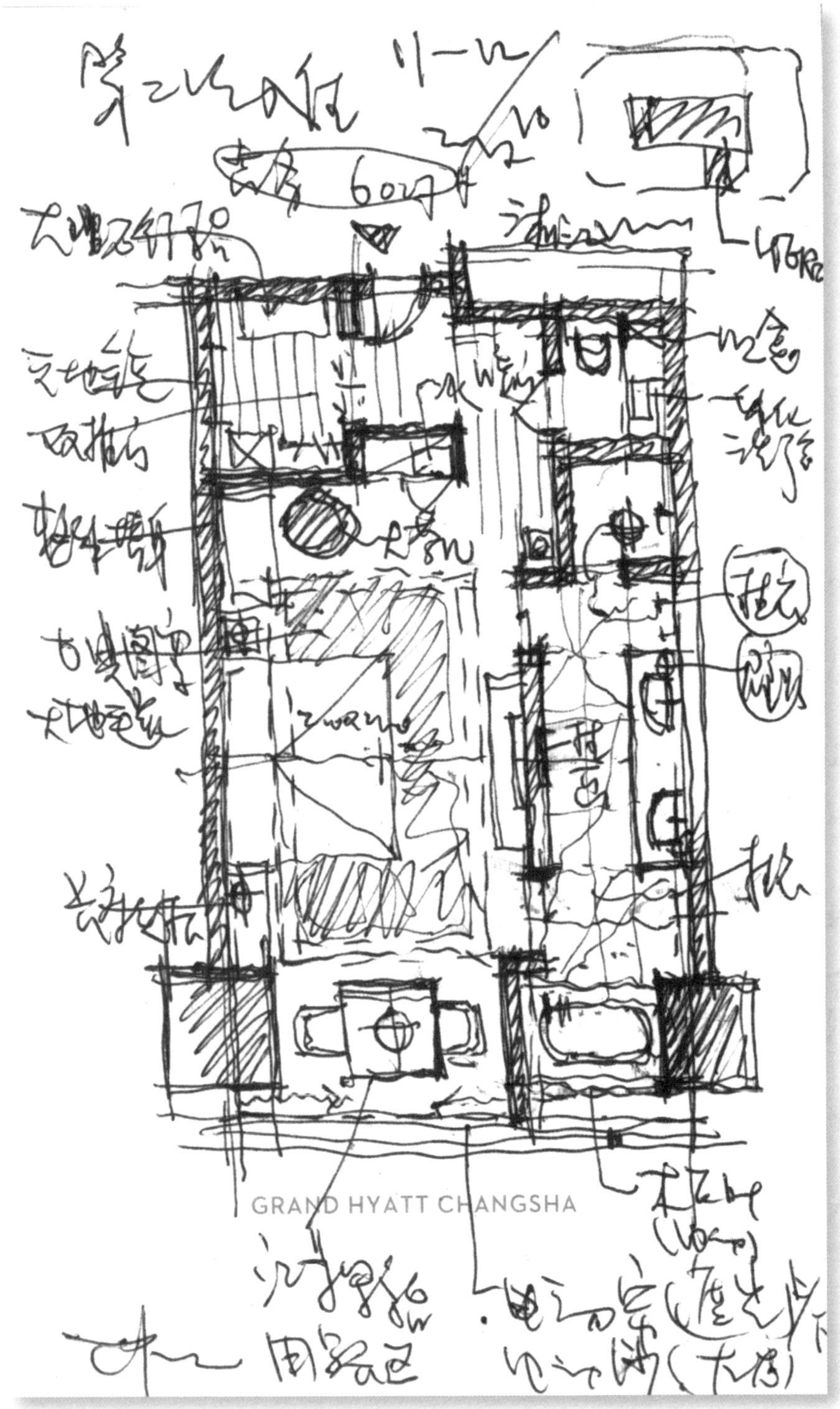

HYATT

GRAND HYATT CHANGSHA

长沙君悦酒店

★★★★★

20 GRAND HYATT CHANGSHA CHINA

Address : No.36 Middle Xiangjiang Road, Tianxin District,Changsha, China
中国湖南省长沙市天心区
湘江中路36号

Telephone : +86 731 8823 1234

Http : //www.hyatt.com

E-mail : Changsha.grand@hyatt.com

长沙君悦酒店·这个就是标准房

Grand Hyatt, Changsha, China

[illegible]

[illegible]

GRAND HYATT CHANGSHA

[illegible]

Try A Standard Room for the Second time

第二次，就住一下标准房吧

雾非常大的一天。早上，窗外几乎都看不清橘子洲头了，望江的大床房还是相当有吸引力的！腾云驾雾的感觉。

What a foggy day! I could hardly see the Orange Island through the window in the morning. The deluxe king-size room which faces the Xiang River was quite attractive with the feeling of the clouds.

还是流行的1.5开间（比瑞吉酒店稍窄一点，应当有6.3米左右吧）,可以让入口、洗手间都有了现代人喜欢的方式：隐私，迂回，厕浴分离，适合长时间在洗手间里面玩手机，双洗手盆设计也相互不影响。这些都是简单而自然的想法，但要做得合理而宜人真的不容易，这个平面可谓“精、细”。

The room was 1.5 metrewide(It was a little narrower than St Regis Hotel, should be around 6.3 meters). The style of entrance and restroom was popular with modern people, privacy and circuity. The toilet and the bathroom were separated. It was suitable to play mobile phones in the toilet for long. The guests didn't affect each other because of the double basins. The idea was simple and natural but not easy to make it reasonable and pleasant. This plan was really delicate and meticulous.

用材上，同上一次体验的弧形大房间一样，看不到一点“没有”装饰的地方，这也许是当下五星级酒店“全装饰”的方式吧（我觉得会流行好一阵子的）。

On the material, it was the same as the curved king-size room I experienced last time. I could not see any place which was not decorated. Maybe it was the style of "Full Inside Decoration" in a five-star hotel at the present. (I thought it would be popular for some time.)

大纹理的木材，仿陶瓷红色板，深浅灰木色（就是与陶瓷相结合），很有“地域风情”（我认为的马王堆文化）。

Big figured wood, imitative ceramics red board and shades of gray wood color (it was combined with ceramics) were full of regional customs. (I found it the culture of Mawangdui.)

最重要的当然是相互之间的统一和谐，相得益彰，LTW公司的林丰年先生的设计确是有其独特之处。

The most important thing was that all of these were so unity and harmony, complement each other. Mr. Lin Fengnian really had his own uniqueness.

标准层平面能做到这么有吸引力也是值得好好学习的！

It was worth learning because the design of a standard plane was so attractive.

Oriental International Hotel,Heyuan
萬綠湖東方國際酒店
中國・河源

21 万绿湖东方国际酒店

★★★★★

ORIENTAL INTERNATIONAL HOTEL HEYUAN CHINA

Address : No.9 Gangjian Road, Xingang Town, Dongyuan County, Heyuan, Guangdong,China
中国广东省河源市
东源县新港镇港建路9号
Telephone : +86 762 229 9999
Http : //www.orientalheyuan.com

Oriental International Hotel,Heyuan
萬綠湖東方國際酒店
中國・河源

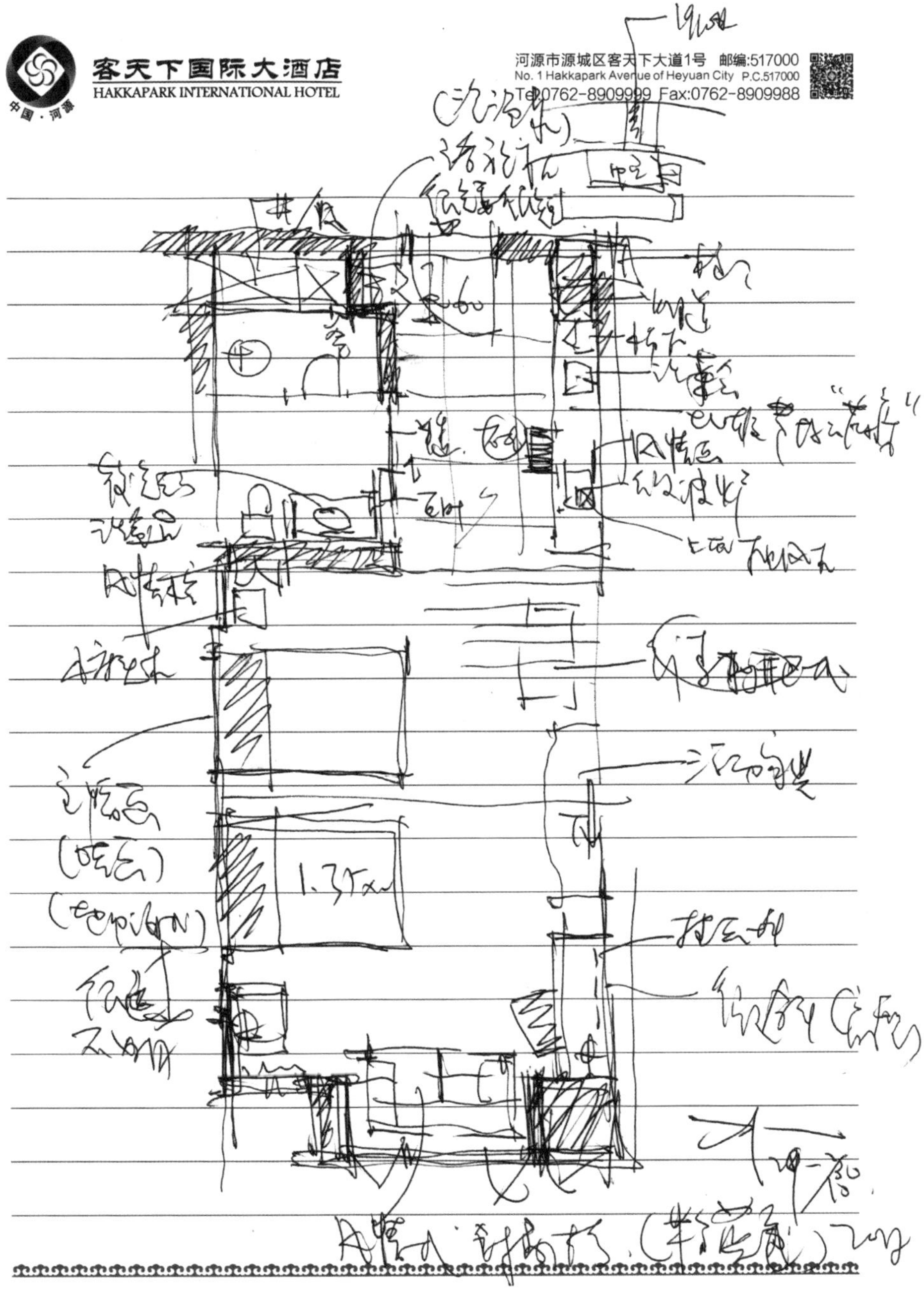

22

客天下国际大酒店

★★★★★

HAKKAPARK INTERNATIONAL HOTEL HEYUAN CHINA

Address : No.1 Hakkapark Averue of Heyuan,China
广东省河源市
源城区客天下大道1号

Telephone : +86 762 890 9999

Fax : +86 762 890 9988

Http : //www.ktxjq.com

客天下国际大酒店
HAKKAPARK INTERNATIONAL HOTEL

河源市源城区客天下大道1号 邮编:517000
No. 1 Hakkapark Avenue of Heyuan City P.C.517000
Tel:0762-8909999 Fax:0762-8909988

中欧行（2017年11月12~22日）

The Trip to Central Europe

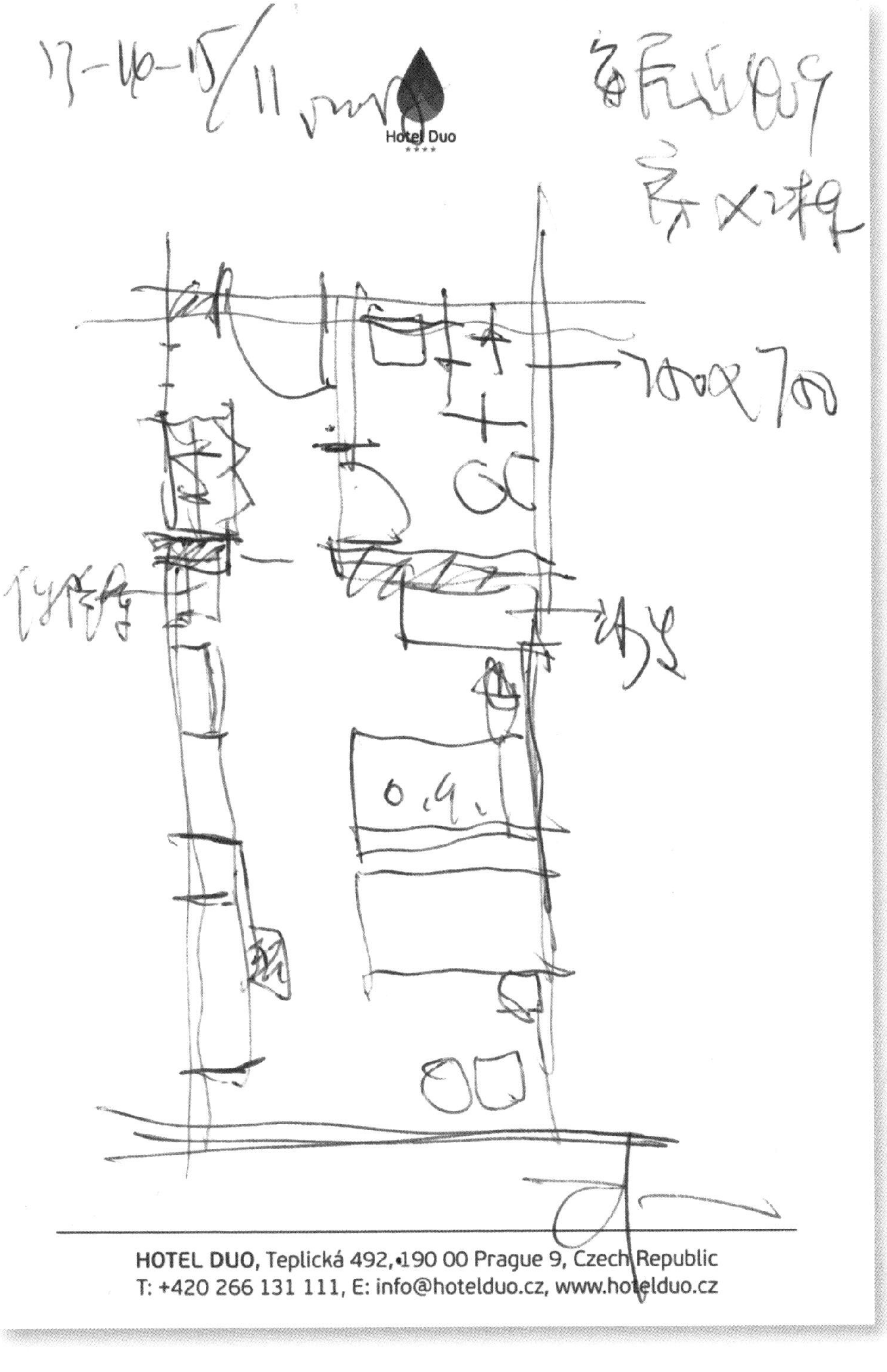

Hotel Duo

HOTEL DUO, Teplická 492, 190 00 Prague 9, Czech Republic
T: +420 266 131 111, E: info@hotelduo.cz, www.hotelduo.cz

23

杜鸥酒店

HOTEL DUO PRAGUE CZECH REPUBLIC

Address : Teplicka 492, 190 00 Prague 9, Czech Republic
Telephone : +420 266 13 111
Http : //www.hotelduo.cz
E-mail :info@hotelduo.cz

杜鸥酒店
Hotel Duo, Prague, Czech Republic

14/11.2017

Hotel Duo

HOTEL DUO, Teplická 492, 190 00 Prague 9, Czech Republic
T: +420 266 131 111, E: info@hotelduo.cz, www.hotelduo.cz

Hotel Duo

HOTEL DUO, Teplická 492, 190 00 Prague 9, Czech Republic
T: +420 266 131 111, E: info@hotelduo.cz, www.hotelduo.cz

Hotel Duo

HOTEL DUO, Teplická 492, 190 00 Prague 9, Czech Republic
T: +420 266 131 111, E: info@hotelduo.cz, www.hotelduo.cz

布拉格的第一天
The First Day in Prague
2017年11月14日
November 14,2017

第三天的上午，吃过早餐后才拿起笔。
On the morning of the third day, I picked up my pen after breakfast.

很久没有和大伙一起出来浪了，这一次选择大家都没有去过的“中欧”三国——捷克、匈牙利、奥地利。行程折腾也顺利，“齐娃娃”（粤语，意为在一起）的时间非常易过：广州——巴黎，隔一个多小时后续飞布拉格——床！！！
I hadn't traveled with my colleagues for long. This time we chose three countries in Central Europe which we had never visited—Czech, Hungary, and Austria. The flight was frustrating but seemed easy to spend because we played happily together. First, we flew to Paris, after an hour's break, we continued to fly to Prague. At last, we arrived —A bed in Prague!

落地时天阴阴的，有一点点的冷，但兴致还是不错。看看查理小桥（伏尔塔瓦河上的），再到罗伦塔教堂、圣维特大教堂，非常哥特式的建筑，虽亦见过，但还是非常令人感到震撼的，特别是彩色水晶玻璃。
The weather was cloudy and a little cold when we arrived. But we were still in high spirits. After visiting Charles Bridge (over the Vltava River), we went to the Loreta Church, St. Vitus Cathedral, both of them were in a classic style of Gothic buildings. Although I saw this kind of buildings before, I was still shocked by its beauty, especially the colorful crystal glass.

下午，天下雨，还是有点意思的！
It rained in the afternoon, but the rainy city was more charming.

一夜无聊，同事们在房间小酌一点，薯片，各种啤酒，九点就睡，半夜一点醒了，是倒时差的开始。
We felt bored in the evening, so we drank a little, ate some chips and tried different kinds of beer. We slept at nine in the evening but woke up at one in the early morning. It was the beginning of the jet lag.

饿着，懒懒床，开始了真真正正的度假模式，酒店一般般，但可以使我慢下来。
Being hungry and lazy in bed, I began my relaxing vacation. The hotel was not very good, but it made me slow down.

到布拉格后睡的第一张床，可以懒！
The first bed in Prague made me relaxed and lazy.

早上饿得慌，但赖赖床还是非常值得的。看看手机，看看股市（虽然“成绩”还是那么的差），再慢慢地洗洗涮涮，吃早餐。

I felt awfully hungry in the morning, but I just wanted to lie in bed lazily. Checked the cell phone and the stock market (though I was not very good at it yet), then I brushed my teeth, washed my face slowly and enjoyed my breakfast.

早餐有不错的选择：肉、蛋、水果、面包，还有浓的咖啡，非常热闹，酒店公共区域不错，房间一般，生意兴隆，这或者也是经营之道。

We had a hearty breakfast including meat, eggs, fruits, bread, and strong coffee. The public area of the hotel was very crowded. Although the guestrooms were not very comfortable, the hotel had a good businese. Maybe it was the way of management.

太阳出来了，新的一天。非常棒的开始！

The sun rose. It was a very good beginning of a new day!

O.'s File

（区生词典）

1.查理大桥：建于1357年，是横跨伏尔塔瓦河上的一条历史著名的古老拱桥，桥长621米，宽10米，有16座桥墩。是欧洲最古老和最长的石桥，桥上有30座巴洛克风格的雕像和雕像群，这些雕像大约建于1700年，但是现在都是复制品了。(维基百科)

Charles Bridge was a historic bridge that crossed the Vltava River in Prague, Czech. It was built as a bow bridge with 16 arches in 1357. It was 621 meters long and nearly 10 meters wide, so it was known as the longest oldest stone bridge in Europe. The bridge was decorated by a continuous alley of 30 statues and statuaries, most of them baroque-style, originally erected around 1700 but now all replaced by replicas. (Form wikipedia)

2.罗伦塔教堂：由意大利建筑师Giovanni Orsi建于1626年。在18世纪初加建由建筑师Christop Dientzenhofer和KilianIgnaz Dientzenhofer设计的巴洛克风格的外墙。(维基百科)

Loreta Church was built in 1626 by the Italian Giovanni Orsi, The baroque facade was designed by the architects Christoph Dientzenhofer and KilianIgnaz Dientzenhofer, and added at the beginning of the 18th century.(Form wikipedia)

3.圣维特大教堂：是哥特式建筑的杰出典范，也是捷克全国最大、最重要的教堂。圣维特大教堂对中欧晚期哥特式风格的发展产生了巨大影响。(维基百科)

St. Vitus Cathedral is a prominent example of Gothic architecture and is the largest and most important church in Czech. It had a tremendous influence on the development of Late Gothic style characteristic for Central Europe. (Form wikipedia)

Hotel Duo

HOTEL DUO, Teplická 492, 190 00 Prague 9, Czech Republic
T: +420 266 131 111, E: info@hotelduo.cz, www.hotelduo.cz

Hotel Duo

HOTEL DUO, Teplická 492, 190 00 Prague 9, Czech Republic
T: +420 266 131 111, E: info@hotelduo.cz, www.hotelduo.cz

小小的网红面包雪糕店，因我们的“蜂拥而来”，立刻手忙脚乱的！这就是强大的购买力。

A small Internet celebrity shop—Trdelink was in a great bustle because we poured in suddenly. How strong the purchasing power was!

布拉格的第二天

The Second Day in Prague

2017年11月15日上午

On the morning of November 15, 2017

在古城“漫走”（慢）的一天，丰盛的早餐后，九点出发，倒时差的“早睡早醒”。先是到古城，在“导游”带着讲解的上午，拿着有特色的“面包雪糕”——拍摄了第一张全家福相片，每个人一个雪糕，满足！

We had a slow walk in the ancient city. After the rich breakfast, we started off at 9:00 a.m. All of us slept and got up early because of the jet lag. The tour guide showed us around the ancient city in the morning. It was fun to taste Trdelink which was the most popular and special ice cream in Prague. We took the first group photo —with ice cream in our hands. Contented!

午餐（中餐），然后自由自在地逛。拍摄——桥上（查理小桥）的风景不错，之后开始了在大街小巷各式精品店无购买式的闲逛、发呆活动。在星巴克买了一个Espresso杯，在水晶店大开眼界，听听整点旧市政厅墙上的天文钟。晚餐的西餐非常有趣味，Olive牛扒、红酒、汤（西红柿）、沙拉，撑“死”了。

Lunch was Chinese food. Then we started to hang out freely and took photos on the Charles Bridge. The scenery was beautiful. There were all kinds of boutiques in the streets and alleys. It was good to do some window shopping. I bought an Espresso cup at Starbucks. The crystal shop was so amazing that it opened my eyes greatly while listening to the ringing of the Astronomical Clock on the wall of the old city hall. The western style dinner was very interesting. We had Olive steak, red wine, tomato soup, and salad. Stuffed!

晚上无所事事，乘出租车去Casino, 未果，再去看看酒店的配套——保龄球满位，只能喝一杯啤酒！

There was nothing to do in the evening. Firstly, we wanted to take a taxi to Casino. But we ended up going to the bowling alley—one of the facilities in the hotel. Unfortunately, it was so full that we had to drink some beer.

惨，十点睡觉，半夜一点就睡醒了，不眠之夜，倒时差有趣，有益身体。离开Prague！

It was so suffering that I slept at ten o'clock and woke up at midnight. Such a sleepless night! Jet lag was fun and good for health. Tomorrow we would leave Prague!

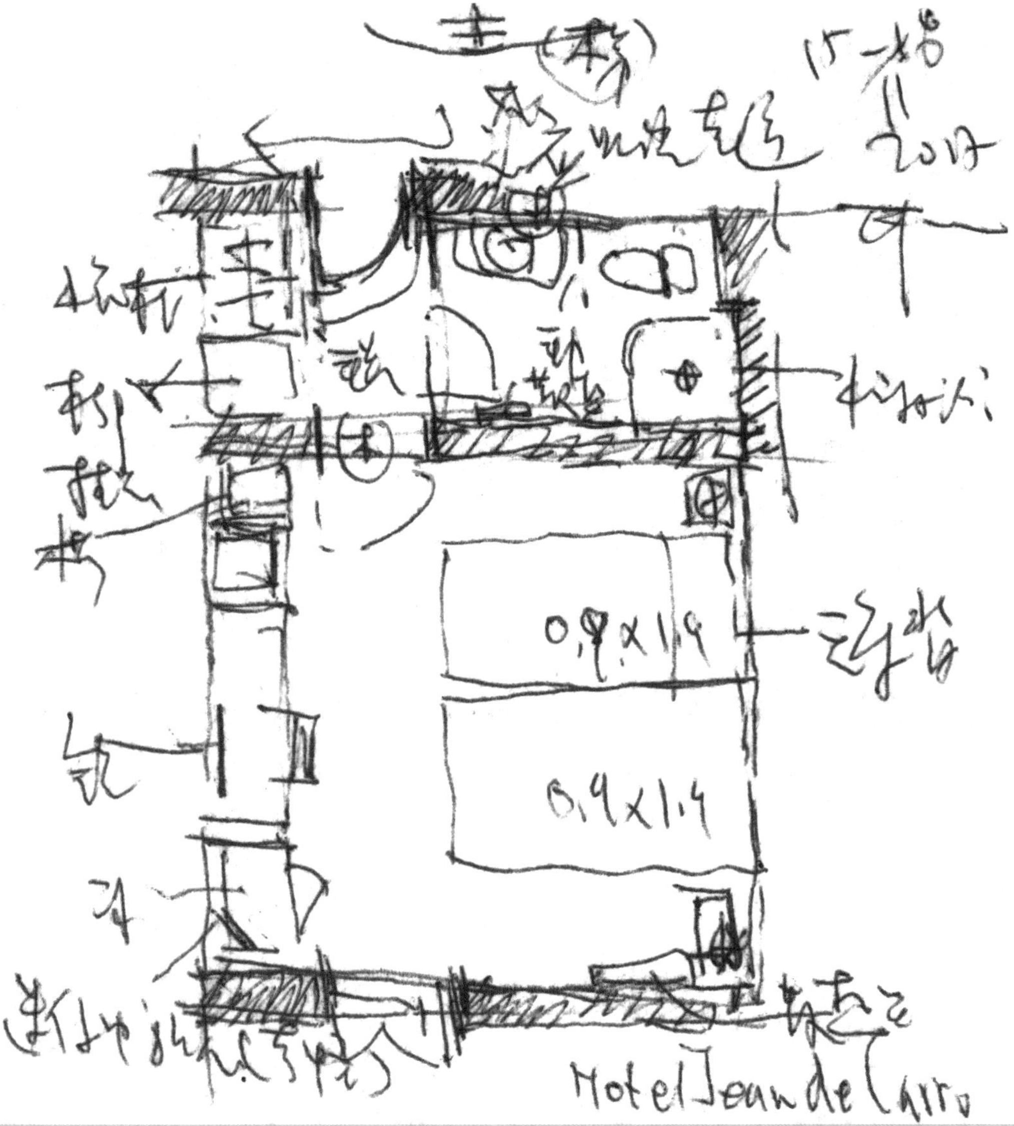

Jean de Carro

珍德卡若健康酒店

24

★★★★

WELLNESS JEAN DE CARRO HOTEL KARLOVY VARY CZECH REPUBLIC

Address : Stezkaleana de Carro
4-6 Karlovy Vary 360 01,
Czech Republic
Telephone : +420 3535 05160
Http : //www.jeandecarro.com

珍德卡若健康酒店
Wellness Jean De Carro Hotel ,Karlovy Vary

第四天（[illegible]第四天）　16/11 2017 早上

[illegible] Fo、[illegible]

洛桑[illegible] 第一次[illegible]洛桑

[illegible]

[illegible]

[illegible]

[illegible] Elephant cafe

（大象咖啡）[illegible]

[illegible]

[illegible]

[illegible]

[illegible] 一件[illegible]

第五天（实际是第四天）

The Fifth Day (Actually the Fourth Day)

2017年11月16日车上去克鲁姆洛夫

November 16 ,2017 on the bus to Krumlov Castle

昨天早餐后，开车，两个小时到达克鲁姆洛夫温泉小镇——卡罗维发利。第一次尝试可以饮的温泉，而且配有各式各样的漂漂亮亮的有纪念意义的温泉杯（我买了一个靛蓝色的），居然可喝，可温暖手，还是挺有趣的！下午休闲逛逛，买水晶杯。乘坐古朴的缆车上山看看城市风景，喝着著名的Elephant cafe(大象咖啡)。住的酒店位于半山，一览全城，牛！

We set off after breakfast. It took us two hours to get to Krumlov by bus. The Spa Town—Karlovy Vary, It was my first time to taste the hot spring. The spring was filled in the cups which were colorful and symbolical(I bought an indigo one). The cup was interesting because it could be used to warm my hands and drink spring as well. I also bought some crystal glasses in the afternoon. We watched the whole city in the old cable car while drinking the famous Elephant Coffee. The hotel we stayed lays in the middle of the mountain, so we could enjoy the beautiful view of the whole city. How magnificent!

湿冷的山区小镇，有一壶可以喝的温泉，幸福。在各式各样的温泉小壶里选购了一个扁扁的靛蓝色的。

It was cold and moist in a small mountain village. I felt happy with a pot of drinkable hot spring. I chose an indigo one from a variety of pots.

晚上聊聊天，喝喝酒，打打扑克，终于在十二点才入睡，当然也不妨碍下半夜醒来，看看手机，发发微信，看看新闻。“愉快的一天”从倒时差开始！

In the evening we talked, drank a little and played cards to kill the time. Finally, I slept at 12:00 p.m, but woke up in the early morning, checked my cell phone, sent messages and read the news. A pleasure day began with the jet lag!

家庭式的酒店早餐， 棒—— 牛角包！

The breakfast in the hotel was family style, the croissants were pretty delidious!

O.'s File

（区生词典）

卡罗维发利温泉：位于捷克共和国波希米亚西部。(维基百科)

Karlovy Vary: It is in western Bohemia,Czech Republic. (Form wikipedia)

25

德沃夏克酒店

★★★★

HOTEL DVORAK CESKY KRUMLOV CZECH REPUBLIC

Address : Radnichi 101 Cesky Krumlov, Czech Republic

Telephone : +420 6080 57057

: +420 2339 20118

Http : //www.krumlovhotels.cz

[illegible]

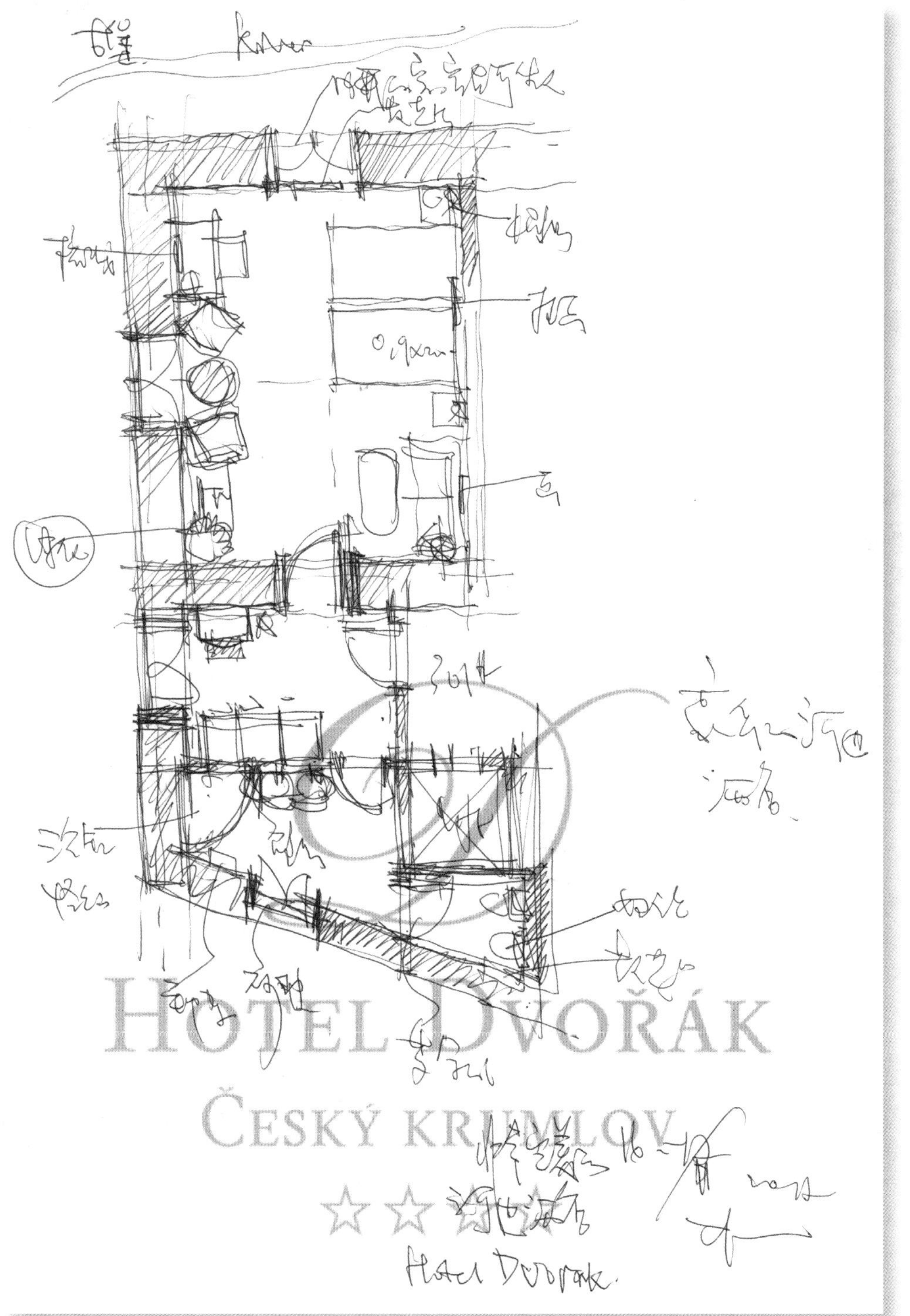
HOTEL DVOŘÁK
ČESKÝ KRUMLOV
Hotel Dvorak.

河边的酒店，我们可以倚窗听听水声，看着山上的古堡，晒晒太阳。

The hotel was beside the river. We could listen to the sound of the water by the window, look at the castle on the mountain and bask in the sunshine.

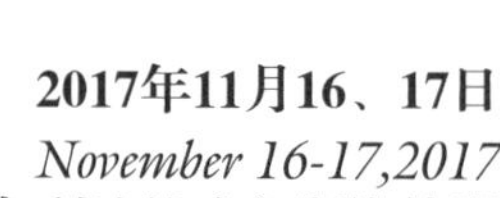

2017年11月16、17日

November 16-17,2017

车上从克鲁姆洛夫到奥利地的哈尔斯塔特湖

on the bus from Cesky Krumlov to Hallstatt Lake in Austria

克鲁姆洛夫，漂亮，一条伏尔塔瓦河环抱，古堡很高，到达，走走逛逛。

The beautiful Cesky Krumlov Castle was surrounded by the Vltava River, it was very tall. We walked around after arriving.

中午饭在广场（中心）旁边的中餐厅（上海菜）吃的，挺好的！

We had lunch in a Shanghai restaurant near the center of the square. The food there was delicious.

午饭后入住这一间在河边、桥边的酒店——非常棒（the Hotel DVOKAK），房间每层只有5间，五层更是有“老虎窗”的特色房间，推窗美景入眼帘，最古典的老宅，阳光/色彩，旋转梯都是“印象派”，棒棒的！

After lunch, we checked in to the hotel which lay beside the river and the bridge. It was pretty cool—the Hotel DVORAK! There were only five guestrooms on each floor. Especially the guestroom on the fifth floor, it was a featured room with a roof window. When I pushed to open the window, the beautiful view came to my eyes. Both the most classical old house with the bright colorful sunshine and the winding stair gave me a deep impression. So terrific!

放下行李，慢慢上山，去用足体验、丈量古堡，风景爆好，鸟瞰小镇，每一张都是明信片。四点太阳下山了，天就冷下来了。下山，逛逛小商店，买买小礼品——木制的“虫虫”。晚餐：西餐，最渴望就是烤猪手了，还是挺美味的。

We left our package at the hotel. And then we climbed the mountain leisurely, treasuring and experiencing the natural beauty of the castle with our feet. We got a bird's-eye view from the top of the mountain; the beautiful city was just like a postcard. The weather got cold when the sun set at 4:00 p.m. We returned to the town. Walked through the shops and bought some souvenirs—wooden worms. We had western food for dinner, hunger for roast pig trotters. Yummy!

土豪的房间，有一点点“穿越”的感觉，我们李总惬意的笑容，自然流露。

The deluxe guestroom has a little feeling of traveling through time. Mr. Li smiled naturally with satisfaction.

享受极奢华的大大洗手间的热水澡（坐在浴缸里）。晚上还是聊聊天，打打牌，尽兴到十二点多，听着河水的潺潺之声，入梦。

I enjoyed the hot bath in the large extremely deluxe bathroom (sitting on the bathtub). In the evening we chatted and played cards happily together till 12 p.m. We went to sleep and had a good dream while listening to the sound of the running river.

早上，早餐在地下一层吃，温暖而小丰富的熏烟肠、面包、咖啡，最爽的就是水果了，酸酸的菠萝，软软的甜瓜，满足，饱撑了！幸福就是吃得好！

In the morning, we had breakfast in the basement. There were warm and juicy smoked sausages, bread, and coffee. I liked the fruits best—sour pineapples and soft melons, I was full and satisfied. Happiness was eating well.

半山上的古堡，艳阳，光影让时光流动。

The castle is on the middle of the mountain. Time flows in the bright sunlight.

O.'s File

（区生词典）

克鲁姆洛夫城堡：城堡的历史可以追溯到1240年，由Witigonen家族建造。目前，该城堡被列为国家遗产。城堡内有世界上保存最完整的巴洛克剧院之一，拥有原有的剧院建筑、舞台技术，等等。(维基百科)

Cesky Krumlov Castle, It dates back to 1240 when the first castle was built by the Witigonen family. Currently, the castle is listed as a national heritage site. It is one of the world's most completely preserved Baroque theatres with its original theatre building, stage technology etc. (Form wikipedia) scalaria™

斯卡拉里亚酒店

26 ★★★★

SCALARIA EVENTHOTEL ST.WOLFGANG AUSTRIA

Address : See 1, St Wolfgang imSalzkammergut St. Wolfgang 5360, Austria

Telephone : +43 6138 8000

Http : //www.scalaria.com

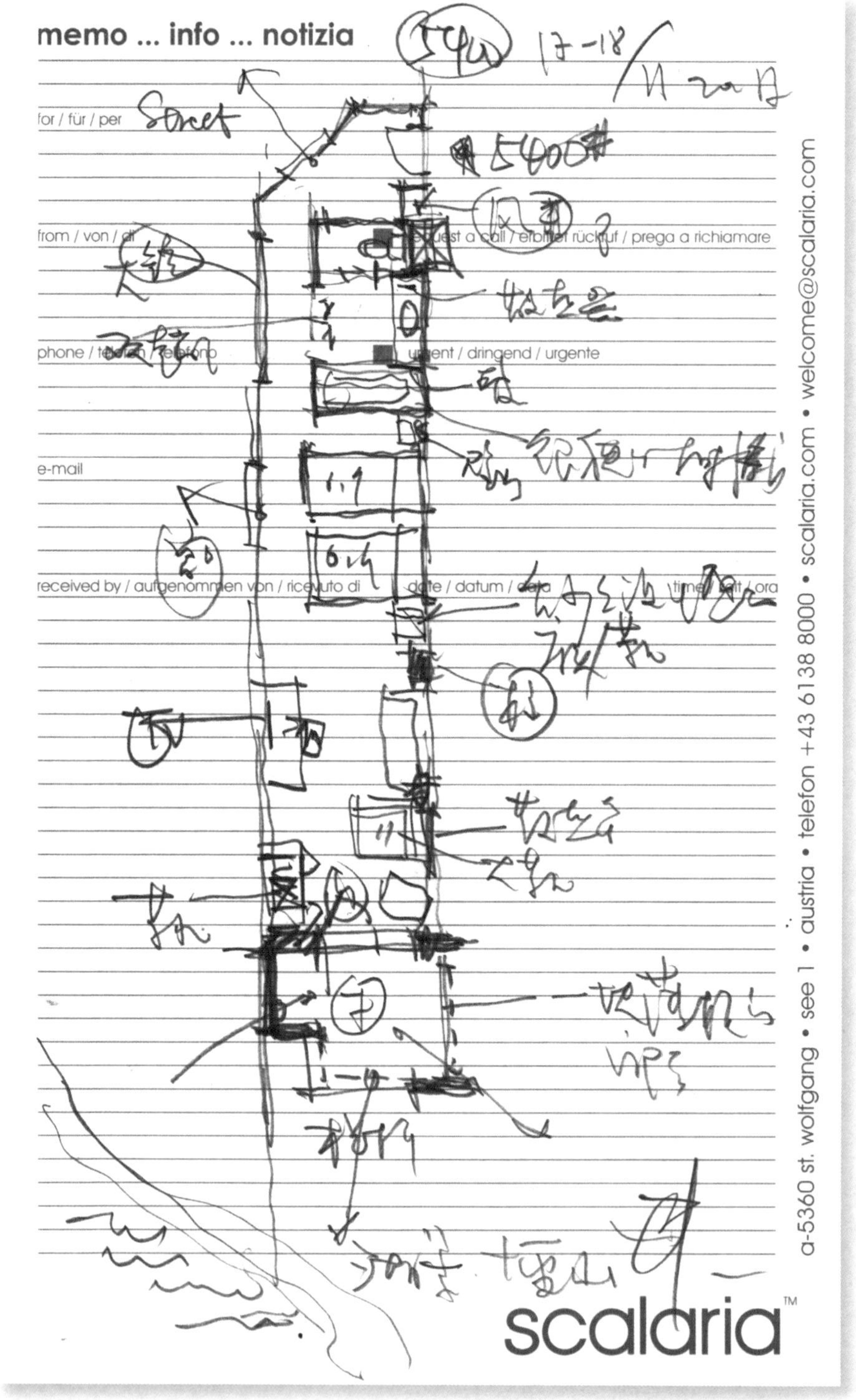

斯卡拉里亚酒店

Scalaria Eventhotel,St.Wolfgang,Austria

memo ... info ... notizia

for / für / per

from / von / di

request a call / erbittet rückruf / prega a richiamare

phone / telefon / telefono

urgent / dringend / urgente

e-mail

received by / aufgenommen von / ricevuto di

date / datum / data

time / zeit / ora

a-5360 st. wolfgang • see 1 • austria • telefon +43 6138 8000 • scalaria.com • welcome@scalaria.com

scalaria™

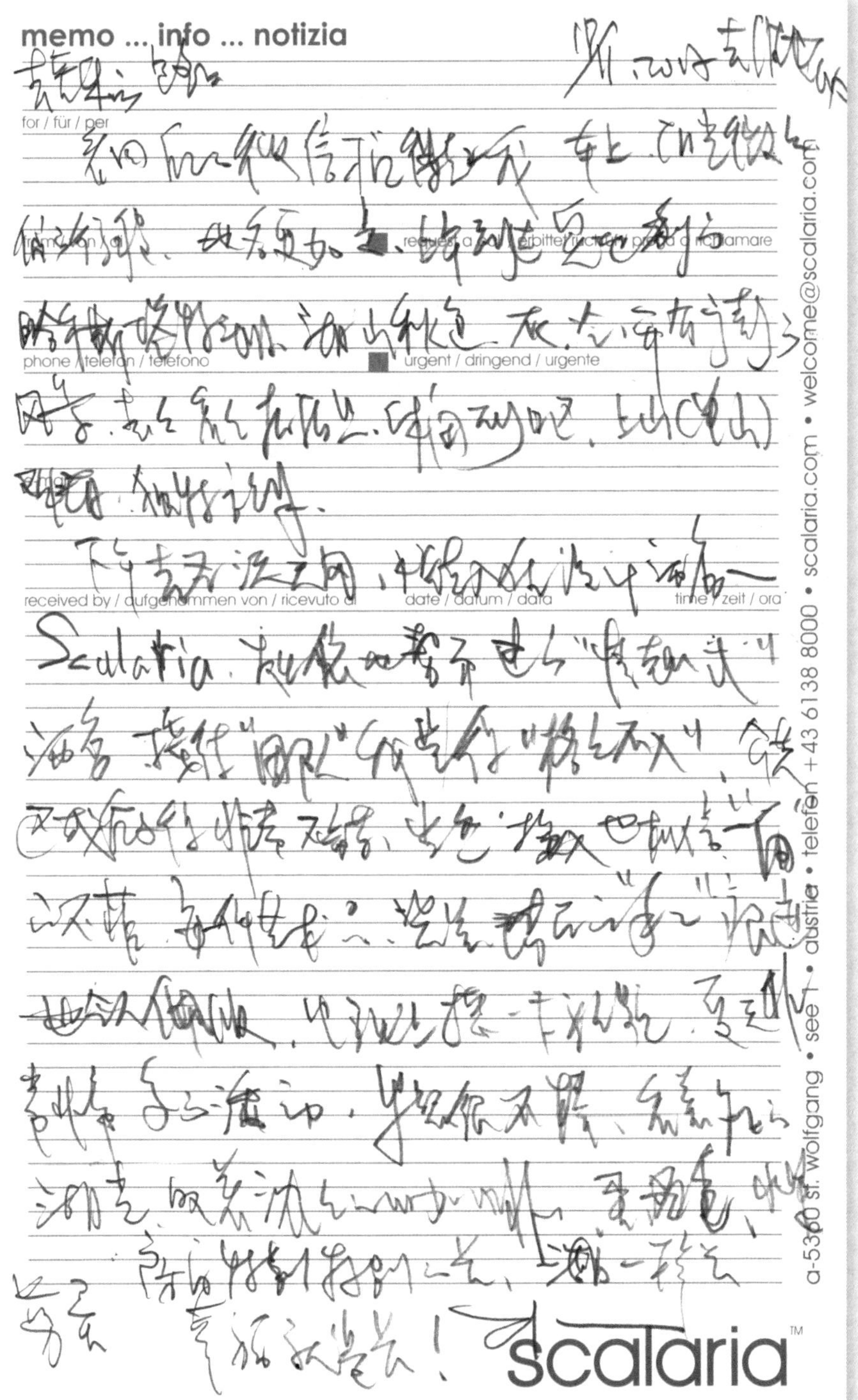
memo ... info ... notizia
for / für / per
phone / telefon / telefono
urgent / dringend / urgente
received by / aufgenommen von / ricevuto da
date / datum / data
time / zeit / ora
a-5360 st. wolfgang • see • austria • telefon +43 6138 8000 • scalaria.com • welcome@scalaria.com
scalaria™

去享“乐”的路上

On the Way to Music

2017年11月18日去维也纳的车上，阳光微微

November 18, 2017 On the bus to Vienna, Sunny

回看自己的微信才记得起我们的行程，地名更加是，上午到达奥地利的哈尔斯塔特湖，湖山秋色，灰、冷而又宁静的风景，走走看看，拍拍照，休闲至极，上山（半山），观湖，景致独特。

Looked back at my Wechat, it reminded me of the trip, especially the names of the places. We visited Hallstatt Lake in Austria, The lake and the mountains were grey and cool but peaceful. Walked around and took pictures leisurely, and then climbed up the mountain. We enjoyed the unique scenery of the lake from the mountainside.

下午去圣沃夫冈，在小镇入住设计酒店——Scalaria, 依山势而建的“情趣式”酒店，接待“团队”，我觉得“格格不入”！公共区域做得非常不错，相信投入不菲，每一件艺术品装置、墙石的“手工痕迹”令人佩服。从电视上播的一个视频得知夏天这里有非常非常多的活动。早餐很不错，看着窗外的湖光，饮着浓浓的咖啡，吃着面包、水果、煎蛋。房间特别特别长，湖一样长，幸福就是长！

In the afternoon, we went to the St.Wolfgang and stayed in Scalaria which was built in the hillside. It was also a "Love Hotel"so living in it for tour groups was out of place. The public area was quite good because it cost a large amount of money to decorate. Not only every work of art but also the trace of handmade on the wall made me feel admired. A show was on TV, there were so many activities in summer here. The breakfast was great. I was enjoying coffee, bread, fruits and fried eggs while looking at the lake view outside. The special guestroom seemed as long as the lake. Happiness lasts forever!

O.'s File

（区生词典）

1. 哈尔斯塔特以其盐的生产而闻名，其历史可以追溯到史前时期，并以哈尔斯塔特文化为名。哈尔斯塔特在1997年被联合国教科文组织宣布为奥地利世界遗产之一的“哈尔斯塔特－达赫斯坦/萨尔茨卡默古特文化景观”的核心。(维基百科)

Hallstatt is known for its production of salt, dating back to prehistoric times, and gave its name to the Hallstatt culture. Hallstatt is at the core of the "Hallstatt-Dachstein/Salzkammergut Cultural Landscape" declared as one of the World Heritage Sites in Austria by UNESCO in 1997. (Form wikipedia)

2. 圣沃尔夫冈小镇：是奥地利中部的美丽小镇，温泉疗养胜地。这个小镇曾被评为世界十大著名小镇。小镇有一座闻名遐迩的圣沃尔夫冈教堂，因有米歇尔·帕赫在1481年完成创作的木刻圣坛——“帕赫圣坛”而吸引了众多观光者。(维基百科)

St. Wolfgang is a beautiful town in central Austria, it is a popular spa resort. This town was once rated as one of the world's top ten famous towns. The town has a famous St. Wolfgang Church, which attracts many tourists because of the wooden altar of Michael Pach, which was completed in 1481, the "Pach altar". (Form wikipedia)

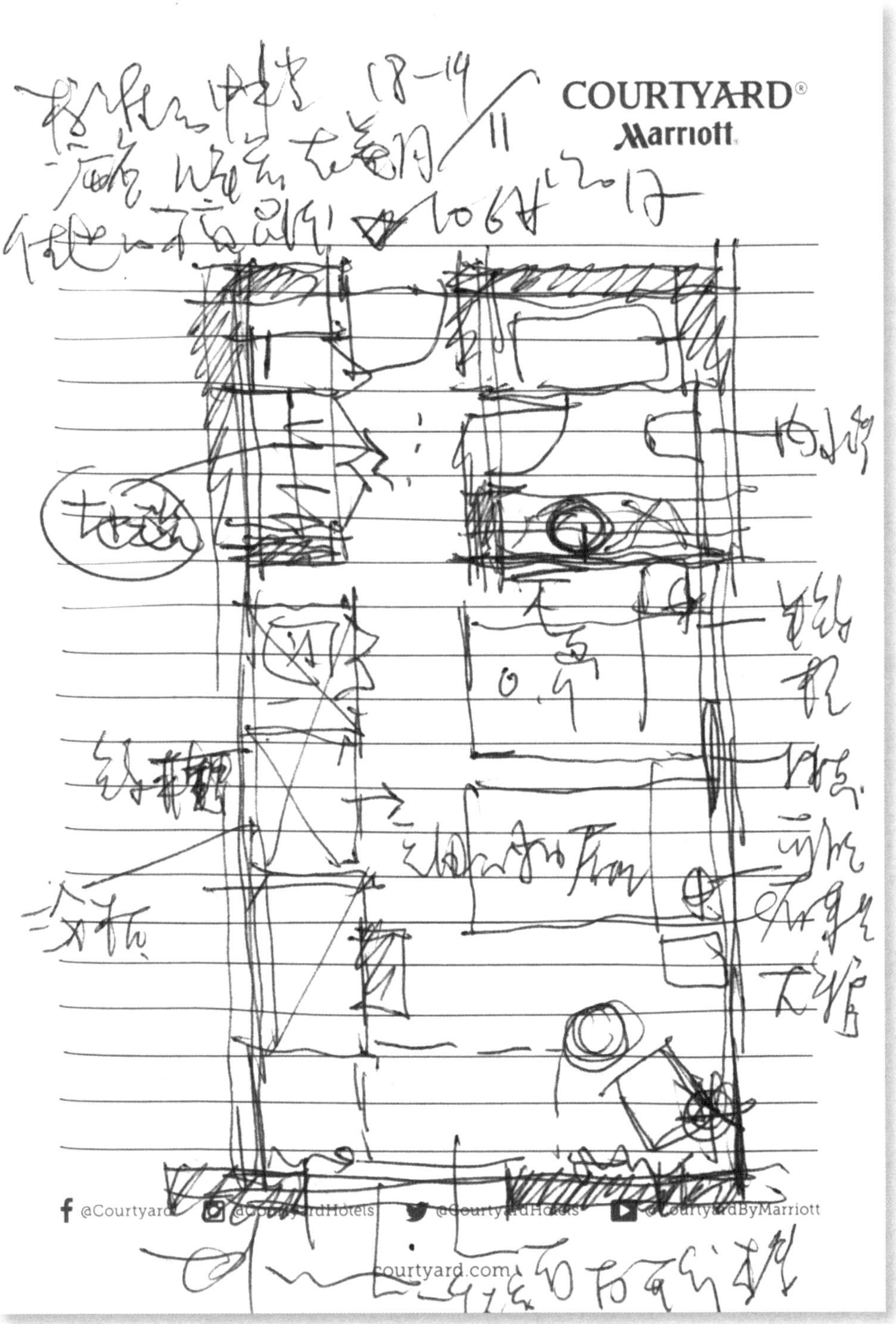

COURTYARD®
Marriott

维也纳美泉宫万怡酒店

★★★★

27

COURTYARD MARRIOTT HOTEL VIENNA AUSTRIA

Address : Schoenbrunner, Schlossstrasse 38-40 Vienna Austria A-1120

Telephone : +43 1810 17170

Http : //www.courtyard.com

维也纳美泉宫万怡酒店

Courtyard Marriott Hotel, Vienna, Austria

COURTYARD®
Marriott

[illegible]

f @Courtyard @CourtyardHotels @CourtyardHotels @CourtyardByMarriott
courtyard.com

COURTYARD®
Marriott

[illegible]

f @Courtyard @CourtyardHotels @CourtyardHotels @CourtyardByMarriott
courtyard.com

2017年11月19日下午,在去布达佩斯的车上

November19 ,2017 on the bus to Budapest

忘记了昨天做了什么，“老年痴呆”！

I forgot what I did yesterday. Alzheimer—perfect!

也许是能倒过来想想做了什么？昨天晚上去听了一个维也纳式的小型音乐会，有演奏，更有独唱，双人男女小品式的独唱，最后以新年音乐会式的乐曲结束，齐鼓掌的名曲——拉德斯基进行曲。

Maybe I could remember what happened if I thought back. We listened to a small Viennese concert last night including music performance, solo singing and male and female duet. At last, the concert ended with the famous symphony—Radetzky March, the audience couldn't help themselves clapping hands by the tempo with the band.

晚上是大大的猪排烧烤，一人一份，太过分（量）了，根本无法消灭。

Our dinner was barbeque pork chops, everyone had a beast size order, it was too huge and I couldn't finish it at all.

格拉茨小镇，不错，走走古城，吃到了有历史的面包店——Hofbackerei Edegger-Tax，还有广场的烧板栗（冬天极品），更有惊喜的格拉茨艺术中心，及其附近的急流水溪中间的金属玻璃“半岛”，饮咖啡——人生夫复何求，路上有阳光，有雪景，有阴天。

The small town of Graz was nice. We tasted delicious bread in a historical bakery —Hofbackerei Edegger-Tax while we were wandering around. We also had the roasted chestnut, the warmest gift in winter. Graz Art Center was an amazing surprise to me, and we enjoyed coffee in the cafe which was made of metal glasses, in the middle of a turbulent creek. I couldn't think of anything that was more beautiful or peaceful than now, just enjoyed the sunshine, snow, and clouds on our way.

喝了咖啡后入住万豪Courtyard，旅游就是够分量的（吃）。

We checked in to the Marriott Courtyard after coffee. Traveling was eating, eating too much!

科林西亚布达佩斯酒店

28 ★★★★★

CORINTHIA HOTEL BUDAPEST HUNGARY

Address : Erzsebet Korut 43-49,
1073 Budapest, Hungary
Telephone : +36 1479 4000
Fax : +36 1479 4333
Http : //www.corinthia.com

CORINTHIA
HOTEL
BUDAPEST

ROYAL LUXURY SINCE 1896

In Residence

CORINTHIA HOTEL BUDAPEST, ERZSÉBET KÖRÚT 43-49, 1073 BUDAPEST, HUNGARY
T +36 1 479 4000 | F +36 1 479 4333 | BUDAPEST@CORINTHIA.COM | CORINTHIA.COM

Secretar

科林西亚布达佩斯酒店
Corinthia Hotel, Budapest, Hungary

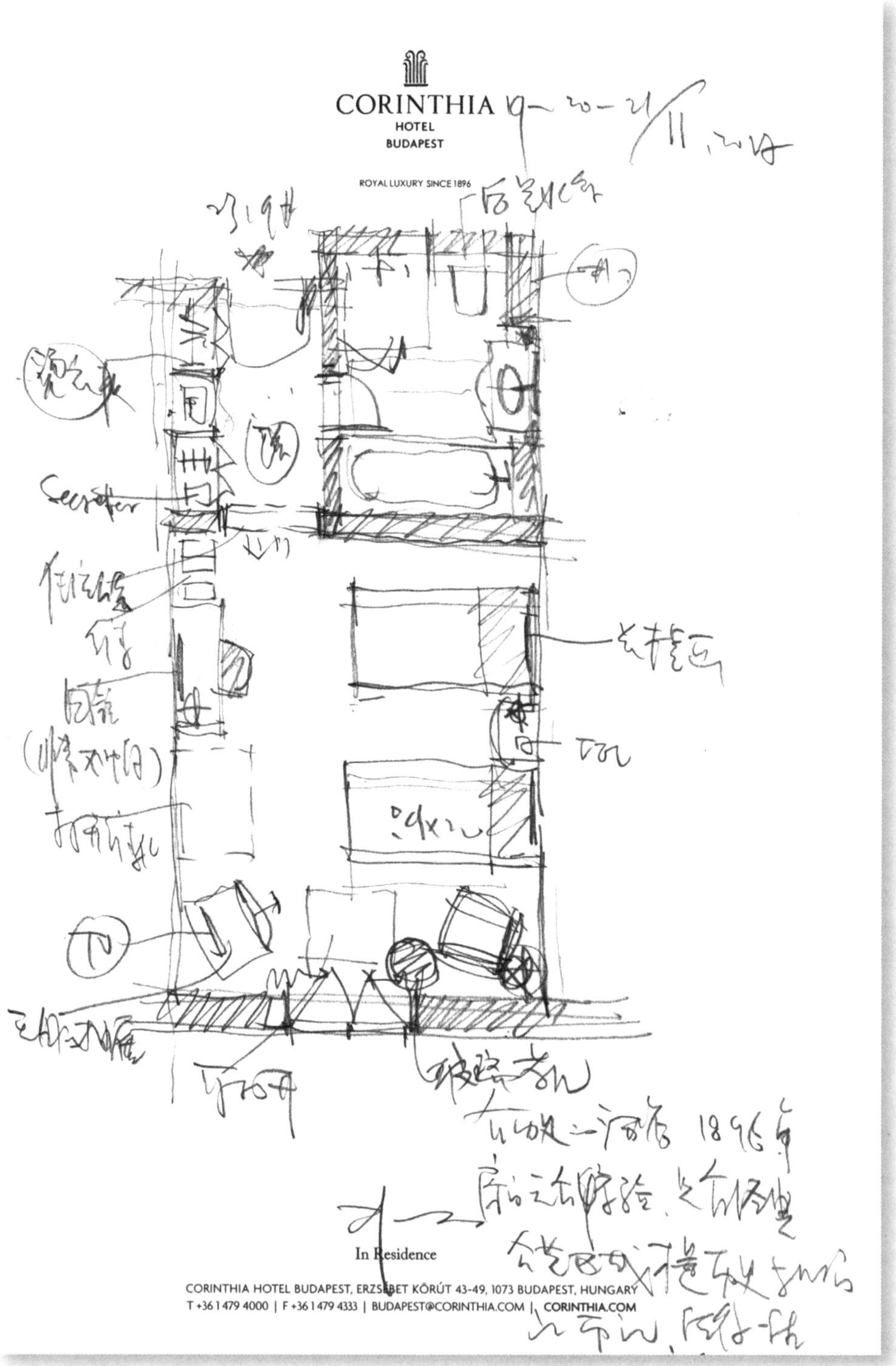

ROYAL LUXURY SINCE 1896

[illegible]

[illegible] Budapest [illegible]

[illegible]

[illegible]

[illegible]

[illegible]

[illegible]

[illegible]

[illegible]

[illegible] the most beautiful cafe in the world

[illegible] cafe [illegible]

[illegible]

[illegible]

[illegible]

In Residence

CORINTHIA HOTEL BUDAPEST, ERZSÉBET KÖRÚT 43-49, 1073 BUDAPEST, HUNGARY
T +36 1 479 4000 | F +36 1 479 4333 | BUDAPEST@CORINTHIA.COM | CORINTHIA.COM

The Right Time is Perfect!

时间合适就好！

2017年11月21日

November 21, 2017

准备离开布达佩斯，在巴黎转机回广州了，我相信同团的同事也想回家了，旅行合适就好（时间），非常完美的一次欧洲行，时间刚刚好，美食刚刚好，美景刚刚好——昨天（这两天）充分感受到。

We were leaving Budapest for Guangzhou, transferred at Paris. I thought my colleagues were also homesick.The time of the travel was important. A very perfect European travel—the right time, the delicious food, the wonderful scenery! We all felt so good during these two days.

多瑙河游船，小阳光，上午半山上的渔人堡建筑群，阳光猛烈，光影强烈，美人美景，晚上的匈牙利“私房菜”堪称一流，有香槟、音乐、沙拉、汤、主菜、甜品、咖啡，满足！

We enjoyed the Danube Cruise in the warm sunshine and then visited the Halászbástya in the middle of the mountain. The sun shone brightly, the light and the shadow were strong. Wonderful view and beautiful people! Dinner was in a Hungarian private kitchen, everything was so perfect, champagne, music, salad, soup, main course, dessert and coffee that we were all satisfied.

半夜再到“the most beautiful cafe in the world”的纽约Cafe，心满意足！

At midnight we went to enjoy coffee in New York Cafe again which was known as the most beautiful cafe in the world. Perfect!

早上休闲早餐，继续“瘫”在床上休息一下，然后逛逛酒店，洗澡，上飞机！

After a leisure breakfast, I had a rest in bed. Nothing to do but going shopping, I wandered around the hotel and took a shower. Then we set off to the airport.

告别欧洲。广州，我们回来了！

Goodbye, Europe. We came back, Guangzhou!

名不虚传的“最美咖啡店”，水晶灯灿烂，极尽奢侈的装饰，值回票价。（不算贵，只是要排队等位）

The most beautiful cafédeserved its reputation. The splendid crystal lamp and the extremely luxurious decoration were worth the price. (Not expensive but had to wait in line)

借着拍摄的技巧，我们跳得“很高”。

Because of the shooting skill, we jumped very high in the sky.

O.'s File

（区生词典）

渔人堡（Halaszbastya）是一个新哥特式和新罗曼风格的观景台，位于匈牙利首都布达佩斯布达一侧多瑙河河畔的城堡山，邻近马加什教堂。它修建于1895年到1902年之间，设计师是弗里杰·舒勒克（Frigyes Schulek）。第二次世界大战期间，渔人堡近乎毁灭。1947年至1948年之间，弗里杰·舒勒克的儿子亚诺什·舒勒克负责了修复工程。(维基百科)

Fisherman's Bastion: Halászbástya is a neo-Gothic and neo-Roman-style viewing platform located on the Castle Hill on the Danube River on the Buda side of Budapest, Hungary, near the Matthias Church . It was built between 1895 and 1902, and was designed by FrigyesSchulek. During the Second World War, Fisherman's Bastion was almost destroyed. Between 1947 and 1948, the son of Frigyes Schulek, Janos Schuleck, was in charge of the restoration project. (From Wikipedia)

BVLGARI
HOTEL BEIJING

北京宝格丽酒店

29

★★★★★

BVLGARI HOTEL BEIJING CHINA

Address : Building 2 Courtyard, No. 8 Xinyuan South Road, Chaoyang District, Beijing,China
中国北京市朝阳区
新源南路8号院2号楼
Telephone : +86 10 8555 8555
Http : //www.bulgarihotels.com/
E-mail : beijing@bulgarihotels.com

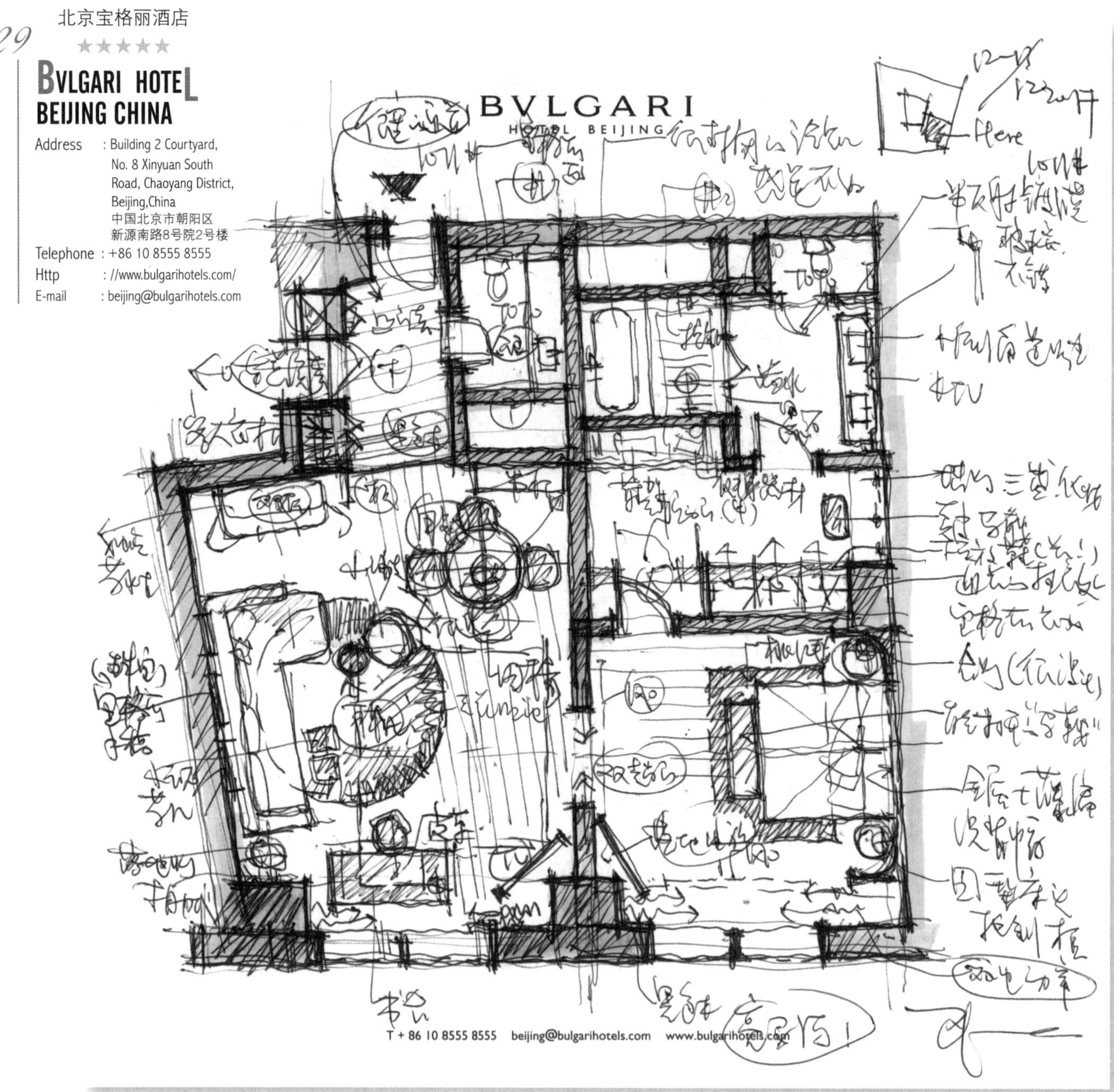

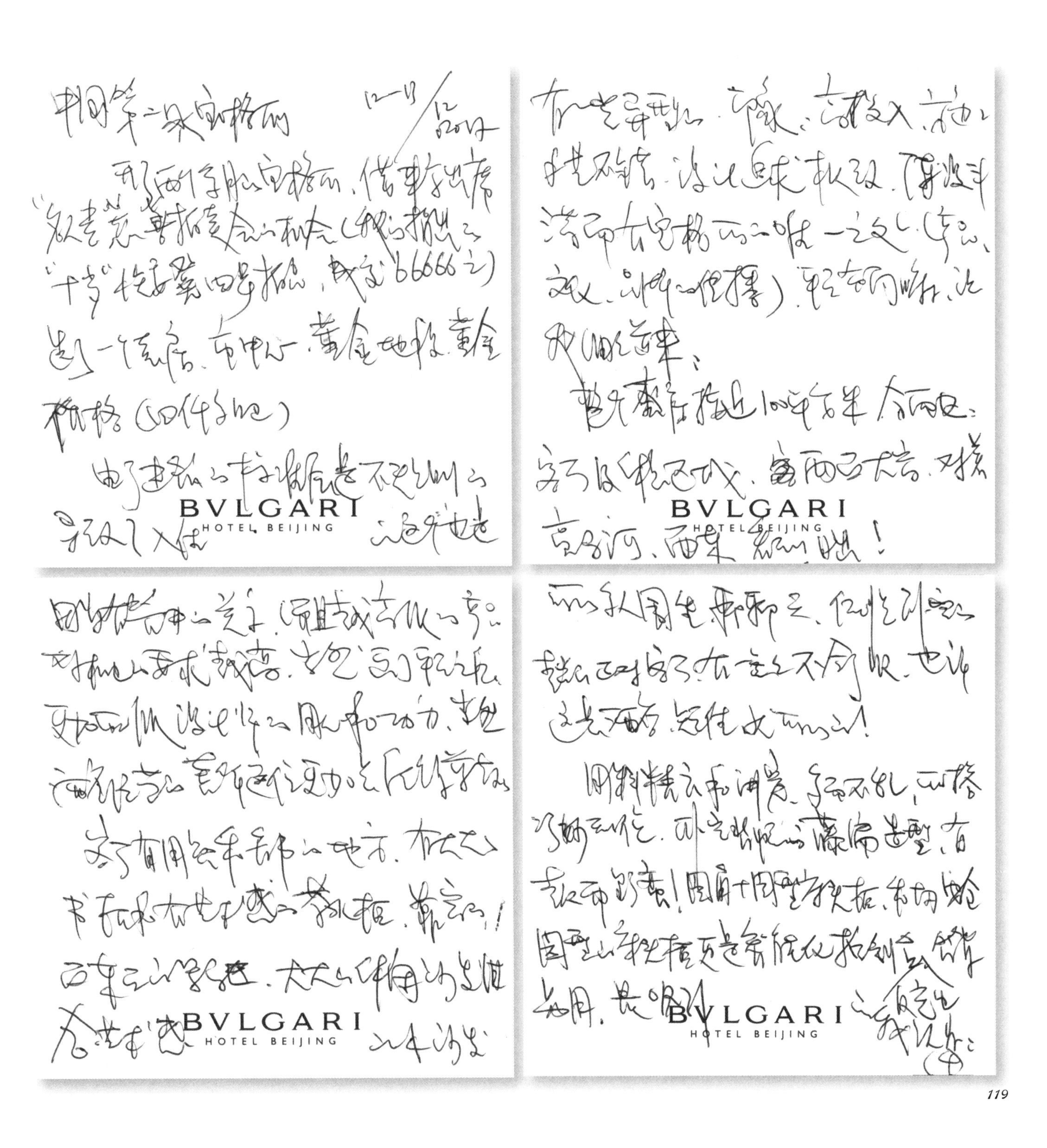

中国第二家宝格丽　　12—13/12 2014

[illegible]宝格丽，[illegible]（[illegible] 66666元）

[illegible]一个[illegible]，市中心，黄金地段，黄金[illegible]（[illegible]）

由于[illegible]

BVLGARI
HOTEL BEIJING

[illegible]

[illegible]宝格丽[illegible]（[illegible]）[illegible]

[illegible]100[illegible]

[illegible]！

BVLGARI
HOTEL BEIJING

[illegible]

[illegible]

[illegible]！

BVLGARI
HOTEL BEIJING

[illegible]也许[illegible]！

[illegible]

BVLGARI
HOTEL BEIJING

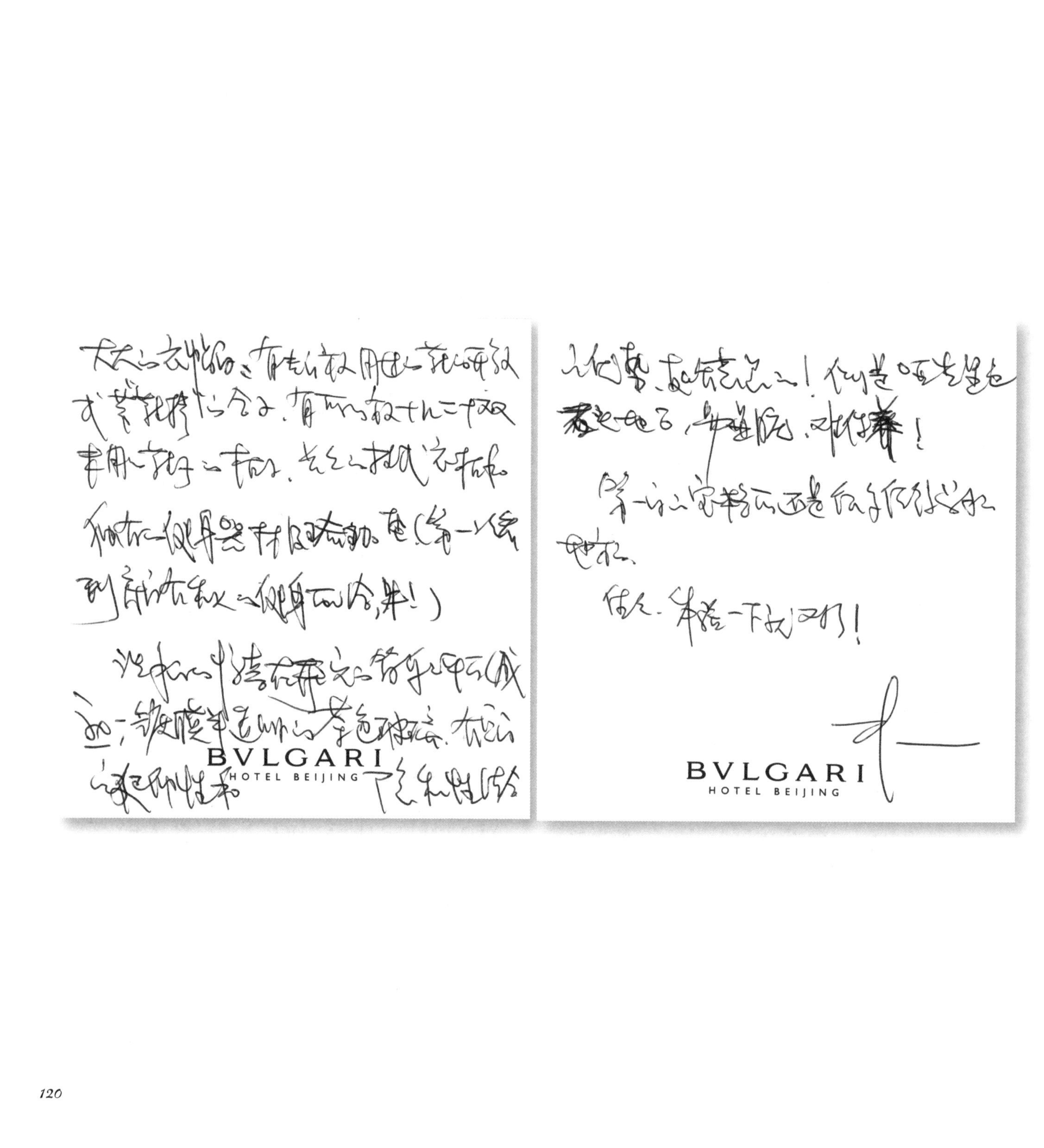

BVLGARI
HOTEL BEIJING

BVLGARI
HOTEL BEIJING

阳光下的早餐

The breakfast in the sun

The First Bulgari Hotel in China

中国第一家宝格丽

在开了两个多月的宝格丽，借来京出席“名人堂”慈善拍卖会的机会（我们捐出了“十岁”小矮凳四号拍品，成交价66666元），选了一个套房，市中心，黄金地段，黄金价格（四千多吧）。

Bulgari Hotel has been opened for two months. Taking the opportunity to attend the charity auction of Hall of Fame(we donated a low stool for kids under ten years old as No. 4 auction item, the final transaction price was 66,666 yuan) , I chose a suite which was in the center of the city, the golden area, golden price as well (more than 4,000 yuan).

由于建筑的标准层平面是不规则的，导致了入住的这个套房也是有一些异形的。印象：高投入，施工手艺不错，设计追求极致，有宝格丽的唯一之处（产品、文化，品牌的传播），平面布局嘛，让我细细道来：

Because the standard floor of the building was irregular, the suite where I stayed was in a strange shape.My impression was high investment and fine workers' craft. Bulgaripursues perfect design and also has its uniqueness including products, culture and the spread of its brand. Let me talk about the plane layout in details.

整个套房接近100平方米，陈设丰满，分两个区：客厅及休息区域，两面大窗，对着亮马河，面东，能看到日出！

The whole suite was nearly 100 square meters. Its furnishings were rich. The suite was divided into two parts: living room and rest area. Two big windows faced the Liangma River and the east, I enjoyed the sunrise.

“厚重”的奢华

The heavy luxury

北京宝格丽酒店

Bvlgari Hotel, Beijing, China

因为有管井的关系（而且越高级的产品对机电的要求越高），等我画了平面之后，更加佩服设计师的用心和功力，当然酒店经营的整体定位更加值得尊敬。

Because of the tube wells, I admired the designers' careful consideration and ability after drawing the plane. (The more advanced products have higher requirements for electro mechanics.) Of course, the overall positioning of the hotel management is more respectful.

客厅有用餐和看书的地方，有大大的书柜和有艺术感的茶水柜，靠窗的面东的写字区，大大的休闲沙发组合，艺术感的小沙发，可以多人围坐，聊聊天。倒是卧室的趟门正对客厅，有一点点不舒服，这是酒店，也许短住还是可以的。

There were places for having meals and reading in living room. It also had a big bookshelf and an artful tea cabinet. The work zone beside the window faced the east. A set of leisure sofa and an artful small sofa provided a place for a few people to sit around and have a chat. But it was a little inconvenient because of the sliding door of the bedroom opposite to the living room. Staying for short might be ok.

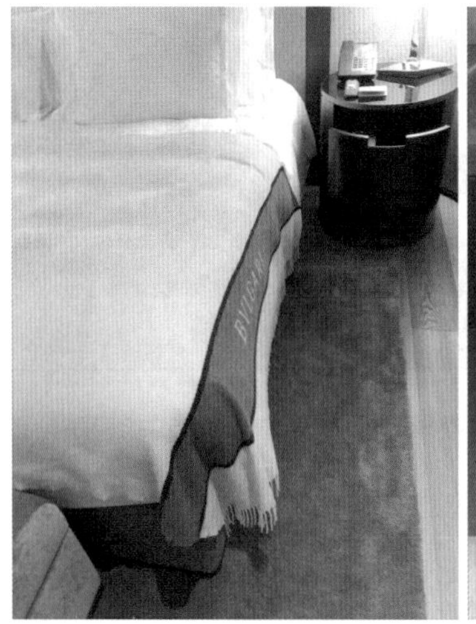

有酒柜和哑铃的房间
The guestroom with a liquor cabinet and a dumbbell

用料精良和讲究，多而不乱，配搭巧妙到位，卧室背幅的藤编造型，有趣而野蛮！圆角+圆形床头柜，和谐安全，圆形的床头柜更是设有智能化控制台及充电设备，简单易用，最吸引我的是大大的衣帽间：有专门放穿过的鞋的开放式带“鞋抽”的盒子，还有可以放十几二十双未穿过的鞋子的柜子，长长的挂式衣柜和独有的健身器材及瑜伽垫（第一次看到房间有私人的健身配给，牛！）。

All the material was excellent and exquisite, rich but not crowded, ingenious match. The rattan modeling on the back of the bedroom was interesting and wild! The round bedside table with rounded corners was harmonious and safe. The round bedside table was also an intelligent control box and charger. It was easy to use. What impressed me most was the big cloakroom. It had some open-plan boxes with a shoehorn to put the used shoes exclusively, and also a cabinet which was used to put ten or twenty pairs of unused shoes. A long hanging wardrobe and the unique fitness equipment,ayoga mat as well. (It was my first time to see the guestroom with private fitness equipment, pretty cool!)

北京宝格丽酒店
Bvlgari Hotel, Beijing, China

北京宝格丽酒店
Bvlgari Hotel, Beijing, China

有酒柜和哑铃的房间
The guestroom with a liquor cabinet and a dumbbell

洗手间的惊喜在于它简单的平面，成熟。镀膜半透明的茶色玻璃，有空间的延伸性和隐私性结合的优势，挺有意思的！倒是哑光黑色石头地面，易显脏，难保养！

The surprise of the bathroom was its simple but mature plan. The coated translucent brown glass made the room extensive and private. It was the advantage of combination of both. So interesting! But the matte black stone ground was easy to get dirty and difficult to maintain.

第一家宝格丽还是有很多值得学习的地方的。

There was still a lot to learn from the first Bulgari.

住住，体验一下就对了！

It's perfect to stay and experience by yourself!

30 上海建业里嘉佩乐酒店

★★★★★

CAPELLA JIAN YE LI SHANGHAI CHINA

Address : No. 480 Jianguo West Road, Xuhui District, Shanghai, China
中国上海市徐汇区
建国西路480号
Telephone : +86 21 5466 6688
Http : //www.capellashanghai.com

小环境

The small environment

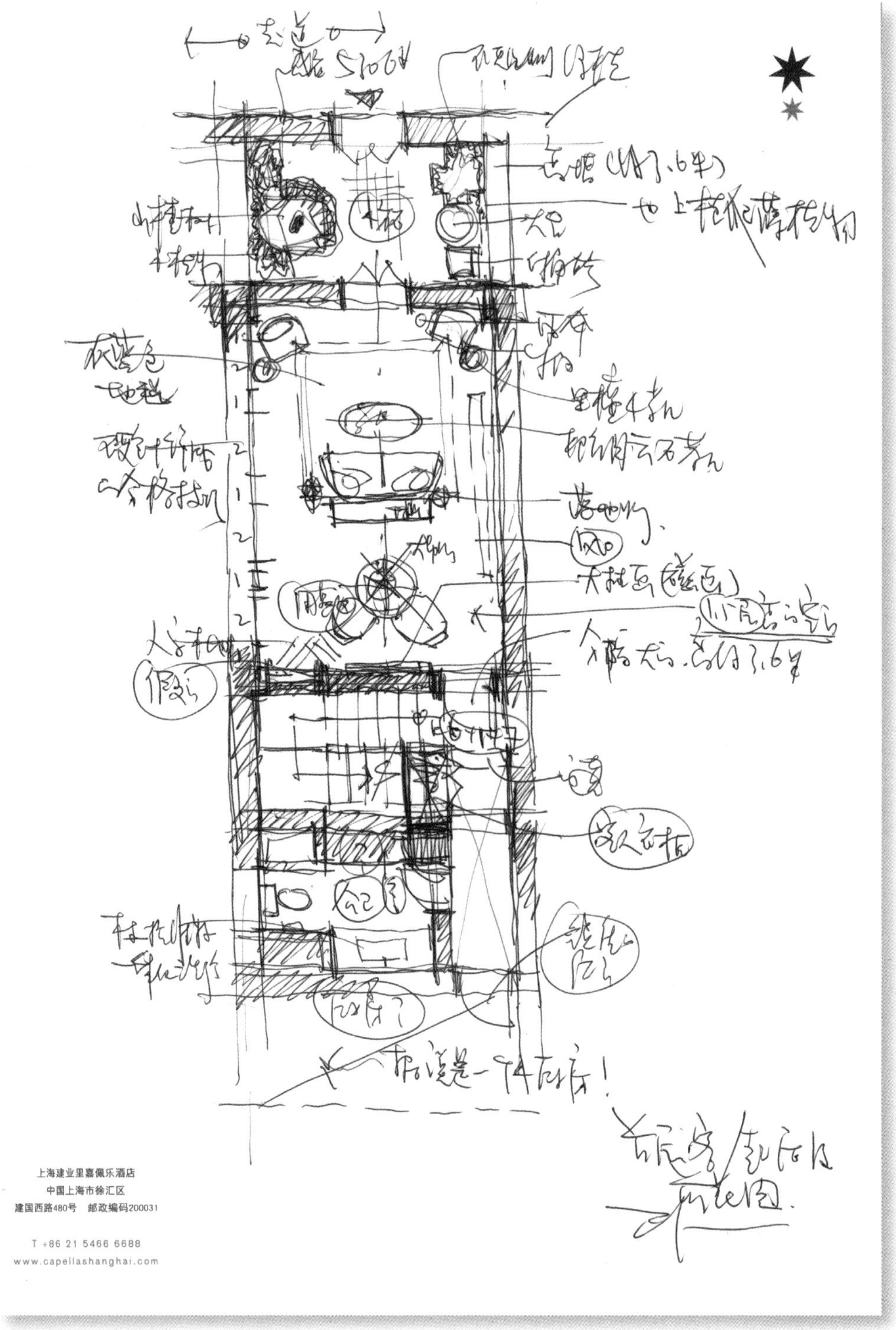

上海建业里嘉佩乐酒店
中国上海市徐汇区
建国西路480号　邮政编码200031

T +86 21 5466 6688
www.capellashanghai.com

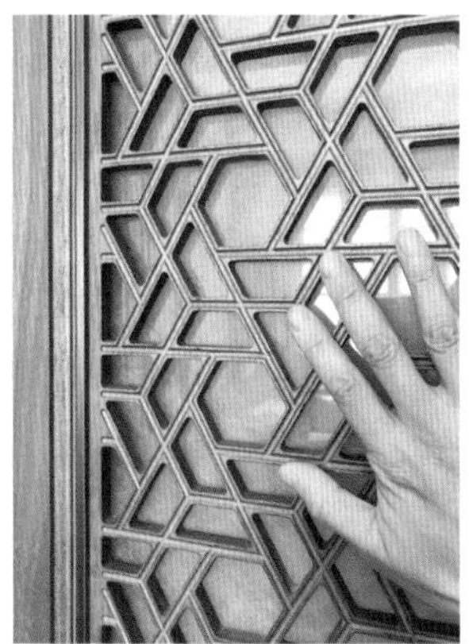

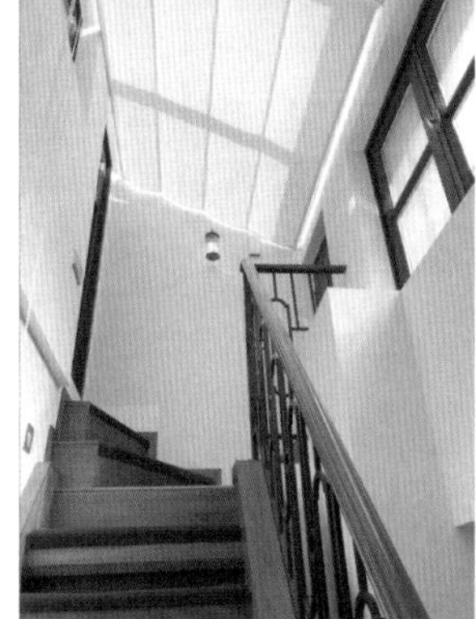

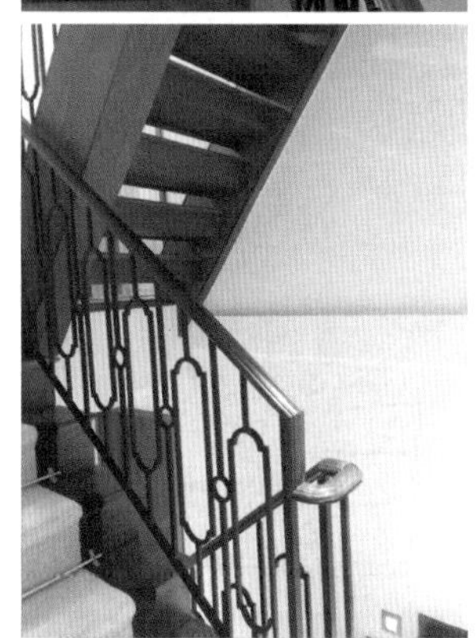

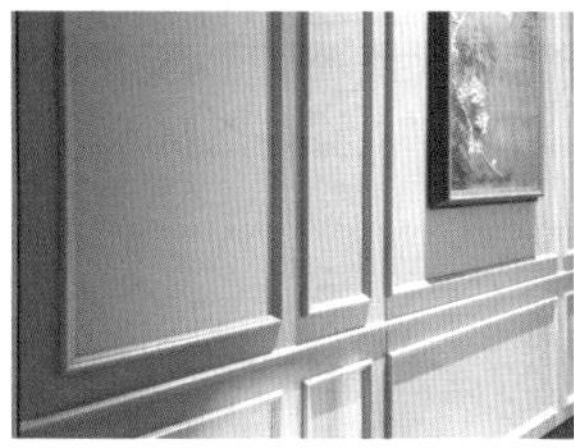

楼梯及墙面的细节
The details of stairs and walls

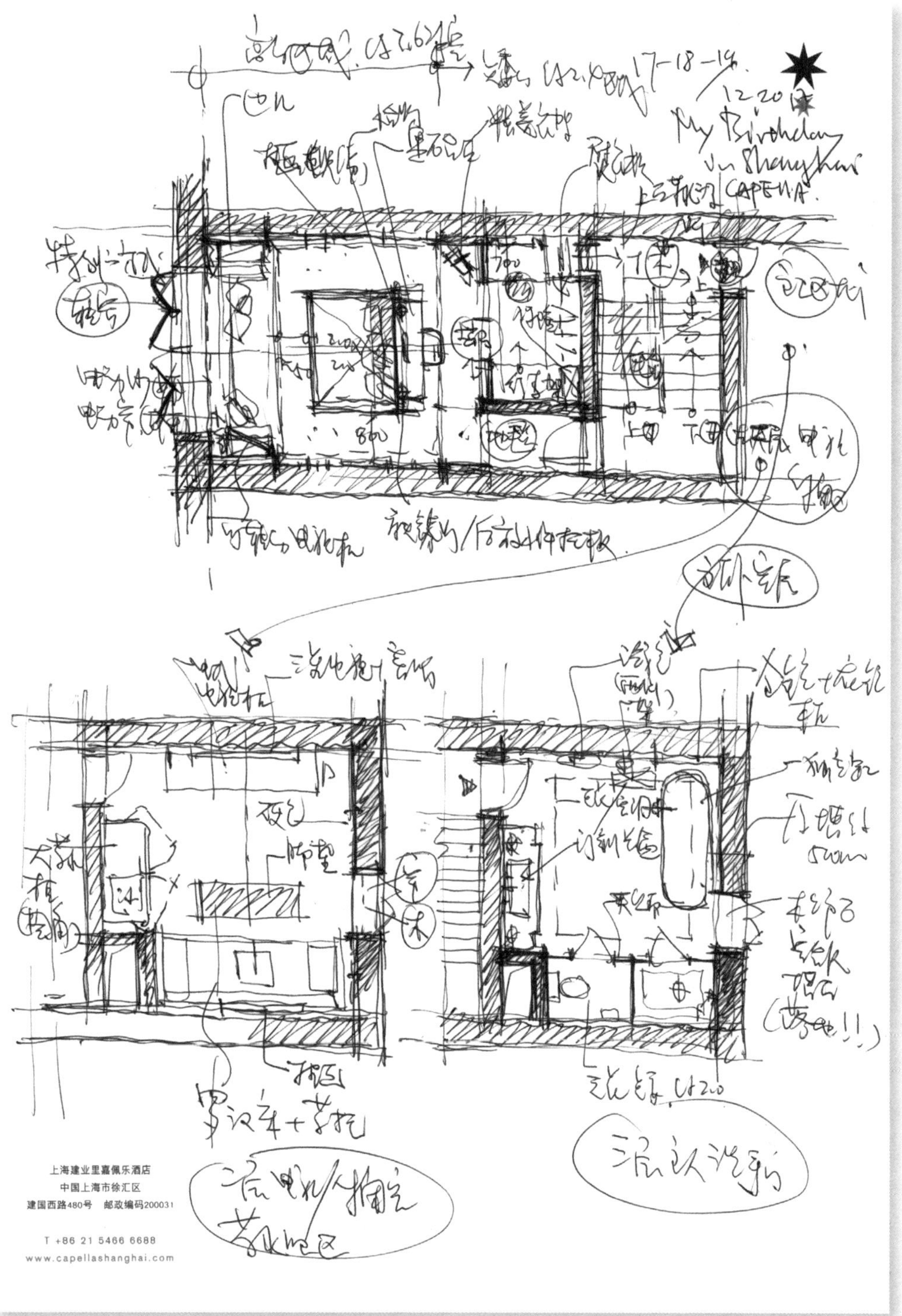

17-18-19.
12.20
My Birthday in Shanghai
CAPELLA.
上海建业里嘉佩乐酒店
中国上海市徐汇区
建国西路480号 邮政编码200031
T +86 21 5466 6688
www.capellashanghai.com

鲜美的早餐
The fresh and tasty breakfast

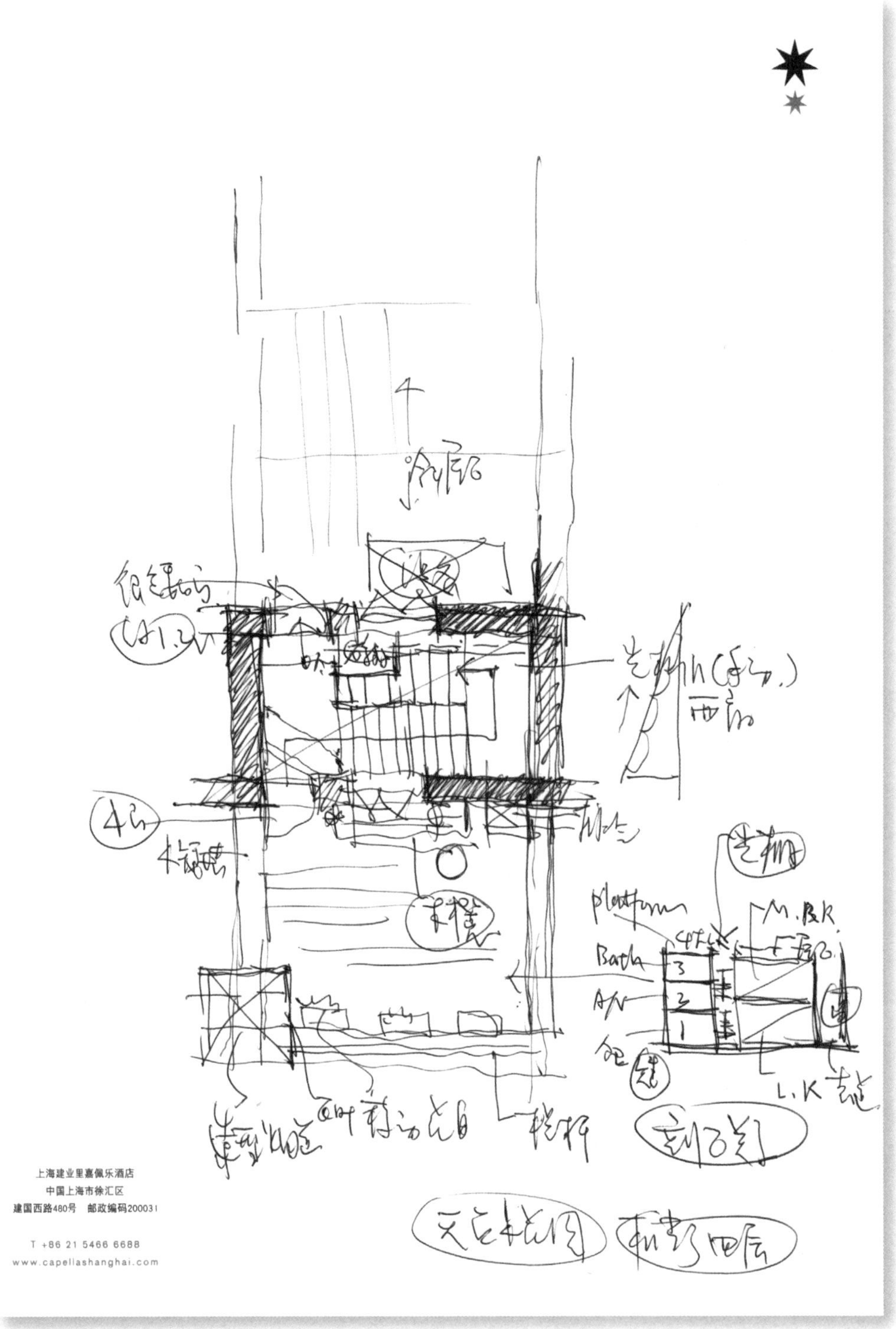

上海建业里嘉佩乐酒店
中国上海市徐汇区
建国西路480号　邮政编码200031

T +86 21 5466 6688
www.capellashanghai.com

量身定做的家具
The customized furniture

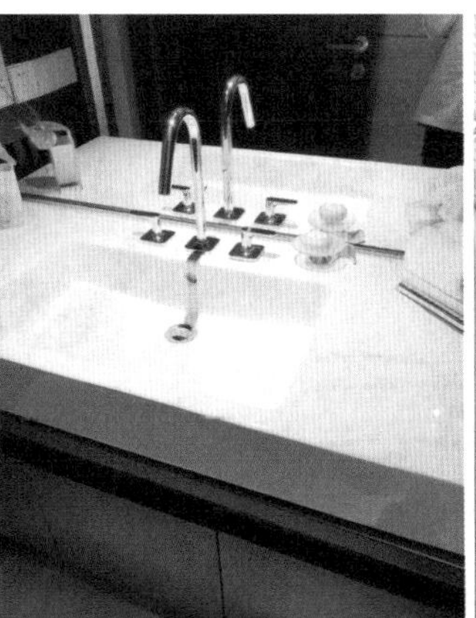

精选的洗手间
The selected washroom

上海建业里嘉佩乐酒店
Capella Jian Ye Li, Shanghai, China

西安凯悦酒店

31

★★★★★★

HYATT REGENCY XI'AN CHINA

Address : No.988 Qujiangchi East Road, Yanta District, Xi'an, Shanxi,China

中国陕西省西安市雁塔区曲江池东路988号

Telephone : +86 29 8579 6082

Http : //www.m.hotels-hyatt.com

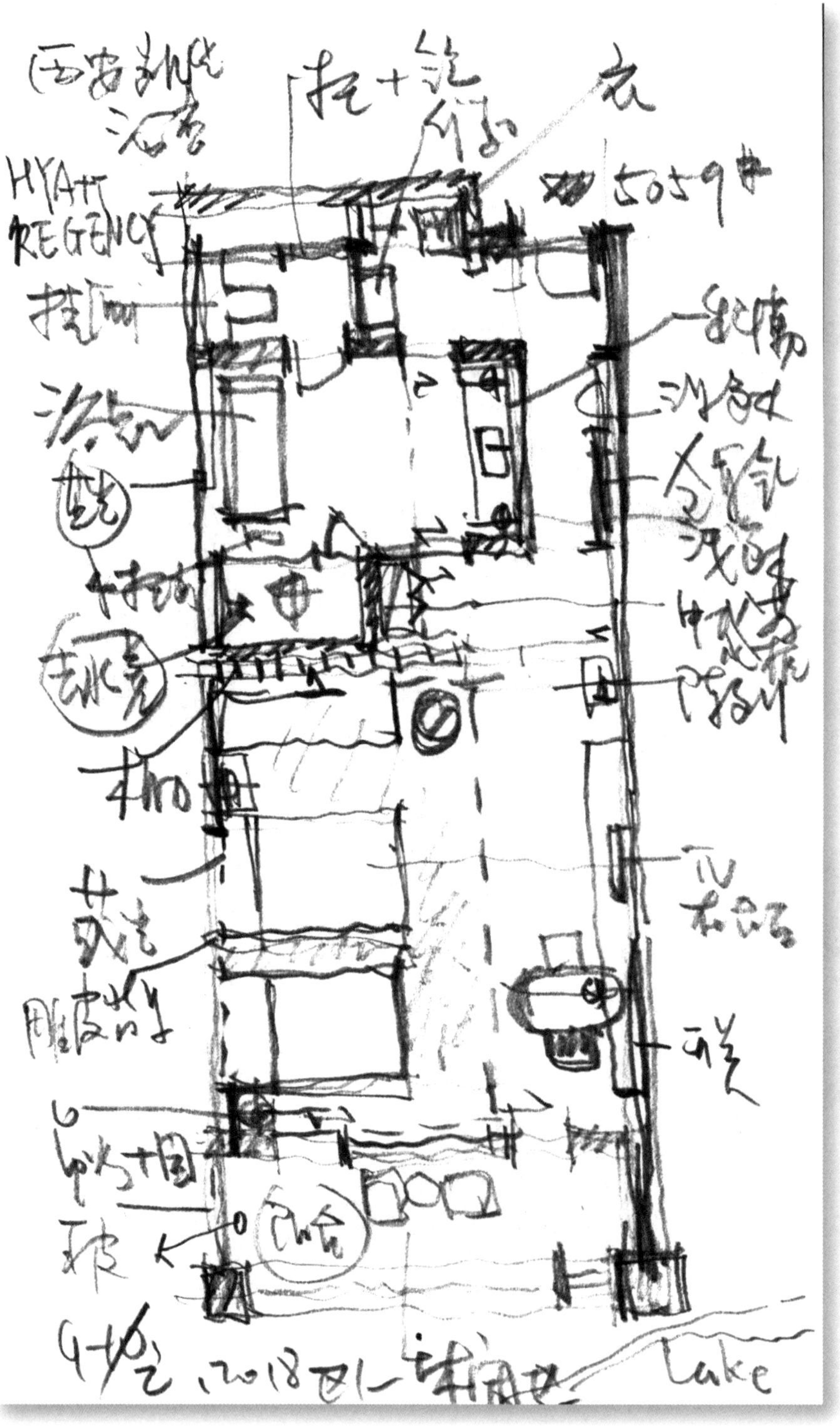

游西安钟楼书店的感 9-10/12，2011

入住西安钟楼饭店，一是地理位置好，二是管理者的主张，[illegible]对我们是我们的客户（不是地方）[illegible]的"[illegible]"。[illegible]

入住是一间标准客房，[illegible]，[illegible]，不需要走一两步就能够到了。（可能是会议式酒店，[illegible]（[illegible]）。[illegible]的[illegible]！

对于这种大面积的书店来说，要做到[illegible]，[illegible]还是不容易的。[illegible]的[illegible]，也是其中一个"亮点"。对于两个[illegible]的使用者[illegible]一点点不如便——[illegible]

的，可能你一[illegible]，所以[illegible]的调整，也可以在不同的层面[illegible]了。大厅、[illegible]等等。

开阔大，[illegible]的[illegible]。[illegible]的[illegible]，[illegible]的[illegible]、空间的[illegible]，[illegible]一个[illegible]的大[illegible]。[illegible]

[illegible]中[illegible]，也[illegible]（[illegible]铅笔），[illegible]一成Logo[illegible]的[illegible]，[illegible]。也不是第一次遇到[illegible]的了！

我觉得，这可能是一个"[illegible]"[illegible]、铅笔有[illegible]、XI'AN.

也[illegible]住在哪里

住哪，在[illegible]一种[illegible]，[illegible]再一次[illegible]的[illegible]，这[illegible]一点[illegible]，不用[illegible]什么[illegible]入住在哪里。

[illegible]的[illegible]，[illegible]你们自己的[illegible]，[illegible]的[illegible]吧

[illegible]
[illegible]/12，18
[illegible]书店·[illegible]

Paper without Label

没有标签的纸

入住西安凯悦酒店，原因一是地段好，二是算最新的五星级。斜对面就是我们的客户（万众地产），六月将开业的“W”，期待。

I stayed in Xi'an Hyatt Hotel because its location was perfect and also a new five-star hotel. Our client, Wan Zhong Real Estate, was in the diagonally opposite. W Hotel is opening in June. Expecting!

入住一间标准房，算是幸运，在电梯口附近，不然要走一百多米就惨了。可能是会议式酒店，层数不多（限高的原因），单层房间有六十多间！

I was lucky to stay in a standard room near the elevator. Otherwise, I had to walk over 100 meters to get to the room. Maybe this Hyatt was a conference hotel, it didn't have many floors (because of height limitation), but there were more than sixty rooms on each floor.

对于这种大开间的标准面积房间，要折腾得好，有小创意还是不容易的，双流线的洗手间，也是其中一个“亮点”，对于两个同性的使用者还是有一点点不好使——经常要关对着卧室的门，可能是统一的设计，所以只是小小地调整，就可以做不同的房型销售了：大床房、标准双床房，等等。

It was not easy to design this kind of big standard guestrooms with creativity. One of the highlights was the double streamlined bathroom. But it was a little inconvenient for two same-sex roommates. You had to close the door opposite the bedroom often; maybe it was a unified design. So if it was adjusted a little, there would be different room types for sale—double room, standard twin room and so on.

开间大，可以有空间的小变化，嵌入式行李架，近床的茶水区，宽阔的写字台，餐区，当然还有一个适合吸烟人士的大露台，吹着冷风也真的是太酷了。

The space was so large that it had a little change in it: the embedded luggage rack, pantry area near the bed, a wide table, dining area. Of course there was also a big terrace for smoking. It was not only enjoyable but also pretty cool blowing the cold wind.

西安凯悦酒店
Hyatt Regency, Xi'an, China

设计中规中矩，找到了笔（只有铅笔），还有没有一丝标识、品牌痕迹的，切割得有点黄金分割比例的小便签纸，好吧，也不是第一次遇到没有标识的便签纸了！

The design was not bad. I found pens (only pencils) and some scribbling paper without logo or brand mark. The scribbling paper was cut by Golden ratio. Ok, it was not the first time to see scribbling paper without logo.

我觉得，这可能是一个潮流。

I think it may be a trend.

幸亏，铅笔有品牌和地址，Xi'an，也知道我住在哪里了！

Luckily, the brand and place can be seen on pencils. Xi'an, I know where I am staying.

住哪，应当是一种记忆，以文字再一次加深记忆，这才有了回忆的点滴，也就不用费神去想什么时间入住在哪里了。

Where to stay? That should be a kind of memory to deepen the impression with its characters again. So I needn't take time to think when and where I stay.

没有标识的纸，是故意让我们用自己的手，留下我们记录生活的痕迹吧。

Paper without logo makes us remember our trace of lives with our hands on purpose.

西安凯悦酒店

Hyatt Regency, Xi'an, China

THE MURRAY

HONG KONG

香港MURRAY酒店

32

★★★★★

A NICCOLO HOTEL IN MURRAY HONG KONG CHINA

Address : 22 Cotton Tree Drive, Central, Hong Kong,China
中国香港特别行政区
中环红棉路22号

Telephone : +852 3141 8888

Http : //www.niccolohotel.com

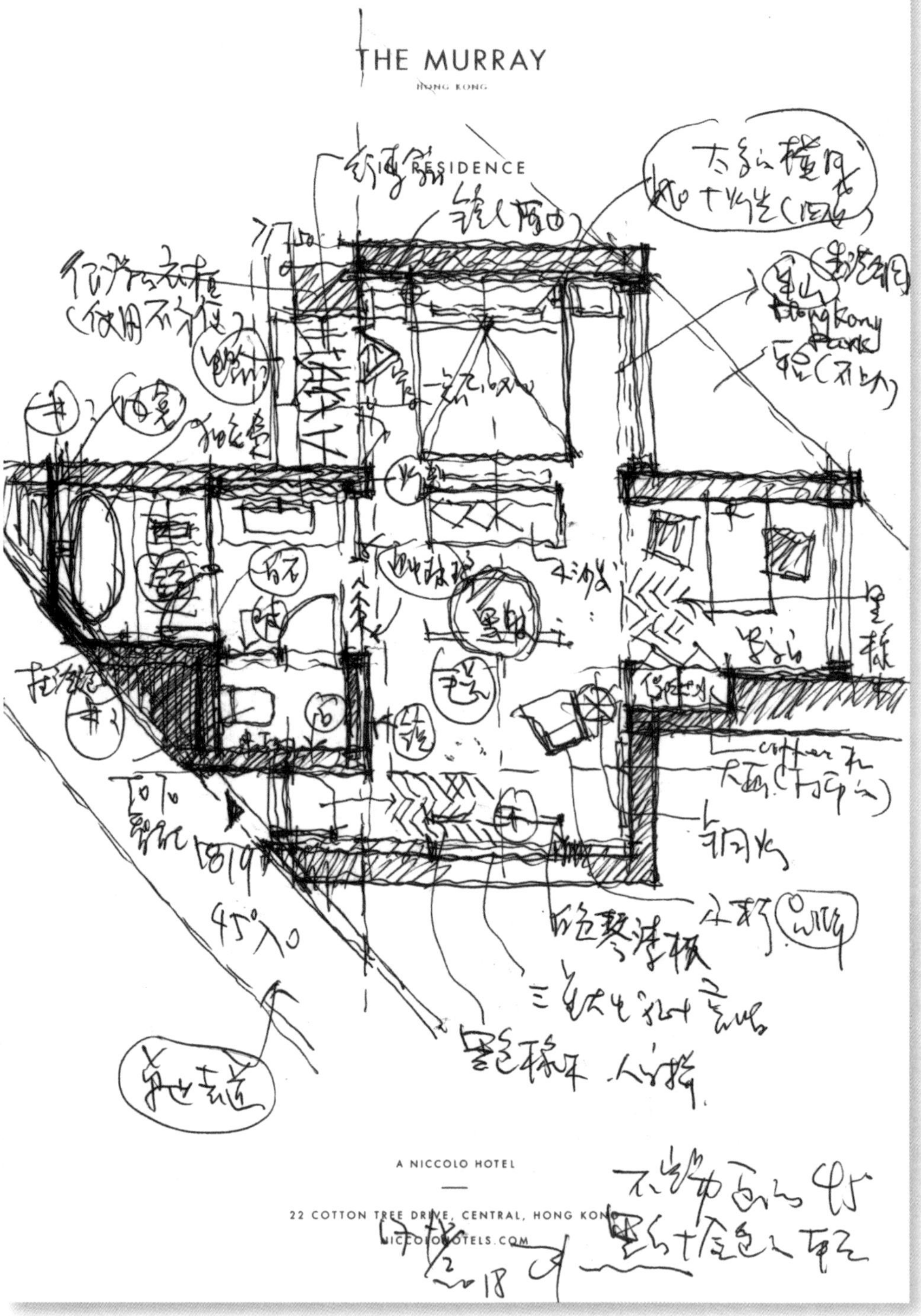

THE MURRAY

HONG KONG

RESIDENCE

A NICCOLO HOTEL

22 COTTON TREE DRIVE, CENTRAL, HONG KONG

NICCOLOHOTELS.COM

THE MURRAY
HONG KONG

17/8
2018

[illegible] Chanel [illegible]

[illegible]

[illegible] Niccolo Hotel [illegible]

[illegible] Norman Foster [illegible]

Andrew [illegible]

[illegible] Coco Chanel [illegible]

A NICCOLO HOTEL

22 COTTON TREE DRIVE, CENTRAL, HONG KONG　T +852 3141 8888　NICCOLOHOTELS.COM

THE MURRAY
HONG KONG

[illegible]

[illegible] Foster [illegible]

[illegible] upper house [illegible]

A NICCOLO HOTEL

22 COTTON TREE DRIVE, CENTRAL, HONG KONG　T +852 3141 8888　NICCOLOHOTELS.COM

THE MURRAY

HONG KONG

[illegible]

A NICCOLO HOTEL

22 COTTON TREE DRIVE, CENTRAL, HONG KONG   T +852 3141 8888   NICCOLOHOTELS.COM

THE MURRAY

HONG KONG

a. [illegible]

b. [illegible]

c. [illegible]

A NICCOLO HOTEL

22 COTTON TREE DRIVE, CENTRAL, HONG KONG   T +852 3141 8888   NICCOLOHOTELS.COM

The Black and White World like Chanel

像Chanel一样的黑白世界

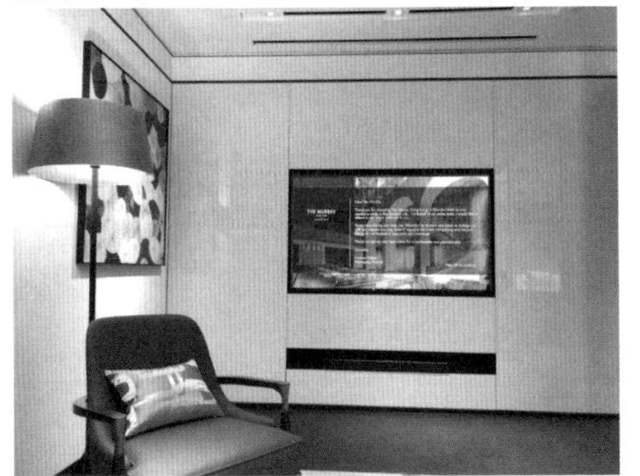

香港MURRAY酒店

A Niccolo Hotel in Murray, Hong Kong, China

突然间会觉得“黔驴技穷”，对于一个酒店房间还有什么会让我激动的?

Suddenly I feel powerless. How can a hotel room make me excited?

过年，来香港发呆一两天，碰巧有新开在美利大厦的尼依格罗酒店，元月十五开张的，位于山脚下的红棉路22号，由诺曼·福斯特爵士主理建筑改造和部分室内空间设计。

During the Spring Festival, I came to Hong Kong to relax for one or two days. It so happened that A NiccoloHotel in Murray opened not long ago, on January 15. It lies at the foot of the mountain, No. 22 Cotton Tree Drive. Its reconstruction and some parts of interior are headed by Sir Norman Foster.

富家公子傅厚民负责房间设计（极致的标准层平面规划）。

Andre Fu who is the son from a rich family is in charge of the guestrooms' design(A Extreme graphic plane of standard floor).

房子不大！黑白如Coco Chanel 小姐。

The guestroom is not large, black and white just like Miss CoCo Chanel.

不过标准房的平面划分规划很考验人。45° 的斜切（不知道是否是原来建筑的结构，我猜是设计大师福斯特的创意吧），让我练习了两次才“守住”了形，而且比例也不容易把握，那还是老老实实地画一下平面吧。家具、饰品都是专门定制设计的，每一样都是精品。

But the graphic plane of the standard floor made me mad. Bevel with a 45 degree angle (I didn't know if it was the structure of the old building. I guessed it was the master Foster's creativity), I had to practise twice in order to hold the shape. And it was also not easy to control the proportion. I just drew the floor plan at last. Furniture and ornaments were custom design and everything was the choicest goods.

房间的尺度只能说是合适，与之前住过的奕居比稍小一些，横向的布局，仅用两个窗户的采光，“当代”的五金龙头，东陶的智能马桶，Claybrook的人造石浴缸，白色大理石+黑色石边，配金色五金（虽然金色不同一），但还是有土豪的调调。

The guestroom was just in a moderate size, a little smaller than the room of Upper house where I stayed before. The horizontal layout, two windows let in natural light. Dornbracht's taps, ToTo's intelligent toilets, Claybrook's artificial stone bathtub, white marble +black stone edge, match golden hardware (the different gold colors), all look extremely luxury.

黑+白配金就是这个酒店的手法（公共区域）也一样，那只能“以点盖全”了。

Black and white and gold is the theme of the hotel, the same as the public area. I had to give a simple generalization.

设计师定好主题、色系，那就不难去营造细节，一个房间的设计整体出来就是这样子的，说一两个值得分享的细节吧。

The designer decides the theme and the color system first so that it is not difficult to make the details. The design of the guestroom will be finished. I would share one or two details of them.

香港MURRAY酒店

A Niccolo Hotel in Murray, Hong Kong, China

a.顶棚有一个小挂板，与踢脚线一致

a. A little plank on the ceiling matches the base shoe

b.地面全黑（橡木）密拼，小木地板“人字拼”

b. Black oak herringbone parquet floor

c.通高玻璃（洗手间的）

c. High glass of the bathroom

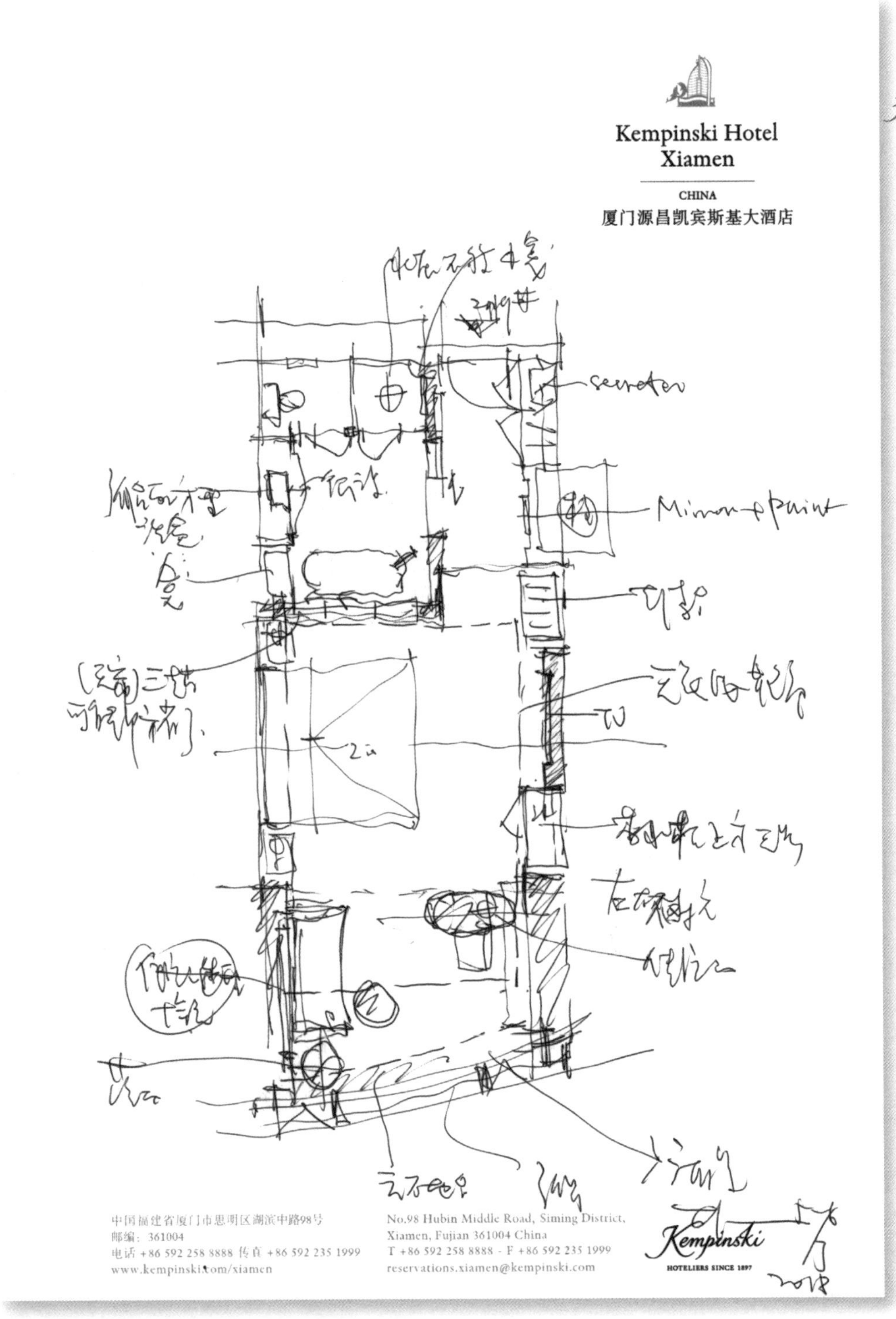

Kempinski Hotel
Xiamen

CHINA

33 厦门源昌凯宾斯基大酒店

★★★★★

KEMPINSKI HOTEL XIAMEN CHINA

Address : No.98 Hubin Middle Road, Siming District, Xiamen, Fujian,China, 361004
中国福建省厦门市
思明区湖滨中路98号
361004

Telephone : +86 592 258 8888

Fax : +86 592 235 1999

Http : //www.kempinski.com/xiamen

Kempinski Hotel
Xiamen

CHINA

厦门源昌凯宾斯基大酒店

中国福建省厦门市思明区湖滨中路98号
邮编：361004
电话 +86 592 258 8888 传真 +86 592 235 1999
www.kempinski.com/xiamen

No.98 Hubin Middle Road, Siming District,
Xiamen, Fujian 361004 China
T +86 592 258 8888 - F +86 592 235 1999
reservations.xiamen@kempinski.com

Kempinski
HOTELIERS SINCE 1897

Kempinski Hotel
Xiamen
CHINA
厦门源昌凯宾斯基大酒店

但又没法也没有拍摄它们的酒店、

记得戴志帅写在当我的《酒乡门门》里，相信我说中心城的酒店都会，这个都没有笔和纸作为"去节日"的初衷在于"留住文化生活"，节日准备放上书刻也是有很高的，可能有意义的意义在于：地点、人物、历史、文化等等等。

这次下榻厦门的凯宾斯基酒店是此行的最后一站，选择这个酒店也是知道它曾经是别人的旧居，但又说不出它的一段，同行的几个人也没有答案和它的经历。但是也许不是结构性的不美在细节。

现在，那时的（酒店管理）高级的投资，这一步都会让自己放下了身段，因为"那个酒店图"的传统变会"卑躬屈膝"，也许这就是原因吧。

酒店单房间，布局依然当初的风格（也许是不完整的成分子）遗憾的酒店！惜！半个世纪的酒店又等不了太久！浅！节从谁上加无从，叹！也许我们不尊重，更被轻待，只在乎这个更为豪华的，从同时酒店中的设计

5-6/1/3
2018

中国福建省厦门市思明区湖滨中路98号
邮编：361004
电话 +86 592 258 8888 传真 +86 592 235 1999
www.kempinski.com/xiamen

No.98 Hubin Middle Road, Siming District,
Xiamen, Fujian 361004 China
T +86 592 258 8888 F +86 592 235 1999
reservations.xiamen@kempinski.com

Kempinski
HOTELIERS SINCE 1897

厦门源昌凯宾斯基大酒店
Kempinski Hotel, Xiamen,China

No Desir to Take A Picture of the Hotel

不想打卡的酒店

记得戴昆帅哥在为我的《住哪? 3》写序时，调侃我什么样的酒店都画，这个和我只把笔和纸作为“分享印记”的初衷有关，“既来之，则安之”，某种程度上对于书刊我也是有取舍的，可能有意义的定义有许多：地点、人物、历史、文化，等等。

I remembered that Mr. Daikun wrote the Foreword for my book Where to Stay? 3. He teased me that I drew all kinds of hotels. This habit was related to my original intention of sharing my mark with pens and paper. Since we come, let's stay and enjoy it. To a certain degree, I will make a choice when my book is edited. Maybe they are based on different meaningful definitions: places, persons, history, culture and so on.

这次入住厦门的凯宾斯基酒店是顺路来看看石材展，选择这间酒店也是想一睹邦胜兄的近作，但入住感觉一般，同行的几个人也没有举手机的欲望，细思也许不单单是设计师的责任吧。

This time, I came to Xiamen Stone Fair and stayed in Kempinski Hotel. It was also because the hotel was Mr. Bangsheng's recent work. But I thought it was just ordinary. A few colleagues even had no desire to take photos like me. Thought carefully, it was not only the designer's responsibility.

五星级，国际品牌（酒店管理），房地产商的投资，这一些都会让设计处于劣势，刚步入“国际酒店圈”的设计公司更会“卑躬屈膝”，也许这就是主要原因吧。

The five-star hotel is managed by international brand and invested by real estate business. Both of them put the design at an inferior position. The design companies which just step into the International Hotels circle are even more servile. Maybe this is the main reason.

说回来房间，布局传统与创新结合（也许是不怎么成功的），设置透明的浴厕，怪！休闲区与写字区组合尺度欠佳，浅！茶水柜上方无灯，缺！也许问题不严重，木色保守，欠亮点，这个更是致命。吸引，从同行开始！加油，同行。

Come back to the guestroom,there is the combination of tradition and innovation(But it was not so successful), the transparent bathroom—Strange! The leisure area and the working area were not in a perfect size—Shallow! There was no light above the tea cabinet—Imperfect! All these might not be serious. Wood color was conservative. No design highlight was even more deadly. Attractive begins with colleagues! Come on! Colleagues!

锦江都城经典南京饭店

34

★★★★

METROPOLO JINJIANG HOTELS CLASSIC SHANGHAI CHINA

Address : No.200 Shanxi South Road, Huangpu District, Shanghai,China
中国上海市黄浦区山西南路200号

Telephone : +86 21 6322 2888

Http : //www.jjmph.com

锦江都城经典南京饭店

Metropolo Jinjiang Hotels Classic, Shanghai, China

METROPOLO
JINJIANG HOTELS
www.jjmph.com

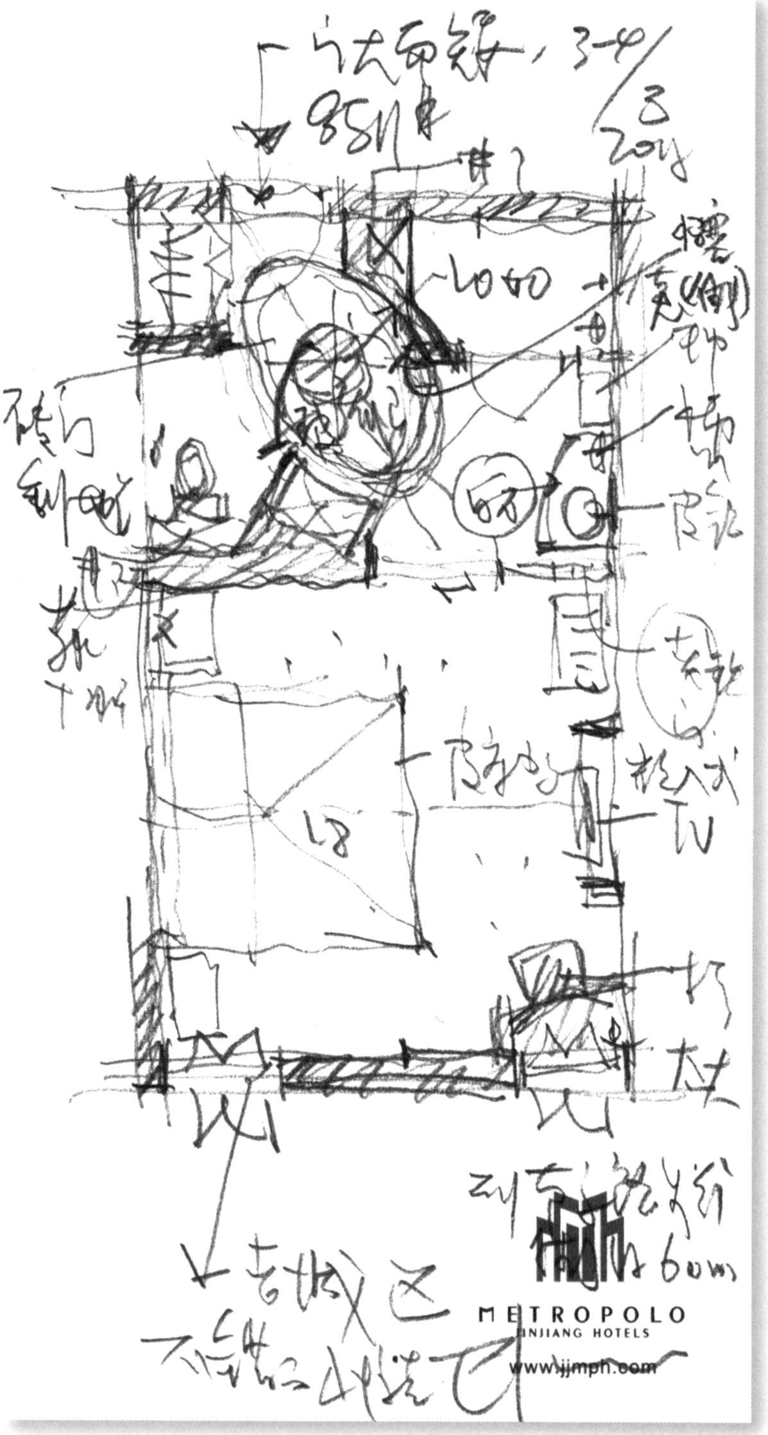

对陈红的怀念 3/3 2018

能在汉口的上海商会的百年下上见到你们怀念，不胜不由。

那支持若是当年陈红的你们这汉宫的"陈红节城"的理念如是样子。大胆的本着的布局（也许的刻意的），好坏是不评价，来中的的投入是不有一种的（对我们这时代的设计生活的信念也是样的的城市的设计的一个相同的的地方的的旅馆的风格，而价旅馆，随便的设计区，怀旧的铁艺装饰，有了几分的新手感的味道，没有色的设计已经[illegible]

METROPOLO
JINJIANG HOTELS
www.jjmph.com

喜欢喜欢，是如此的，但是真实（你的）的[illegible]生活味的设计的感觉，相信

当时的风格过往的记忆。你们的[illegible]的生活了，我们这些新生活，新的记忆，新的陈红对未来的新的也许是也是一定的的节奏。

每一次都有怀念，那这个[illegible]

下一次还有[illegible]

METROPOLO
JINJIANG HOTELS
www.jjmph.com

An Unforgettable Surprise

难忘的小惊喜

能在改造的小酒店空间及平面上感到有小惊喜，确实不易。

It's really not easy to feel surprised at the space and floor plan of a small converted hotel.

朋友推荐免费试住的南京饭店改造成的“锦江都城”就是这样的，大胆的小房间布局（也许是刻意的），好坏先不评价，与中档的投入是不相一致的（至少我们设计师第二天上午吃早饭时讨论是这样说的）。开放式的区域以一个椭圆、砖的水刀切割商标拼花巧妙组织：厕浴分离，椭圆的洗手盆区、怀旧的铁艺趟门，有了几分老克拉的味道，深灰色调调的主卧区，喜欢喜欢，见仁见智；倒是真皮（仿古）的床背靠有人情味，有神往的感觉！

My friend recommended me Jinjiang Metropolo which was converted by Nanjing Hotel. I stayed in it for free. This brand is just like that. The bold layout of the small guestroom was inconsistent with the mid-range investment(Maybe it was deliberate). I didn't evaluate it good or bad first(At least we designers thought so when we talked about it at breakfast the next morning). The open area was separated by the pattern of Logos which was cut by an oval-brick water knife. The toilet and the bathroom were also apart. The oval sink area and the nostalgia iron sliding door had a little feeling of Old Carat, the color of the master bedroom was dark grey, I liked it very much, but different people had different views. The leather antique backrest was warmly human and fascinated me.

窗外的风景还是很棒的，一座小小的矮矮的老房子，我们的惊喜来自生活，来自自己的感受，来自对陌生、对未来的祈求，也许住也是其中一点点的部分。

The view outside the window was perfect. A low old house could be seen. Our surprise comes from life, comes from our feelings, comes from the prayer for the strange future, and maybe staying in a hotel is also part of it.

每一次都有小惊喜，那这个品牌就更厉害了。

If there is a little surprise every time, this brand will be much greater.

下一次还可以再找找。

I will try to discover surprise next time.

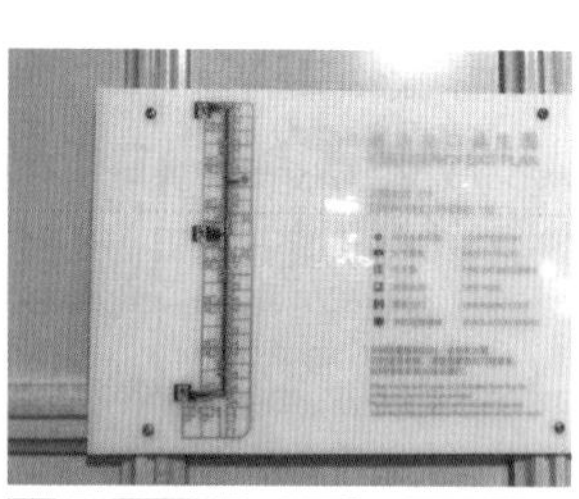

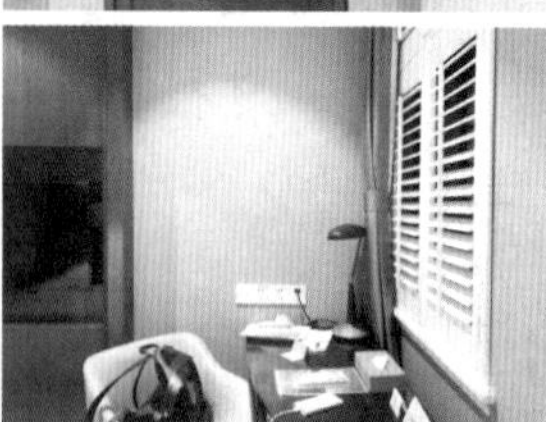

锦江都城经典南京饭店
Metropolo Jinjiang Hotels Classic, Shanghai, China

O.'s File

（区生词典）

老克拉或称为“老克勒”，指的是老上海阅览深、收入高、消费前卫的都市男性族群。(以上内容转自《百度百科》)

Old Carat means those old gentlemen,who live in Shanghai, with rich experience, high-income and consumption fashion.

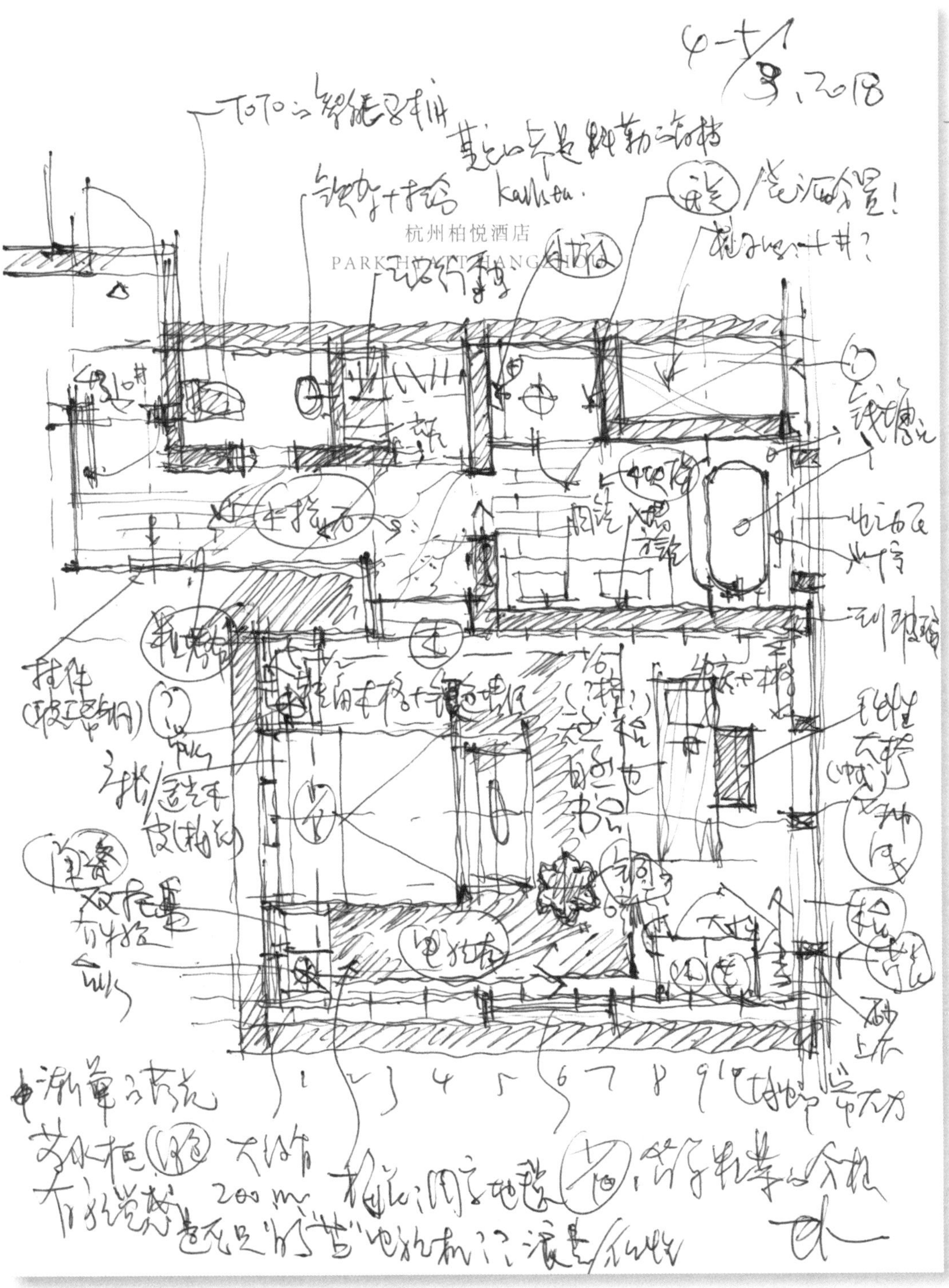

HYATT

杭州柏悦酒店

35

★★★★★

PARK HYATT HANGZHOU CHINA

Address : No.1366 Qianjiang Road, Qiangjiang New Town, Jianggan District, Hangzhou,China
中国浙江省杭州市江干区
钱江新城钱江路1366号
Telephone : +86 571 8696 1234
Fax : +86 571 8693 1234
Http : //www.hyatt.com
E-mail : hangzhou.park@hyatt.com

玻璃地台
The glass wall cabinet

荷花与石头
The lotus and stones

杭州柏悦酒店

PARK HYATT HANGZHOU

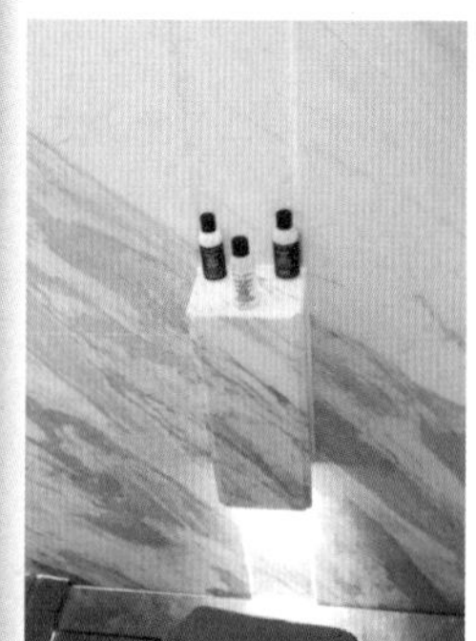

细节处理

Details treatment

灯
Lamps

Shadow
“影子”

入住在杭州算是最好的酒店，柏悦，有熟悉的感觉，大大的大堂，全江景大堂酒吧，单一的木色和“花花绿绿”的大理石（肆意地用）。

I stayed in Park Hyatt which is the most expensive hotel in Hangzhou. I had a familiar feeling with it. The large lobby and the lobby bar faced the wonderful river view. A single wood color and colorful marble were used recklessly.

一查百度，乔治・亚布的室内设计，佩服，又卖掉了一个矿山（像广州W酒店一样），因为很多的项目模仿他们的用材，所以这种石头很热卖，很快就卖光了，设计推动市场的力量不容小觑。

Looked up in Baidu, the interior was designed by George Yabu. I respected him. Of course he sold a mine (Just like W Hotel in Guangzhou). Because many programs copied their material, this kind of stones sold well and sold out quickly. The power of the design which pushes the market couldn't be underestimated.

杭州柏悦酒店
Park Hyatt, Hangzhou,China

方形的传统塔楼，我选了一个江景的近转角位的大房间，平面布局成熟而有独到之处，厕浴分离，成熟而有序。夜晚睡觉更加感受到：声音、灯光的影响（平面是关键），挺棒的，比广州的W有条件，虽然有大师的影子（其实每一个人都有自己固有的一套体系），面江的浴区，心意合一，洗手盆不对称的设计也是当下的一种时髦（如季裕棠的广州文华东方也采用这种设计）。卧室区非常深，写字台原木独特，在一侧的染色荷花茶水柜成为焦点，大大的双床头柜，配金色木椭圆挖孔，奢侈，但夜光太亮了！而更土豪的在于全固装的木格墙身，转角的小格配银质墙的细节，简约的大面块配精细的细节，这也是“影子”！这就是大师给我的。

多个系统的灯光
Multi-system lighting

The traditional tower was square. I chose a big guestroom with river view at the corner. The layout was mature and has something special. The toilet and the bath were separated maturely and orderly. You could feel it better while sleeping at night: the sound and light had a perfect effect. It was better than W Hotel in Guangzhou. Although we found out the master's design in it. (In fact, everyone has his own pattern of system.) The bath area facing the river view combined heart and soul. The design of the irregular sink was fashionable at the moment. (Like Tony Chi's Mandarin Oriental in Guangzhou.) The sleeping area was very deep. The wooden table was unique. The tea cabinet, which its side was dyed lotus, became the highlight. Double big beside tables with golden wooden elliptical holeswere luxury. But the night light was too bright. The most luxury was the whole wood frame wall which was fixed decoration.Small squares at the corner matched silver wall, simple big plate fit refined details, and both of these details were the shadow of the master. That was what I learned from the master.

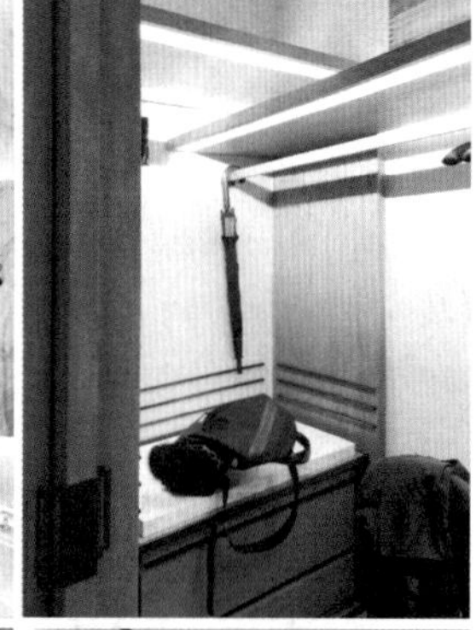

荷花酒吧柜
Liquor cabinet with lotus

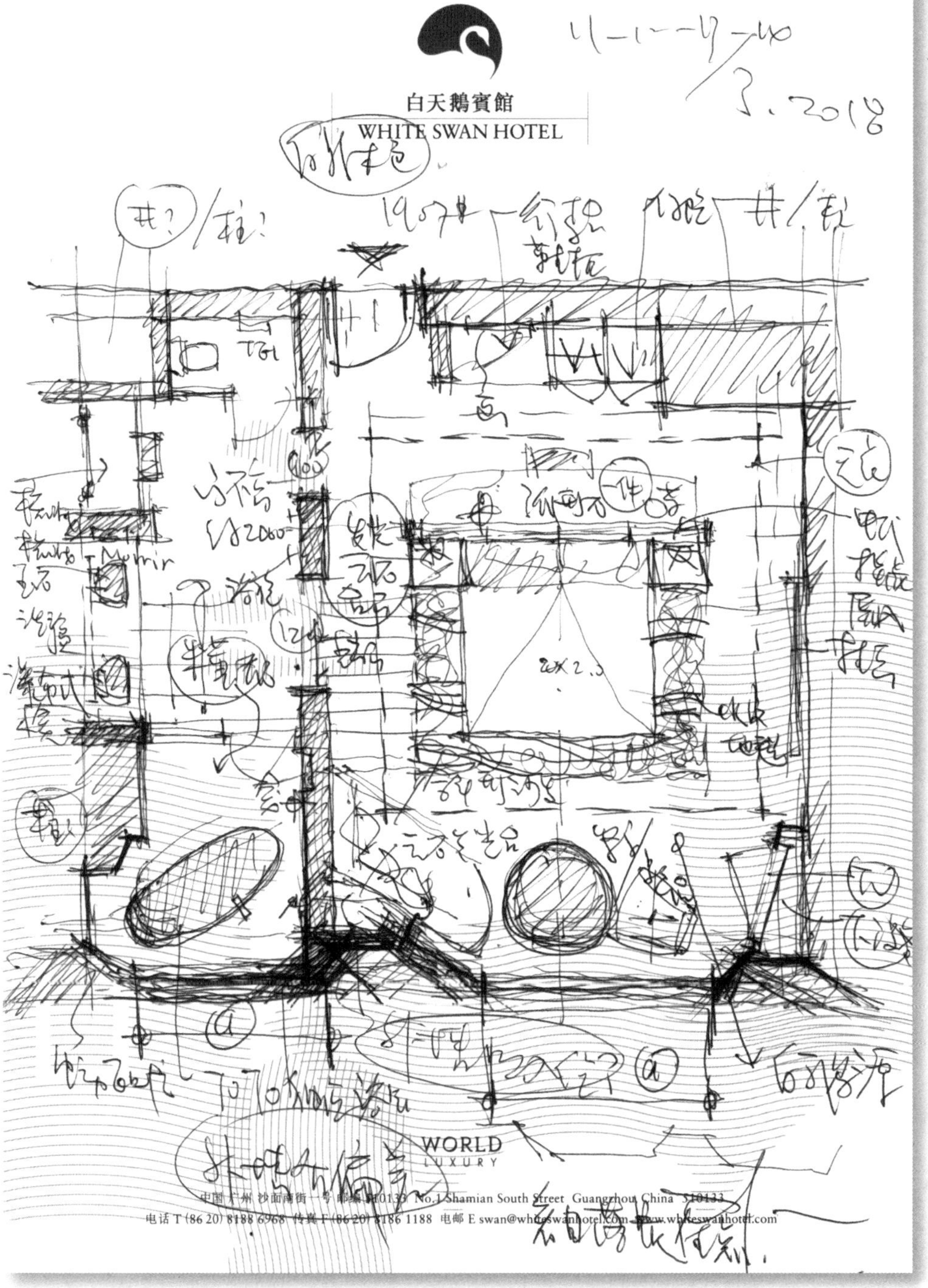

白天鹅賓館

36

广州白天鹅宾馆

★★★★★

WHITE SWAN HOTEL
GUANGZHOU CHINA

Address : No.1 Shamian South Street, Guangzhou, China
中国广州
沙面南街1号
Telephone : +86 20 8188 6968
Fax : +86 20 8186 1188
Http : //www.www.whiteswanhotel.com
E-mail : swan@whiteswanhotel.com

广州最美的白天鹅宾馆 [illegible] 15/2 2018

白天鹅賓館
WHITE SWAN HOTEL

当然能够住在[illegible]近三十年仍[illegible]。（佩服[illegible]！）[illegible]。但能一睹[illegible]“[illegible]”也是[illegible]一段[illegible]。[illegible]入住[illegible]一些的[illegible]70m²[illegible]房间（[illegible]基本[illegible]，也难怪[illegible]HBA大师们）[illegible]，一[illegible]入住时的[illegible]”。

这次[illegible]入住[illegible]白鹅潭[illegible]房，60m²，[illegible]。[illegible]，中[illegible]大[illegible]。[illegible]入住[illegible]不适。[illegible]。[illegible]也[illegible]的[illegible]人。[illegible]：

[illegible]。从[illegible]。[illegible]，从[illegible]到[illegible]。[illegible]下一[illegible]也不会[illegible]。[illegible]最美的。[illegible]。

WORLD LUXURY
中国广州沙面南街一号 邮编 510133 No.1 Shamian South Street Guangzhou China 510133
电话 T (86 20) 8188 6968 传真 F (86 20) 8186 1188 电邮 E swan@whiteswanhotel.com www.whiteswanhotel.com

The Most Beautiful Guestroom to Watch the Sunset in Guangzhou

广州最美的日落房间

虽说白天鹅经历了近三年的全停业改造，重新开业，但能一睹其客房的“芳容”也是在前一段时间，当时是朋友推荐入住经济一些的望沙面的30平方米的小房间（就是基本按原来的间隔的小房间，也是难为HBA大师了），评价特别的差，一群设计师在入住房间时这样评价。

Although the White Swan Hotel has been closed to reform for almost three years and it reopened, I haven't got a chance to visit it until not long ago. I recommended some economic rooms which face Shamian to my friends. (these rooms basically maintain original interval—small size ones, it made difficult for the Master HBA.) The comments were really bad—A group of designers were critical of the hotels when they stayed in it.

这次好了，入住望着白鹅潭的望江房，60平方米，相当于两间合并的产品，浅木色+采光的洗手间，锯齿形的原外立面也给室内设计师提出难题，中置面江的大床式布局，半透的木质屏风分区可能会令长者不适，全木地板的应用也属不错，但这些都不足以打动人，倒是：

It was better to stay in the guestroom facing the Pearl River. The room was 60 square meters which equaled two original rooms. Light wood color and natural lighting bathroom, the zigzag original facade set the designers a very difficult task. The king bed was in the middle of the room facing the Pearl River. The subtransparent carved wooden screen division might make the elderly feel uncomfortable. The wood floor was also perfect. But all of these didn't touch me except the beautiful sunset.

广州白天鹅宾馆

White Swan Hotel, Guangzhou,China

广州白天鹅宾馆
White Swan Hotel, Guangzhou, China

日落，江水倒影，从下午五点左右到六点多，倚窗，品茶，读书，从半空光芒四射的艳阳到锐气渐收的半夕阳，再到慢慢收敛光华的半空的“蛋黄”，渐渐的、慢慢的过程。

The sunset reflected in the Pearl River from five to six in the late afternoon. Stood by the window, drank tea and read books. The beautiful bright sun was fading to the half circle of the setting sun in the sky inch by inch. The "golden Yolk" was falling down and grew weak slowly.

原来，落日是这样的，不一定能看到全落下的过程，也不会全落下。

The setting sun was like this, perhaps you couldn't see the sun fall behind horizon, it wouldn't fall down completely, either.

这里，可谓是最美的日落。

The White Swan Hotel is the most beautiful place to see the sunset.

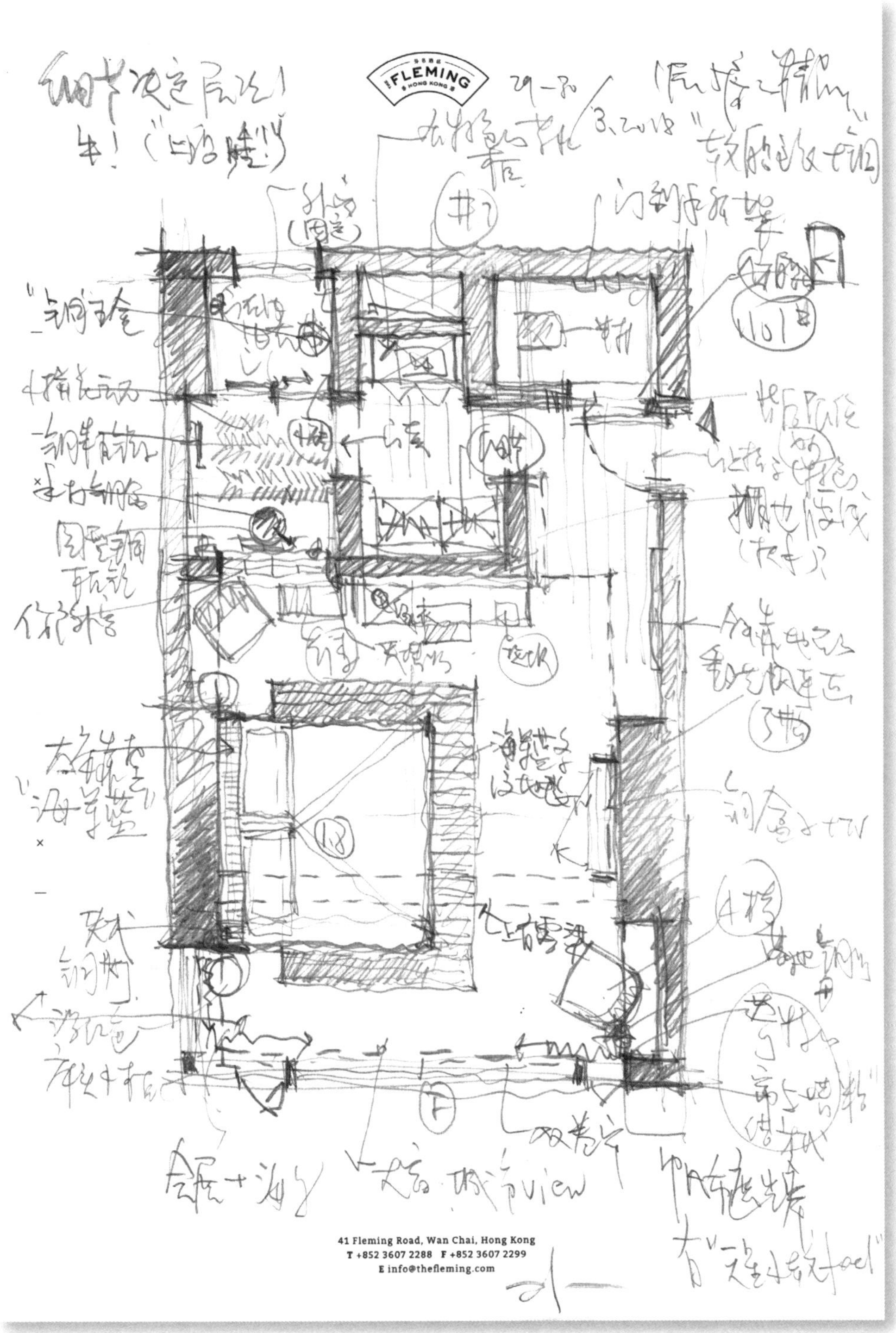

41 Fleming Road, Wan Chai, Hong Kong
T +852 3607 2288 F +852 3607 2299
E info@thefleming.com

37

香港芬名酒店

★★★★

FLEMING HONG KONG CHINA

Address : No.41 Fleming Road, Wan Chai, Hong Kong,China
中国香港特别行政区
湾仔菲林明道41号

Telephone : +852 3607 2288

Fax : +852 3607 2299

Http : //www.thefleming.com

E-mail : info@thefleming.com

外观和网红餐厅

The appearance and the Internet celebrity restaurant

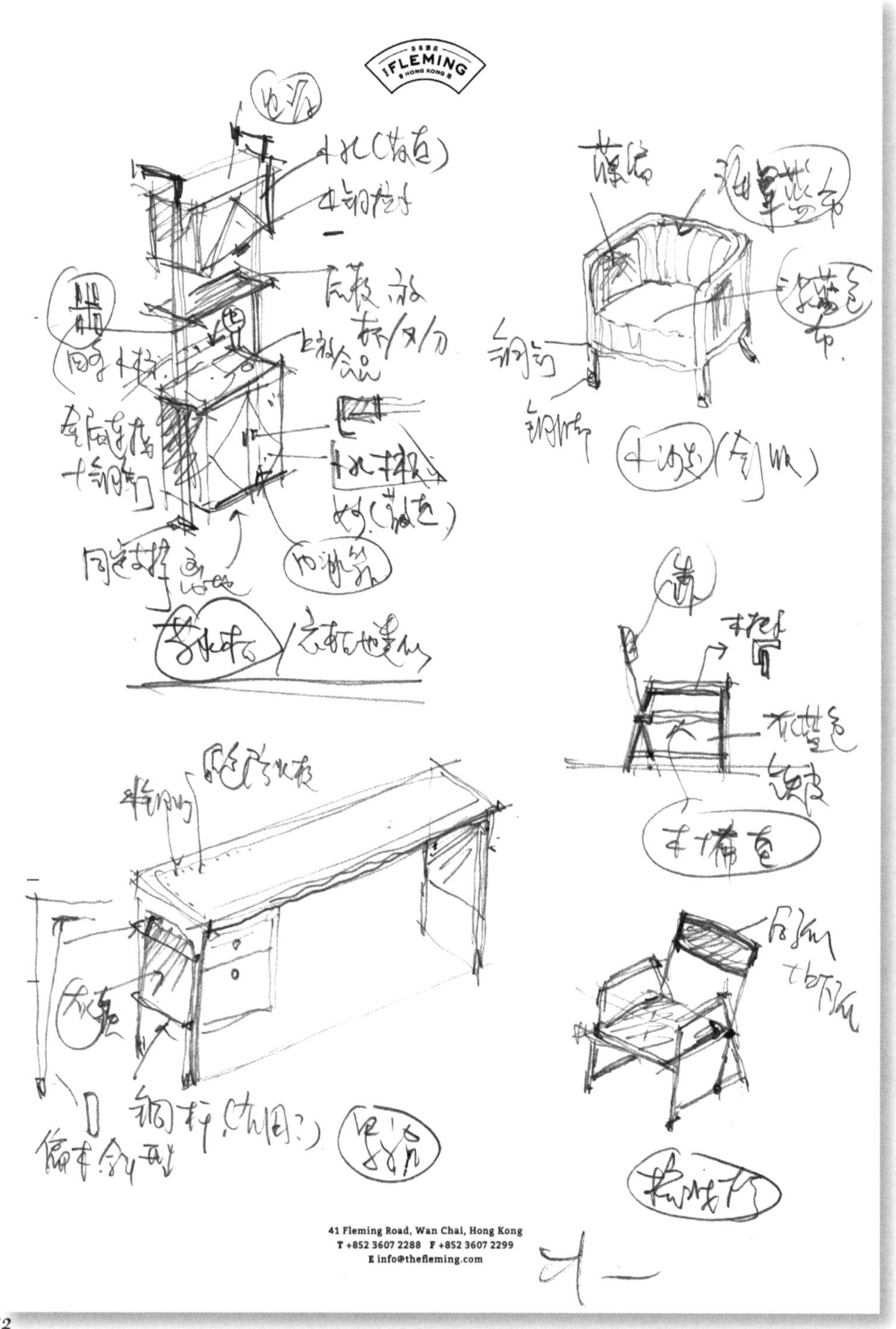

THE FLEMING
HONG KONG

41 Fleming Road, Wan Chai, Hong Kong
T +852 3607 2288 F +852 3607 2299
E info@thefleming.com

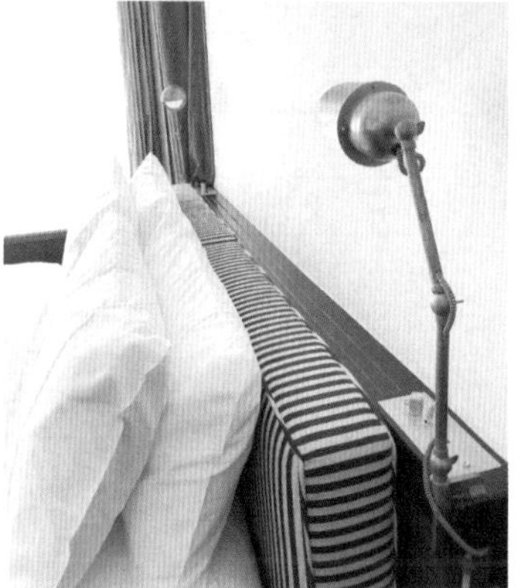

细节
Details

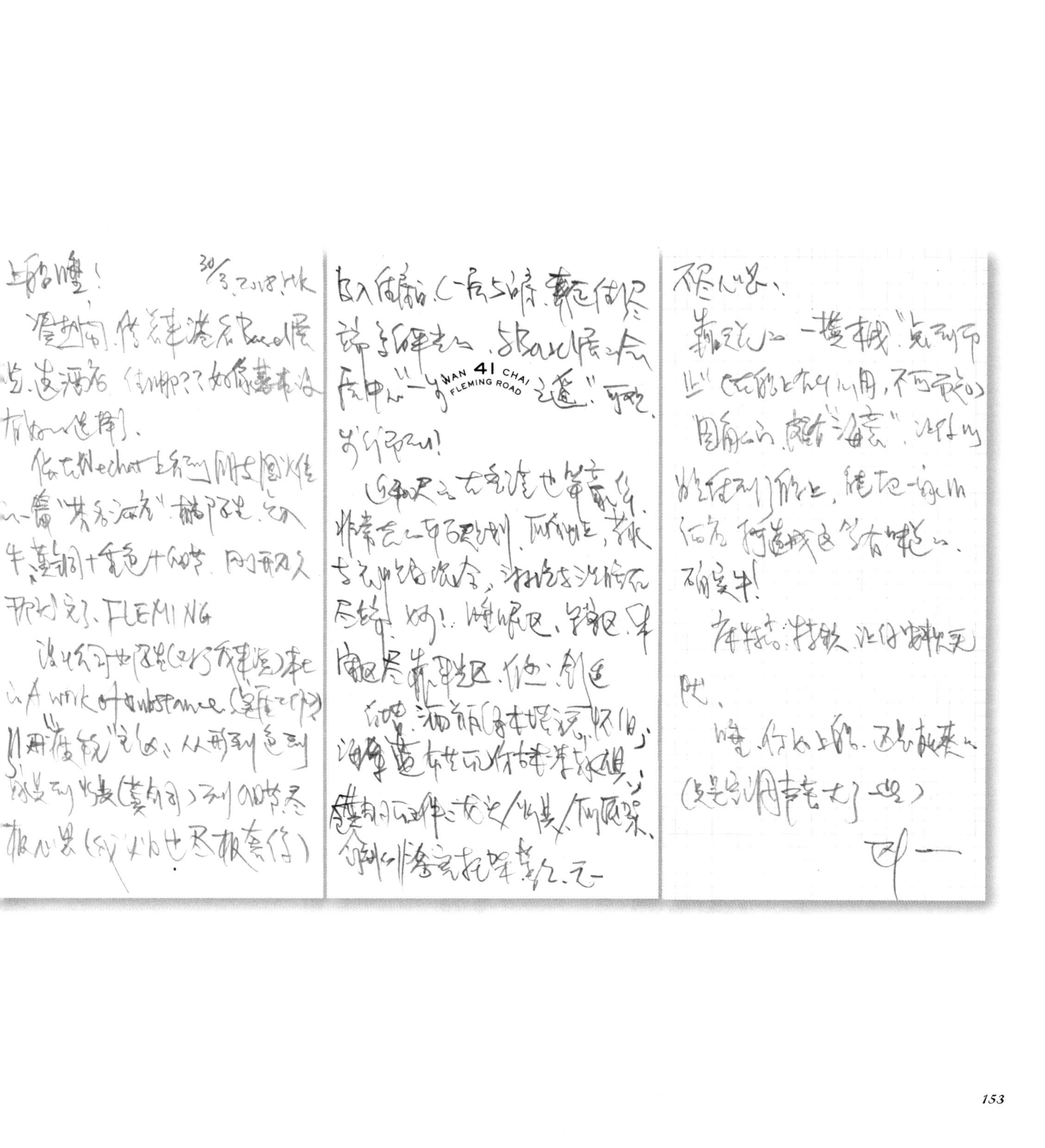

上路嘍！　　　　30/3，2018，HK

[illegible]Bar[illegible]……

你在Wechat上[illegible]……一篇"[illegible]"，[illegible]……FLEMING

[illegible]A work of substance（[illegible]）[illegible]……

WAN 41 CHAI
FLEMING ROAD

[illegible]……

[illegible]……

不忍心呢。

[illegible]一樣[illegible]……（[illegible]，不可[illegible]）

[illegible]……確實牛！

[illegible]……

（[illegible]）

[illegible]

Welcome Onboard Hotelship

欢迎登船

凑热闹！借着来香港参加巴塞尔艺术展览，选酒店，住哪？好像基本没有好的选择了。

To join in the fun, I came to Hong Kong to visit the Basel Exhibition. I had no idea where to stay when I wanted to choose the hotel. It seemed no good choice.

偶然在微信上看到朋友圈疯传的一家“芬名酒店”，相当陌生，点入链接，牛！黄铜+重色+细节，刚开业不久，就它吧，芬名酒店！

I occasionally found the Fleming which was hot on the Internet. It was very strange to me. I clicked into the website to have a look. Cool! Brass, heavy color things and details, it opened not long ago, That was it—Fleming!

设计公司也陌生（对于我来说），为香港本土的公司A Work of Substance。采用“渡轮”主题：从形到色到家具到灯具（黄铜）到细节，极尽心思（我认为也极尽奢华）。及后入住房间（一层只有5间房），幸运入住尽端多向采光的房间。酒店与巴塞尔展的会展中心“一步之遥”，步行而至！

The design company was also strange to me. A local design company—A Work of Substance. The concept of design came from the Ferry: from shape to color, furniture and brass lamps, the designers made all the details perfect with heart. (I thought everything was luxury as well.) After checking in, I found that there were only five rooms on every floor. I was lucky to stay in a room with multi-natural lighting which lay at the end of the corridor. The hotel was only a step away from the convention center, so I could walk there.

香港芬名酒店
Fleming, Hong Kong,China

房间
Room

近400尺在香港也算奢华，非常态的平面规划，厕所独立，茶水间与衣帽间结合，淋浴间与洗脸区在尽端，妙！睡眠区、写字区、休闲区尽靠采光面，自然，舒适。白墙，酒瓶绿木墙裙，怀旧；海军蓝布艺配仿古木漆家具，全黄铜配件——龙头、灯具、厕纸架、订制的浴室托架，等等。

It was luxury in Hong Kong that the hotel room was nearly 400 feet. The floor planning was unusual, independent toilets, the pantry was also the cloakroom, the shower room and the sink are both at the end of the room, so smart! All the sleeping area, the working area and the leisure area were close to the lighting surface, natural and comfortable. The white walls, bottle green and the wooden dado were nostalgic; Navy blue fabric sofa matched antique wood lacquer furniture, all the accessories were brass: taps, lamps, toilet paper holders, customized bathroom brackets and so on.

无一不尽心思：靠顶棚的一横木线“点到即止”（这个在船上有什么用，不得而知）。圆角的门颇有“海意”，让你以为是住到了船上，能把一家旧酒店打造成这等有味道的，确实牛！

The designers spent enormous efforts on everything: a wooden line which was close to the ceiling (Nobody knew what it was used for on the ship).The door with rounded corner gave out the feeling of the sea to make you consider of staying in the ship. It was really amazing to rebuild an old hotel like this!

床特高，特软，让你安枕无忧。

The bed was so high and soft that you could sleep without anxiety.

睡，仿如上船，还是挺爽的。（只是空调声音大了一些，权当是海浪的声音吧！）

Sleeping was just like in a ship, so cool! (The noise of the air conditioner was a little loud, imagine the sound of waves.)

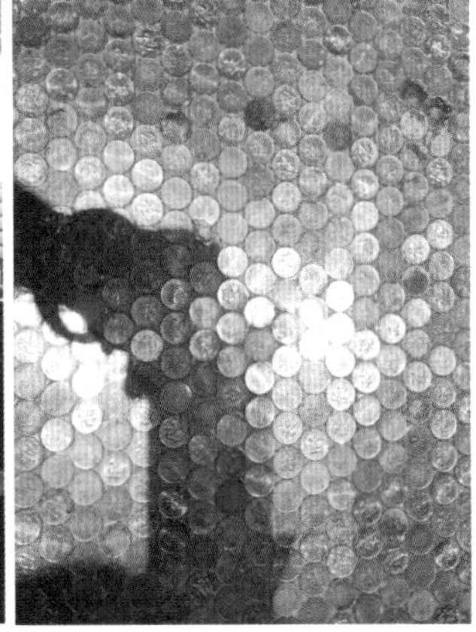

香港芬名酒店

Fleming, Hong Kong,China

METROPOLO
JINJIANG HOTELS

38 上海锦江都城青年会经典酒店

★★★★

METROPOLO JINJIANG HOTELS CLASSIC YMCA SHANGHAI CHINA

Address : No.123 Tibet South Road
Huangpu District,
Shanghai,China
中国上海市黄浦区西藏南路
123号

Telephone : +86 21 5160 1931

Http : //www.jjmph.com

METROPOLO
JINJIANG HOTELS
www.jjmph.com

6.2014

(1929年)

[illegible]

[illegible]

METROPOLO
JINJIANG HOTELS
www.jjmph.com

[illegible]

METROPOLO
JINJIANG HOTELS
www.jjmph.com

Something Old, Something New

旧而弥新

入住近90年楼龄的（建于1929年）锦江都城青年会经典酒店，可谓再一次向“古典致敬”。

I stayed in a classic hotel named Metropolo Jinjiang(built in 1929), it really "honors Classic" again.

酒店位于西藏南路，旺地，交通方便，更是食肆林立，居然走几步就到了我们喜爱的上海菜“新陶陶”的云南南路。

The hotel lies on Tibet South Road in busy area with convenient transportation and lots of restaurants. It took us only a few steps to get to a Shanghai restaurant named "XinTaotao" which was on Yunnan South road.

旧就是靠时间养的，最奢侈的就是时间了。设计师嘛，总爱挑刺，多入住几间“锦江都城”就会比较一下，老毛病，画完了这个平面后总觉得自己会做得比这个好！真的可以吗?

Old style needs time for nourishment. Nothing is more luxurious than time. Designers are always critical. Compared with a few "Jinjiang Hotels" which I have stayed in, I think that my design will be better after drawing the floor plan. It is my old habit. Can I really make it?

房间位于走廊的尽端，“L”形的，转角窗，睡眠区占据了最好的位置，对着大马路，而洗手间像其他同类酒店一样放在不太重要的位置，也许这就是我挑毛病的地方呢，洗手间不可以更好一些吗?

My room was at the end of the corridor with "L" shaped and a corner window. Sleeping area occupied the best place facing the main road. The bathroom was in a less important place like the same kind of hotels. Maybe this was my criticism, why couldn't the bathroom be better?

入门、大堂、餐厅、电梯等等都“修旧如旧”。房间更是旧味十足，挺吸引人的。小休闲区，大大的写字台（可吃饭），反而床的位置挺“紧”的，让人有一点坐立不安。不安的或不足的地方还不少：窗框、帘都不如意，我觉得还不如直接一框对一个窗，不强求对称和遮挡原有不严格的窗、窗台及结构柱；地毯褪色老旧而显脏，对了，“旧”但不能脏，特别是感觉，油漆和墙面就有这样的强烈感觉，旧但不能烂，毛巾还破了一个洞，更加不用说洗了过百次的感觉了。

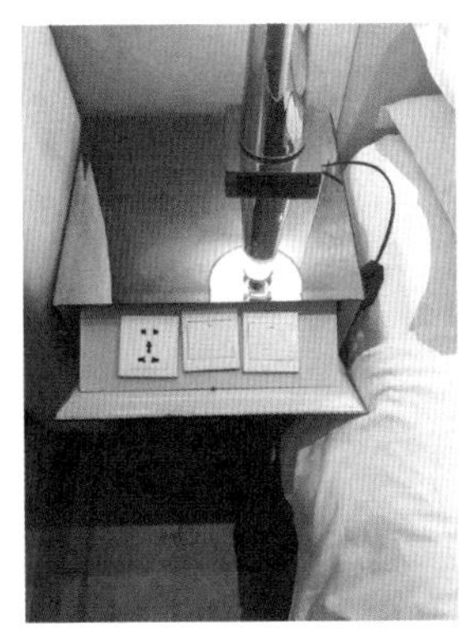

The entrance, lobby, restaurant and the elevator were all repaired to be old. The guestroom was full of old style and charms: a small leisure area, a very big table (for meals). While the sleeping area was so narrow that I felt a little upset. Here were some disadvantages: the window frame and curtains were unsatisfactory. I found it might be better that one frame matched one window instead of the pursuit of symmetry and covering the irregular window, windowsill and structural pillars. Faded carpets were old and dirty. By the way, it might be old but not dirty, especially the feeling. The painting and the surface of the wall also gave out this strong feeling. Towels might be old but not worn-out. Seeing a hole in it, you couldn't help thinking it had been washed for hundreds of times.

终于懂了，原来“旧”也要有性格和坚持的底线，当然更需要高成本的保证：要整洁。旧是一种价值，更应该用价值去维护，我们入住的时候应当用一种“纯洁”的心情去体验这近90岁的老建筑的风韵。

I understood at last. Old must have character and the baseline of persistence, of course the guarantee is high cost. Old needs to be kept clean. Old is also a kind of value, worth being protected. When we stay in it, we should experience the charm of this ninety-year-old building with our most pure heart.

酒店可以怎么“旧”，让它有尊严、有魅力的“旧”，这才是我们应做的！

How old can a hotel be? We should keep its dignity and charm. That's what we should do.

我会继续住下去的，找找旧建筑，或更旧的建筑。

I will keep staying in these kinds of old buildings and searching for them, or even older ones.

上海锦江都城青年会经典酒店
Metropolo Jinjiang Hotels
Classic YMCA, Shanghai,China

METROPOLO
JINJIANG HOTELS

上海锦江都城经典南京东路外滩酒店

39 ★★★★

METROPOLO CLASSIC BUND CIRCLE SHANGHAI CHINA

Address : No.98 Nanjing East Road, Huangpu District, Shanghai,China
中国上海市黄浦区南京东路98号

Telephone : +86 21 5160 1931

Http : //www.jjmph.com

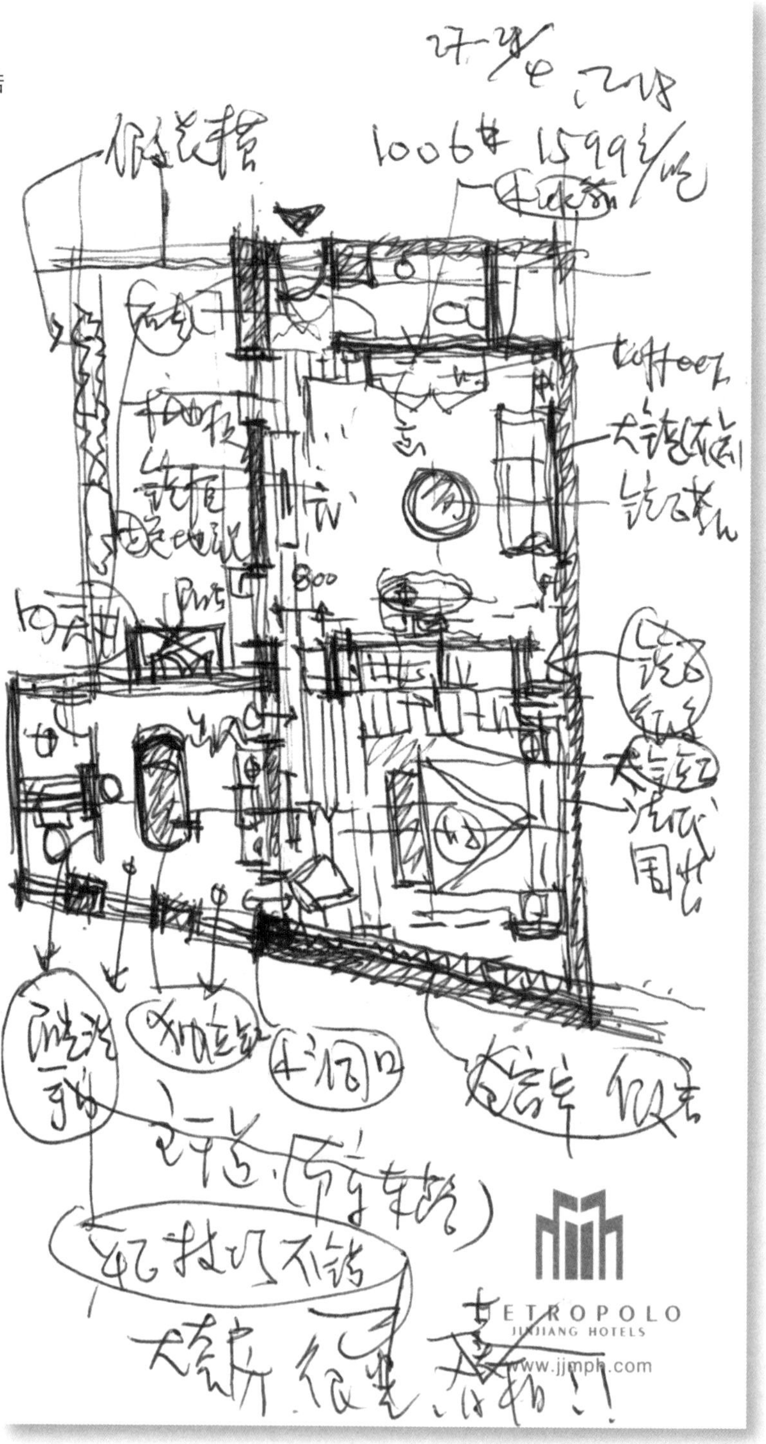

METROPOLO
JINJIANG HOTELS
www.jjmph.com

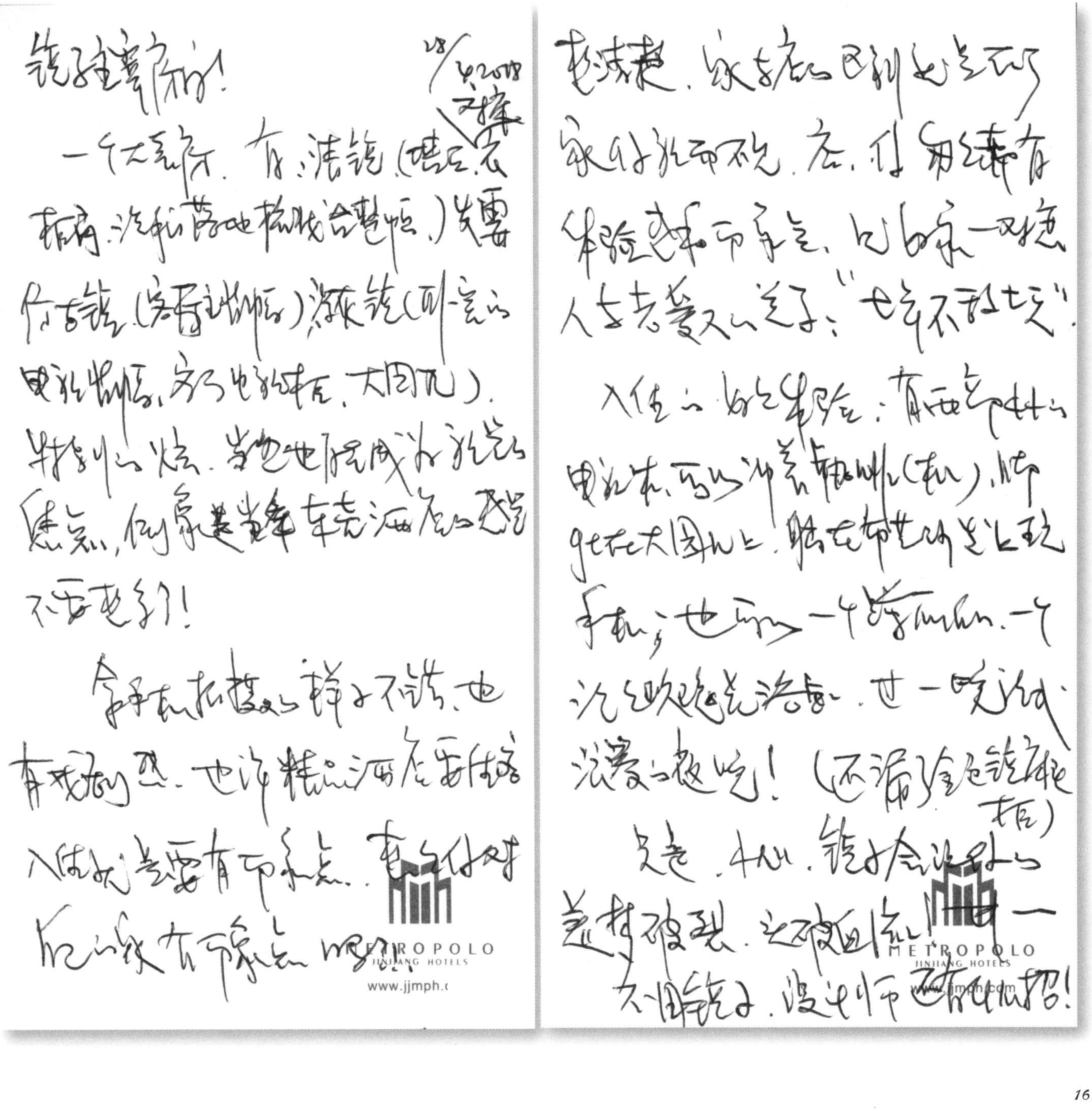

锦江都城你好!

28/8/2018
文辉

一个大气派,有:法馆(塔尖、花园、洋松落地格栅门窗整修后)、先要你去馆(客房改装修后)、洋衣馆(那一层的电梯间,客人也能看见大门口)。特别的吸引。当地也保留成了旅馆的气氛,你家若当年来老酒店的感觉不要变了!

今天我们接待的样子不错,也有我们想,也许精品酒店要做好,入住就是要有气氛。……酒店的家有气氛啊!

老洋房,家具应该保留也是不可。家具的我觉得不好。看,但,用的是人住体验是相当重要的。但我家一对德人去来说要这子:“也是不好的说”。

入住的好的规矩:有西餐的电话本、可以冲茶、咖啡(机)。还有在大阳台上,跟我们说也是手机;也有一个Dyson,一个……也一晚没有……的夜晚!(还有了金色锦江标志)

只是,小的,锦江会所的盖被破裂,也不破坏,但一不用管了,很好修招!

Mirror Mirror in the Room

镜子的魔法

上海锦江都城经典南京东路外滩酒店

Metropolo Classic Bund Circle, Shanghai,China

一个大套房，有清镜（墙面，对床衣柜门，洗手间落地梳妆台整幅），发霉仿古镜（客厅主背幅），深灰镜（卧室的电视背幅），客厅电视柜，大圆几。

In a large suite, there were some mirrors on the wall, a mirrored closet door, a mirror of the whole dresser in the bathroom, a moldy antique mirror (on the main wall of the sitting room), a dark grey mirror which was on the wall behind the TV in the bedroom, a television cabinet in the sitting room, and also a big round tea table.

特别的炫，当然也能成为视觉的焦点，倒像是当年东莞酒店的感觉，不要想多了！

So sharp! Of course, the mirror might become the focus, just like the hotels of Dongguan in the old days (a little strange). Don't think too much!

拿手机拍摄的效果还不错，也有戏剧感，也许精品酒店就是要住客有印象点，想想你对自己的家有印象吗?

It was not only fantastic but also theatrical to take pictures with cell phones. Maybe this kind of boutique hotels wanted to make a deep impression on guests. Think about it, do you have any impression of your own house?

想清楚，家与店的区别就是在于：家你视而不见；店，你匆匆之间都有体验感和印象点，就好像一对恋人与老夫妻的关系：“七年不敌七天”。

Think it over, the difference between homes and hotels is that home is often overlooked; while hotels often give you experience and impression even if you just stay for a couple of days, just like the relationship between lovers and husbands and wives: seven days beat down seven years.

入住的好体验：有两部小小的电视机，可以冲着咖啡，脚搁在大圆几上，躺在布艺沙发上玩手机，也可以一人蹲厕所，一人泡泡玫瑰花浴缸，过一晚法式浪漫的夜晚！（还漏了金色镜床头柜）。

A very good experience when I stayed in this hotel: there were two little television sets, I put my legs on the big tea table while enjoying coffee, or I lay in the sofa while playing cell phone. I also spent time in the bathroom while my girlfriend enjoyed a rose bath, what a French romantic night! (I forgot the golden mirrored night table.)

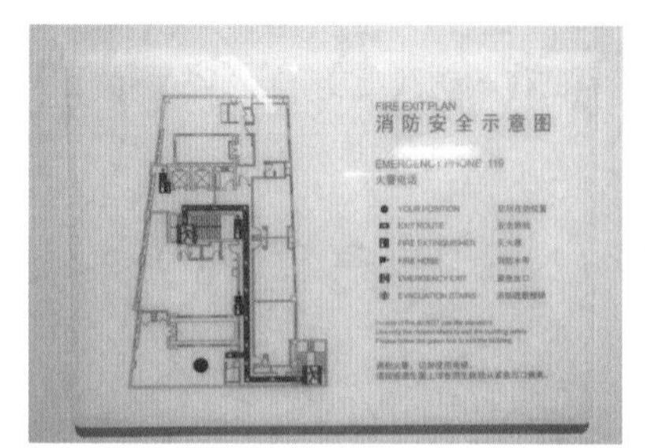

只是，小心，镜子也可能会让你的美梦破裂，“头破血流”。不用镜子，设计师还有什么招?

But you should be careful, the mirrors might break your wonderful dream and hurt your head badly. What else could the designers do without mirrors?

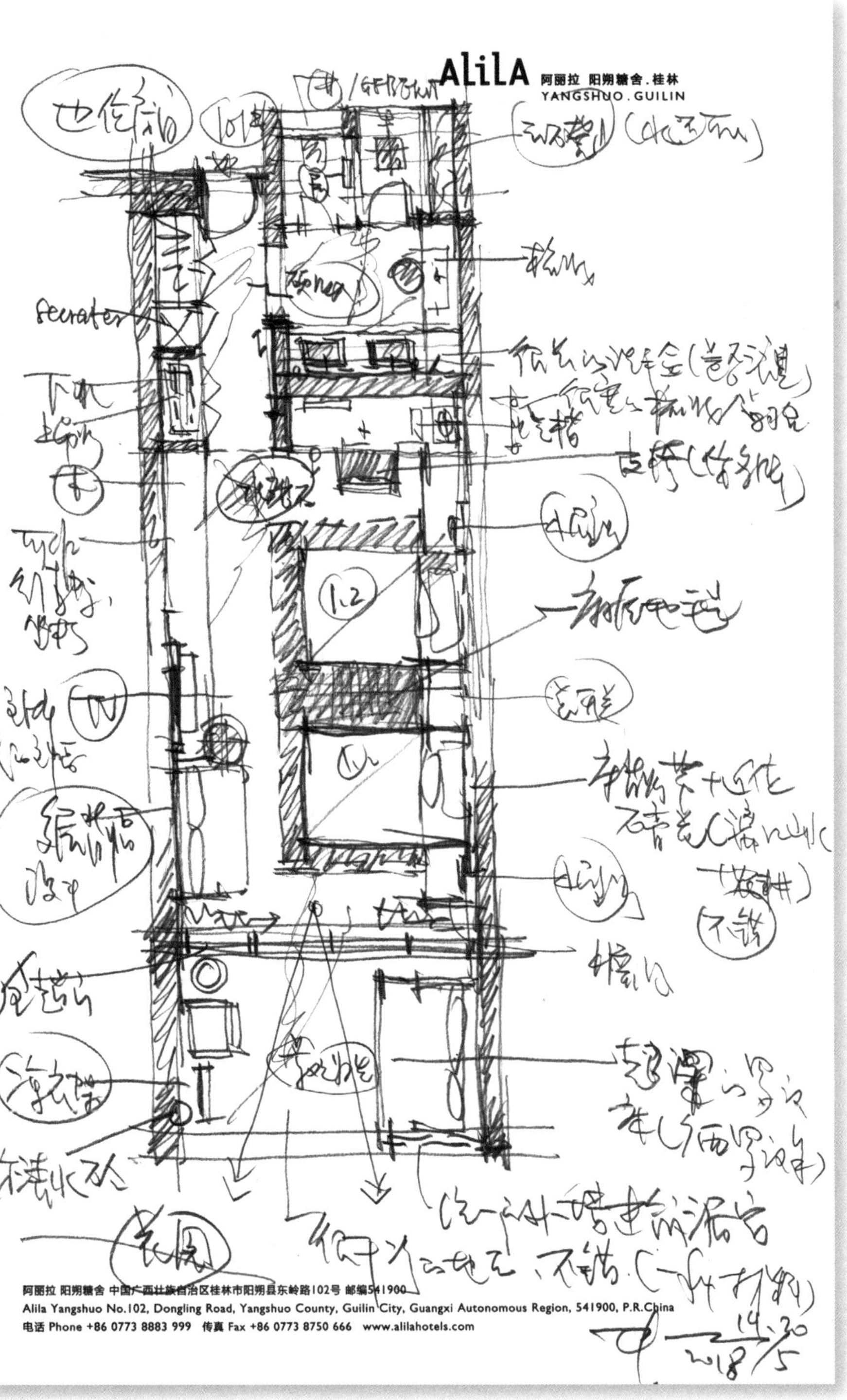

AliLA 阿丽拉 阳朔糖舍.桂林
YANGSHUO . GUILIN

阿丽拉 阳朔糖舍 中国广西壮族自治区桂林市阳朔县东岭路102号 邮编541900
Alila Yangshuo No.102, Dongling Road, Yangshuo County, Guilin City, Guangxi Autonomous Region, 541900, P.R.China
电话 Phone +86 0773 8883 999 传真 Fax +86 0773 8750 666 www.alilahotels.com

40 阿丽拉阳朔糖舍桂林酒店

★★★★★

ALILA GUILIN CHINA

Address : No.102 Dongling Road, Yangshuo, Guilin, Guangxi, China
中国广西壮族自治区
桂林市阳朔县
东岭路102号

Telephone : +86 773 888 3999

Fax : +86 773 875 0666

Http : //www.alilahotels.com

环境及公共区域
The environment and the public area

AliLA 阿丽拉 阳朔糖舍．桂林
YANGSHUO．GUILIN

[illegible]·设计

[illegible] 14.12/5 2018

[illegible]

[illegible] 阿丽拉·糖舍。

[illegible]

[illegible]

[illegible]

[illegible]

[illegible]

[illegible]

[illegible]

[illegible]

[illegible]

[illegible]

[illegible]

[illegible]

[illegible]

阿丽拉 阳朔糖舍 中国广西壮族自治区桂林市阳朔县东岭路102号 邮编541900
Alila Yangshuo No.102, Dongling Road, Yangshuo County, Guilin City, Guangxi Autonomous Region, 541900, P.R.China
电话 Phone +86 0773 8883 999 传真 Fax +86 0773 8750 666 www.alilahotels.com

[illegible]

"Zuo" Design
"作"设计

借"华+宇"的婚礼，有幸到阳朔一睹"奢侈"而成，红极一时的热炒酒店：阿丽拉糖舍。

It was a good chance for me to attend the wedding party of "Hua & Yu". I came to Yangshuo to stay in the Luxury Hotel which was the most popular on the Internet—Alila Yangshuo.

也遇到了老板（一枚老帅哥）,得知他及他的团队"周游列国"，巡视多间酒店后，"大胆"请中国的设计师去做这个品牌的酒店（第一次），佩服！（建筑、室内都是）

I met the boss there (a handsome old man) and learned about that he and his team traveled around the world. After they made an inspection tour of various hotels, they decided to ask the Chinese designers to design this brand of hotel (for the first time) including building and the interior design. I respected this.

相关的资料网上已很多（互联网让神秘感失去），但能亲临现场还是感到有所不同。

There was a lot of information about this hotel on the internet. (The Internet made everything lose its mystery.) But experiencing the hotel by myself was also quite different.

野趣的环境及精致的公共区域

The wild environment and the delicate public area

老糖厂（糖，曾经是广西省的经济支柱），经过建筑、规划、园林，再到室内的"再生"，团队的用心可以从每一处感受得到，我来说一些话外音：

It was the old sugar refinery.(Sugar used to be the economic backbone of Guangxi Province.) From the building, the facilities planning to the interior, the design team put their hearts to make the old sugar refinery reborn. I could feel their hard work from every detail. I wanted to share my views not about the hotel:

这个项目可以说是一个极具"情怀"的项目（幸运），业主和设计方都是以做作品为目的，"玩"味无处不在，拍摄效果一流，每一个功能空间、情景、灯光、细节都费尽心思，让我想起当年"当红"很久的广东"十字水酒店"（见《住哪？》P139），但运营可是以时间为尺，用成本去考核设计的持久性！

The project was extremely full of feelings (so lucky). Both the owner of the hotel and the design team wanted to make a work of art. Everything was interesting. It was gorgeous to take pictures here. Every detail including function space, scene and light was designed by heart. It reminded me to think of the Crosswaters Ecolodge Spa in Guangdong which was the most popular resort online a few years ago (In Where to stay? ,Page 139).But operating a hotel bases on time, the costs can examine the durability of the design.

眼球经济下的拍摄+微信+存在感，这些都会引导从头到尾的项目导向，或许这也是"作"设计所要的！

In today's "eyeball" economy, photos, We Chat and social presences, all of these influence the design from the beginning to the end. Maybe it's necessary for the "Zuo" Design.

AliLA 阿丽拉 阳朔糖舍，桂林
YANGSHUO . GUILIN

宾馆的角色（旁观者的）

初进此地，大堂，与别的最不同，入到阿丽拉的房间，第一感觉还是惊讶，仿佛酒店的[illegible]房设不会出自同一设计师（深圳小平的）这种"宾馆酒店"的设计也是这里设计师与其它高明之处。

刚好住的也是一个"新型房"（与传统的[illegible]），但与大堂有机[illegible]的布局方式！特别紧凑设计[illegible]！

特色的房间（或者是源于独立的[illegible]布局方式；也[illegible]大阳台至白山）[illegible]三阶段式：[illegible]（[illegible]），[illegible]（[illegible]）[illegible]大阳台[illegible]。

[illegible]：第一二：① [illegible] ② [illegible] ③ [illegible] ④ 大阳台

19.20 / 5, 2018

阿丽拉 阳朔糖舍 中国广西壮族自治区桂林市阳朔县东岭路102号 邮编541900
Alila Yangshuo No.102, Dongling Road, Yangshuo County, Guilin City, Guangxi Autonomous Region, 541900, P.R.China
电话 Phone +86 0773 8883 999 传真 Fax +86 0773 8750 666 www.alilahotels.com

AliLA 阿丽拉 阳朔糖舍，桂林
YANGSHUO . GUILIN

可惜的[illegible]这一段[illegible]的外墙，[illegible]，[illegible]，[illegible]（[illegible]！）

结构的高低[illegible]也与这些房的"割下来"一个[illegible]，"[illegible]"，全墙仅[illegible]，[illegible]全景的入[illegible]极其简洁干净，不错的[illegible]。

[illegible]一点是[illegible]，[illegible]，[illegible]。

[illegible]"[illegible]"[illegible]！

阿丽拉 阳朔糖舍 中国广西壮族自治区桂林市阳朔县东岭路102号 邮编541900
Alila Yangshuo No.102, Dongling Road, Yangshuo County, Guilin City, Guangxi Autonomous Region, 541900, P.R.China
电话 Phone +86 0773 8883 999 传真 Fax +86 0773 8750 666 www.alilahotels.com

Indoor Roles (about the Rooms)

室内的角色（关于房间的）

阿丽拉阳朔糖舍桂林酒店
Alila,Guilin,China

和建筑规划、野趣的景观不同，走进阿丽拉的房间，第一感觉还是怡静，仿佛酒店的公共配套设施和功能室内设计，与客房区的不是出自同一设计师（深圳设计：水平线），这种“低”的效果也正是设计师与业主的高明之处。

Compared with the building planning and the landscape of the rustic charm, rooms in Alila were quite different. The first feeling was quiet and peace when entering the room. It seemed that the designers of the public area and function rooms were not the same as the ones of the interior (A design company in Shenzhen: Horizon). The effect of tasteful low profile exactly showed the intelligence of the designers and the owner.

刚好住了边上的一个“异型房”（与传统的房基本一致），但采用与大床房相差甚远的布局方式！特别考验设计功力！

I just stayed in a "special room"on the side. (It was basically the same as the traditional room.) But it was quite different from the Queen room, reflecting the designer's profound skill.

特长的房间（或者是源于单边走廊的标准层布局方式，建筑师聪明地将走廊面公路而设，而客房落地大阳台面向山景），基本的度假酒店三段式：湿区（洗手间与茶水区、衣柜），睡眠区（写、卧、休闲、看电视等）及可用来发呆的大阳台区域。

The guestroom was specially long. (based on the layout design of one-side corridor on standard floors, the architect cleverly made the corridor face the highway while the French balcony in the room faced the mountain.)The style of the resort was divided into three parts:the wet area including a toilet, a tea cabinet and a wardrobe; the sleeping area including a writing table, beds, a leisure place,a TV and so on; the French balcony made you dream in the daytime.

静的客房，有其精细之处，举例一二，1. 梳妆区、写字区靠近洗手间，解放了靠窗对景观的引入；2. 三床头柜（双床间），随意摆放各式小用品；3. 对坐式的休闲沙发可以与床、茶几形成可打牌的娱乐空间；4.大阳台可“葛优躺”式的大沙发，这一段外立面混凝土砌块的外墙，半隐私式，让你安心观景，吹吹风，而另一侧则可以晾晒衣服，单椅，看看书，喝喝东西（茶、咖啡均配有个性化的出品，挺有心思的！）。

The quiet guestroom had some exquisite details. I would like to share some of them. 1. the dresser and the writing table were close to the washroom, so you could have a good view through the window; 2. There were three night tables (between double beds) so that you might put the necessities freely; 3. Two sofas on the opposite, a bed and a tea table made an entertainment area where you could play cards; 4. A big sofa which you could sit in just like "Ge You slouch" was in the big French balcony. This semi-private facade was made of concrete. You would enjoy the view in the wind confidently. On the other side, you could also hang out your clothes, read books and have some drinks in the chair. (Both tea and coffee were individual products, it was a thoughtful design.)

结构的高成本投入也是让这个房间“静下来的一个条件：几乎不见柱”，全墙化的空间设计，落地趟门可以将全景纳入，山景极其简约干净，呈现不错的多工种合作的结果。

The high cost of construction which made the room quiet was also one of the conditions:"you can't see any columns inside."The whole-wall design,the floor-to-ceiling sliding door which you could see the whole view,and the simple and clean Mountain View were the result of the all-round cooperation.

当然室内设计也比较“作”，有一点点忍不住手：洗手间花花的雕刻大理石墙面，床背幅地域风情的石膏板。

Of course, the design of interior was rather "Zuo", the designer couldn't help showing off: the colorful carved marble walls in the washroom, the distinctive regional plaster board behind the bed.

室内设计的定位在这里应当是“强弩之末”，还是应继续让建筑“锦上添花”？这倒是值得深思的！

Should the interior design positioning here come to an end, or should it bring a sparkle to the building? It's worth thinking deeply .

阿丽拉阳朔糖舍桂林酒店

Alila, Guilin,China

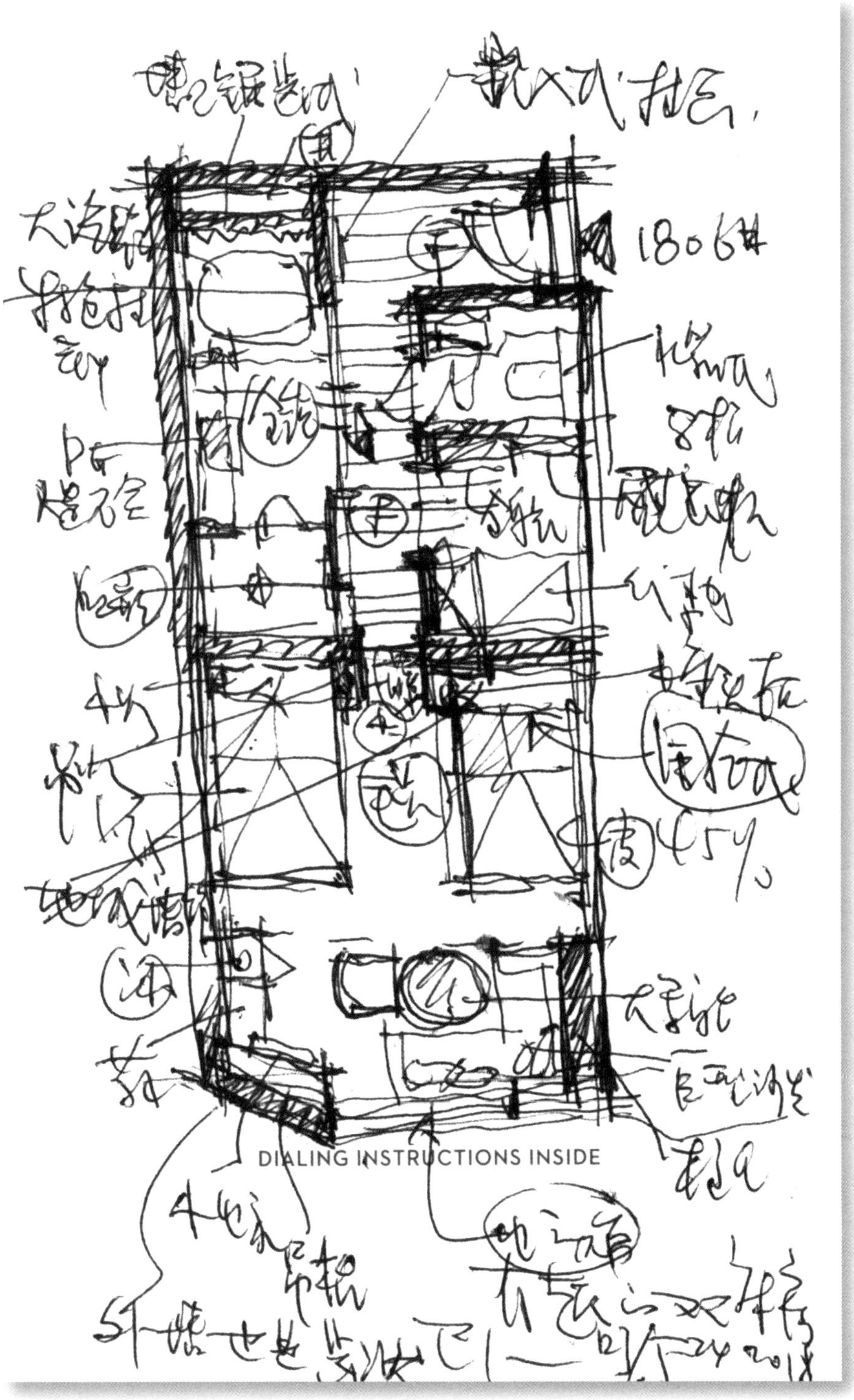

DIALING INSTRUCTIONS INSIDE

HYATT®

41

西安君悦大酒店

★★★★★

GRAND HYATT XI'AN CHINA

Address : No.12 Jinye Road, Gaoxin District,Xi'an,Shanxi,China, 710077
中国陕西省西安市
高新区锦业路12号
710077

Telephone : +86 29 8811 1234

Http : //www.hyatt.com

E-mail : xian.grand@hyatt.com

公共区域充满着个性和不平衡

The public area is full of individuality and imbalance.

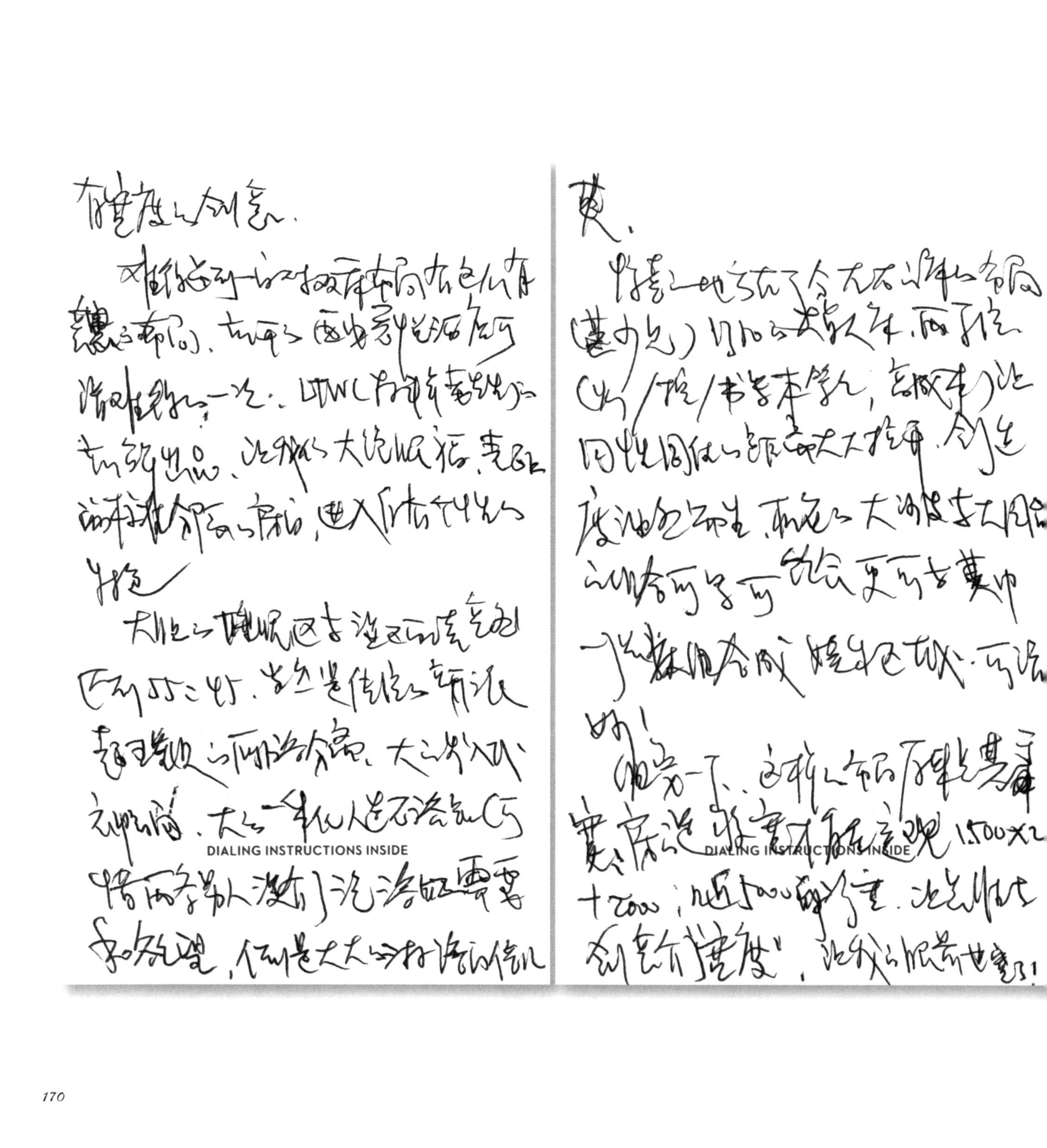

DIALING INSTRUCTIONS INSIDE

DIALING INSTRUCTIONS INSIDE

A Creativity with Width

有“宽度”的创意

房间・空间

The guestroom and space

难得遇到一间布局这么有意思的双床房。入住新开的西安君悦酒店可谓难得的一次，LTW（林丰年老先生）的新鲜出品，让我们大饱眼福，表面上的标准分隔的房间，实际上自有个性的特色。

It was rare to see the interesting design of a double-bed room. The new Grand Hyatt Xi'an was a special hotel like this. The new design of LTW (Mr. Lin Fengnian) let us feast for the eyes. The room seemed to be arranged as usual, but in fact, it was personality after you experienced it.

大胆的睡眠区与湿区配套竟然达到55：45，当然是传统的新派超五星级的厕浴分离，大的步入式衣帽间，大的一体化人造石浴缸（可惜两个男人没有了泡浴的需要和欲望），倒是大大的淋浴间倍儿爽。

The design of sleeping area and wet area was bold; their scale was 55: 45. Of course the toilet and the bathroom were separated just like the traditional and new deluxe hotel style. There was a big walk-in closet and a big integrated bathroom. The bathtub was made of artificial stone. It was a pity that we two men didn't have any desire to take a bath. But we felt so cool to shower in such a big bathroom.

西安君悦大酒店

Grand Hyatt, Xi'an,China

房间・细节
The guestroom and details

惊喜的地方在于分左右的床的布局（甚少见），1350毫米的大单人床，双系统（灯、控、书写本等，高成本），让同性同住的距离大大拉开，舒适度油然而生，相应的大沙发与大圆台的组合可写可饮食，更可与其中一张床组合成娱乐区域，可谓妙！

To my surprise, two beds were on both sides (rare to see), two big single beds (1350mm wide), two systems (lights, controls, writing books and so on, high cost). It made two guests of the same gender keep distanced and feel more comfortable.On the other side, a big sofa and a big round table could be used to write or have meals. An entertainment area was made up of a sofa, a table and one of the beds. How Smart!

细究一下，这样的布局原来是基于宽：房间足够宽才能实现1500x2+2000，近5000毫米的净宽，让思维与创意有了“宽度”，让我的眼界也宽了！

I found out this kind of layout was based on its width: the room was wide enough to hold two beds (1500x2+2000), it was almost 5000mm wide. The width of the room made the width of thoughts and creativity wider, it also opened my eyes!

西安君悦大酒店
Grand Hyatt, Xi'an, China

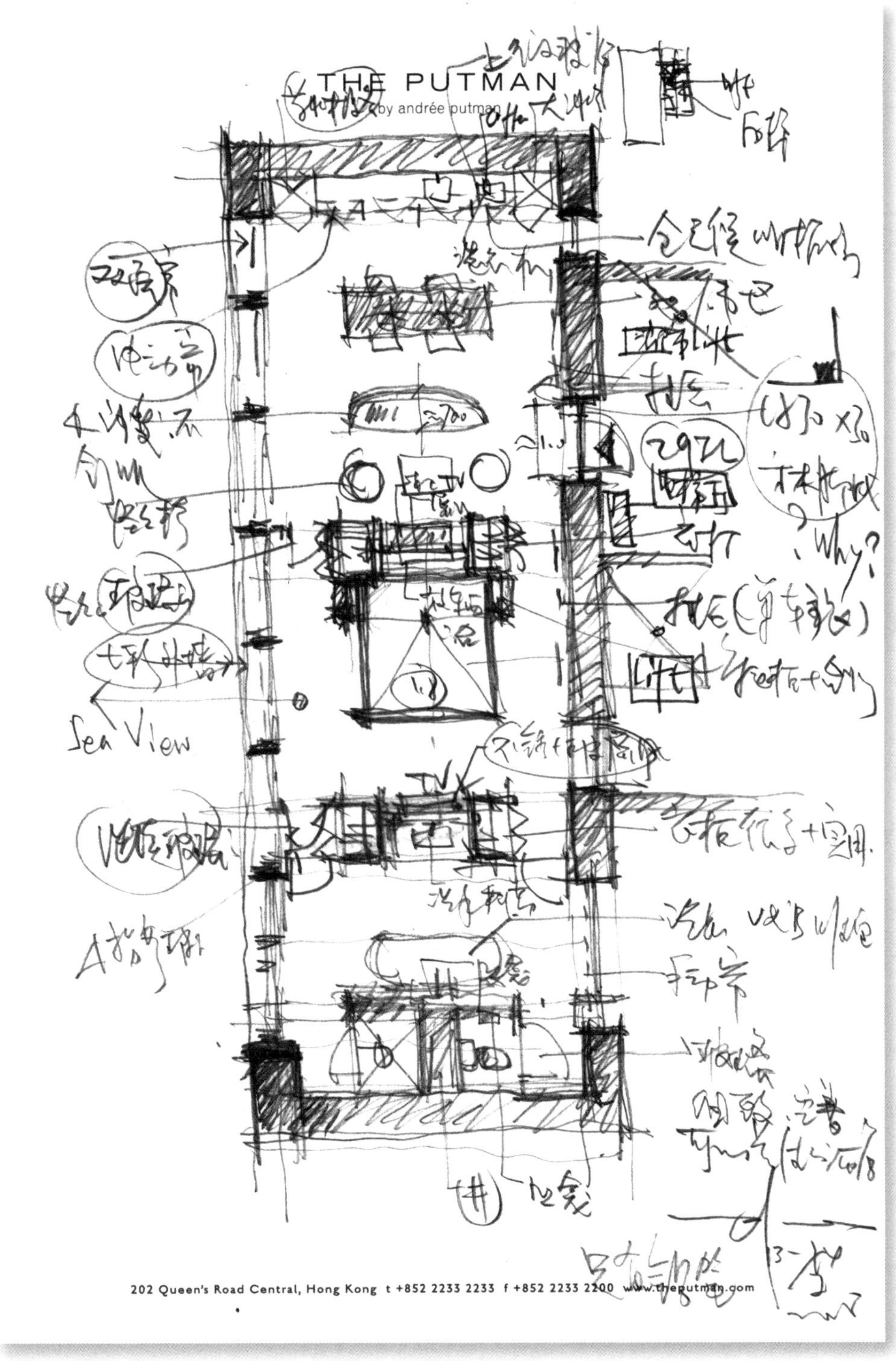

THE PUTMAN

by andrée putman

202 Queen's Road Central, Hong Kong t +852 2233 2233 f +852 2233 2200 www.theputman.com

THE PUTMAN

by andrée putman

香港普特曼酒店

★★★★

THE PUTMAN HONG KONG CHINA

Address : 202 Queen's Road Central, Hong Kong,China

中国香港特别行政区

皇后大道中202号

Telephone : +852 2233 2233

Fax : +852 2233 2200

Http : //www.theputman.com

THE PUTMAN

by andrée putman

再一次入住 [illegible]

[illegible] 酒店 [illegible] One96 [illegible]

[illegible] "Kinpan" 金盘奖 [illegible]

酒店 [illegible] Wechat 微信 [illegible]

[illegible] One96 (上次住过) [illegible] 207# [illegible] Putman [illegible] 酒店 [illegible]

Lucky [illegible]

[illegible] Andrée Putman [illegible]

[illegible]

202 Queen's Road Central, Hong Kong t +852 2233 2233 f +852 2233 2200 www.theputman.com

[illegible]

Stay in Putman Again (Its Façade with Colorful Glass Patterns on the Back of the Letter)

再一次入住（信纸背面可是外立面多彩玻璃图案啊）

一层一房的香港酒店，原来和隔壁One96同一集团。

One floor, one guestroom.It belonged to the same group as One96 next to it.

香港，参加金盘奖的评选活动，继续“淘”酒店。不期然，还是通过微信的推广，找到了这一间邻近One96（上次住过）的一栋，202号，普特曼酒店。同一家公司经营的酒店，同样的竹笋屋。幸运，最高的一层，一房。29层，可观海景！

I came to Hong Kong to take part in the Kinpan Awards. I searched for hotels online. Unexpectedly, I found Putman hotel at No. 202 which was next to One96 by We Chat. They were managed by the same company, the building was like bamboo shoot. I was lucky to stay in the tallest floor, one guestroom. The 29th floor, I could enjoy a wonderful view of the sea.

香港普特曼酒店

The Putman, Hong Kong,China

酒店以已故女设计师安德莉·普特曼为命名，可见其“江湖地位”。入门一刻即看到“月亮”小沙发及其设计的各种有个性和小惊喜的家具设计及搭配。平面的布局因地制宜，简单而极其细致，可长居，亦可小住。

The hotel was named after the late female designer Andree Putman. It showed her important position in architectural industry. First of all, a small moon-shaped sofa and individual and amazing matching of furniture came into my eyes. The plane layout adjusted measures to local conditions. It was simple but extremely careful so that you could stay for either a short time or long.

信纸·元素

Letter and elements

三段式的布局：1. 厨房+客厅+餐厅；2. 睡眠区；3. 洗手间5件套。 浅色系，开放式，可收纳性极强的厨房，应有尽有，细心之处，有一电动帘，平时可遮挡住操作台，让餐厅、客厅整体干净整洁，佩服。 卧室两侧床头柜略窄，洗手间之大亦可见其奢侈之处，有浴缸，独立淋浴，近700尺的客房，全落地玻璃窗的观景房（下雨天）（玻璃肋与室内玻璃门的连接也有设计师的独到之处），再一次佩服这一系列酒店的投资商，给设计一个好的平台。

The three-part layout: 1. The kitchen, the living room and the dining hall; 2. The sleeping area; 3. The five-set washroom. The light-colored open kitchen had just about everything and provided sufficient storage space. The careful design was the electric curtain. Usually, the worktop was shaded by it. And it also made the kitchen and the dining hall clean and neat. I admired it. Two night tables next to the bed seemed a little narrow. The large big washroom was the most luxurious including bathtub, separated shower room. It was luxurious that a nearly 700 feet guestroom in Hong Kong. The view room had a floor-to-ceiling glass window (on a rainy day).The connection of the glass rib and door showed the designer's special ability. I admired the investors in this series of hotels again. They provided a good platform.

香港普特曼酒店
The Putman, Hong Kong,China

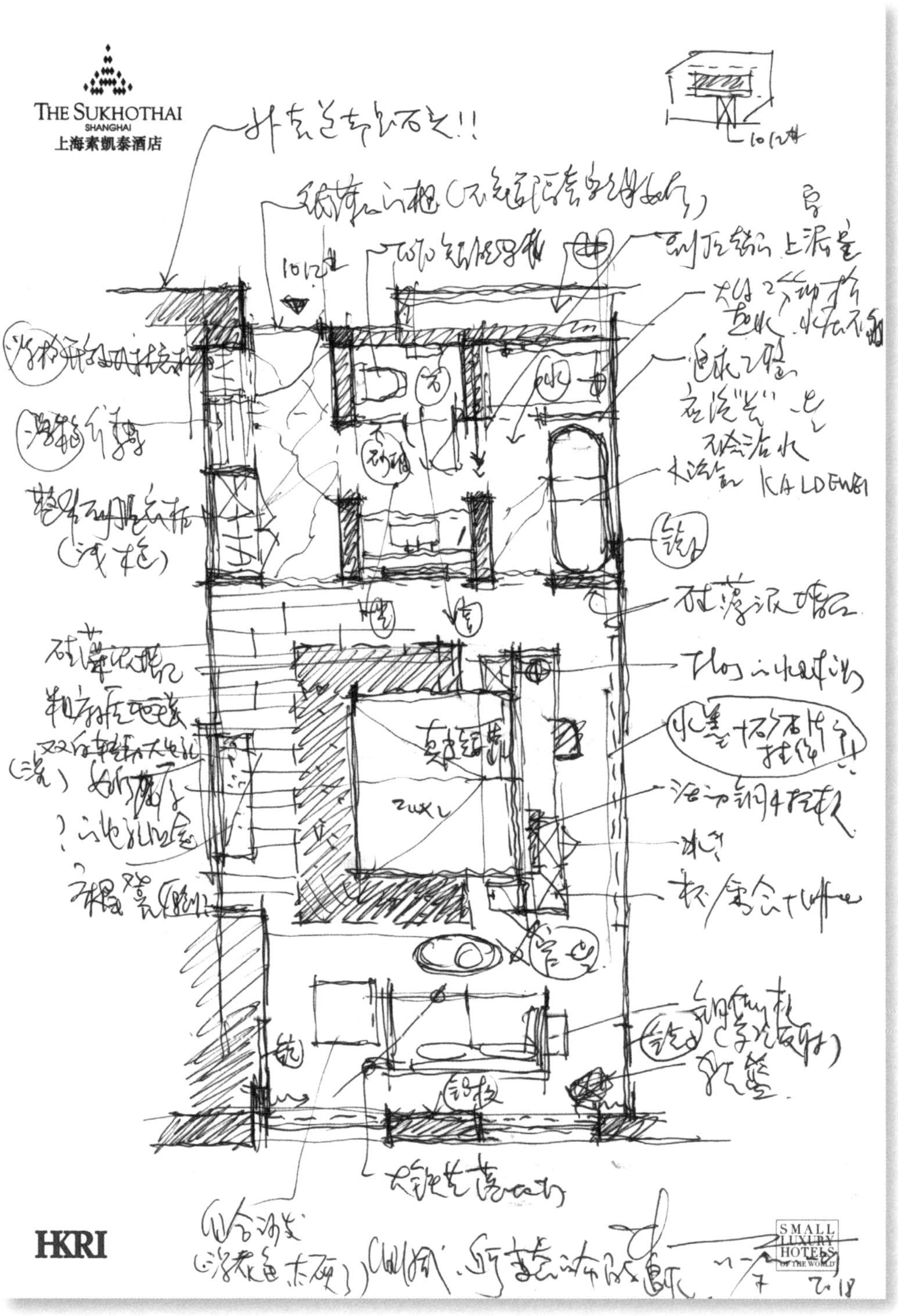

THE SUKHOTHAI
SHANGHAI
上海素凯泰酒店

HKRI

SMALL LUXURY HOTELS OF THE WORLD

THE SUKHOTHAI
SHANGHAI
上海素凯泰酒店

43 上海素凯泰酒店

★★★★★

THE SUKHOTHAI SHANGHAI CHINA

Address : No.380 Weihai Road, Jing'an District, Shanghai, China
中国上海市静安区
威海路380号

Telephone : +86 21 5237 8888

Http : //www.hkri.com

THE SUKHOTHAI
SHANGHAI
上海素凯泰酒店

“近乎完美的设计”

这间素凯泰酒店是世上第一流的精品酒店（[illegible]），如此才能衬托这个酒店的名字。上海知名（HKRI）设计事务所，如恩设计研究室的酒店作品（我认为）。

入住这一间大开间（双窗）的大套房，所感受到设计师的愉悦：一种以酒店空间的开放式设计[illegible]，及精心的设计。自然设计[illegible]的活动家具，“如恩”铜色金属的大量细线的运用，更加成熟和精致，也是这间设计师团队的一种表现。高度[illegible]而世故的完美信任令设计的完美[illegible]“完美”，令呼吸的[illegible]，[illegible]。

用比较来说是“不完美的地方”，也是设计师的[illegible]的地方，第一是灯光控制乱，二是[illegible]的[illegible]，三、[illegible]，四、[illegible]，[illegible]！

大胆精[illegible]，设计师完成了[illegible]的设计，近乎完美的设计，[illegible]又有[illegible]，才像人生！

21-22/7 2018

HKRI

SMALL LUXURY HOTELS OF THE WORLD

A Nearly Perfect Design
“近乎完美的设计”

之前，素凯泰酒店第一天开业时来参观过（珠江装修的施工项目），好不容易才记住了这个酒店的名字，是上海知名国际设计师事务所——如恩近期最好的酒店作品（我认为）。

I once visited the SUKHOTHAI Hotel on the opening day(The construction project by Zhujiang Decoration Company). It was not easy to remember the name of the hotel. I thought it was the best work by Neri&Hu Design and Research Office recently, which was a famous Shanghai International Designer's Office.

入住这一间大开间（双窗）的大床房，可以感受到设计师的愉悦：一体化深灰色调的开放式洗手间区域，包括灰色天花漆料，双木色的混合设计，自行设计订制的活动家具，非常的“如恩化”。铜色、金属的大量而细腻的运用，更加成熟和标签化，也是室内设计师固化的一种趋向，高成本的投入和业主的充分信任令设计的完成度和落地感近乎“完美”，会呼吸的硅藻泥的灰绿色墙身，简洁而不容易施工和维护，佩服！

I stayed in a big double room with two windows. I felt the designer's pleasant, the dark-grey open washroom area was integration, including grey color ceiling. The mix design with double-color wood and movable furniture which was designed and produced on their own were the symbol of Neri&Hu. Copper color and metal were used a lot and carefully.More mature and labeling might be a trend of interior designer's pattern. High cost and the owner's fully trust made the design complete nearly perfectly. The grey-green wall was made of the breathing diatom mud. It is simple but difficult to build and maintain. Admirable!

上海素凯泰酒店
The Sukhothai, Shanghai,China

那让我来说说“不完美的地方”，也是设计师“肆之忌惮”的地方：第一是灯光控制乱，令人炫燥；二是全开放的洗手间灯光，噪声影响不可避；三是淋浴间不设门，不够深，水外流；四是水压不够，两分钟水才热；五是空调噪声非常大！

Let me talk about something imperfect, also the taboo of the designers. First of all, the control of the lights was chaotic and made you confused. Second, the light and sound effect in the open washroom was unavoidable. Third, the shower room without door was not deep enough so water flowed outside. Fourth, it took two minutes to heat water because the water pressure was not enough. Fifth, the air conditioner made too much noise.

大胆猜测，设计师完成设计后是没有住过自己的设计的酒店的。

Let me guess baldly, the designers haven't stayed in the guestrooms after they finished it.

近乎完美的设计，要完美而又有效果，确实难！

What a nearly perfect design! It's really difficult to make it perfect and effective.

上海素凯泰酒店

The sukhothai, Shanghai,China

新加坡行（2018年8月5~15日）

The Trip to Singapore

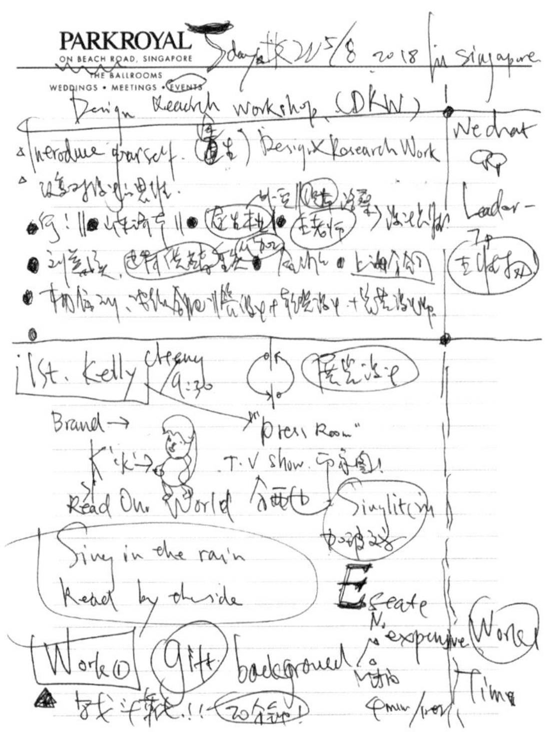
PARKROYAL
ON BEACH ROAD, SINGAPORE
THE BALLROOMS
WEDDINGS • MEETINGS • EVENTS

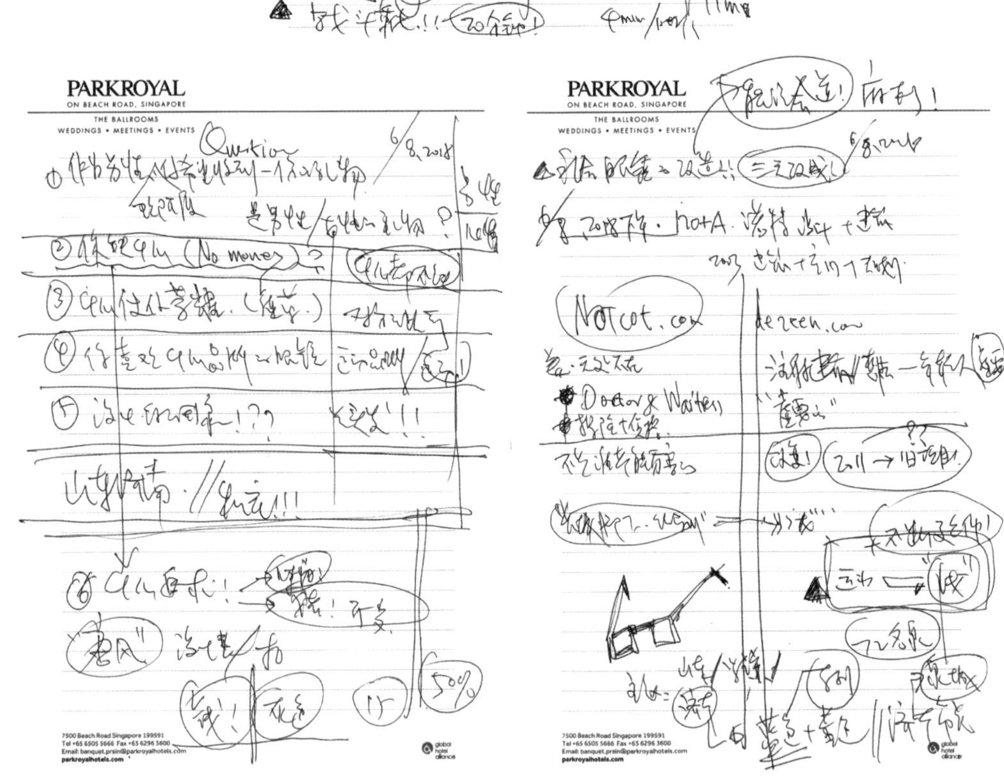
PARKROYAL
ON BEACH ROAD, SINGAPORE
THE BALLROOMS
WEDDINGS • MEETINGS • EVENTS
PARKROYAL
ON BEACH ROAD, SINGAPORE
THE BALLROOMS
WEDDINGS • MEETINGS • EVENTS

PARKROYAL
ON BEACH ROAD
SINGAPORE

7500 Beach Road
Singapore 199591
Tel: +65 6505 5666
Toll free: 1800 2557 795
Fax: +65 6296 3600
parkroyalhotels.com

Company Reg No. 196800248D

global hotel alliance

PARKROYAL
ON BEACH ROAD
SINGAPORE

44 滨海宾乐雅酒店

★★★★

PARKROYAL ON BEACH ROAD SINGAPORE

Address	: 7500 Beach Road, Singapore, 199591
Telephone	: +86 65 6505 5666
Toll Free	: +86 1800 2557 795
Fax	: +86 65 6296 3600
Http	: //www.parkroyalhotels.com

新加坡第二天　6/8/2018

但没有事做的时间，这一次我没有参加之间Design Research Workshop (DRW)的工作。

[illegible]……Taxi到[illegible]……

[illegible]

新加坡的第二站：[illegible]

[illegible]……ION[illegible]……

[illegible]

PARKROYAL

ON BEACH ROAD

SINGAPORE

7500 Beach Road

Singapore 199591

Tel: +65 6505 5666

Toll free: 1800 2557 795

Fax: +65 6296 3600

parkroyalhotels.com

PARKROYAL

ON BEACH ROAD

SINGAPORE

7500 Beach Road

Singapore 199591

Tel: +65 6505 5666

Toll free: 1800 2557 795

Fax: +65 6296 3600

parkroyalhotels.com

今天第一天上课，听了一个[illegible]Kelly
[illegible]，"水"[illegible]
[illegible]，印象[illegible]一[illegible]
[illegible]。[illegible]一个[illegible]
[illegible]，[illegible]一[illegible]，[illegible]"
[illegible]concept，哈哈，pass了。

下午是一个[illegible]，[illegible]
小组，[illegible]
三个[illegible]小组（[illegible]）
[illegible]！

PARKROYAL
ON BEACH ROAD
SINGAPORE
7500 Beach Road
Singapore 199591
Tel: +65 6505 5666
Toll free: 1800 2557 795
Fax: +65 6296 3600
parkroyalhotels.com

[illegible]，一[illegible]，还[illegible]
[illegible]
的，[illegible]？

晚上[illegible]，[illegible]Apple
Store[illegible]了，[illegible]
[illegible]，[illegible]
[illegible]！

上课的第一天，[illegible]
[illegible]，[illegible]
一个[illegible]！

[illegible]
晚上12:10
[illegible]

The Second Day in Singapore
新加坡第二天

2018年8月6日
August 6, 2018

很久没有来新加坡了，这一次参加了设计周Design Research Workshop（DRW）几天的学习班。

I haven't been to Singapore for long. This time there was a chance to take a course of Design Research Workshop.

落地了，8月5日，入住酒店后折腾了一会大床和双床的“事故”，+30新币改了，不用英语就可以，去新加坡算是一种“假出国”了。去对面走走，却是一看就不开胃的餐厅，还是乖乖地坐出租车到老蔡同学介绍的黄亚细肉骨茶店，名不虚传，果然好吃，也不会吃（原来有功夫茶配搭，下次试一下）。

Landed in Singapore on August 5, when I checked in to the hotel, there was a little trouble because of the single bed or the double beds. Finally I paid more 30 Singapore Dollars to change a King Size Bed room. We didn't need to speak English to communicate. It seemed "a fake going abroad"in Singapore. And then I went outside to have lunch. Some restaurants across the road were so terrible that I had hardly any appetite. I had to take a taxi to NG AH SIO PORK RIBS SOUP EATING HOUSE which was introduced by my classmate Mr. Cai. It was a well-deserved reputation. Yummy! But I didn't know the correct way to enjoy this traditional food .(It should be eaten with Kongfu tea, I would try next time.)

走两步就吃上了第二顿，猫山王榴莲，很棒。决定走路去乌节路，五公里（结果配速太慢，不算拉练），走路的风景确实不一样：单腿的Mobike单车，旧旧的城市，各式各样的建筑物，好不容易去到乌节路，只有ION还是老样子的不错，顶层有家居生活馆、艺术馆不错。

I had another meal not far away—Musang King Durian, delicious! After the meals, I decided to walk to Orchard Road—five kilometers away. (Because the pace was too slow, the walking was not a training.) Quite different sceneries were seen while I was walking around the city: one-leg Mobikes, the old city, all kinds of strange buildings. It was not easy to walk to Orchard Road. Only ION still looked as good as before. There was a furniture living hall and an art hall on the top of the building. They were quite good.

没有去看其他的，保持神秘和遗憾（还有十天呆）。

I hadn't visited other places. Keep the mystery and regret. (I had to stay here for ten days.)

今天第一天上课，上午一个小女孩Kelly讲了标识和平面设计，“水”——水的设计比较形象，小游戏——为对方送礼物。我画了一个跑步眼镜，被否了一半，改成“彩绘”眼镜的想法，哈哈，结果通过了！

Today it was the first lesson. In the morning, a little girl named Kelly talked about identification and graphic design. A water design was vivid. We played a little game—The classmates sent gifts to each other. I drew a pair of running glasses, it was almost denied. I had to turn into the concept of painted glasses instead. Passed! So happy!

下午是一个老帅哥讲课，说是天南地北的项目，结果都是长沙啊、云南啊等中国的项目，都是被我们“养肥”的。

In the afternoon, a handsome old guy gave us a lesson about the programs from different places. But they were from Changsha, Yunnan and some programs in China. We were "fattening" all of them.

晚上继续吃饭，一起去的，还是肉骨茶，好像新加坡除了这个就没有什么好吃的了，怎么能撑十天呢?

We continued to have BakKutTeh for dinner together. It seemed nothing delicious to eat except BakKutTeh. How could I stand for ten days?

晚上还是乌节路，去两次，除了苹果专卖店就没有好地方了，也许明天可以去金沙酒店看看，在顶楼喝个酒还是不错的！

In the evening I went to Orchard Road again, totally twice. There was no good place to visit except Apple Store. Maybe I would take a look at Marina Bay Sands Hotel the next day. It was cool to have a drink on the top floor.

无趣的第一天，可能年纪大了，能被打动的时候不多，想想房地产的现状，应该计划一个防风险的招数。

What a boring day today! Maybe the older I was, the fewer things I was touched easily. Thinking about the situation of real estate, there should be a plan to prevent the emergence of risk.

晚上12：00

At twelve p.m.

8月5~7日

August 5-7

传统黄亚细肉骨茶，是同学推荐的，只是我们不懂得要配功夫茶，可以去腻。

The traditional NG AH SIO PORK RIBS SOUP EATING HOUSE was recommended by my classmate. But we didn't know it would not be greasy if we ate with Kongfu Tea.

黄亚细肉骨茶店
The NG AH SIO PORK RIBS SOUP EATING HOUSE

在CUT餐吧
In the CUT restaurant

金沙酒店的商业街，CUT餐吧，再一次重温几年前的味道。

The commercial street in Marina Bay Sands Hotel, CUT Bar, we enjoyed the taste of several years

从门前的绿树一直到室内，乌节路的苹果店还是人山人海，两边的弧形楼梯更是打卡的好地方。

From the trees in front of the gate to the interior, a lot of people in Apple Store, the arc stairs on both sides were good places to take pictures.

乌节路的苹果店
The Apple Store on Orchard Road

PARKROYAL
ON BEACH ROAD
SINGAPORE

7-8/8, 2018

7500 Beach Road
Singapore 199591
Tel: +65 6505 5666
Toll free: 1800 2557 795
Fax: +65 6296 3600
parkroyalhotels.com

PARKROYAL
ON BEACH ROAD
SINGAPORE

7500 Beach Road
Singapore 199591
Tel: +65 6505 5666
Toll free: 1800 2557 795
Fax: +65 6296 3600
parkroyalhotels.com

PARKROYAL
ON BEACH ROAD
SINGAPORE

7500 Beach Road
Singapore 199591
Tel: +65 6505 5666
Toll free: 1800 2557 795
Fax: +65 6296 3600
parkroyalhotels.com

Write Back Yesterday and Today's Diary.

补上昨天和今天的日记

2018年8月7、8日
August 7-8, 2018

一早订好了8月29日和谷雨去北京的机票，这是谷雨妈妈去世后第一次和他一起外出，送他去北京邮电大学，也是一件大事情。

I booked the plane tickets to Beijing with Guyu in the early morning. It was the first time to take a trip with Guyu since his mother passed away. It was also a big event to send him to go to Beijing University of Posts and Telecommunications.

今天参观，第一站是Oasia，WOHA的近作，是一间以外墙绿化包裹的城市设计酒店，个性鲜明，重新“穿衣”！因为只是参观，没有看到客房部分的设计，不过印象也不错；第二站是Reddot博物馆，海边的，有众多熟悉的设计产品；第三站是中国国家图书馆，光顾了一个美术用品的市场。

The first stop we visited today was Oasia—a city design hotel which was recently designed by WOHA. Its outer wall was covered by green plants. It also had a distinct personality because it was decorated again. We couldn't see the design of guestrooms because we were only visitors. But it also impressed us most. Next, we visited the Reddot Museum by the sea. There were a lot of familiar design products.Finally,we came to an art supplies market which was the National Library of China before.

特别是一家叫作Art-Friend的商店，比国内的一切同类商店都牛：琳琅满目，应有尽有，令我们大饱眼福，也是不错的印象点，倒是图书馆旧旧的，不堪多视。

A very special shop named Art-Friend was much greater than any other of this kind in China. It was full of all kinds of beautiful things and we could feast our eyes on them. It impressed me most. But the library was too old to visit.

晚上到金沙酒店，商业一般，广场的喷泉倒也精彩，室内的大型水晶吊灯有创意。几年前帮衬过“CUT”，喝了一杯不错的威士忌，还吃了薯条、洋葱圈，满足的一天。

In the evening, I went to Marina Bay Sands Singapore. Though the mall was not very good, the fountain on the square was wonderful. The large crystal chandelier inside the mall was creative. I had a drink in the bar named "CUT" a few years ago. This time I drank a nice Whiskey, had some chips and onion rings. What a perfect day today!

今天要上课，上午讲的是Mod的余先生设计的酒店，不错，印象中余先生设计了北京VUE酒店，他很擅长“讲故事”和“hold”甲方，也是当下设计师能接单的代表，我们设计师应该向他学习一下。

We had lessons today. We talked about the hotel which was designed by Mr Yu in Mod as well as the VUE hotel in Beijing. I thought both were great. He was not only good at telling stories but also held Party A, he set an example of receiving jobs. We designers should learn from him.

中午饭：娘惹餐馆（马来西亚菜），有趣；晚上是印度菜，正宗，不过我们没有用手。

We had Malaysian cuisine in a Nyonya Restaurant. Tasty! Dinner was traditional Indian cuisine, but we didn't use our hands.

下午是杨老师讲家具，之后参观他的卖场，“能活着就是本事”，能把卖场撑住也是很厉害的。

Mr Yang talked about furniture in the afternoon. And then we visited his furniture store. "Keeping alive must have great ability." It was great of him to support the business.

两天也就这么过了，继续！

Two days had passed. Go on!

写写画画，挑选一下酒店和跑步。

I wrote and drew, chose the hotel and ran.

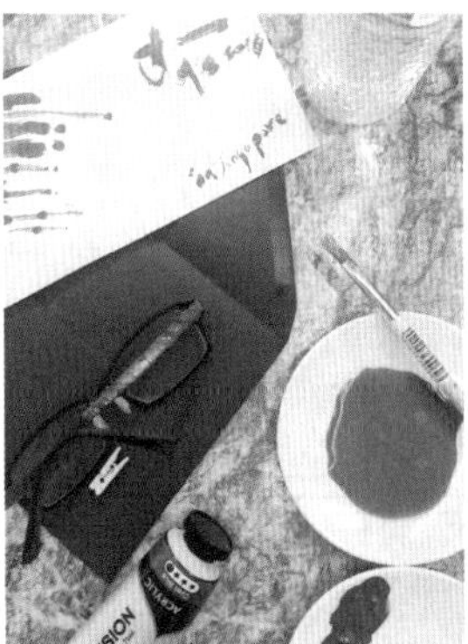

彩绘眼镜，五块钱的课堂作业，送给团友的礼物。

A pair of painted glasses.My class work which cost five yuan was sent to a group member as a present.

Art Friend 文具店，琳琅满目，无敌！

Art Friend, which was a stationer, was full of various things,second to none!

娘惹餐厅，纯正的马来西亚菜，好不好吃，你试过就是。

Nynonya Restaurant severs pure Malaysian food. Whether it tastes good or not, just come to try!

Oasia酒店，城市的地标建筑，当年的垂直绿化技术全球领先，顶层的酒店泳池区亦是自然通风，空间也颇特别。

Oasia Hotel is the landmark of the city. The vertical greenery technique was the top in the world of those years. The swimming pool on the top of the hotel was natural ventilation and the space was also unique.

PARKROYAL
ON BEACH ROAD
SINGAPORE

7500 Beach Road
Singapore 199591
Tel: +65 6505 5666
Toll free: 1800 2557 795
Fax: +65 6296 3600
parkroyalhotels.com

9-10/8, 2008

[illegible] Lasalle [illegible] Sota [illegible]

PARKROYAL
ON BEACH ROAD
SINGAPORE

7500 Beach Road
Singapore 199591
Tel: +65 6505 5666
Toll free: 1800 2557 795
Fax: +65 6296 3600
parkroyalhotels.com

[illegible] Design. Centre.) [illegible]

[illegible] space [illegible]

[illegible] Design Resources Workshop !!

Try to Remember the Name(s)!
努力把名字记起来!

2018年8月9、10月
August 9-10, 2018

循例的酒店早餐：coffee，两个煎鸡蛋，牛角包，两根鸡肉肠，足矣。

I had breakfast in the hotel as usual, coffee, two fried eggs, a croissant and tow chicken sausages, that was enough!

去Lasalle拉萨尔艺术学院，人少（因为是50周年新加坡国庆日，放假），逛逛，以喝咖啡结束。接着去看SoTa艺术学院，私人地方，不允许拍摄，倒是前广场很丰富精彩。中午饭一般，天台花园精彩而有艺术特色。下午去参观大坝Vivo，逛逛商场。

We went to visit Lasalle College of the Arts. There were not many people.(Because it was the National Day of Singapore, people there had a holiday.) Walked around and enjoyed coffee. And then we went to the School of the Arts Singapore (SOTA), we weren't allowed to take pictures there because of the private place. Luckily, the front square was wonderful. Lunch was so so. The garden on the rooftop had its unique artistic features. We visited the Vivo Dam and some shops in the afternoon.

晚上回来做完老师的作业，一个彩绘的眼镜，希望帅哥喜欢！

After coming back to the hotel, I completed the task assigned by the teacher, a pair of painted glasses. Hoped the handsome guy will love it.

10日，学习的最后一天。第一站上课地点：国家设计中心（National Design Centre），算是最好的上课地点了，旁边有教堂，其本身的建筑空间、布局、室内设计，包括保留有旧教堂的我们上课的阶梯课室，太绝了，真的是无与伦比。首层的卖场兼餐厅（中午饭）的地方，有小资情调，让上课少了寂寞！下午讲解作业，练习口才，结果是王总（图森）的一首诗，绝对胜出，牛，实至名归！

Today was the last day of the course. We had the first class at the National Design Center which was the best place, I thought. There was a church beside it. The architectural space, the layout and the interior design including the ladder classroom in the church where we took the class were all so perfect and second to none. The first floor was the place which was not only a market but also a restaurant with some petty bourgeoisie sentiment, so we felt less lonely while having classes. In the afternoon, we presented our homework and practised our speaking skill. Mr. Wang, (the CEO of Tusen Company), won with a poem by great superiority. Cool! He deserved it.

晚上算是告别餐，也斯文，有虾蟹，安安静静的，蛮不像国内设计师的风格。

Dinner was the farewell party. All the people were quiet to have sea food. It was quite different from the domestic designers' style.

这几天是有收获的，也许吧。

Maybe I learnt something during these days.

下午参观Space，再一次看看新加坡（澳大利亚人投资）的顶级家具卖场，氛围不错，有上升能力和标杆作用，卖，实体店真不容易。就像这一次课程，从一开始招生就可以看到他们的不容易，加油，Design Research Workshop！！！

In the afternoon, we visited Space again which was the top furniture shop in Singapore. It was invested by Australians. The commercial atmosphere there was so good that it kept rising and also should be the flag in the same industry. It was really not easy to sell things in a physical store. Just like our course, enrolling students was not easy from the very beginning. I felt their pressure.

Come on, Research Workshop!!!

国家设计中心

National Design Centre

国家设计中心，新旧结合的综合楼，上课的阶梯课室原来是一个小教室。

National Design Centre is a multiple-use building that stitched together old and new.The lecture theatre we had classes was used to be a small classroom.

学习中心

Learning Centre

互动性的学习，我们这些“老学生”也挺认真的。

Interactive learning.The "old" students like us were quite serious.

SPACE家具店

SPACE家具店，丰富的国际一线品牌家具卖场。

SPACE—A top furniture shop with plenty of international famous brands.

CAPELLA
SINGAPORE
1 THE KNOLLS
SENTOSA ISLAND
SINGAPORE 098297

T +65 6377 8888
F +65 6337 3455
www.capellasingapore.com

CAPELLA
SINGAPORE
1 THE KNOLLS
SENTOSA ISLAND
SINGAPORE 098297

T +65 6377 8888
F +65 6337 3455
www.capellasingapore.com

2018年8月11日

August 11, 2018

今天正式结束学习，各奔东西。当然也有像我们这样的多停留一些日子的。懒洋洋地吃过早餐，逼自己画了住了6个晚上的酒店房间平面，也试着找到这间与我设计水平相当的客房的“优点”。

We drifted apart at the end of the course today. Of course, there were also some people staying for a few more days like us. After breakfast, I made myself draw the plane of the guestroom which I have stayed for six nights. While drawing, I tried to find out the advantages of it. I thought the level of the interior design was the same as mine.

真的不容易：中档项目“比上不足，比下有余”，要有效果，确要花心思（比贵的项目难度更加大，我认为），退房，走走逛逛旁边的安达仕酒店（傅厚民的室内设计），倒是它旁边的新古典风格（Art Deco）的大厦首层大堂是久违的豪华与精彩，想起了三十年前手绘时期的主流风格，坚持也是不容易！当然像陈杨所说的，要投入足够才能有这样的效果，像洛可可一样要时间和金钱！

It was really not easy: the mid-range hotel was not the best, but better than many others. If you wanted some effects, you must pay more attention(I thought it was much more difficult to design expensive ones). And then I checked out and walked around Andaz nearby (the interior design was made by Andre Fu). Beside it, the launch of the building on the first floor was in the neoclassical style—Art Deco. It looked luxury and wonderful I had never seen this style for long. It reminded me of the mainstream style of hand-painted time thirty years ago. Persistence was not easy! Just like Chen Yang's words, enough money could make the effects like Rococo—it needed a lot of time and money as well.

入住圣淘沙的嘉佩乐酒店，多年前诺曼·福斯特做的建筑设计，新旧结合，已故的设计师贾雅先生做的室内设计，倒是室内设计见多了，没有了新鲜的感觉，旧了，怎样才有延续的吸引力，这也值得思考！！

I checked in to Capella in Sentosa which was designed by Norman Forster many years ago.It was the combination of old and new. The interior design was made by JAYA who passed away. Maybe I had seen too many interior designs, I couldn't feel surprised again. How can old things keep attractive? It was worth thinking about.

12-13/8 2018

住这两天，[illegible]，[illegible]一到就不想出去了，一家酒店能[illegible]了。某种程度上[illegible]设计[illegible]，这个酒店[illegible]国际[illegible]！Kerry Hill[illegible]的手法。

我们昨天下午去看了Kerry Hill设计的新酒店Warehouse Hotel，那是一个[illegible]仓库改造的[illegible]。之后[illegible]（Scott设计）和JW万豪酒店South Beach区的Norman Foster设计的Phillipe Starck的室内设计：精彩，很，好！

CAPELLA
SINGAPORE
1 THE KNOLLS
SENTOSA ISLAND
SINGAPORE 098297

T +65 6377 8888
F +65 6337 3455
www.capellasingapore.com

关于[illegible]：[illegible]设计[illegible]的设计细节。每一个[illegible]需要设计[illegible]，这就是[illegible]不[illegible]。[illegible]“[illegible]设计”[illegible]，[illegible]（钱）[illegible]。

[illegible]

CAPELLA
SINGAPORE
1 THE KNOLLS
SENTOSA ISLAND
SINGAPORE 098297

T +65 6377 8888
F +65 6337 3455
www.capellasingapore.com

今天的话题：“8.13 [illegible]，[illegible]酒店[illegible]（钱）这个[illegible]酒店[illegible]。

市场→设计→运营→推广

[illegible]

[illegible] 200～300万[illegible]

south 5%～10%（1000万[illegible]）

CAPELLA
SINGAPORE
1 THE KNOLLS
SENTOSA ISLAND
SINGAPORE 098297

T +65 6377 8888
F +65 6337 3455
www.capellasingapore.com

2018年8月12、13日

August 12-13, 2018

休息两天，写写东西发发呆，早餐是最后一刻才去吃的，一家酒店住两天足够了，基本能体验到设施和服务，上午把酒店房间的平面图和文字写好了，为《住哪？4》积累足迹。

During two leisure days, I was in a daze and wrote something, had breakfast at the last minute. Staying in the same hotel for two days was enough to experience the facilities and service. I also finished the plane of the guestroom and my comments in the morning to accumulate footprints for Where to stay? 4

倒是昨天下午约了Kelly钟小姐去看看酒店Wart house Hotel，聊聊小晖和她的合作意向，开心。之后去了洲际酒店（SCDA设计）和JW万豪的新酒店，South Beach区诺曼・福斯特设计的建筑，菲利普・史塔克做的室内设计，狂，狠，花！

It was happy to make an appointment with Kelly, Miss Zhong to have a look at Wart Bouse Hotel, and talked about the cooperation with Xiao Hui. After that, we not only visited Inter Continental Hotel which was designed by SCDA and the new Marriott Hotel but also the Norman Forster's building in South Beach, the interior design was made by Philippe Starck.Crazy, determined, Garish!

关于公司定位，看了几个新加坡项目及背后的设计公司，每一个项目业主都需要设计公司，为什么是我，不找你？这个是我们的定位问题。再看到类似国内"启迪设计"这样的公司，水平不一定高，但是收费高，姿态好！发展得不错，而我们还是几千万的营业额，不容易扩展，基础（钱）不牢固。

About the positioning of the company, I observed a few Singapore grogrammes and the design companies behind them. I found out that the owner of each programme needed design company, but why me, not you? It depended on the positioning of a company. I watched the design company in China like "Qidi Design", its professional level was not high but the price was high, the attitude was also great. This kind of company ran well. Our turnover was still about tens of millions. It was not easy for us to develop our business because of the weak foundation (especially money).

45 嘉佩乐酒店

CAPELLA SINGAPORE

Address : 1 the Knolls Sentosa Island, Singapore, 098297
Telephone : +86 65 6377 8888
Fax : +86 65 6337 3455
Http : //www.capellasingapore.com

CAPELLA
SINGAPORE
1 THE KNOLLS
SENTOSA ISLAND
SINGAPORE 098297

T +65 6377 8888
F +65 6337 3455
www.capellasingapore.com

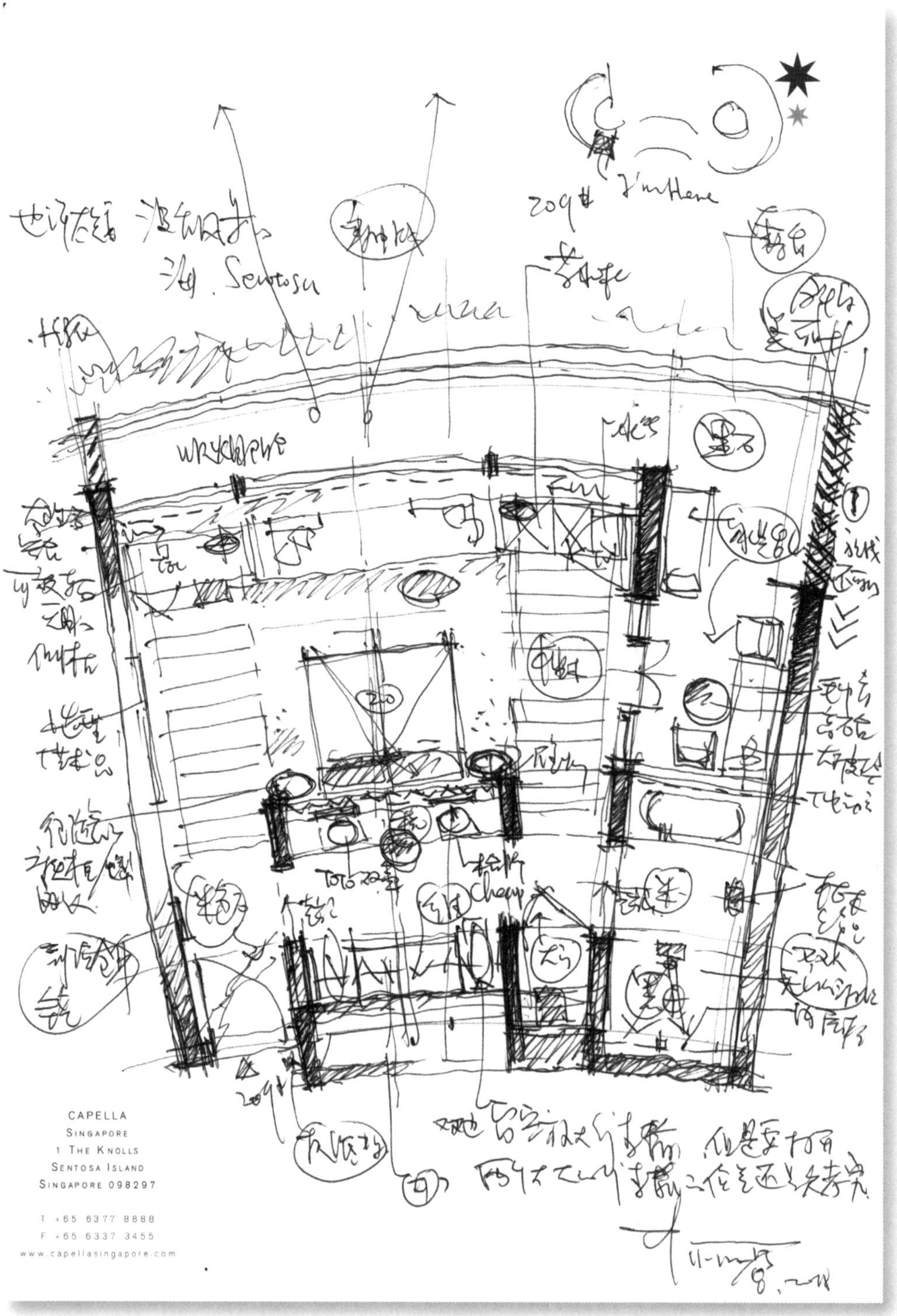

O.'s File

（区生词典）

贾雅(Jaya)

自幼深受印度尼西亚爪哇文化熏陶，加上西方教育的洗礼，养成了贾雅对东方与西方元素融合的独到见解与功力。贾雅的风格偏爱使用自然材料，在他设计的酒店中看不到大理石或者很具跳跃性的色彩，他的坚持，也是他给予品牌的不二法门。贾雅的代表作：北京颐和安缦酒店，杭州富春山居，上海璞丽酒店。(以上内容引自《诠释TRENDS》杂志)

Influenced by the Javanese culture in Indonesia and baptized by western education since his childhood, Jaya has developed his unique views and skills on the integration of eastern and western elements. Using natural materials became his style. There is no marble or jumping colors in the hotels which were designed by Jaya. His persistence is also the key to his brand and success. His representative works include: Aman at Summer Palace in Beijing, Fuchun Resort in Hangzhou, the Puli Hotel in Shanghai. (From《诠释TRENDS》)

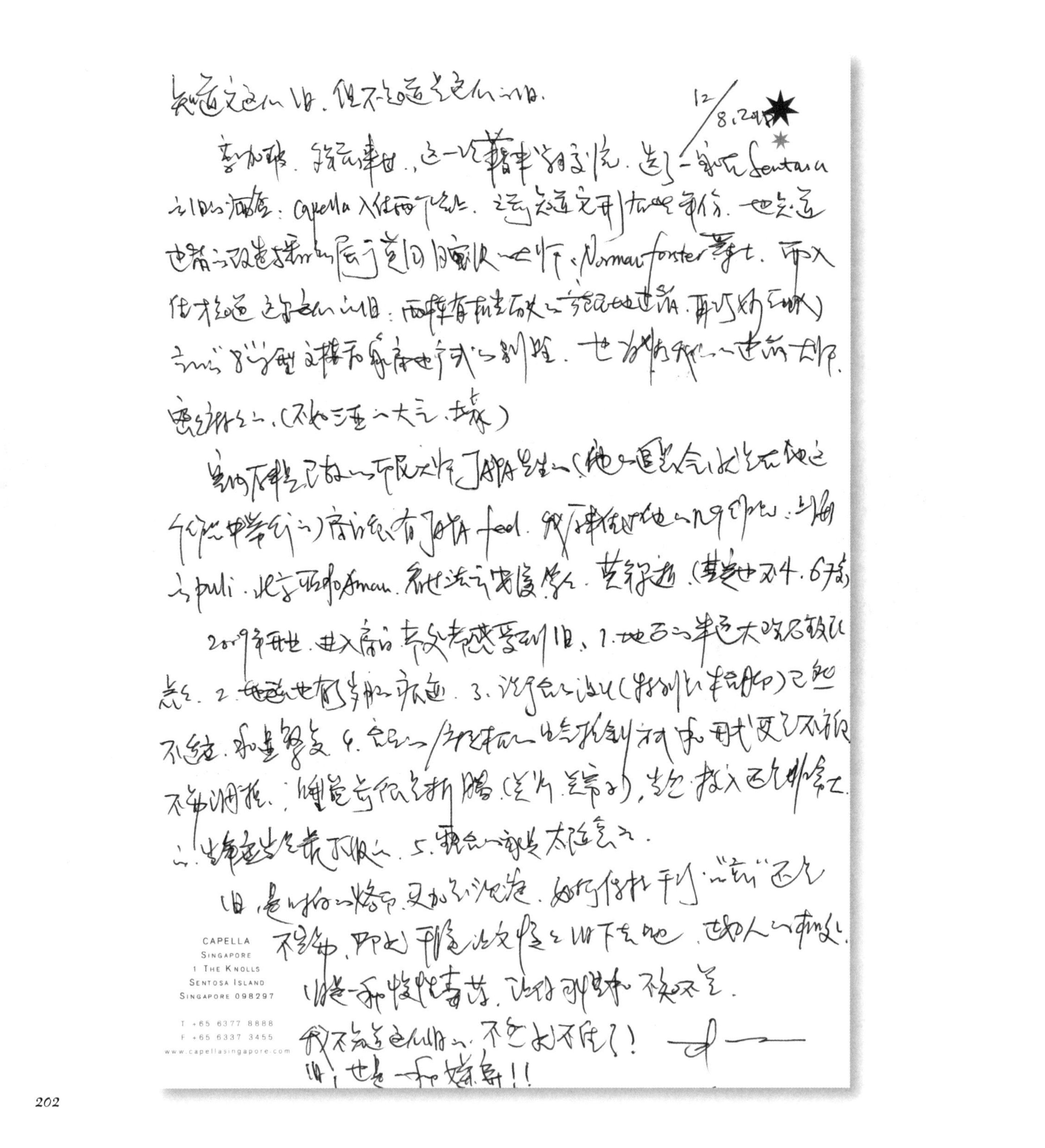

12/8, 2013

[illegible]

CAPELLA
SINGAPORE
1 THE KNOLLS
SENTOSA ISLAND
SINGAPORE 098297

T +65 6377 8888
F +65 6337 3455
www.capellasingapore.com

I Know It is Old, but I Have Never Thought It is So Old

知道它这么旧，但不知道它是这么的旧

2018年8月12日
August 12, 2018

新加坡，多年前来过，这一次借着来学习交流，选了一家在圣淘沙的旧酒店，在嘉佩乐入住两个晚上。之前知道它开了有些年份，也知道建筑的改造与翻新属于英国国宝级的大师——Norman Forster爵士，而入住才知道这里这么的旧。两栋有相当历史的殖民地建筑，再巧妙融入新的“8”字形主楼和类似房地产式的别墅，也为难了我们的建筑大师，密密麻麻的（不如三亚的酒店大气、奢华）。

I came to Singapore many years ago. I came again for an study exchange this time. I chose an old hotel in Sentosa—Capella to stay for two days. I knew it had been opened for some years. I also knew that the building was transformed and renovated by the national treasure master—Sir Norman Forster. Actually, I found it was so old after staying in it. Two colonial style buildings with a long history were tactfully integrated into the new main building in the shape of "8" and the villas like real estate. This stumped our great master. The dense layout was not so luxurious as the hotels in Sanya.

室内原来是已故的印尼大师贾雅先生做的设计（他的追思会就是在这个作品中举行的），房间很有JAYA feel，我原来住过他的几个作品：上海的璞丽，北京颐和安缦，还看过法云安缦等等，英年早逝（其实也不算“早”，67岁）。

The interior was designed by deceased Indonesian master JAYA. (His memorial was held in this place), it was full of JAYA feel. I had stayed in some hotels designed by him: Puli in Shanghai, Beijing YiheAmanHotel and FayunAman Hotel. He died young. (not really young, already 67 years old.)

2009年开业，进入房间，到处都能感受到旧。1. 地面的米色大理石斑痕花花；2. 地毯也有了岁月的痕迹；3. 洗手台的设计（特别是木台脚）已然不适应和显繁复；4. 台面的、床头柜的电器控制方式和形式更是不方便，不易调控，睡觉前很是折腾（关灯，关帘子），当然投入还是非常大的，当年应当是顶级的；5. 露台的家具太随意了。

The hotel opened in 2009. I felt old everywhere when I entered the guestroom. 1. The beige marble floor was mottled; 2. The blankets had years' trace; 3. The washbasin was unsuitable and seemed too heavy and complicated in modern time; 4. The controls of electrical appliances on the table or on the night table were not easy to use (it took some time to turn off the lights and the curtains before sleeping). Although a large amount of money was spent on it. The system should be the top at that time; 5. The furniture in the balcony seemed too casual.

旧，是时间的烙印，更加是沉淀，如何保持干净的“新”还是不容易，那就干脆让它慢慢旧下去吧，正如人的相处，旧是一种慢性“毒药”，让你习惯和不知不觉。

Old is a mark of time and sedimentation as well. It is not easy to keep things clean and new. So let it be. Just like getting along with others. Old is one kind of a slow "poison", it makes you get used to it when you are unaware of.

我不知道这么旧的，不然就不住了！

I didn't know it was so old. Otherwise, I wouldn't choose to stay here.

旧，也是一种“嫌弃”！

Old is also a kind of "disgust"!

O.'s File

（区生词典）

诺曼・福斯特

英国人，是国际上最杰出的建筑大师之一，被誉为“高技派”的代表人物，第21届普利兹克建筑大奖得主。诺曼・福斯特特别强调人类与自然的共同存在，提倡那些适合人类生活形态的建筑方式，一生获奖无数，被册封为爵士。福斯特建筑事务所的代表作品包括中国首都国际机场新航站楼、苹果总部，等等。(百度百科)

Norman Forster is one of the most outstanding British architects in the world. He is also known as a representative of the "High technology school", as well as the winner of the 21st Pritzker Architecture Prize. Norman Forster emphasized the co-existence of human and nature, and advocated those architectural styles which were suitable for human life. In his lifetime, he won a lot of awards and was knighted. The representative works of Foster+Partners include New terminal of China Capital International Airport, Appleheadquarters,and so on.

(From Baidu Encyclopedia)

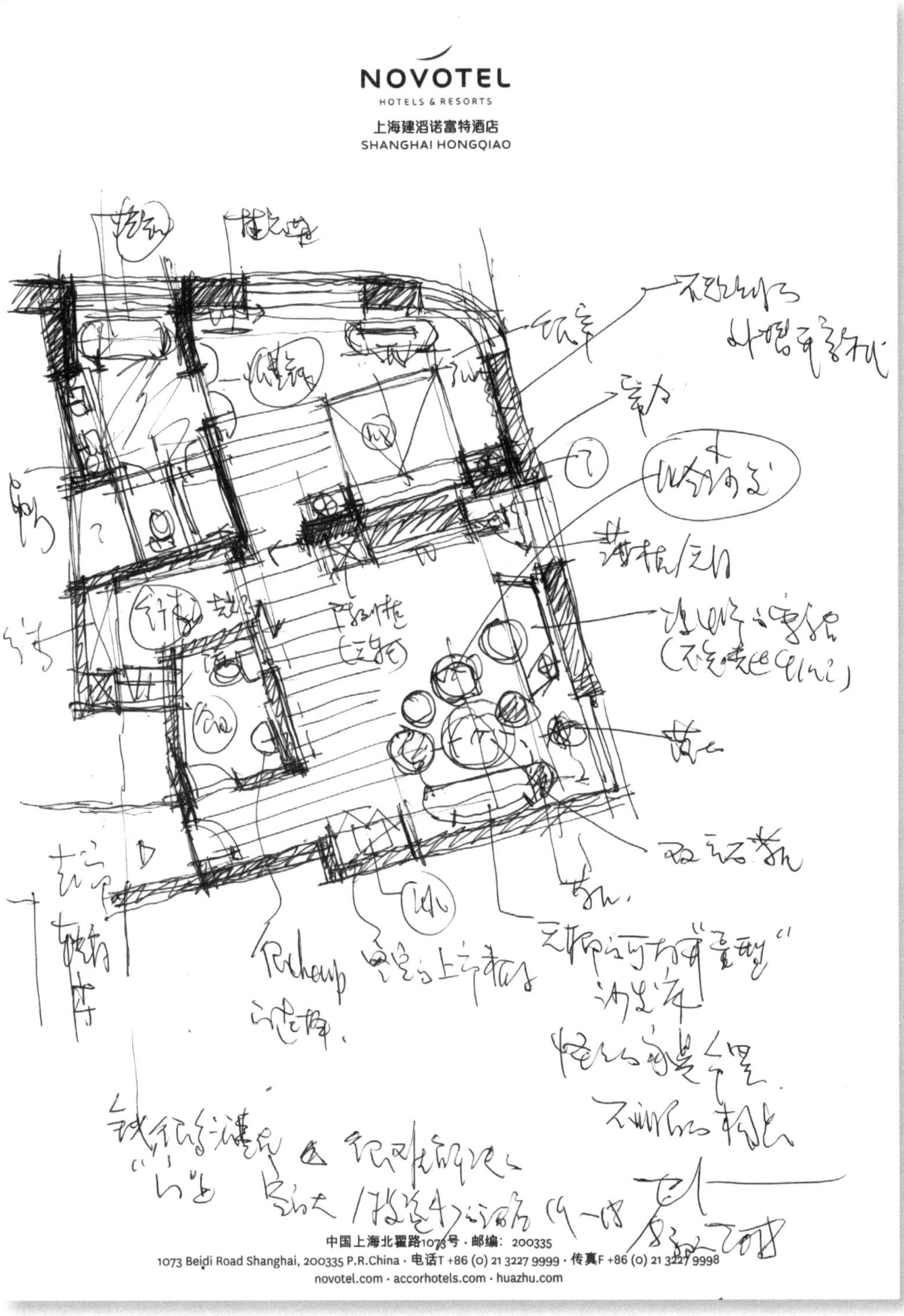

NOVOTEL
HOTELS & RESORTS
上海建滔诺富特酒店
SHANGHAI HONGQIAO

上海建滔诺富特酒店

46

★★★★

NOVOTEL HOTELS & RESORTS SHANGHAI CHINA

Address : No.1073 Beidi Road, Minhang District, Shanghai, China
中国上海市闵行区北翟路1073号 200335
Telephone : +86 21 3227 9999
Fax : +86 21 3227 9998
Http : //www.novotel.com/accorhotels.com/huazhu.com

NOVOTEL
HOTELS & RESORTS
上海建滔诺富特酒店
SHANGHAI HONGQIAO

中国上海北翟路1073号 · 邮编：200335
1073 Beidi Road Shanghai, 200335 P.R.China · 电话T +86 (0) 21 3227 9999 · 传真F +86 (0) 21 3227 9998
novotel.com · accorhotels.com · huazhu.com

NOVOTEL
HOTELS & RESORTS
上海建滔诺富特酒店
SHANGHAI HONGQIAO

中国上海北翟路1073号 · 邮编：200335
1073 Beidi Road Shanghai, 200335 P.R.China · 电话T +86 (0) 21 3227 9999 · 传真F +86 (0) 21 3227 9998
novotel.com · accorhotels.com · huazhu.com

上海建滔诺富特酒店
Novotel Hotels & Resorts, Shanghai, China

How to Spend Money on A Guestroom?
一个客房的钱应当怎么花?

入住这家今年四月份开业的诺富特，一方面是我习惯的品牌，另一方面，它的外立面有一些设计感。

I stayed in Novotel Hotel which opened in April. On one hand, it was my familiar brand, on the other hand, its facade had a sense of design.

入住的感觉特别不错。热情（可能还在推广期间吧）。公共区域装修特别豪华，全白石，有大的流水瀑布墙等，比原来类五星的酒店的定位提升不少。

My stay was quite good and I felt warm (maybe during the promotion). The decoration of the public area was deluxe with all white marble,and there was a big waterfall wall and so on. The position of this five-star hotel was upgraded a lot.

端头的套房，考验设计师的“客房骨头”，这一间也就是这样，入住后，第一感觉是“简陋”，客厅区域很大，很多浪费钱的木饰面、饰品柜（空空的）、大理石主幅，而软性的陈设如画、灯、家具、饰品等都欠缺或干脆就没有，第一感觉特别不好！

The suite at the corner was irregular, really challenged the designer, so did this room. After entering it, my first feeling was simple. The living area was very big. A lot of wood veneer, empty ornament cabinets and the marble wall were a waste of money. While there wasn't any soft furnishing at all, such as paintings, lights, furniture or ornaments.My first feeling was quite terrible.

入住卧室睡眠区，全采光的感觉不错，包括洗手间浴缸的城市风景也还可以——工地+车水马龙，可能是结构的因素，房间不规则且窄，有不合理的设计与家具配置，短板突显，钱继续浪费在无聊的暗藏趟门、双掩门等地方，包括双洗手盆等梳妆区的设计，挺不切实际的，这才是刚刚好的五星级啊，也难为了设计师了。

Entering the sleeping area in the bedroom, the natural light felt good.You could enjoy the beautiful city view, building site and busy traffic in the bathtub. Because of the construction, the guestroom was irregular and narrow. The disadvantages of some unreasonable designs and furniture configuration were obvious. The useless hidden sliding door, double door and others wasted a lot of money, double basins and dressing area as well. It was quite impractical. This five-star hotel was just good enough but hard for designers.

想想，一个房间的花费该怎么分配，也是值得研究和关注的！

We should think over how to spend money on the guestroom.It is worthy of study and paying attention.

北京南铜锣巷亚朵酒店

47

★★★★

ATOUR HOTEL BEIJING CHINA

Address : 1 Xiyangwei Hutong A, Dongcheng District, Beijing,China
北京市东城区西扬威胡同甲1号

Telephone : +86 10 5750 6999

Http : //www.yaduo.com

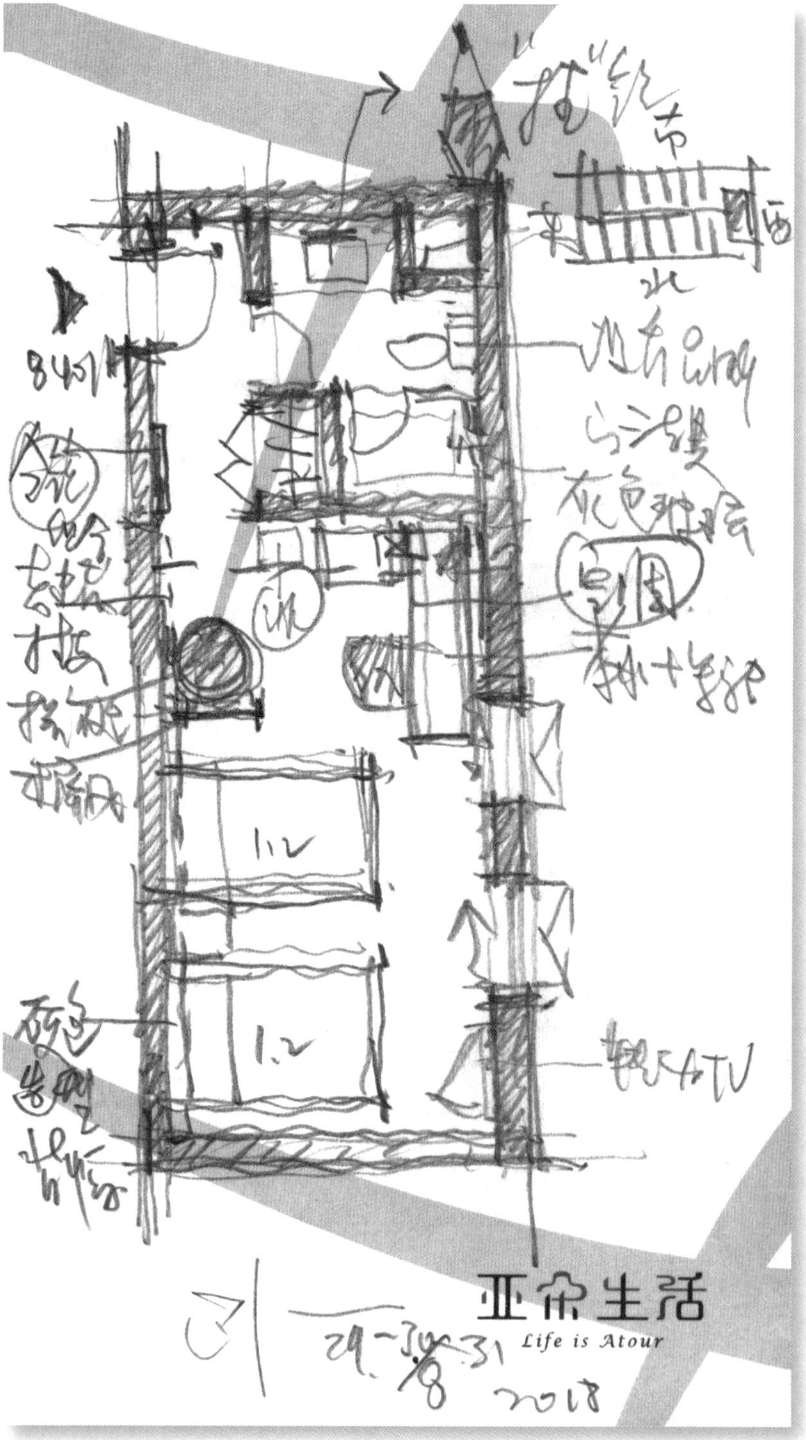

亚朵生活
Life is Atour

Charming of Refined Systematic Hotel Chain

精细化的系统连锁酒店的魅力

入住这间高消费的亚朵Atour酒店（1200元），因为在老区，是南铜锣巷的改造老店。周边都是一层的京味十足的居民四合院，大树却有20多米高，极端的“不和谐”。

Staying in the expensive Atuor Hotel (1200 yuan), because it lay in the old district, was a transformed house in South Luogu Lane. There were traditional Beijing Courtyards around it. A tall tree was more than twenty meters. Extremely disharmony!

亚朵的发展有目共睹，所以来体验一下。标准的旧楼改造项目，现代中式味道，设计布局颇为老到。入住的是端头的房型，配套标准化（因为有三间房，不同房型作对比）而有经验，更是有它自家的手段和特色。

The development of the Atour was for all to see. So I hoped to experience it. It was a standard old building renovation project. The hotel was modern Chinese style. Design layout was the best. I stayed in the corner guestroom with standardized supporting and experience(Compared three different types of guestroom which we stayed in).

1. 标配的三件套洗手间、管井、布局、间隔（灰色玻璃）；细节：个性挂镜和化妆镜，合适的毛巾台下柜，双手纸架，双淋浴去水地漏。

A three-set standardized washroom, the tube well, layout and space by grey glass; details: personalized hanging mirror and makeup mirror, appropriate towel cabinet, double toilet paper holders and double shower floor drains.

2. 全仿旧中式（类榆木）的家具。立面上间隔屏风（疏了一点点），大到台、椅、床头柜、仿古（砖）木地板，小至书签架、配品夹子，可谓尽心思和控成本。

All the furniture were old Chinese style (like Elm wood furniture). The facade was spaced with a screen which was a little crude. The tables, chairs, bedside tables, antique (brick) wood floor, even the bookmark rack and accessories clip can be said to do the best and cost control.

亚朵的配色系统还是不错的。房间为浅草绿色+灰色床背，配木色。草黄色床上用品与再造纸相一致，有环保与温馨的感觉！赞！

The color matching system of Atour was still good, the light grass green room and the grey bed back matched the wood color. Straw yellow bedding was consistent with recycled paper. The room was full of the environmental and cozy feeling! Wow!

软的也不错，入住有茶饮，退房有水送。（当然因为送小朋友上大学，也在酒店买了被子和枕头，可抵房费。精明的酒店！）

The service was also good. Guests could enjoy welcome tea while checking in, and free water while checking out. (Of course, I bought a quilt and pillow in the hotel to cover the cost of my rooms because I sent my child to university. What a smart hotel!)

相信亚朵也是一匹大黑马，会成为盈利的连锁品牌。

I believe Atour is also a dark horse, a profitable chain brand.

北京南铜锣巷亚朵酒店
Atour Hotel, Beijing, China

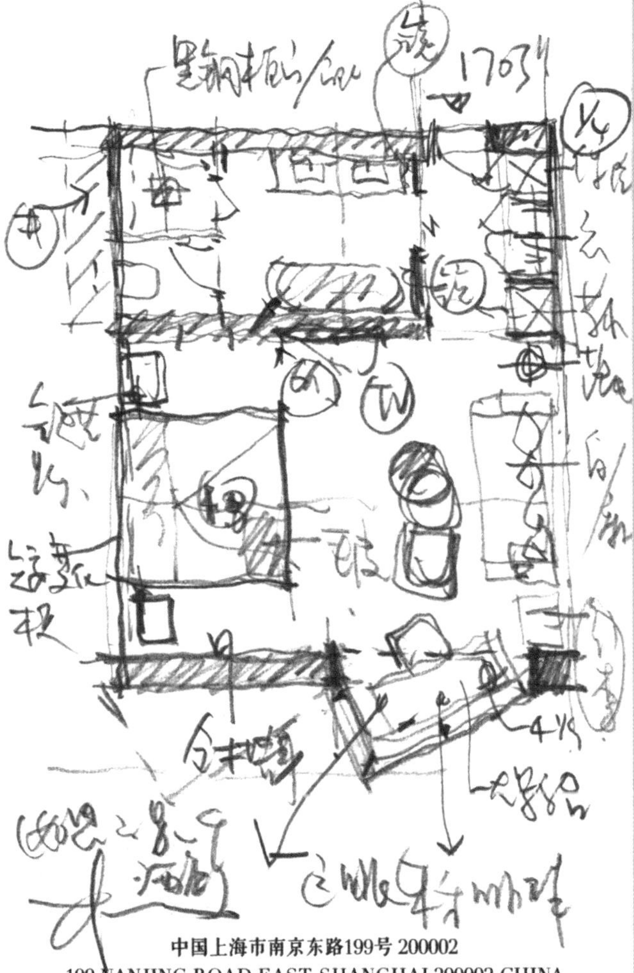

中国上海市南京东路199号 200002
199 NANJING ROAD EAST, SHANGHAI 200002 CHINA
电话 PHONE +86 21 5368 9999 传真 FAX +86 21 5368 9998
WWW.EDITION-HOTELS.CN

上海艾迪逊酒店

48 ★★★★★

THE SHANGHAI EDITION CHINA

Address : No.199 Nanjing Road East, Shanghai, China
中国上海市南京东路199号
Telephone : +86 21 5368 9999
Fax : +86 21 5368 9998
Http : //WWW.EDITION-HOTELS.CN

THE SHANGHAI
EDITION
上海艾迪逊酒店

copy一下，客户……[illegible] 上海Edition艾迪逊酒店 [illegible]

中国上海市南京东路199号 200002
199 NANJING ROAD EAST, SHANGHAI 200002 CHINA
电话 PHONE +86 21 5368 9999 传真 FAX +86 21 5368 9998
WWW.EDITION-HOTELS.CN

THE SHANGHAI
EDITION
上海艾迪逊酒店

每个人的创意……[illegible] 我copy……[illegible]

中国上海市南京东路199号 200002
199 NANJING ROAD EAST, SHANGHAI 200002 CHINA
电话 PHONE +86 21 5368 9999 传真 FAX +86 21 5368 9998
WWW.EDITION-HOTELS.CN

THE SHANGHAI
EDITION
上海艾迪逊酒店

[illegible] copy……[illegible]

中国上海市南京东路199号 200002
199 NANJING ROAD EAST, SHANGHAI 200002 CHINA
电话 PHONE +86 21 5368 9999 传真 FAX +86 21 5368 9998
WWW.EDITION-HOTELS.CN

上海艾迪逊酒店
The Shanghai Edition, Chine

Would You Mind a Copy of Design?

一个复制的设计你介意吗?

之前入住上海素凯泰酒店，有小惊喜和学习的地方，这一次借来上海看家具展览的机会，选了一个位于南京路步行街旁的旧改项目——上海艾迪逊酒店，闹中取静，有出入方便的优势，当然入住房间可以眺望黄浦江和东方明珠塔也是“卖点”之一。

I had a little surprise and learned a lot when staying in the SUKHOTHAI Hotel in Shanghai before. This time I came to the Shanghai Furniture Exhibition, so I took this opportunity to choose Shanghai Edition Hotel which lay beside Nanjing Road Pedestrian Street. It was also a transformed project. It was an oasis of serenity with the advantage of transportation. Of course, one of the attractions was to overlook the Huangpu River and the Oriental Pearl Tower.

房间位于楼龄30多年的主塔楼，而副楼是楼龄近80年的老建筑，都是配套的酒吧、会所，及餐厅（粤菜、西餐）。

Guestrooms were in the main tower building with 30-year-old history while the annex was an 80-year-old building with included bars, clubs and restaurants (Guangdong and western cuisine).

每个人的创意也许都有局限，这间酒店同样是出于如恩设计之手（同素凯泰，牛！），用材有雷同，白石、哑光、黑铁、木饰面等等。 客户，我复制一个我之前的案例，你介意吗？当然，不要太小看了这家设计公司，有创意的室内家具、全木墙身的设计有别于之前的设计，封闭的洗手间，也许玩不出什么花样，倒是有几点也体现出这一间设计公司的功力。

A designer's creativity might be limited. This hotel was from Neri&Hu Design and Research Office (the same as the SUKHOTHAI Hotel, Amazing!). They used similar material, such as white stones,matt, black iron,wooden veneer and so on. Dear client, if I copy a previous design for you, do you mind? However, don't look down upon this design company. The creative indoor furniture made of wood was a little different from before. They didn't have any new ideas about a closed washroom. However, the following highlights showed the special abilities of this design company.

整体：四门柜位于入口走廊一侧，一字展开，其中最靠近卧室的是一收纳式的茶水柜，当然如果门可收入侧边，那就更加完美了。横向的开门布局，让床与休闲区对着，正好利用斜窗作为写字区，妙！有设计公司的个性和固化的“手法”，但应当看到它更加有其研究过的、对品牌的独特要求的解决能力。

On the whole, a four-door cabinet lay up on a side of the corridor.One of them was a stored cup cabinet next to the bedroom. It would be perfect if the door was put aside. Horizontal door design made the bed face the leisure area. The slanting window was just used to be a writing area. Perfect! All of these could demonstrate the personality of the design company and their solidified skills. You might notice while staying in it. They also had ability to solve the unique requirements for their studied brands.

Copy，不是和盘而是合理取舍。

Copy should be a reasonable trade-off but not the whole.

BVLGARI
HOTEL SHANGHAI

上海宝格丽酒店

49 ★★★★★

BVLGARI HOTEL SHANGHAI CHINA

Address : No.33 North Henan Road, Shanghai,China
中国上海市河南北路33号

Telephone : +86 21 3606 7788

E- mail : shanghai@bulgarihotels.com

Http : //www.bulgarhotels.com

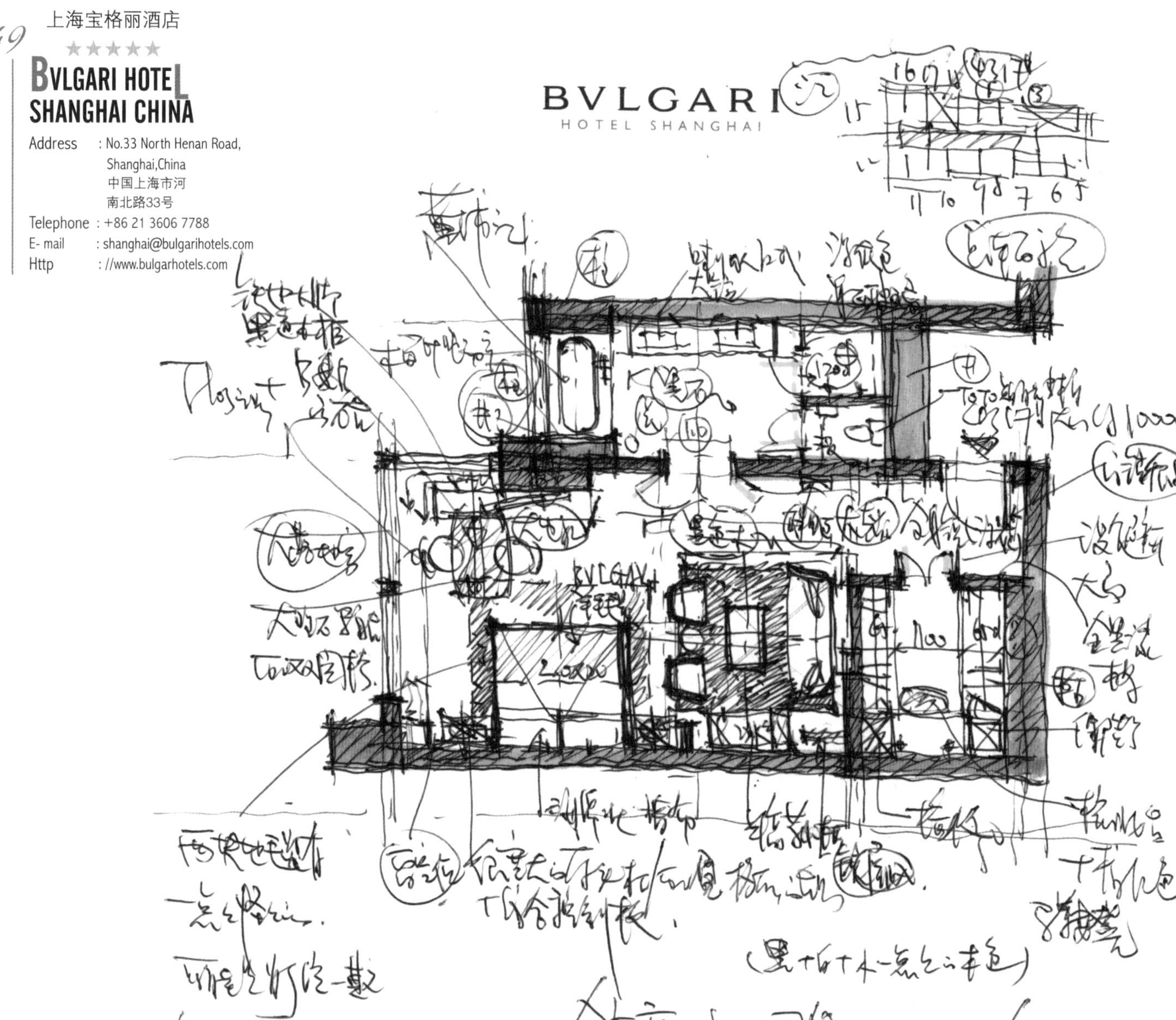

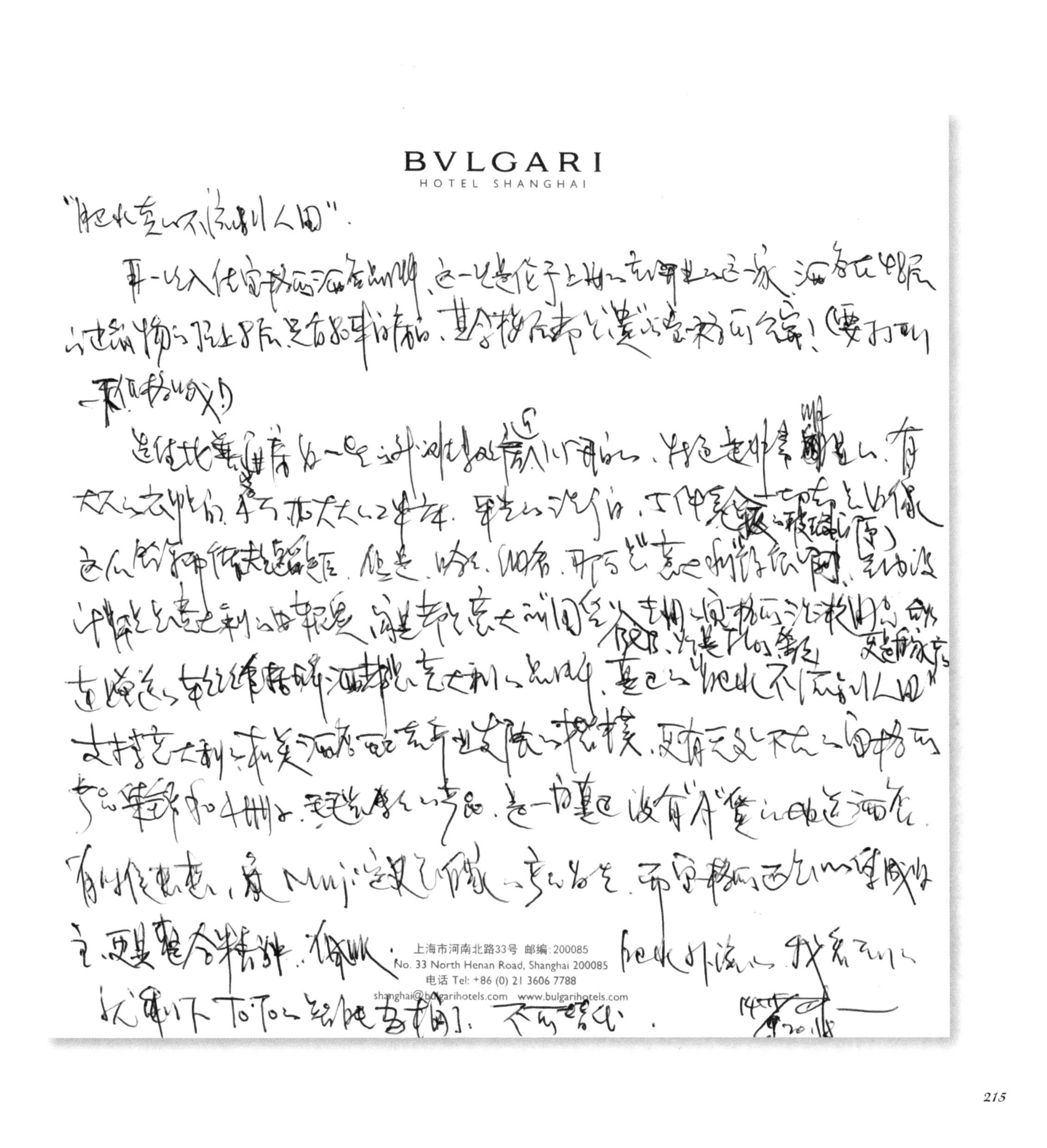

BVLGARI

HOTEL SHANGHAI

[illegible]

上海市河南北路33号 邮编：200085
No. 33 North Henan Road, Shanghai 200085
电话 Tel: +86 (0) 21 3606 7788
shanghai@bulgarihotels.com www.bulgarihotels.com

Save the Best for Oneself

“肥水不流外人田”

再一次入住宝格丽品牌酒店，这一次是位于上海的新开业的这一家，酒店在48层建筑物的顶上8层，只有80来间房间，其余楼层都是“贵”的宝格丽公寓！（要打听一下价格哦！）

I stayed in a new hotel in Shanghai which belonged to the BVLGRI Brand. The hotel was on the top eight floors of a 48-floor building. There were only around 80 guestrooms. The others were all expensive BVLGARI apartments. (The price should be inquired.)

选住比普通房好一些的外滩景观房，近1.5开间的，特色是非常明显的。有大大的衣帽间，小客厅和大大的2米床，采光的洗手间，5件套，单反的玻璃间隔，一切都是好像这么简单而循规蹈矩。但是，哈哈，细看，那可是“意大利”得很啊！室内设计师是意大利的安东尼奥，家具都是意大利的“国货”B&B，灯具是Flos，等等。专用的宝格丽洗漱用品更是自家产品，连赠送的软饮包括啤酒都是意大利的品牌，这是真正的“肥水不流外人田”，可谓支持意大利的相关酒店配套产业发展的楷模。更有无处不在的宝格丽产品集锦和小册子、毛毯等的产品，是一间真正没有“A”货的地道酒店。有时候想想，像无印良品定义的是自家的产品为先，而宝格丽还是以集成为主，更具整合精神，佩服！肥水不流外人田，就剩下东陶的智能马桶了，不可替代。

I chose a better guestroom which I could enjoy the bund. It was nearly 1.5 bay.The feature was obvious. There was a big cloakroom, a small sitting room, a two-meter-wide bed, a washroom with natural skylights including five sets divided by glass, all these seemed easy and followed the rules. However, watched carefully you would find everything was from Italy. The interior designer was Antonio from Italy. The furniture was B&B made in Italy. The lights were Flos and so on. The exclusive BVLGARI toiletries were made by their own company; soft drinks on the house as well as beer were Italian brands. It was a good example that Benefits should always be kept for their own people to support the development of Italian hotel (supporting) industry. BVLGARI product collections, booklets, blankets and some other things could be seen everywhere. It was really a typical hotel without fakes. Sometimes I think that Muji gives priority to its home products, while BVLGARI prefers collecting different boutiques. BVLGARI should have more integration spirit. I admired it. Benefits should always be kept for their own people. Nothing can be taken place of but only ToTo's intelligent closestools.

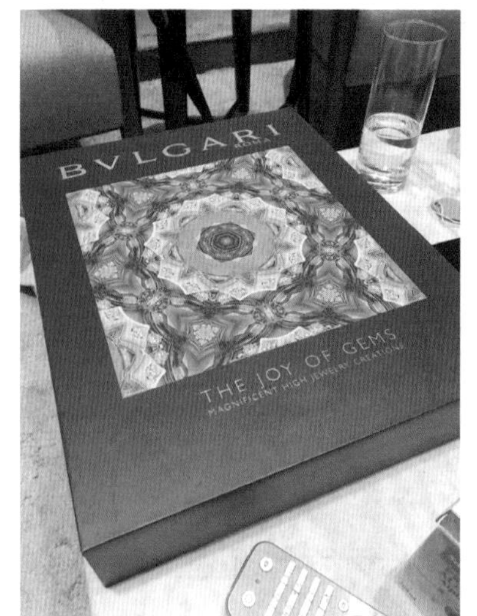

房间都是品牌的推广・上海宝格丽酒店

BVLGRI in the room · Bvlgari Hotel, Shanghai, China

[illegible]

The 24-Hour Breakfast (Cost is Said to Incur)

24小时的早餐（成本据说这样子产生了）

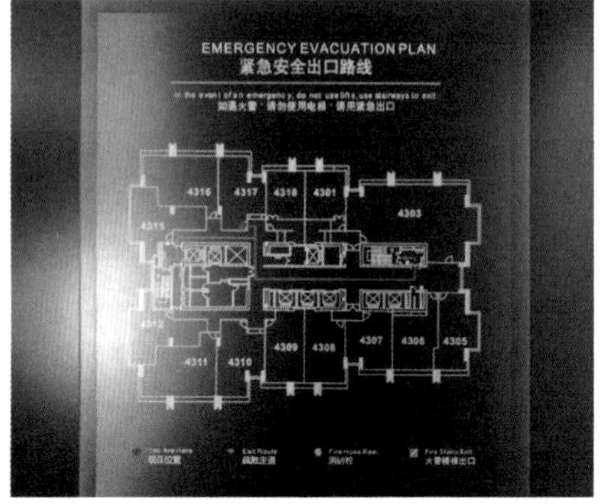

少见的标准层平面

A rare standard floor plan

当天下午5点多才能入住。第二天上午十点多吃早餐，被“隆重”告知：其实你们可以24小时随时来吃“早餐”！ OMG！！牛！

I didn't check in until more than five o'clock in the afternoon. The next day, I had breakfast at about ten in the morning. I was told seriously that breakfast was available at all hours of the day and night. OMG! Fancy that!

五千元一天的房费，当然“贵得有道理”！ Check in，有书看，有水喝（凉茶或柠檬水）。因为延误了我们的入住，还免费送了现打的果汁和咖啡作为补偿。

The guestroom cost 5000 yuan one day. The high price was really quite reasonable. You could read books, drink tea or lemon water while checking in. Fresh juice and coffee were given free to us as compensation for delaying our check-in.

房间大，1.5开间，全面黄浦江，采光的洗手间，宝格丽标配大大的可住一个月阵势的衣服间，小冰箱软饮任饮。硬件投入不容置疑，一次性的。而最令我印象深刻的还是配套的东西。

The big room was 1.5 bay facing the Huangpu River, and a washroom with natural skylights. The cloakroom which was standard in BVLGARI was so big that you could stay here for a month. Soft drinks in the small fridge were all free for guests. It was no doubt for the one-off hardware investment. What impressed me most was the facilities that came with it.

尽情享受免费软饮・上海宝格丽酒店

Enjoy free soft drinks · Bvlgari Hotel, Shanghai, China

“24小时”早餐，算是我第一次听闻和遇到，绝没有哪一个顾客“半夜吃早餐”，但这样子的“吆喝”让我服了。

The 24-hour breakfast, it was the first time I had heard and met, I didn't think anybody would have breakfast at midnight. But I was convinced by its hard sell.

我喜欢，当下的酒店，没有对比，就是伤害，土豪，成本就是这样子来的?

I like it. Hotels at the present must be compared with each other, if not, it will hurt. Rich men, was the cost made by this?

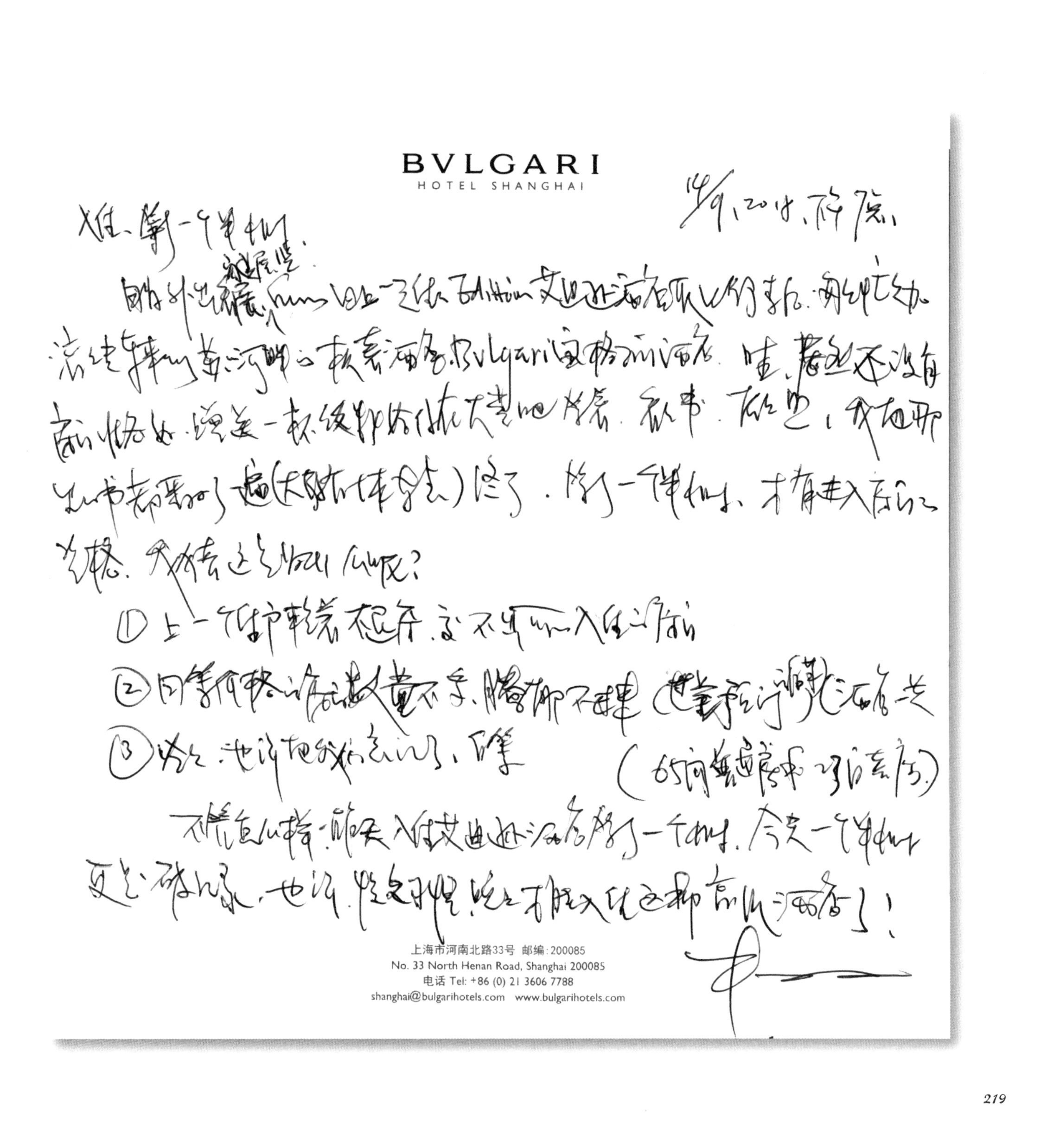

BVLGARI
HOTEL SHANGHAI

上海市河南北路33号 邮编：200085
No. 33 North Henan Road, Shanghai 200085
电话 Tel: +86 (0) 21 3606 7788
shanghai@bulgarihotels.com www.bulgarihotels.com

A Long Checking in

漫长的入住

因为外出看家具展览展，所以回上一天住的艾迪逊酒店取回行李后，匆匆忙忙叫滴滴专车来到苏河畔的极奢酒店宝格丽酒店。哇，居然还没有房间准备好，赠送一杯饮料给你在大堂等着，看看书，拍拍照。我把那里的书都翻了一遍（大约有十本左右）。终于，等了一个半小时，才有进入房间的资格，我猜这是为什么呢?

Because I went to the Furniture Exhibition in the morning, I had to go back to Edition Hotel which I stayed in the day before to get my luggage.And then I hurried to the deluxe BVLGARI Hotel along the Suzhou riverside by Didi Taxi. Wow! The guestroom hadn't got ready yet!I had to wait in the lobby and enjoyed a free drink, read books and took photos. I looked over all the books (about ten books). Finally I was eligible to check in after one hour and a half. I guessed why?

1. 上一个住户赖着不退房，交不出可以入住的房间。

The last guest stayed on, so no guestrooms were available.

2. 同等价格的房间数量不多，腾挪不过来（过量预定的结果，酒店一共65间普通房和23间套房）。

There were not many guestrooms of the same prize. As a result of overbooking, they didn't have enough guestrooms to offer. (There were sixty-five regular rooms and twenty-three suites in total.)

3. 哈哈，也许把我们忘记了，白等。

Aha, maybe we were forgotten. We waited for long in vain.

不管怎么样，昨天入住艾迪逊酒店等了一个小时，今天一个半小时，更是破纪录，也许，慢慢习惯，晚上入住这种高级酒店了！

Anyway, I waited for an hour to check in at Edition Hotel the day before, and one hour and a half at BVLGARI today. It broke my record. Maybe I have to get used to checking in this kind of deluxe hotels after dark.

风景如画的房间・上海宝格丽酒店

A picturesque room · Bvlgari Hotel, Shanghai, China

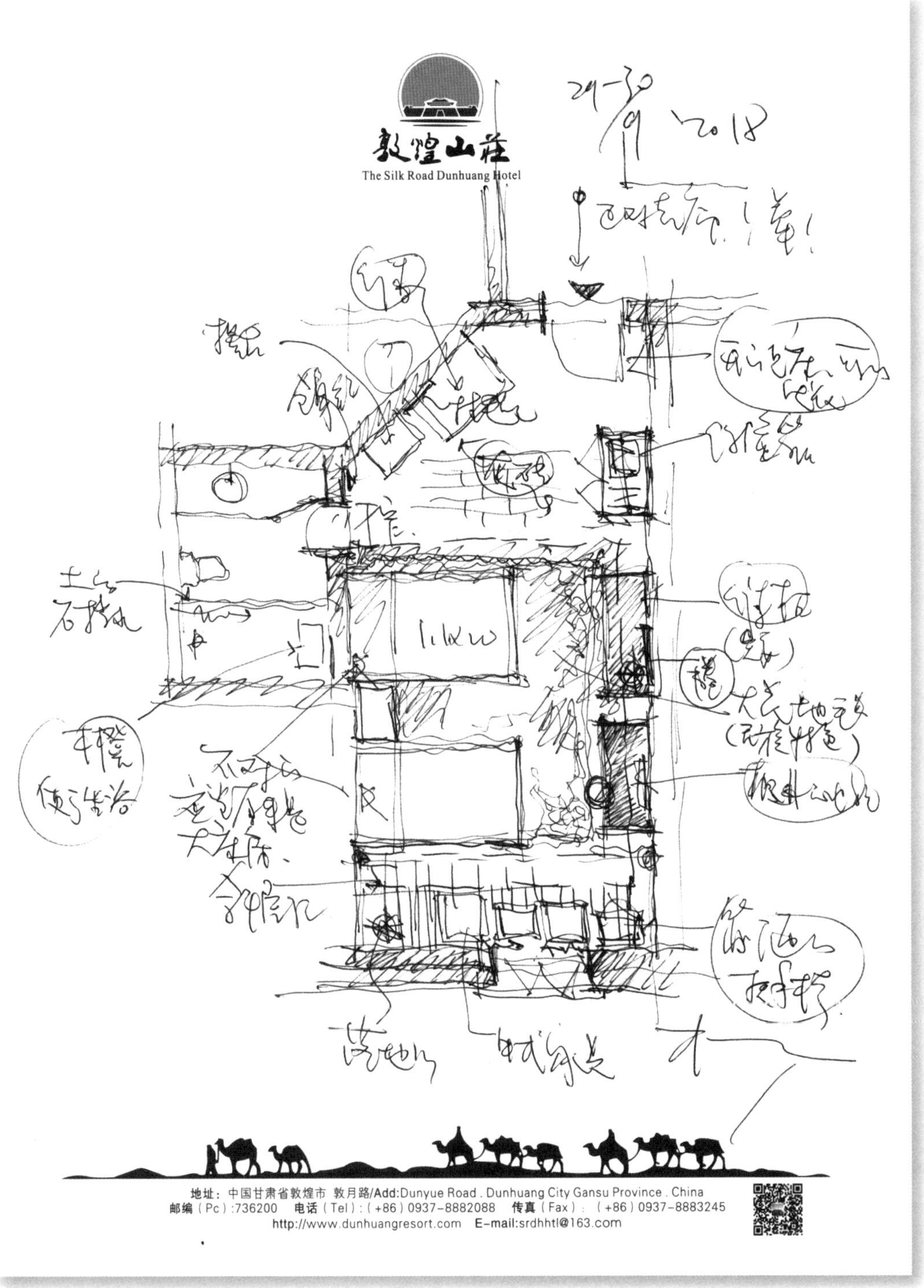

敦煌山莊
The Silk Road Dunhuang Hotel

地址：中国甘肃省敦煌市 敦月路/Add:Dunyue Road . Dunhuang City Gansu Province . China
邮编（Pc）:736200 电话（Tel）:（+86）0937-8882088 传真（Fax）:（+86）0937-8883245
http://www.dunhuangresort.com E-mail:srdhhtl@163.com

敦煌山庄

50

★★★★

THE SILK ROAD DUNHUANG HOTEL GANSU CHINA

Address : Dunyue Road, Dunhuang, Gansu, China
中国甘肃省敦煌市
敦月路
Telephone : +86 937 888 2088
Fax : +86 937 888 3245
E-mail : srdhhtl@163.com
Http : //www.dunhuangresort.com

"全副武装"的"戈壁战士"
A heavily armed Gebi fighter

敦煌山庄
The Silk Road Dunhuang Hotel

[illegible]

地址：中国甘肃省敦煌市 敦月路/Add:Dunyue Road . Dunhuang City Gansu Province . China
邮编（Pc）:736200 电话（Tel）:（+86）0937-8882088 传真（Fax）:（+86）0937-8883245
http://www.dunhuangresort.com E-mail:srdhhtl@163.com

Another Shouldn't-See the Bed While Opening the Door

又一个开门见床的“不应”

敦煌山庄

The Silk Road Dunhuang Hotel, Gansu, China

参加旭辉地产的戈5戈壁133公里，四天的徒步，组委安排入住当地最好的酒店——敦煌山庄。个人觉得非常有心思的西域地区特色的建筑群设计，包括门廊、大堂、公共区域都十分有印记，房间也颇有味道，用材地域配饰、配色，特别是挂件、地毯、窗帘非常有地域风情，倒是平面难以完美。

I took part in a four-day hike which was organized by CIFI Group. It was the fifth hike in Gobi desert. It was totally 133km. The organizer arranged us to stay in the best local hotel—the Silk Road Dunhuang Hotel. Personally I thought the buildings were special with regional characteristics of Western Regions, including the porch, lobby, public area. All of them had marks. Guestrooms had some kinds of feelings, local material and color, especially pendants, carpets, curtains with regional characteristics, while the plane was hard to be perfect.

我们的房间位于大堂右侧正正的一条长走廊的尽端，房间门正对走廊。而且，关键是开门见床，颇没有考虑心理感受。那可以改吗？我试着优化一下，感觉有难度，但不是不可行。这样就要有取舍了!让入口做一个小过厅，床易边后可远离洗手间，同时解放床背幅的长度，让双床的感觉更加合适，当然，应当巧妙地布置衣柜，摆台及行李台也要花一点心思，整体确实是利大于弊。

Our guestroom was at the end of the long corridor on the right side of the lobby. The door exactly faced the corridor. The key was that seeing the bed while opening the door. Maybe the designer never thought about the feelings of guests. So could it be corrected? I tried optimizing it and found it difficult. Of course it could. However, something had to be chosen. Make a small hall at the entrance. Change the location of the bed far from the washroom. At the meanwhile, the back of the bed should be long enough to make the double bed comfortable. Of course, the wardrobe should be arranged skillfully, the table and baggage desk as well. On the whole, the disadvantages gave way to benefit.

相信十有九入住的人都会有我这种感觉，不明白设计师为什么不去修改一下，你的设计应当从心理感受开始！

I believe most people have the same feeling as mine. I don't understand why the designer didn't correct it. Your design must start from your feelings!

51 上海CitiGo酒店

CITIGO HOTEL SHANGHAI CHINA

Address　: NO.1093 Xinzha Road,
Jing'an District, Shanghai,China
中国上海市
静安区新闸路1093号

Telephone : +86 21 6318 1588

上海CitiGo酒店
Citigo Hotel, Shanghai, China

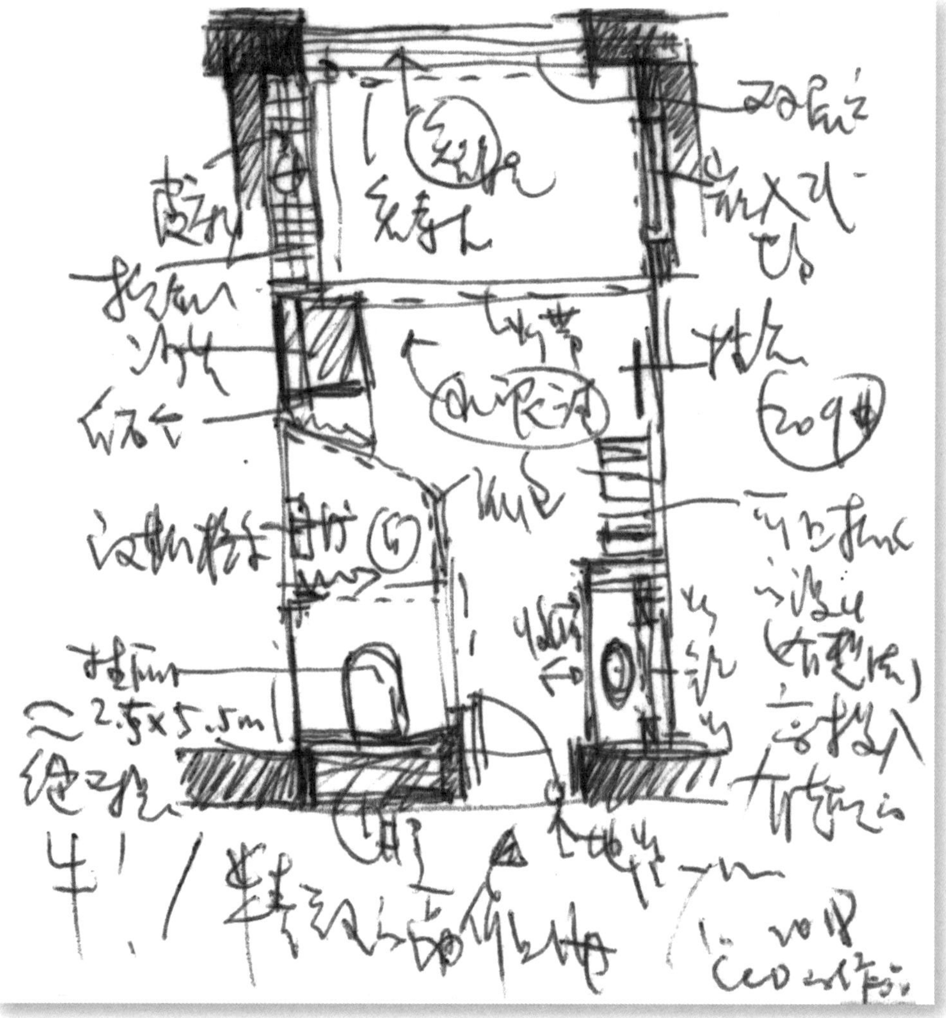

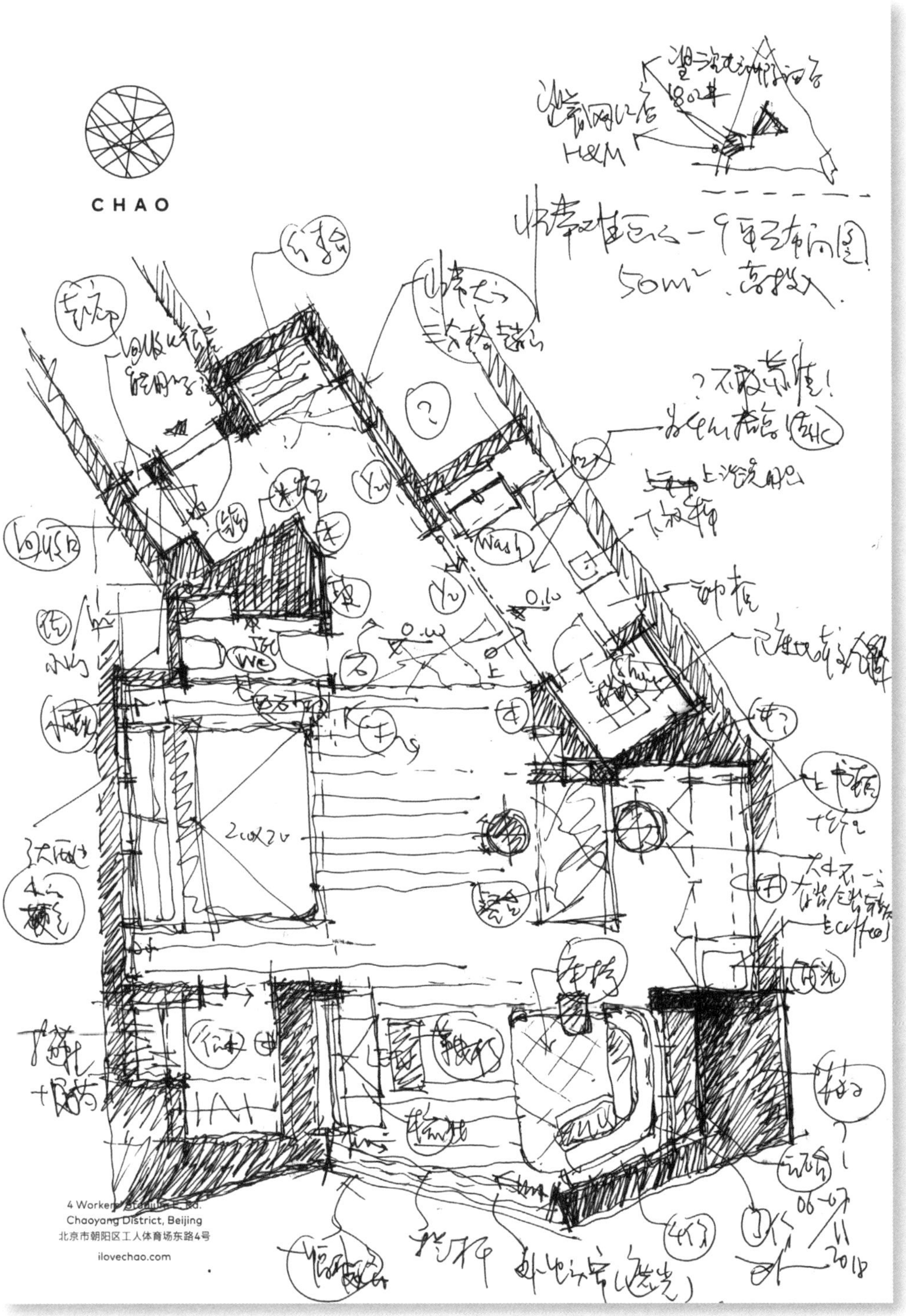

CHAO

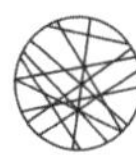

CHAO

潮酒店

52 ★★★★★

ChaO
BEIJING CHINA

Address : 4 Workers' Stadium E. Rd.
Chaoyang District, Beijing,China
中国北京市朝阳区
工人体育场东路4号

Telephone : +86 10 5871 5588

Http : //www. ilovechao.com

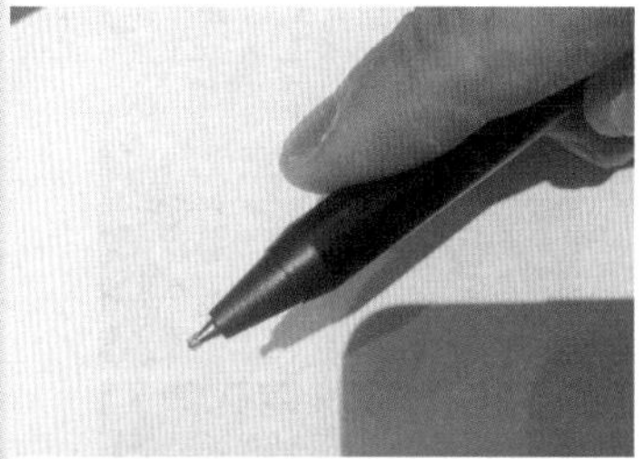

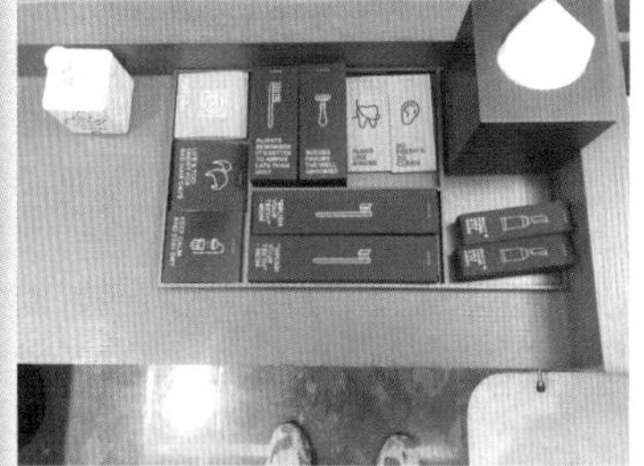

潮酒店

Chao, Beijing, China

灯光・潮酒店
Chao, Beijing, China

难得一见的"年轻"! 真难!

这次北京(东莞回)入住北京的"chao"潮之酒店. 据说是由在个酒店(不管他的)房间真之出乎. 三更八角的! 服了建筑师和室内设计师.

房间特大. 50m²左右吧. 算它在一个体方. 又有大大的卫生区、用餐区、整体的摆设思路. 更说打破常规设计师们、也许因为是"创意"—chao. 浴缸洗手台不接、是镜子的洗手盆还是多功能的"旅途"我会一样的所. 一样快的设计. (这在香港的套房的格局里成功地完整地有趣味. 过瘾啦! (特别像"起点")可参一览.

当然, 极大地颠覆的创作和酒店风格. 从灯光到家具. 它们都体现了设计感和个性. 他们胆也许才让我们行走在酒店设计之布局!

追求"潮"的酒店设计师真的是"永不停止"啊

一个有设计的细节. 房间的笔也是卷入本子的造型设计. 这是我的第一个用的. 真支持!

—— [签名] 6-7 / 11 2018

4 Workers' Stadium E. Rd.
Chaoyang District, Beijing
北京市朝阳区工人体育场东路4号
ilovechao.com

A Rare "CHAO" Plane

难得一见的"潮"平面

2018年11月6、7日

November 6-7, 2018

途经北京（去美国），入住北京的"Chao" 潮酒店，才知道标准层是一个三角形（不等边的）房间。真的是牛，三尖八角的！服了建筑师和室内设计师。

On the way to America, I stayed in "Chao" Hotel in Beijing. I just found the guestroom was so great because its standard floor was a scalene trigale. I admired the architect and the interior designer.

房间非常大，大概50平方米左右吧，有一个小厅，又有大大的茶水区、用餐区，整体可谓费尽思量，真的是折腾"死"设计师了。也许过度追求"创意"——Chao。洗手间厕浴分离不提，关键是洗手盆区与淋浴间"被迫"提高了一步台阶——极少见的设计（香港奕居的抬高区域比较完整和有理据，这里嘛！你来住就知道了），可谓一绝。

The guestroom was very big, about 50 square meters, I thought. There was a small living room but a big tea area and dining area. The designers really took great pains over the whole plan. It really tortured the designers. Maybe they chased creative—Chao excessively. I didn't want to mention the separated toilet and bathroom. The point was that the basin area and the bathroom had to be lifted to a step up—a rare design. (Once I stayed in the Upper House in Hong Kong. The raised area was reasonable and completed. But here, you would understand the feeling after staying.) "Chao" was quite unique.

潮酒店

Chao, Beijing, China

当然投入是相当的到位和淋漓尽致，从灯具到家具、立面都是相当有设计感和个性，"倒腾"也许才让我记住了这间酒店的平面布局！

Of course, the investment was really put in place and vividly from lamps to furniture and the facade. They were all well-design and individual. Creativing made me remember the plane of this hotel.

追求"潮"与舒适好用真的是"水火不相容"啊！

Chasing fashion and comfort would be totally in incompatible.

一个有趣的细节：房间的笔，还套着小小的塑料套子，应当我是第一个用的，"老克拉"！

An interesting detail, the pen in the guestroom was still covered with a plastic cap. I must be the first "old Carat" to write with it.

美国西海岸学习团（2018年11月7~17日）

The West America Learning Tour

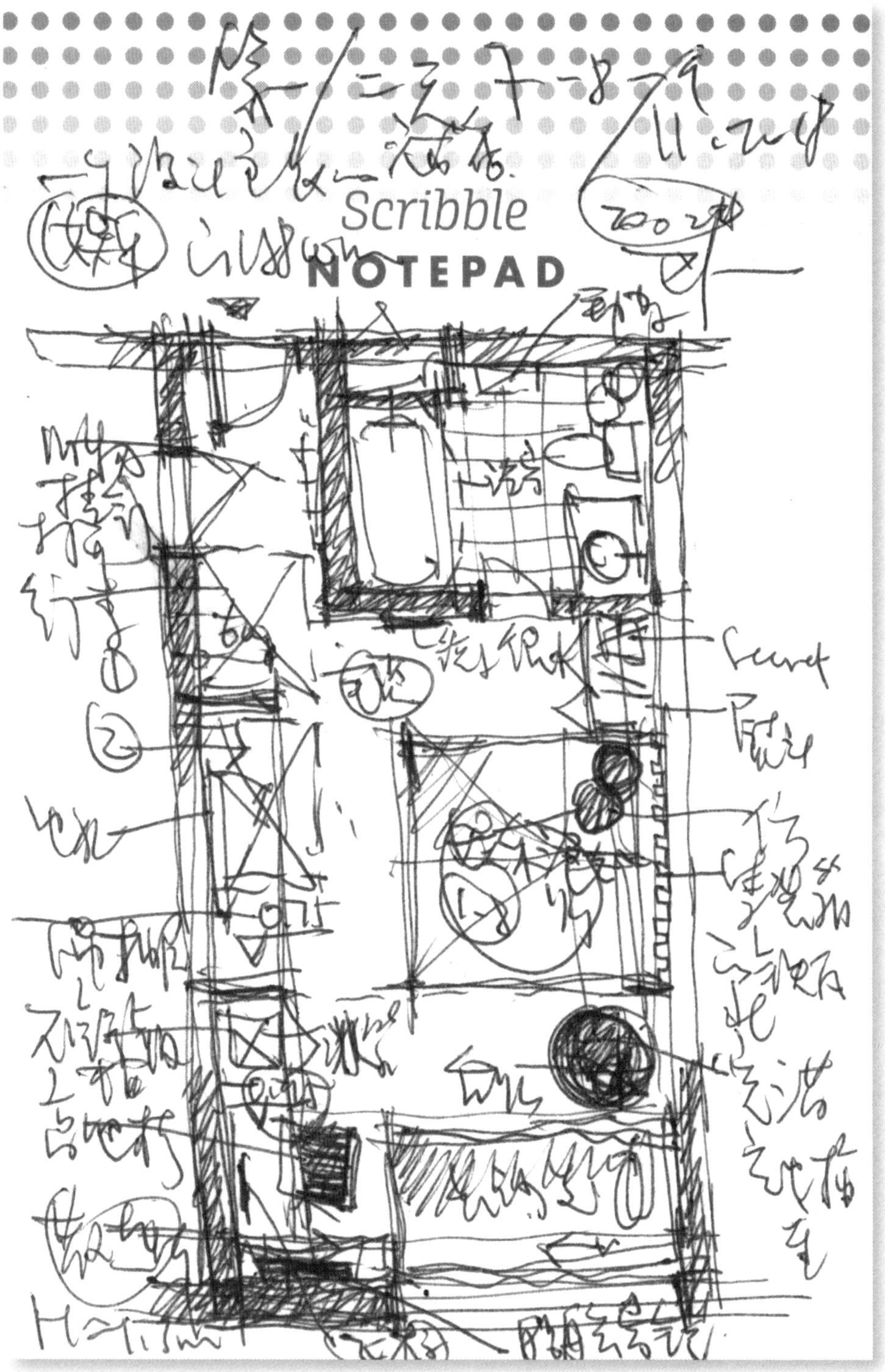

Scribble
NOTEPAD

旧金山和风酒店

53

★★★★

HOTEL ZEPHYR SAN FRANCISCO AMERICA

Address : 250 Beach Street, Fisherman's Wharf, San Francisco, CA 94133, America

Telephone : +415 617 6565

Fax : +415 986 7853

Http : //www.hotelzephyrsf.com

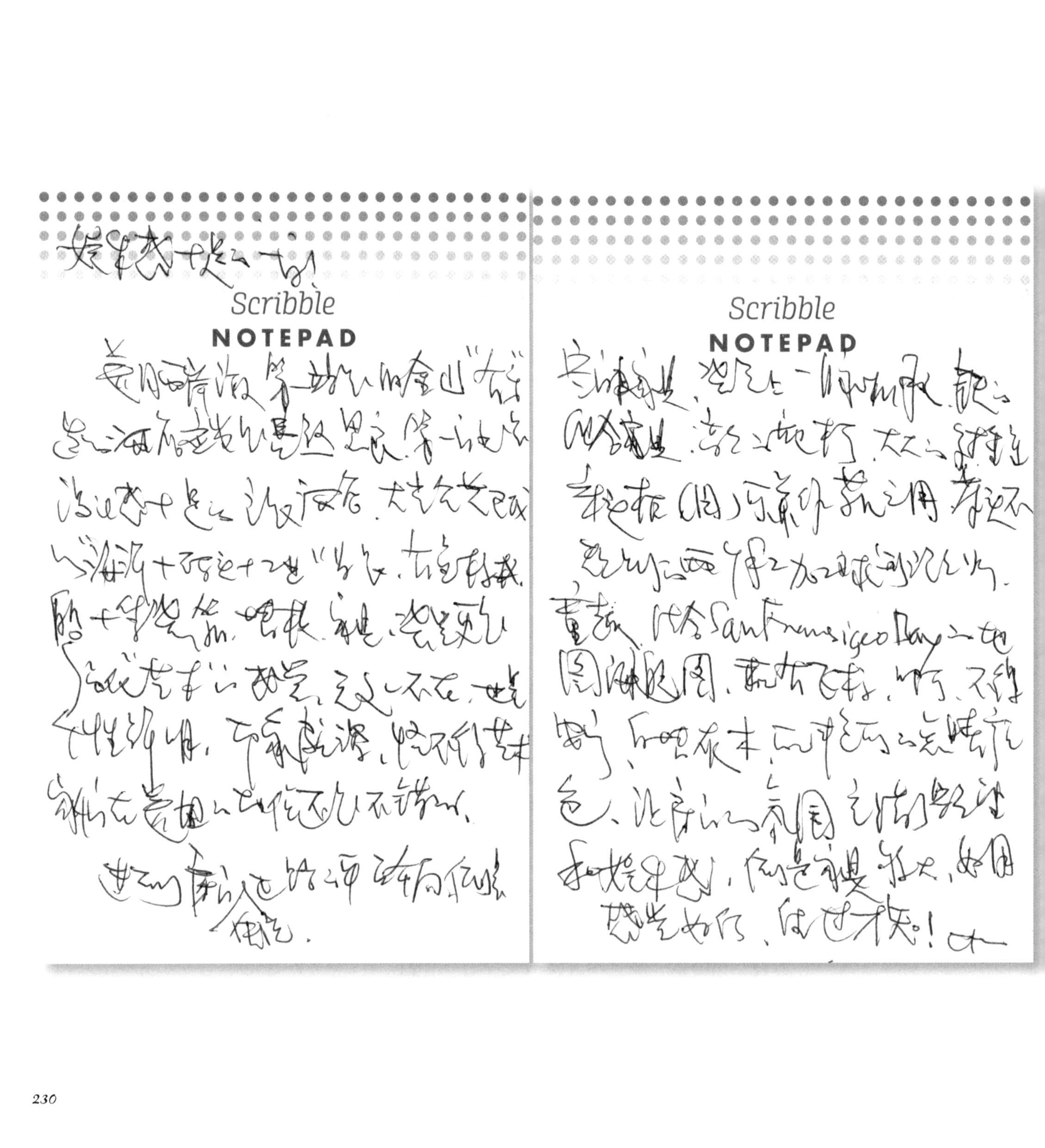

Scribble
NOTEPAD

Scribble
NOTEPAD

A Fun Hotel

多“Fun”酒店

2018年11月7、8日
November 7-8, 2018

美国西岸游的第一站是旧金山。“有方”选的酒店应当是费煞思量。第一间就是设计感十足的主题酒店，大堂公共区域的“海洋+ 码头+工业”为题，有鱼标枪+船+集装箱，墙板、家具、装饰更是强化“艺术”的感觉，无处不在，也是个性鲜明，印象颇深，怪不得艺术家们在美国的地位还是不错的。

The first stop of the West America Tour was San Francisco. "Position" must spend a lot of time choosing the hotels. The first one was a theme hotel full of entertainment. The public area in lobby showed the topic of "Ocean+Wharf+Industry". There were spear guns, ships and containers. The wallboard, furniture and decorations had a strong artistic feeling with bright individual. You could feel it everywhere and had a deep impression on it. No wonder the artists were in a high position in America.

进到房间传统正路的平面布局，倒是空间及家具、装置上一脉相承，铁的组合家具，高高的吧椅，大大的多插座床头柜（圆），可兼作茶几之用，床头不规则的两个手工加工球型泡泡灯，富有童趣，结合三藩市海湾的地图航海图，配有飞镖，啊！不得“安宁”。白墙灰木，配艳丽的可以点睛的颜色，让那个房间的氛围充满了娱乐感，倒是家具够大，好用。

The layout inside was the traditional style. The design of space and furniture came down in one continuous line, such as iron modular furniture, high bar stools, a big night table with multiple sockets which was also used as a tea table. Two irregular hand-made round bubbles above the bed were full of childlike. There was a sailing map of San Francisco Bay with darts. Oh! You would never stop to rest. The bright color was the highlight of white wall and grey wood. It made the whole room full of desire and entertainment. The furniture was quite large and easy to use.

感觉如何，住过才知！

How do you like the hotel? You won't know the feeling until you experience it yourself.

重写的现代——美西建筑

徐千禾

前言

> 然而，如果传统——传承的唯一形式仅是盲从或怯懦地复制前一代成功的方法，那“传统”断然早已失传。这样天真的潮流迷失在时代沙河里的情况屡见不鲜；新颖总是比重复要好。传统是一件具备更为广泛意义的事情。它不是固守就能得到传承的，你必须得付出巨大的努力……[1]
>
> ——T.S. 艾略特

好莱坞、硅谷、嬉皮士、牛仔、拉斯维加斯喧嚣的赌场……一望无际的沙漠、被风沙侵蚀的山体、昏黄的落日、一匹马与骑坐在马背上的人，或是一辆奔驰着的皮卡穿过寂静的小镇、广袤无垠的城市奔向无尽的远方……提起美国西部，脑袋里闪现出来的联想多少都带点“离经叛道”、一骑绝尘的味道。无数的电影作品已经将美国西部特有的自然条件和城市肌理深深地烙印在我们的脑海里。而谈到建筑，美西却不似东岸，始终无法用一个明确的形象来概括。在这片土地上发生的一切就像是在一张可重复书写的羊皮纸上不停描画，被抹去的依旧留下了可辨读的痕迹。

一万多年前，在这一片辽阔的土地上曾孕育出多个古老的文明，玛雅、托尔科特和阿兹特克等，在前哥伦布时期这里已经有着丰富的宗教和农业文化，玛雅建筑的烤砖构造和灰泥涂料（stucoo）的强烈视觉特征中隐藏的神秘感经常被运用在电影的视觉艺术中，《银翼杀手》（*Blade Runner*）中泰勒公司的总部外形就是一个放大的玛雅金字塔，耸立在电影里 2019 年阴雨绵绵的洛杉矶，为复制人最终“弑父”的场景增添了一层

希尔顿美洲休斯敦酒

Hilton Americas Houston, Ameri

Day 1 November 7, American Time

第一天美国时间 11月7日

到达旧金山的时间是当地12点多，多折腾。出关抽检行李。惨，被问了很多问题，练口语……

It was over 12 P.M.(local time) when arriving in San Francisco. It was so frustrating that my luggage was chosen to check at customs. So terrible! I was also asked a lot of questions. I have to practise my oral English……

第一个参观的项目是一个有特殊结构的混凝土结构的教堂（圣母升天主教堂），到处都是“十字”——十字结构的平面、十字的采光彩色玻璃、吊着十字架、斜切的石栏杆……一切都让你向着主。当然高的投入，如选址（阳光灿烂）、木材、地砖……让我们看到了追求精细的第一件建筑，不错的“开局”。

The first building we visited was a church with a special concrete structure—Cathedral of Saint Mary of the Assumption. Cross was everywhere—the crossed plane, the crossed lighting stained glass, the hanging Cross, the beveled stone balustrades…All led you to the Lord Jesus. Of course, the investment was high, the location (in the sun), as well as the wood and tiles. It was a good beginning that we saw this magnificent building.

圣母升天主教堂

Cathedral of Saint Mary of the Assumption

看似简单的建筑形体，产生戏剧般的空间和光影，仪式感更是非常之到位。

The shape of the building looked simple but made the dramatic space and light. It was full of ritual sense.

第二个参观的建筑是笛洋美术馆，全铜的不同图案的拼板外墙，灵动的平面布局，极致追求的“裂缝”！

The second building was de Young Museum. Its all-copper outer wall was made up of different patterns, a vivid plane and the abnormal “cracks”!

广场地面故意模仿地面裂缝，独立的观光台（塔），展品也有不错的，如达利的几幅作品——小福利，看了对面的皮亚诺设计的“科学中心”，只看看外观，是早期的生态建筑，棒棒的！

The floor of the square was like cracks intentionally. Independent sightseeing tower and the exhibits were all amazing, including a few Dali's paintings—a small bonus. Across the museum, we visited the Science Center which was designed by Renzo Piano, but just looked outside. It was an early ecological building. Great job!

笛洋美术馆

de Young Museum

追求一条“缝”，相信不容易。

I'm sure it was not easy to pursue a “crack”.

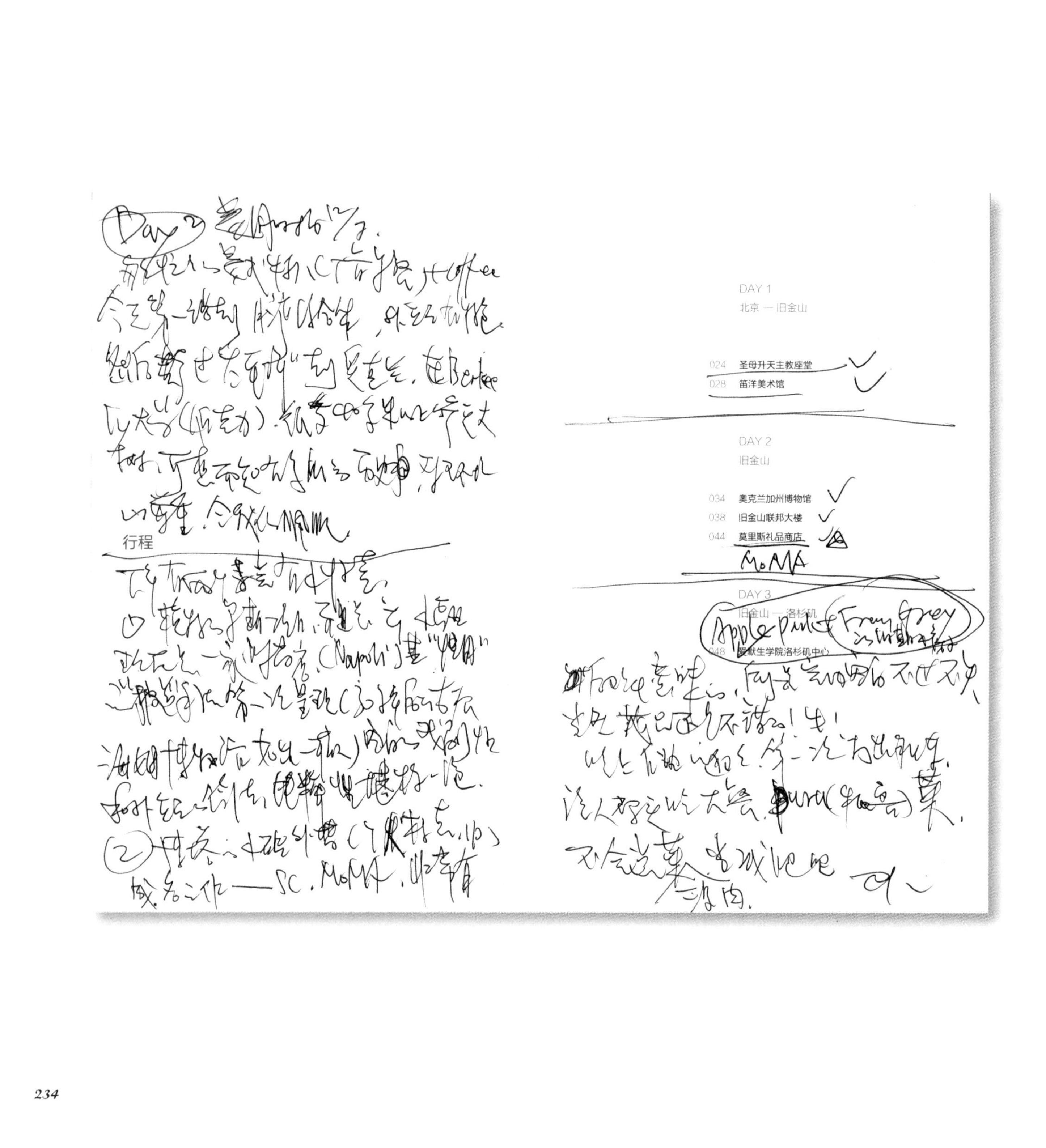

行程

DAY 1

北京 — 旧金山

024 圣母升天主教座堂

028 笛洋美术馆

DAY 2

旧金山

034 奥克兰加州博物馆

038 旧金山联邦大楼

044 莫里斯礼品商店

DAY 3

旧金山 — 洛杉矶

048 爱默生学院洛杉矶中心

Day 2 November 8, American Time

第二天美国时间 11月8日

匆匆忙忙的美式牛扒（T骨扒）+coffee。

A hurried breakfast—T bone steak and coffee.

今天第一站去了月湾综合馆,外立面有特色。然后跨过太平洋去了奥克兰，在伯克利大学，有很多40多米以上的参天大树，可想而知有多久的历史，对环境的尊重，令我们佩服。

The first stop was the Moon Bay gallary which had a distinctive facade. And then we crossed the Pacific to Oakland. There were lots of over 40-metre-tall trees in Berkeley. Such a long history and the respect for the environment made us admire.

（T骨扒）+coffee/ 伯克利大学参天大树

T bone steak and coffee/ Tall trees in Berkeley

下午有两个景点，有小惊喜。

In the afternoon, we visited two places of interest and felt a little surprised.

1.赖特的早期项目，是一个小商业，现在是一家时尚店（Napoli）。其“惯用”的“坡道”手法，第一次呈现（与30多年后的古根海姆博物馆如出一辙），空间的戏剧性和外立面的简洁、纯粹性堪称一绝。

1. Frank Lloyd Wright's early work was a business, now it was a fashion shop (Napoli)with his usual "slope" skill. Here was his first design. (the same as the Guggenheim Museum which was built after more than thirty years.) The dramatic space and the simple and pure facade were absolutely unique.

赖特早期项目-Napoli *T bone steak and coffee*

很喜欢的一个赖特的作品-*It was my favorite work of Wright's.*

2. 博塔的小砖外墙（个人标志，IP），成名之作——SC.MOMA，非常有后现代意味，倒是室内空间无过无失，当然藏品还是不错的，牛！

2. Mario Botta's brick outer wall (Personal logo, IP). His masterpiece was SC.MOMA with quite postmodern feeling. The interior was not bad and the collections were quite good. Great!

晚上自由地逛逛，第一次打出租车，在渔人码头吃大餐，秘鲁菜，不会点菜，没肉，当减肥吧。

We walked around freely in the evening. It was my first time to take a taxi and then have a big meal —the Peruvian food at Fisherman's Wharf. Ididn't know how to take an order, only ate some vegetables as a diet.

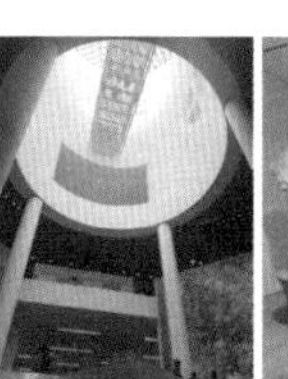

博塔设计的MOMA

Mario Botta's—SC.MOMA

红砖外墙的特色，博物馆最喜欢的展品就是：蜘蛛。

The facade was red brick. My favorite exhibit in the museum was the spider.

DAY 3 舊金山 9/1 2018，從舊金山→洛杉磯

[illegible]早餐，舊金山[illegible]，在舊金山真不錯！

一如往常早餐後，去Apple Park看它的外觀，它的建築很美，也可以看到它的[illegible]設計！

蘋果 設計 很不錯

飛往洛杉磯，居然行李都丟了，要辦手續，[illegible]，只好[illegible]。

下午到住在Hollywood + Sunset 的西[illegible]酒店，[illegible]一個小時[illegible]也要[illegible]，外在[illegible]，[illegible]（外場的建築）[illegible]。

[illegible]

傍晚，去到Frene Grey的[illegible]，[illegible]

Day 3 November 9
(From San Francisco to Los Angeles) American Time

第三天美国11月9日（旧金山—洛杉矶）

Apple Park

苹果总部接待中心，精彩，AR模型区，有未来感。

The reception center of Apple Headquarters was wonderful. The AR model area was futuristic.

照旧吃的牛扒早餐，等了有半个小时，在美国急不得！

Steak was for my breakfast as usual. I had waited for half an hour patiently. There was no rush in America.

一小时的车程，去Apple Park 看看外观。远观主建筑体，也可谓向乔布斯致敬！

We took an hour-drive to have a look at the appearance of Apple Park. Watching the main building from a distance was also to salute to Steve Jobs!

苹果确实与众不同！

Apple was quite special!

飞洛杉矶，居然行李超重，要部分手提，在机场吃饭更是费时，不得不跑步上飞机。

I took a flight to Los Angeles. My luggage was even overweight so I had to carry some of them by hand. It took so much time to have meals at the airport that we ran to get on the plane.

下午到达Hollywood+日落大道，堵车近一个小时到达爱默生学院。外立面建筑空间十分别致，对光影、新技术（外墙电磁板）的应用更是榜样。

We reached Hollywood and Sunset Strip in the afternoon. It took nearly an hour to get to Emerson College because of the traffic jam. The facade of the building was quite unique. The light and shadow as well as new technology (outer wall's electromagnetic board) were a good example.

傍晚，去看了弗兰克·盖里的早期作品“小房子”，看到了他年轻时的“嫩”，谁没有！

In the evening, we visited a small house which was Frank Gehry's early work. I saw the immaturity in his youth. Who had never been young?

洛杉矶比尔特摩尔千禧酒店

54

★★★★

MILLENNIUM BILTMORE HOTEL LOS ANGELES AMERICA

Address : 506 S. Grand Ave.,
Los Angeles,
California 90071, America

Telephone : +1 213 624 1011

Http : //www.MillenniumHotels.com

MILLENNIUM
BILTMORE HOTEL
LOS ANGELES
You are the Center of Our World

506 S. Grand Ave., Los Angeles, California 90071 T +1 213.624.1011
www.MillenniumHotels.com

A Member of Millennium Hotels and Resorts

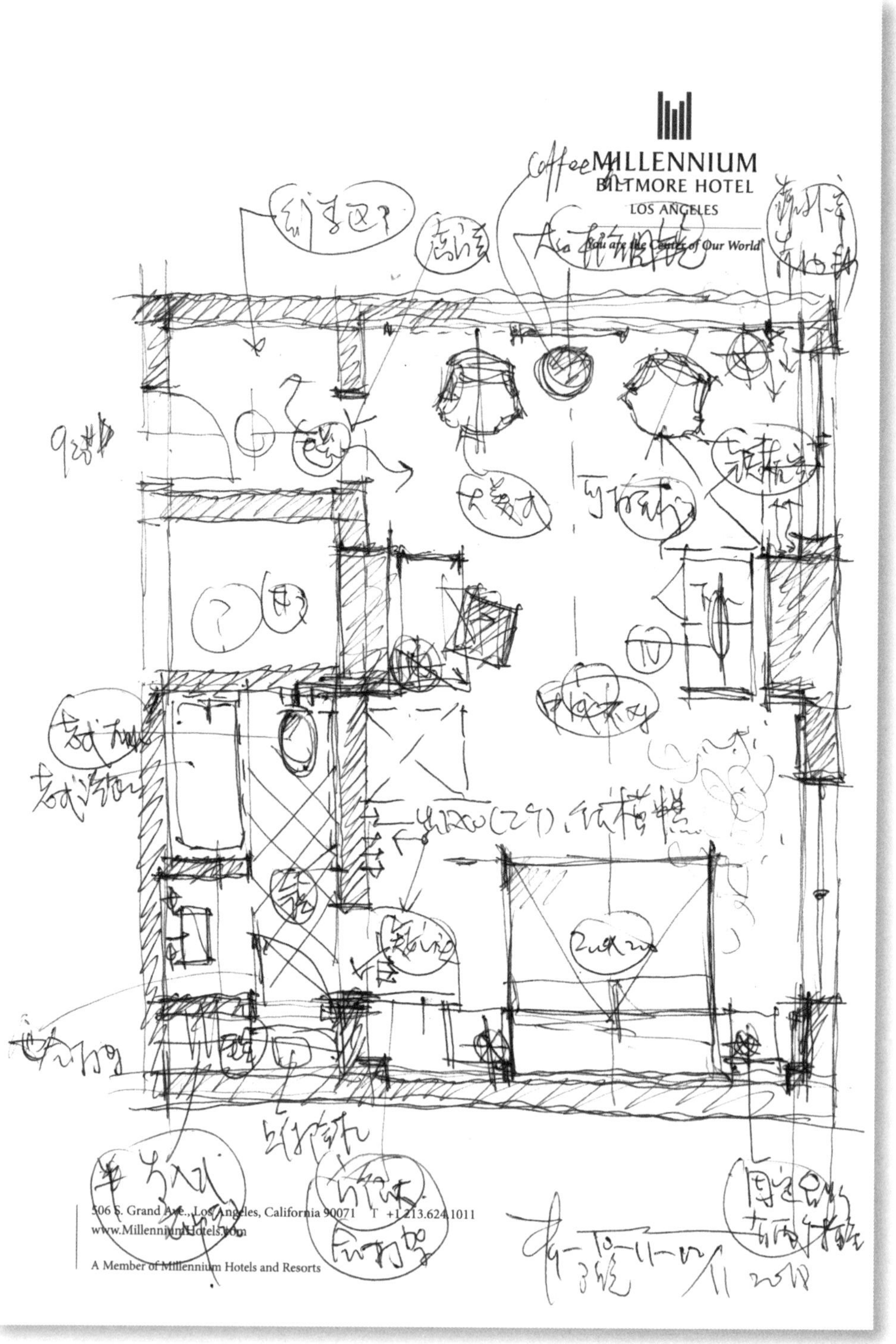

O.'s File

（区生词典）

比尔特摩尔千禧酒店位于洛杉矶，是家4星级酒店。酒店拥有19世纪的典雅奢华装饰，是洛杉矶市中心最具地位的豪华酒店，无与伦比的历史性地标性建筑。奥斯卡奖杯的设计者Cedric Gibbons就是在酒店用餐时突发灵感，设计了奥斯卡奖杯，这间宴会厅是无数电影的拍摄地及历史活动的举办地，如《真实的谎言》《独立日》等，现代的舒适和精良的设施使比尔特摩尔千禧酒店成为理想的社交活动、婚礼和会议场地。（百度百科）

Millennium Biltmore Hotel is a four-star hotel located in Los Angeles. With the 19th century elegance and luxury decoration, the hotel is one of the most prestigious luxury hotels in downtown. It is an unparalleled historic building. Cedric Gibbons got his inspiration to design the Oscar trophy while having dinner at the hotel. The banquet hall is also thesite of filming location and historical events, such as "True Lies""Independent Day" and so on. The comfortable and excellent facilities make the hotel an ideal venue of social activities, wedding parties and conferences. (From Baidu Encyclopedia)

Day 4 Millennium Biltmore Los Angeles

第四天比尔特摩尔千禧酒店

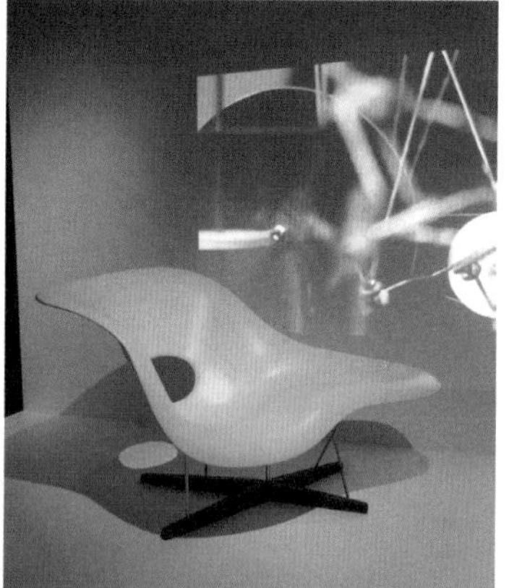

EAMES的家具

幸运地看到了EAMES的家具藏品展览，不枉此行。

I saw the EAMES Furniture Collection Exhibition luckily, it deserved the trip.

95年前建成的酒店，古典，浓重，丰满，是我喜欢的样式，想起当年手绘的光景。

The hotel was ninety-five years old. It was classic, strong and full which was my favourite style.It reminded me about the time of hand drawn.

迪士尼音乐厅，弗兰克・盖里的大作，“无曲不成欢”。山上面向太平洋的EAMES自宅,我个人觉得一般般，当然有87年（1932~2018年）的历史，超前的布局和家具摆设，还是非常赞的。更牛的是1927年的Nentra研究住宅二号，令人叹为观止。多层次、多空间、多园景极品之作，不枉此行。辛德勒自宅工作室（曾经在赖特事务所工作过），比赖特的更加矮的尺度（不理解的尺度），有趣，非常态的“四人同居”，意识和行为的艺术建筑师。

Disney concert hall was Frank Gehry's masterpiece. "No joy without curving or music". EAMES private residence was on the hill and faced the Pacific. But I found it was just so so though it had a 87year history, from 1932 to 2018. Both advanced layout and furnishings were great. Nentra's Research House II which was built in 1927 was much cooler. Totally stunning! It was the perfect one with multi-level space and more gardens. What a worthwhile trip! Schindler House(Schindler's scale was shorter than Wright's and once worked for him)was interesting. That was a sick act that "four people live together". Schindler was a consciousness and behavior artisticarchitect.

反而赖特在比华利上的小商业，我认为很一般。

I reckoned that Wright's business on Beverly Hills was not so good.

平静而充实的一天！

What a peaceful and rich day!

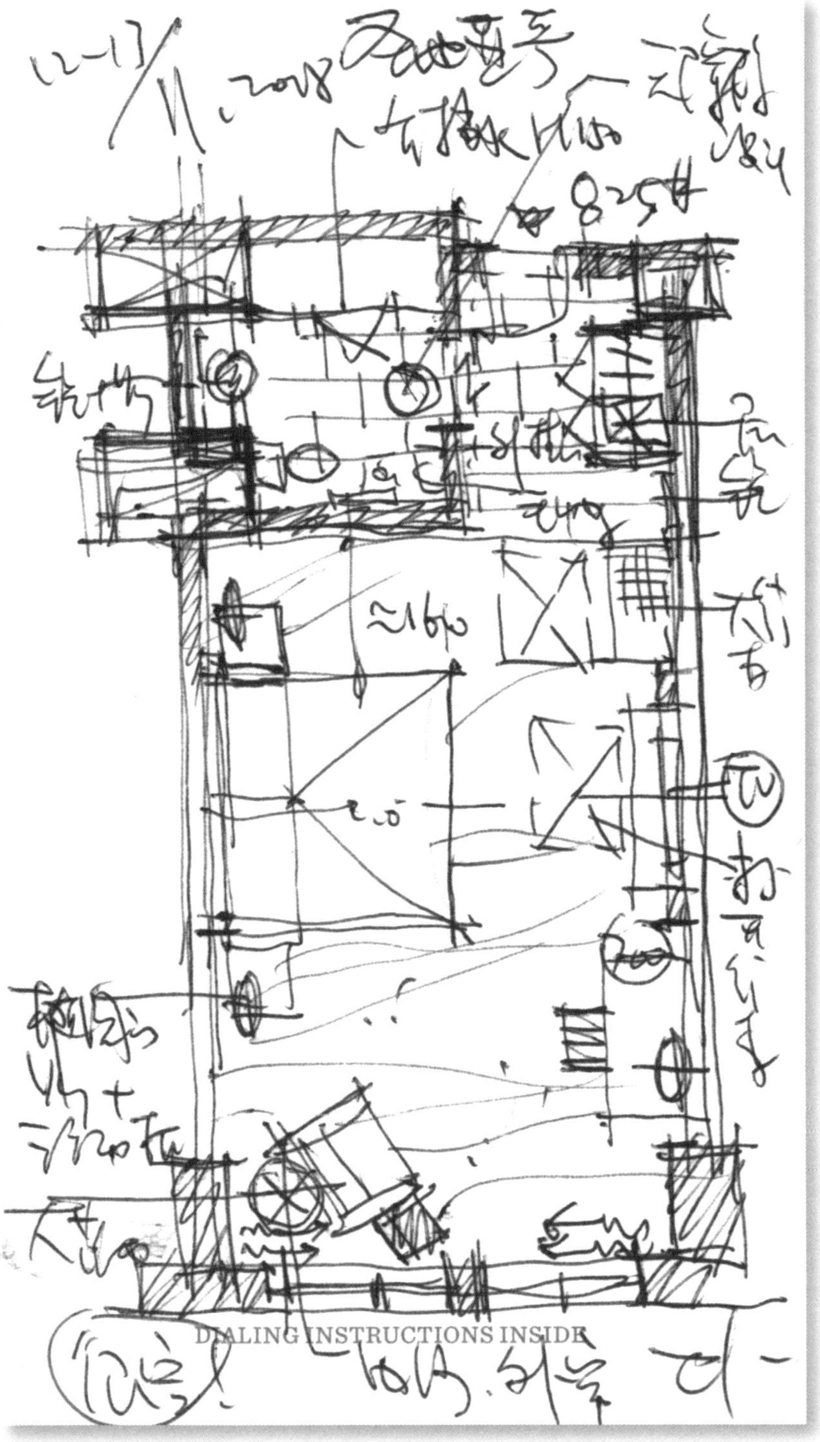

55 阿文蒂诺凯悦拉霍亚酒店

★★★★

HYATT REGENCY LA JOLLA AMERICA

Address : 3777 La Jolla Village Dr, San Diego, CA 92122

Telephone : +1 858 552 1234

Http : //www.hyatt.com

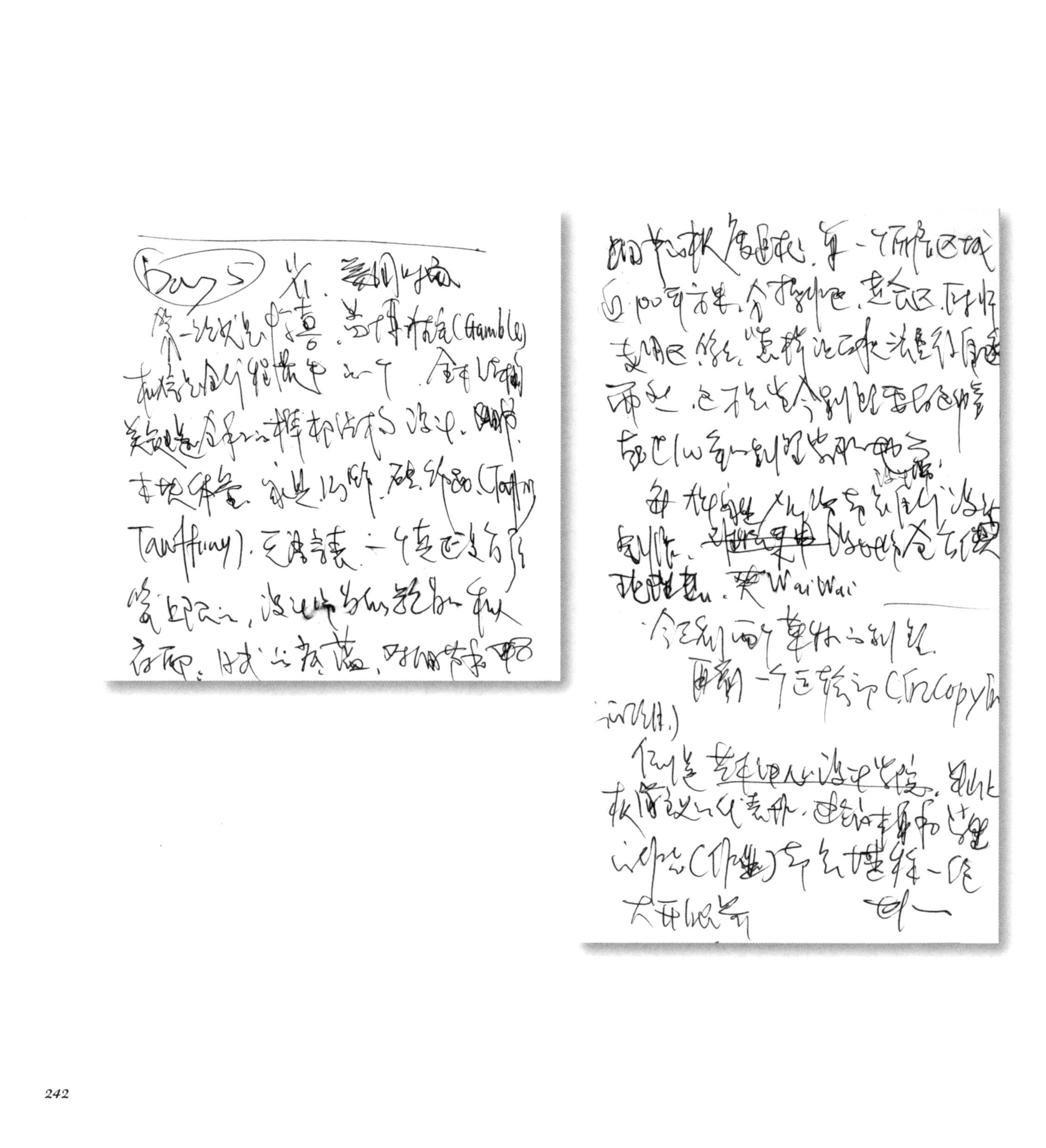

Day 5 November 11, American Time

第五天美国时间 11月11日

第一站就是惊喜，盖博住宅（Gamble），相信是全行程最牛的一个，全木结构，关键是全手工的榫卯结构设计，细节，木块的体量，家具，灯饰，砖，饰品（蒂芙尼）……无以言表，一个真正没有预算上限的设计师“为所欲为”的私人府邸：日式的底蕴，对细节和平面细节的极度追求，单一个厨房区域就近100平方米，分为操作区、煮食区、厨师专用区，等等。“怎样让面积浪费得自然而然，这才是当今别墅要向这幢超过100年的别墅学习的地方。”

The first stop, Gamble House, gave me surprise. I believed it was the most superior building during this journey. Gamble House was made of wood and the point was mortise and tenon joint structures which were all made by hand. The details, block size,the mass of wood, furniture, lightings, bricks and Tiffany decorations were so perfect that I couldn't use words to describe.The endless budget let the designers do whatever they wanted—Japanese background, the extreme pursue of details and graphic minutiae. Only the kitchen was nearly 100 square kilometers. It was divided into different zones, including operating, cooking and Chef's. "How to waste the space naturally was what we should learn from this over 100-year villa."

每件家具、灯饰都是设计师自行设计制作，设计师全方位实现理想，爽歪歪。

Every piece of furniture and lights were all designed and manufactured by the designer. The designer achieved his dream completely. How fantastic!

今天看到了两个赖特的别墅，还看了一个运输部（自己复制自己的项目）。

We visited two villas made by Wright. And then a transportation department (a duplicate project of their own).

倒是艺术中心设计学院，半山上极简主义的代表作，建筑本身和学生的作品（作业）都堪称一绝，让人大开眼界。

The Art and Design Institute was at the mid-level of the mountain. It was a master piece of minimalism. The building and the students' works were so phenomenal that they opened my eyes.

OMNI HOTELS & RESORTS
fort worth

56 沃斯堡奥姆尼酒店

★★★★

OMNI HOTELS&RESORTS FORT WORTH AMERICA

Address : 3777 La Jolla Village Dr, San Diego, CA 921221300 Houston St, Fort Worth, America TX 76102

Telephone : +1 817 535 6664

Http : //www.omnihotels.com

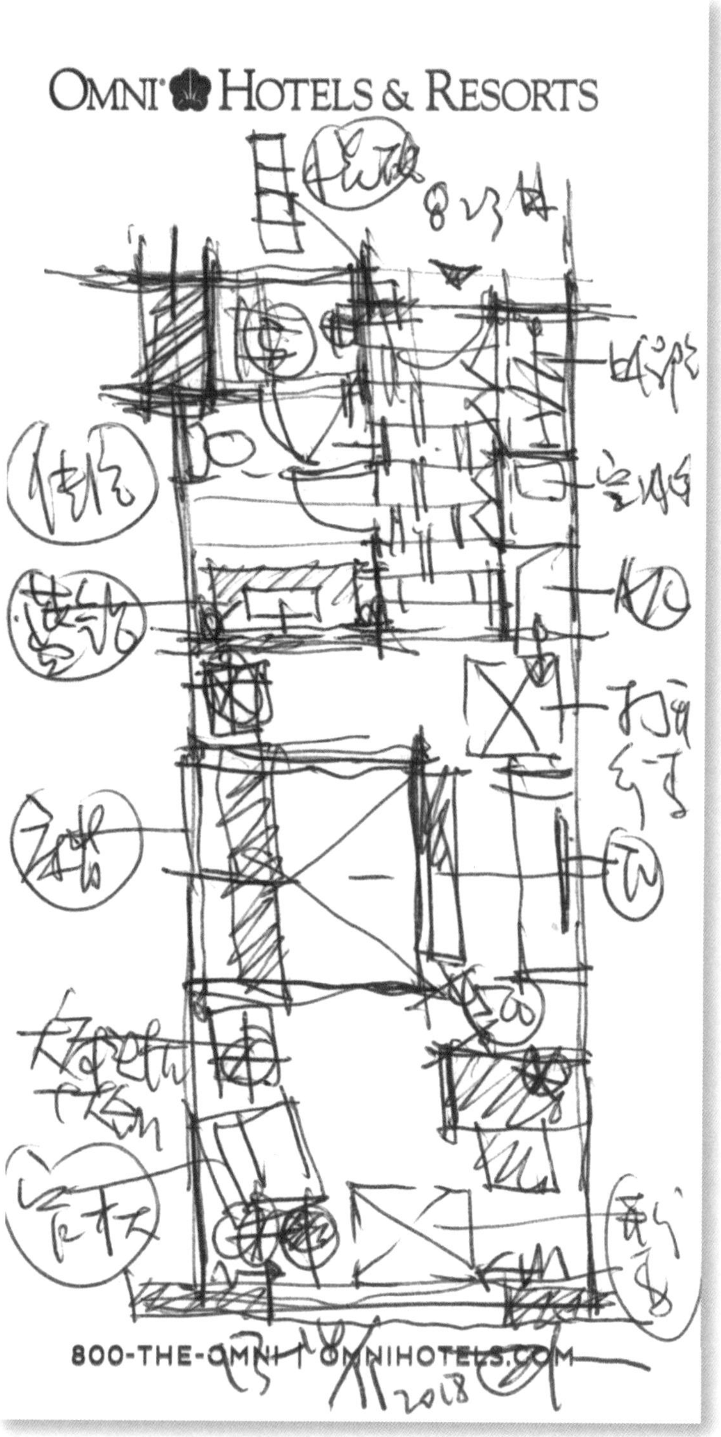

OMNI HOTELS & RESORTS

800-THE-OMNI | OMNIHOTELS.COM

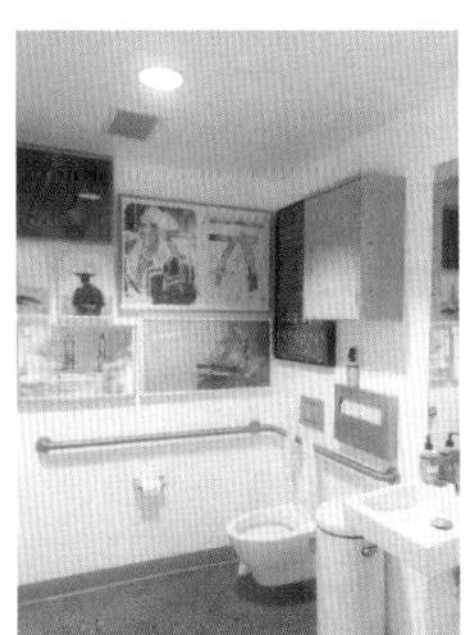

Morphisis建筑事务所

与一位日裔的建筑师交流可以练习练习口语，这一家公司的环境真不错，作品也非常出色：Morphisis。

I talked with a Japanese architect and practised my spoken language. The environment of this company was really perfect as well as its work: Morphisis.

Day 6 November 12, 2018 American Time

第六天美国时间 2018年11月12日

非常充实完美的一天行程。上午拜访Morphisis建筑事务所，牛！

What a wonderful Day! In the morning, we had a visit to an architecture firm—Morphisis. Impressive!

“0”耗能的孤立式庭院办公室，一层半，有花园，私家太阳能停车场，吸引我们的东西实在太多了，艺术品一样的各式模型，实验室一般的制作车间，先进的顶级3D打印设备，琳琅满目的叠积如山的模型，有工业风的铝板台面会议台，洗手间也是绝，厕所内有各式作品挂画及员工签名活动照片。当然项目的品质才是这家公司的杀手锏，从体到壳（蛋）+(XY)+Z，再到用“挖”的理念去搭建建筑创意思想体系——无往而不利！

The office was a zero-energy isolated courtyard with one and a half stories, a garden and a private solar parking lot. So many things interested us. Different models were like art works and the manufacturing workshop was like a lab as well. There were also advanced top 3D printing devices, the variety models piled up like a mountain, a conference table was aluminum plate with industrial style. The restroom was amazeballs because lots of paintings and photos with staffs' signature were hung on the wall. Of course the master skill of this company was good quality. From shapes to eggshell, XY and Z, till setting up the architecture creative system by the concept of excavation—it would be successful in doing anything!

参观一间艺术学院（放假），人员稀少，自由逛逛，几幢“砼”（混凝土）建筑，光影不错！

And then we visited an art institute. There were only a few people because of the holiday. We walked around freely and saw several concrete buildings. Light and shadow were perfect!

去圣地亚哥的路上——看到半山上面向太平洋的赖特的孙子设计的玻璃小教堂，心思很棒，细嚼慢咽，回味无穷。

On the way to San Diego, there was a small glass church on the mid-level hill, facing the Pacific. It was designed by Wright's grandson.The idea was so perfect that I watched the church carefully and thought a lot. It was very memorable.

再一组教堂，全玻璃幕与迈耶的建筑，再与现代主义的早期教堂组合，互相辉映，富有乐趣（更有趣的是：据说是基督教教会破产了，卖给天主教的项目，让我大开眼界。）

Another group of churches had a full glass façade and Meier's building. It was harmonious and interesting to combine with an early modernism church. (Something more interesting was that the Christian Church was sold to Catholicism because of bankrupt. It really opened my eyes.)

继续旅程的路上看了小建筑，墨西哥大餐，难食，分量大；

乖乖地到圣地亚哥，凯悦丽晶（酒店），后现代代表作，格雷夫斯的作品。

We continued to visit a small building and then had a big Mexican meal. It tasted terrible and too much for me.

Then we arrived in San Diego obediently. After checking in at Hyatt Regency Hotel, we went to visit Michael Graves' masterpiece of the Post-Modernism.

Day 7 美国加州 17/11/2018

今天是[illegible] Louis Kan[illegible]的项目[illegible]
"[illegible]"，[illegible]十[illegible]"。[illegible]
[illegible]，[illegible]。[illegible]（[illegible]
[illegible]（[illegible]）[illegible]（Salk Institute [illegible]
[illegible]）[illegible]（[illegible]
[illegible]"唯一"的[illegible]。

[illegible]
[illegible]（[illegible]）[illegible]一[illegible]、
[illegible]"[illegible]"（[illegible]）[illegible]
[illegible]，[illegible]
[illegible]！

下午飞去[illegible]，[illegible]大[illegible]

[illegible]！

[illegible]一[illegible]。[illegible]牛[illegible]
[illegible]一下[illegible]
[illegible]牛[illegible]
[illegible] Uncle Jesse
[illegible]牛[illegible]！
[illegible]！

~~[illegible]（[illegible]）~~ [illegible]
~~[illegible]~~
[illegible]！！[illegible]

Day 7 November 13, 2018 American Time

第七天美国时间 2018年11月13日

今天只有路易斯·康的项目值得一提，“面朝大海，春暖+花开”，选址还是相当重要的，全混凝土是主体，地上四层，使用隔层（夹层），设备（Salk Institute /萨尔克生物研究所实验室）相当有前瞻性（我认为是康的“唯一”的杰作）。

Today, only Louis Kahn's project deserved mention. "Face the sea in warm spring and flowers come out."Location was quite important. The main building was all concrete with four stories. The building was interlayer using and the equipment (which was in the Salk Institute Lab) was prospective. (I considered it was the only masterpiece of Kahn's.)

路易斯·康萨尔克生物研究院
我认为是“康”最值得看的项目。
I think this Louis Kahn's project was the most worth visiting.

未经证实的小插曲：景观设计师在半围合的广场（庭园）不植一物，坚持空白给远方的“海”（太平洋）。真是英明，为创造了最美的拍摄地立下伟绩！

An unproven episode was that the landscape designer had never grown any plants on the half-closed square. He insisted that the space was for the faraway "sea" (the Pacific). How brilliant! He made great contributions to create the most beautiful place for taking photos!

下午飞达拉斯，在房间叫餐，大大的分量。

但一般，好像真的是除了牛扒还可以吃一下，其他都不行。

真怀念芝加哥的厚的炭烧烤牛扒和纽约第五大道边上Uncle Jack的35天熟成牛肉！

洗洗，准备休息了！

We flew to Dallas in the afternoon. I called room service and the meal was too much for me. But it tasted so so. It seemed nothing was good except steak.

I always thought about the thick charcoal grilled steak in Chicago and the 35 days' air-dried beef at Uncle Jack on New York Fifth Avenue.

I took a shower and rest!

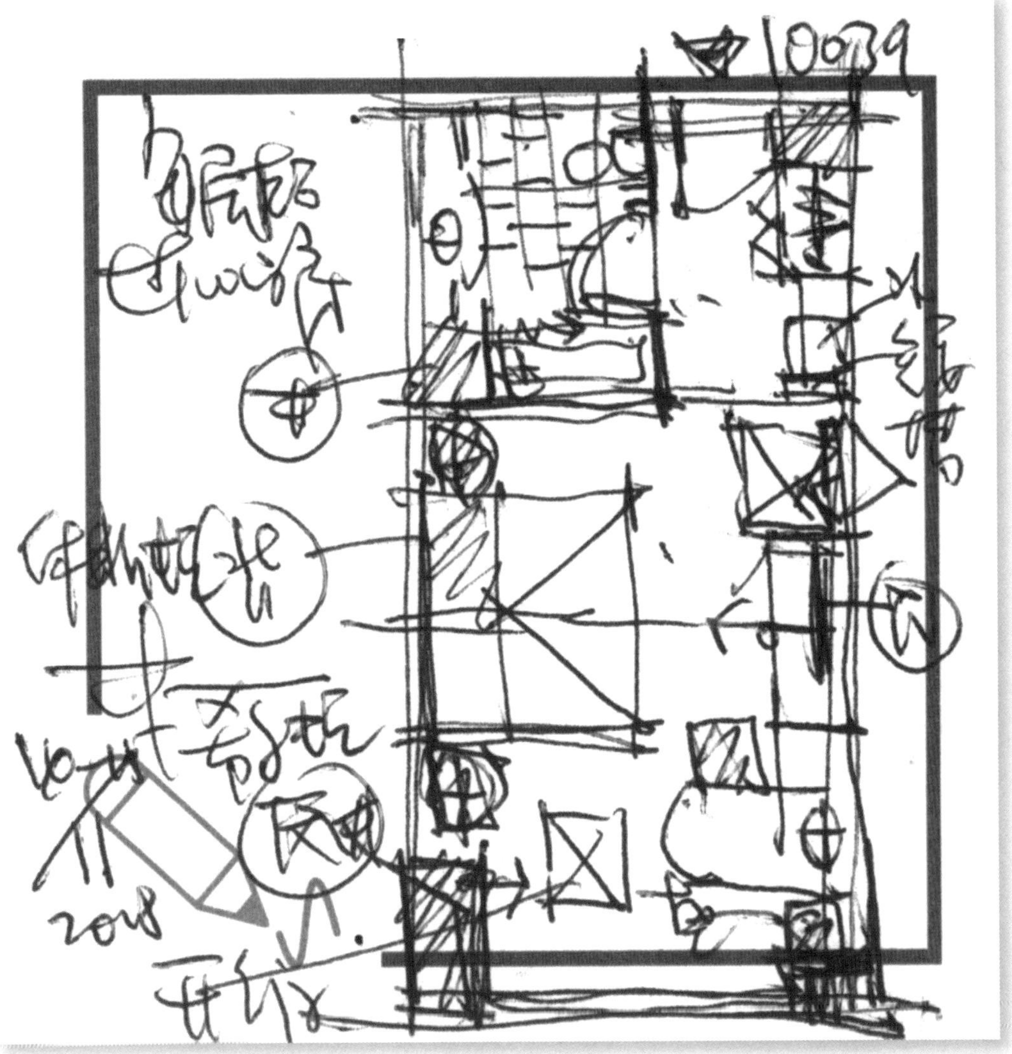

57 希尔顿美洲休斯敦酒店

★★★★

HILTON AMERICAS HOUSTON AMERICA

Address : 1600 Lamar St,
Houston, TX 77010
Telephone : +1 713 739 8000
Http : //www.hilton.com

LAURIE SIMMONS
BIG
CAMERA
LITTLE
CAMERA

October 14, 2018–January 27, 2019
MODERN ART MUSEUM OF FORT WORTH

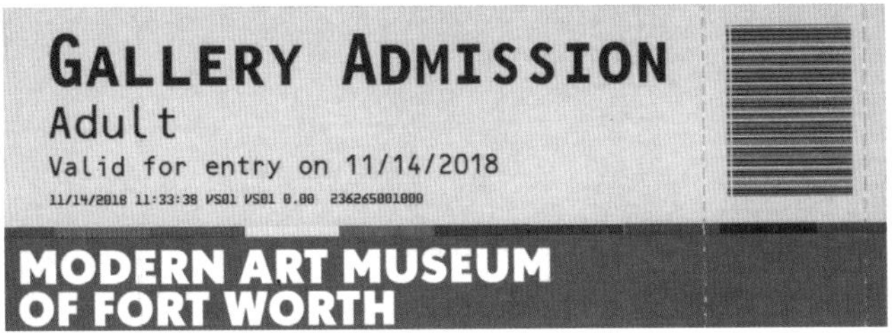

GALLERY ADMISSION
Adult
Valid for entry on 11/14/2018
11/14/2018 11:33:38 VS01 VS01 0.00 236265001000

Welcome!
Nov 8 2018
EAMES ADD ON $4
OAKLAND MUSEUM OF CA
The museum of us.
PEEL HERE

Day 8 [illegible] 16/11.2017

[illegible] "[illegible]" [illegible]

[illegible] "Water Gardens" [illegible]

[illegible]

[illegible] Louis Kahn's Kimbell [illegible]

[illegible]

[illegible] piano [illegible]

[illegible]

Day 8 the last day but one, November 16, 2018 American Time

第八天倒数第二天 美国时间 2018年11月16日

一早，吃昨天晚上的“客房服务”剩下的一点东西，匆匆忙忙去酒店旁边的“Water Gardens”走走，下沉的多层级水池，“四水归堂”的感觉，可以逐级而下，滑，危险，得匍匐前进（据说曾淹死了几个人），设计师的非人性化设计害死人了！

In the early morning, I finished the food which was left last night. Walked around the "Water Gardens" beside the Hotel hurriedly. The sunken pond with multi layers was similar to"Four Water to Church". Travellers stepped down but had to crawl because it was too slippery and dangerous. (It was said that a few people had been drowned.) The designer's inhumanized design killed people.

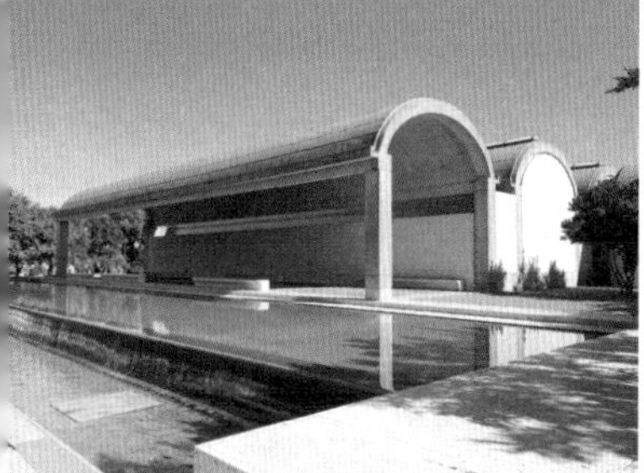

金伯莉美术馆

佩服这么精巧的“薄壳”结构，与室内的功能需求“完美组合”,真的是领先。

I admired such a skillful thin-shell structure. It combined perfectly with the indoor function and indeed led the way.

继续参观路易斯·康的金伯莉美术馆，细腻，“玩”光、影，是结构及构造的“小把戏”，经典；藏品不错。

We went on to visit Louis Kahn's Kimberly (Kimberly Gallery). The delicate light and shadow as well as the trick of the structure and construction made it classic. The collections were awesome.

三馆齐下——金伯莉的扩建部分，皮亚诺的近作，附近更加有安藤忠雄的现代艺术馆美术馆，尽兴。

Three museums got together—the extension of Kimberly Gallery,Renzo Piano's latest work and the Modern Art Museum of Tadao Ando. We really had a great time.

4个小时的车程到达最后一站休斯敦，回房间，整理，看看股市，通过微信工作一会。

Arrived in the last station Houston after four hours' drive.Checked in, cleaned up, looked at the stock market and worked on We chat for a while.

洗洗，睡！

Took a shower and slept!

新西兰之旅（2018年11月20~28日）

The Trip to New Zealand

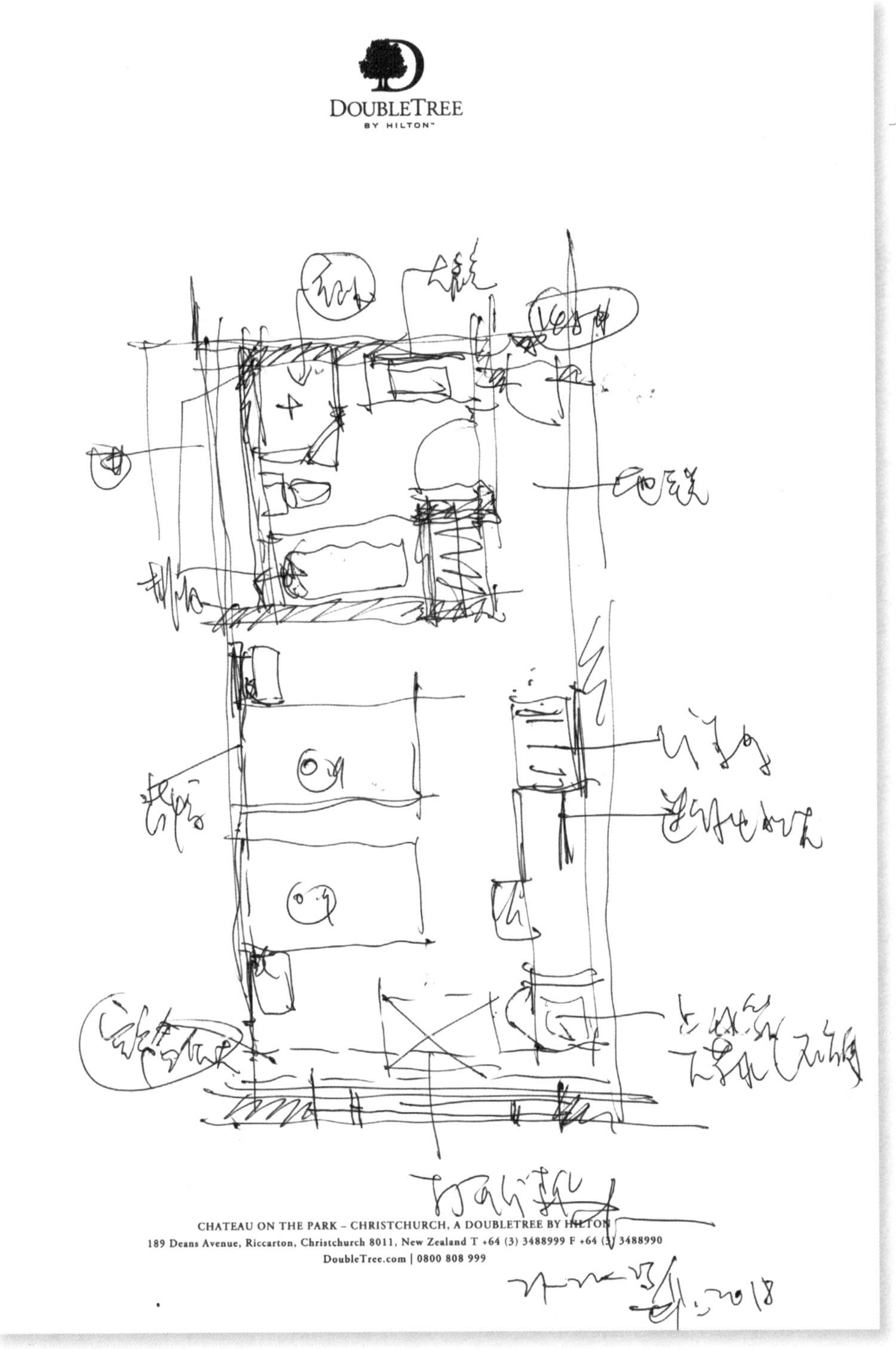

希尔顿逸林酒店

58 ★★★★

DOUBLETREE BY HILTON NEW ZEALAND

Address : 189 Deans Avenue, Riccarton, Christchurch 8011, New Zealand
Telephone : +643 3488 999
Fax : +643 3488 990
Http : //www.DoubleTree.com

CHATEAU ON THE PARK – CHRISTCHURCH, A DOUBLETREE BY HILTON
189 Deans Avenue, Riccarton, Christchurch 8011, New Zealand T +64 (3) 3488999 F +64 (3) 3488990
DoubleTree.com | 0800 808 999

Day 2/3

DOUBLETREE
BY HILTON

新西兰时间 28日 凌晨[illegible]

今天继续前往Neverland的第三天

是Day 2，[illegible]

① 入住的酒店太[illegible]！

② [illegible]

我们[illegible] golf [illegible]

③ [illegible] 144人的Killer游戏，

[illegible]

④ [illegible]

[illegible]

DoubleTree.com | 0800 44 8002

[illegible]

DOUBLETREE
BY HILTON

[illegible]

[illegible]

[illegible]

[illegible]

[illegible]

[illegible]

[illegible]

DoubleTree.com | 0800 44 8002

The Second or Third Day

第二/三天

新西兰时间11月23日凌晨1:31分

New Zealand Time, November 23, 1:31 a.m.

希尔顿逸林酒店

Doubletree By Hilton, New Zealand

分不清来到新西兰算是第三天还是第二天，有趣的事情总是很多！旅游嘛！

I was confused if it was the second or third day in New Zealand because of the jet lag. There were always so many interesting things. Traveling was just like that!

1. 入住的酒店不错，花多！
2. 旁边是大草坪城市广场，牛！我们尝试了简单高尔夫（用脚），也是乐趣。
3. 第一次玩19个人的“杀人”游戏，来劲！
4. 第二天，长途去观抹香鲸，只有一条懒洋洋的，反而看到了上千条的海豚，来劲！中餐、晚餐有大餐，龙虾三食，鲍鱼，石斑鱼，等等。

晚上继续“杀人”游戏，可以让大家聚在一起，放下手机，是不错的游戏。

酒店的酒吧已经打烊了！

我们继续，一路继续，感恩大家，今天是节，“感恩节！”

1. The hotel where we stayed was perfect with so many beautiful flowers.

2. The City Square next to our hotel was such a large publicarea that we couldn't help playing golf with our feet happily like children. So funny!

3. For the first time, we nineteen people played the game of "killer" together. How exciting!

4. The next day, we travelled a long way to see the sperm whale, but there was only one swimming languidly. Luckily, we saw thousands of dolphins instead.

We had delicious seafood in Chinese restaurant for dinner, including fresh lobsters, abalones, groupers and so on.

After dinner we continued to play the game of "killer" so that we all put down our cell phones to enjoy the time we stayed together. I thought it was a great game to keep off the phone.

The bar in the hotel was closed.

We would go on our trip with grateful heart, thanks for all on this special day just like Thanksgiving Day!

59 瓦纳卡卓越服务式公寓

★★★★

DISTINCTION WANAKA SERVICED APARTMENTS NEW ZEALAND

Address : 150 Anderson Road, wanaka, new zealand, 9305

Telephone : +64 3 443 2325

Http : // www.distinctionhotels.co.nz

A DISTINCTION HOTEL

www.distinctionhotels.co.nz

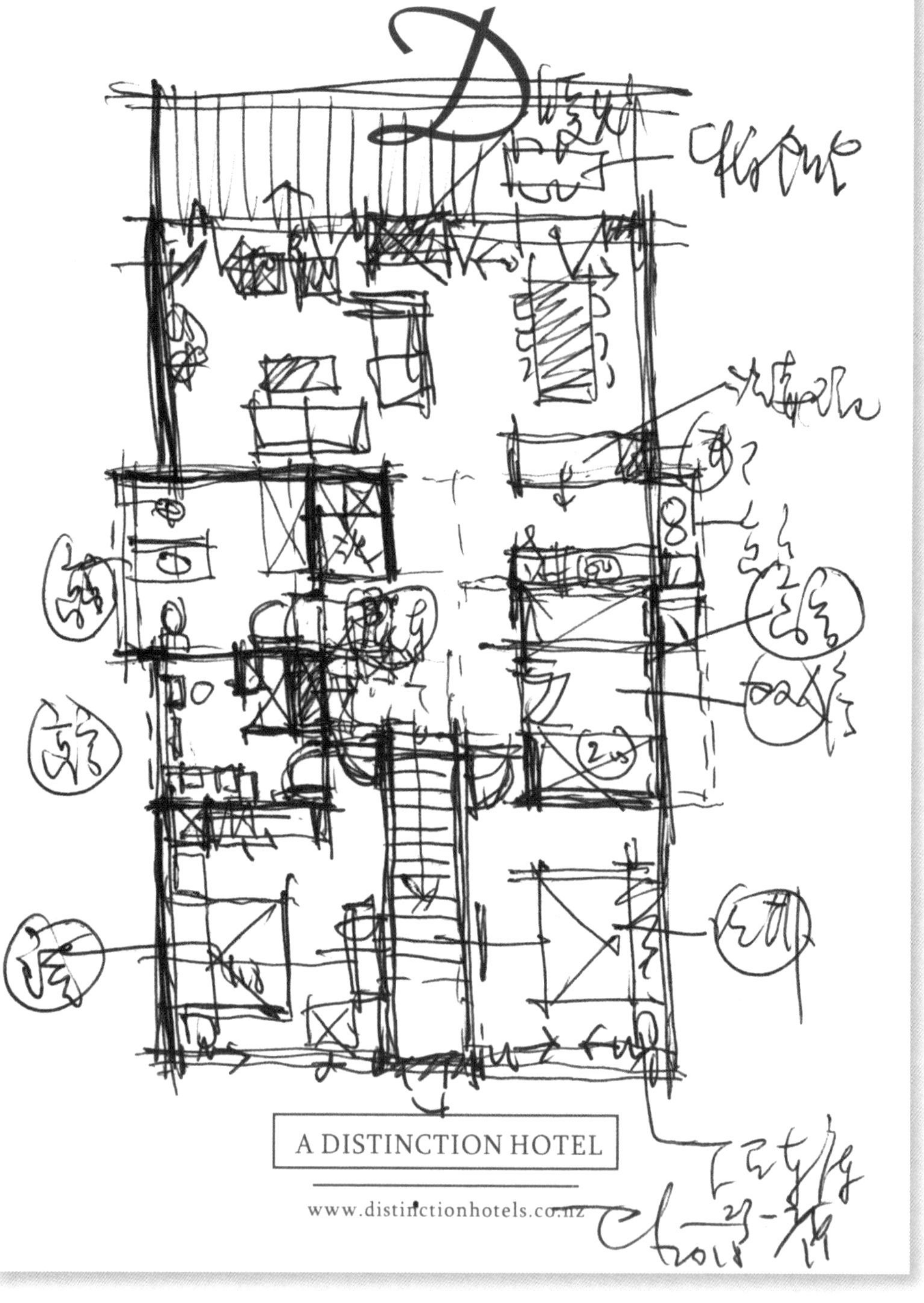

D

A DISTINCTION HOTEL

www.distinctionhotels.co.nz

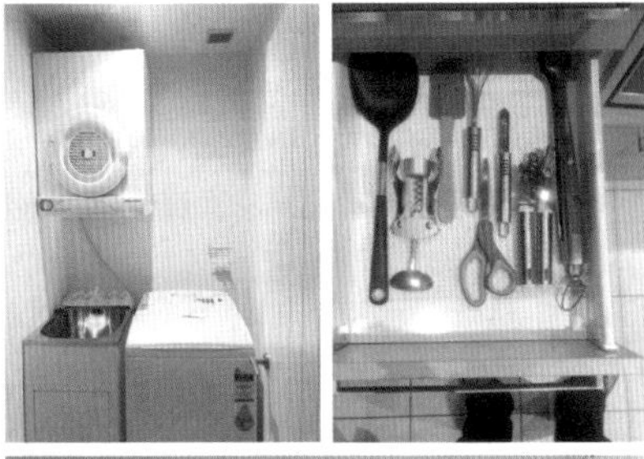

D

A DISTINCTION HOTEL

www.distinctionhotels.co.nz

D

A DISTINCTION HOTEL

www.distinctionhotels.co.nz

The First Experience in Serviced Apartment

酒店式公寓之初体验

2018年11月24日
November 24,2018

这次在新西兰(南岛)之旅，有机会入住公寓式酒店，久未有此体验（上一次应当是在斐济的经历）。也许是度假式的国度，游客对经济性比较敏感，我认为更可能是外国人的人工贵，做“低”服务的酒店式公寓或称公寓式酒店可以最大限度地去掉“人”的依赖，不用太多的软性服务，亚洲人最土豪。

During this trip to the south island in New Zealand, I had a chance to stay in a serviced apartment. It had been a long time for me since I stayed in the serviced apartment in Fiji several years ago. Maybe New Zealand is a travel country, so visitors are sensitive to the price. But in my opinion, because the salary in foreign countries is too high, the service in the hotel has to be downgraded. The advantage of the serviced apartments (or apartment hotels) is that tourists can stay here more freely and needn't depend on too much service. They are provided only some limit service; of course they don't need to pay more money for the high quality service. Except in Asia, crazy rich Asians.

这就是一个小别墅，讲求功能配套齐全，可长期居住，经济实惠。

First, we stayed in a villa with multiple facilities. This kind of serviced apartment was also good for long-termed staying because it was economical and practical.

首层有两个车位，我们住二层，有一个公共的客厅，餐厅，阳台，一个大套间，一间卧室，一间小的房（却是双人床），当然还有公共卫生间和洗衣干衣间，“五脏”俱全。厨房更是齐全，从杯杯碟碟到刀刀叉叉，洗，消毒，冰箱，应有尽有。

There were two parkings on the 1st floor. We stayed on the 2nd floor with a public living room, a dininghall, a balcony, a grand suite, a bedroom and another smaller bedroom with double beds, of course a public washroom and a laundry as well. All the necessary components you need could be found here. There was also a kitchen with all kinds of things, including cups, plates, knives, forks. You could wash, cook and sterilized in it. The fridge had just about everything you could wish for.

心里想，单日租客一定很贵，而长租一定很便宜，最要命的是对我们这些仅住一晚的人来说：洗手间简陋，热水要轮流使用。追求目的不同，自然有不同的体验结论，用平常心去体验就好！就如你不会用酒店的心态和要求去评判自己的家一样。

I thought the serviced apartment was convenient for tourists to spend their holiday; the cost for the single night must be more expensive than that for the long term. The worst thing was that the bathroom was too simple, especially the hot water, we had to take turns and waited for a long time to take a shower. On the other hand, different requirements led to different cost and different experience and conclusion, just enjoyed your trip with a common heart, like nobody compared their home and hotels with the same standard. Since we came for holiday, just relaxed ourselves, just enjoyed ourselves, that's all!

有空去住住这种公寓也不错，平衡片面追求奢侈的心情。

It is also an interesting experience to stay in this kind of serviced apartment. Keep balance to purse a luxurious feeling.

OAKS

HOTELS • RESORTS • SUITES

盛橡酒店及度假村

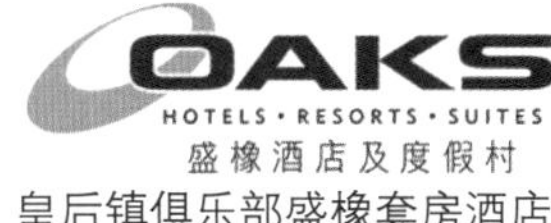

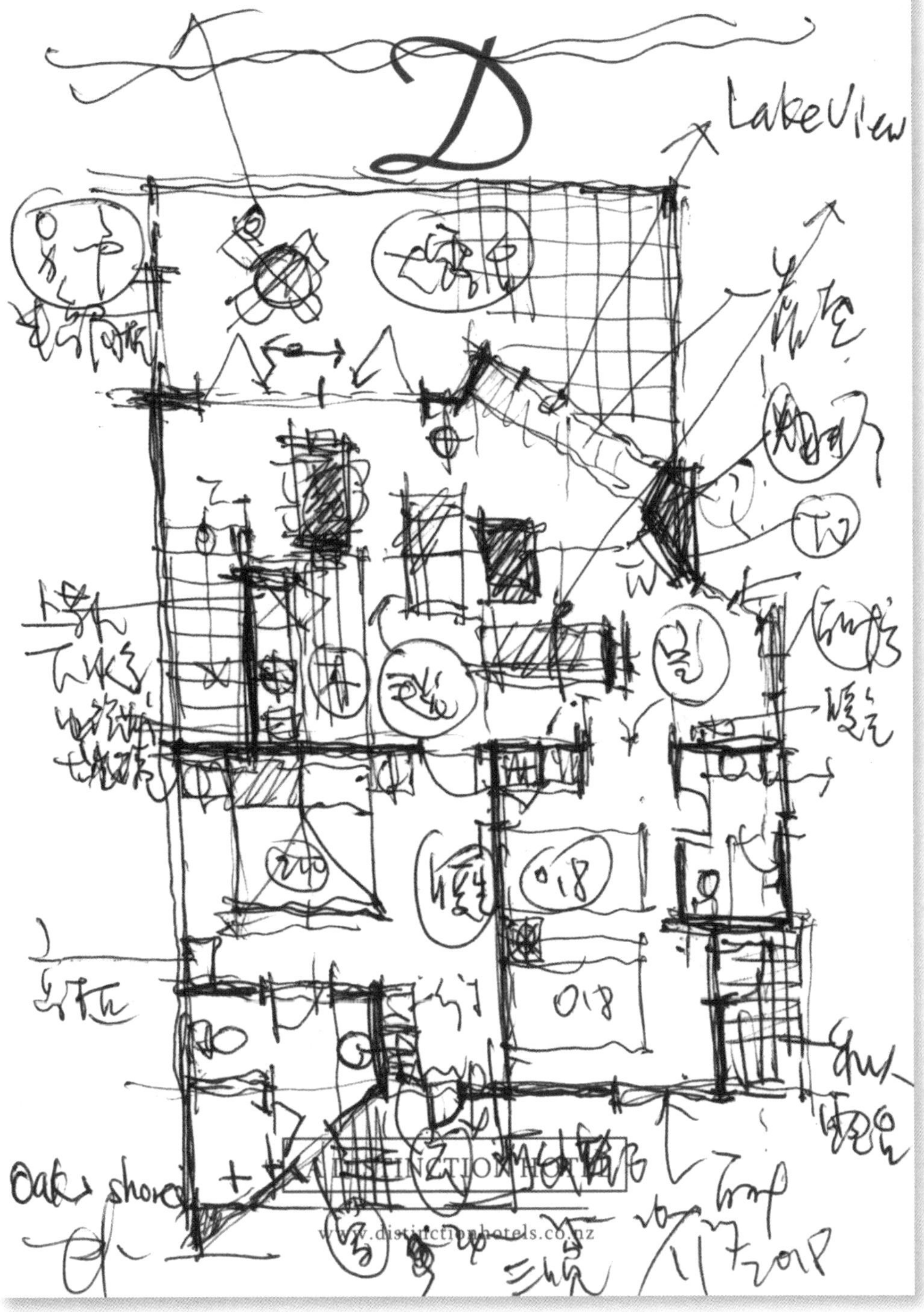

皇后镇俱乐部盛橡套房酒店

60

★★★★

OAKS QUEENSTOWN SHORESRESORT NEW ZEALAND

Address : 171-179 Frankton Road, Queenstown STH 9300

Telephone : +64 3 450 2700

Http : // www.distinctionhotels.co.nz

E-mail : club@theoaksgroup.co.nz

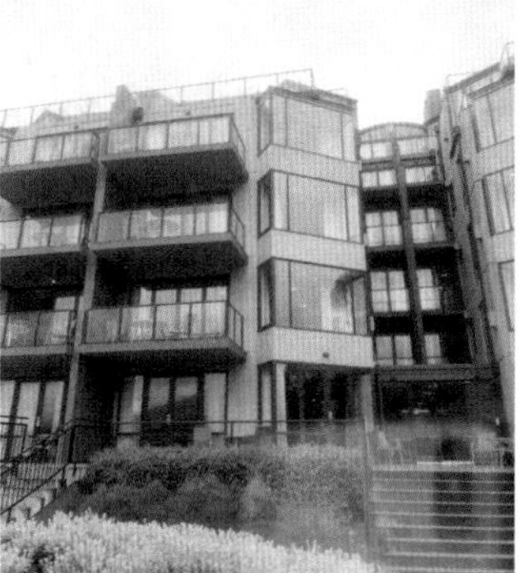

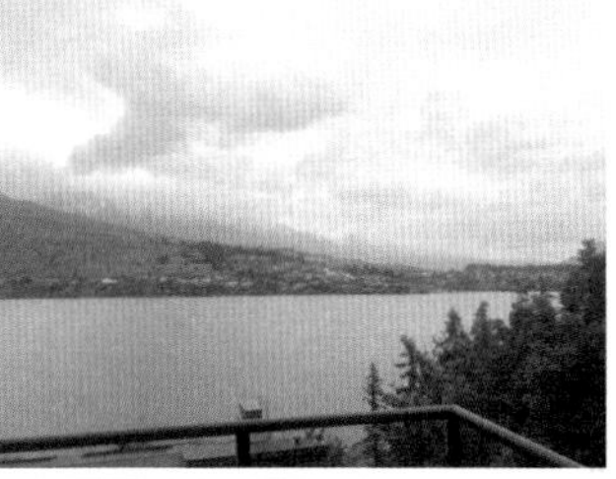

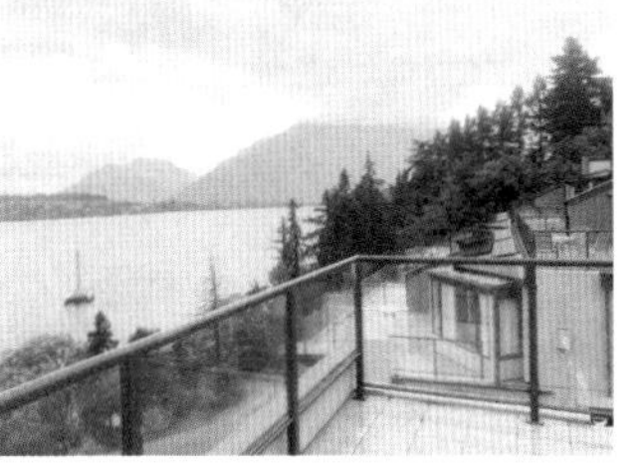

D 2018

A DISTINCTION HOTEL

www.distinctionhotels.co.nz

D

By HILTON; HYATT.com;

Zephyr Fishermans wharf

A DISTINCTION HOTEL

www.distinctionhotels.co.nz

D

A DISTINCTION HOTEL

www.distinctionhotels.co.nz

皇后镇俱乐部盛橡套房酒店

Oaks Queenstown Shoresresort, New Zealand

About the Identification of Pens

关于笔的识别性

2018年12月1日

December 1, 2018

继续《住哪？4》，继续收藏笔。可以一览全世界各个酒店品牌对配品、耗品的态度，实在太多了，每一本《住哪？》近100支不同酒店的笔（包括部分未入选出书的酒店），要回过头来一一对应识别是什么地方，住过什么酒店，百分之八十是对得上号的，于是我就细看它们的可识别性。

Keep writing Where to stay? 4,keep collecting pens. It shows different hotels' attitude to accessories and consumables. I have too many pens. I have collected almost 100 pens for every book, including some of them which haven't been chosen in my books. I look back to recognize the pens one by one, and think back the hotels which I have stayed in. Eighty percent of pens can be matched, so I check the identification carefully.

品牌几乎是百分之百可识别的，如HILTON、HYATT、COM……没有地区那就头疼了，国内或一些小众精品的酒店会印上城市的唯一性，如Zephyr Fisherman’s Wharf，这样我觉得会更有纪念性和收藏性。

Pens with brands are easy to recognize, for one hundred percent, such as HILTON, HYATT, COM, and so on. It's difficult to recognize if there is no area. Some domestic hotels and minority boutique hotels print the only symbol of the city on pens, such as Zephyr Fisherman's Wharf. I think these kinds of pens are more memorial and collectible.

计划用三十年左右时间，出版十本《住哪？》。难得会回头看看各种收获，笔——我相信是“最丰富的一笔”，希望我还能一一想起来每一支的出处。

About thirty years, ten books of where to stay? It's difficult for me to look back at different kinds of my collections. Pens —I believe they will be the richest ones. I hope I will remember where every pen comes from.

当然，有城市特征的，或有个性特征、提示的我更喜欢。

Of course, I prefer pens with city characteristics, personality and logo.

笔，我能一直识别的。

Pens, I can always recognize them.

澳门新濠天地

61 ★★★★★

MORPHEUS MACAU CHINA

Address : Morpheus At City of Dreams Estrada Do Istmo, Cotai, Macau, China
新濠天地摩珀斯
澳门路氹连贯公路

Telephone : +853 8863 3088

Http : //www.CITYOFDREAMSMACAU.COM

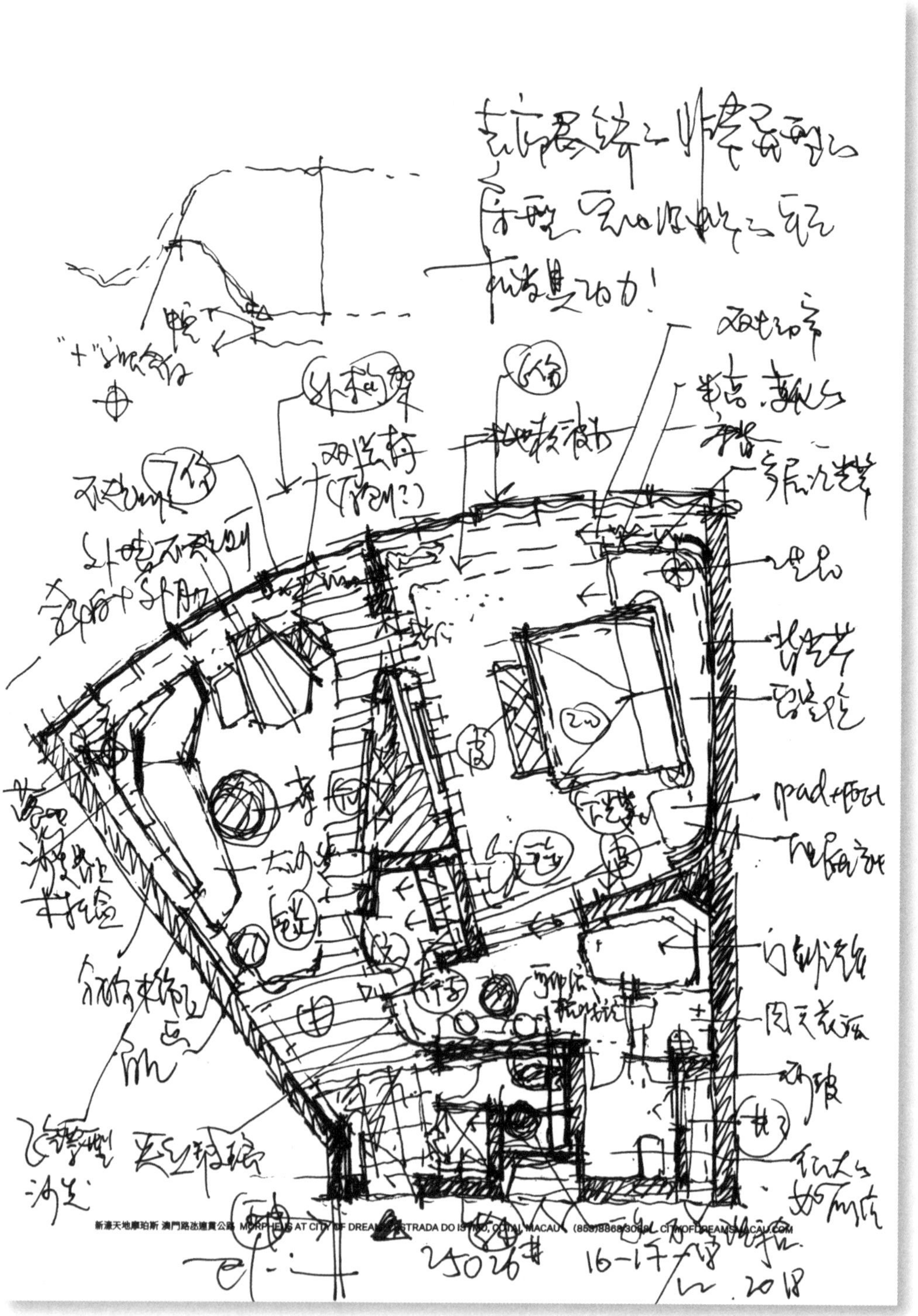

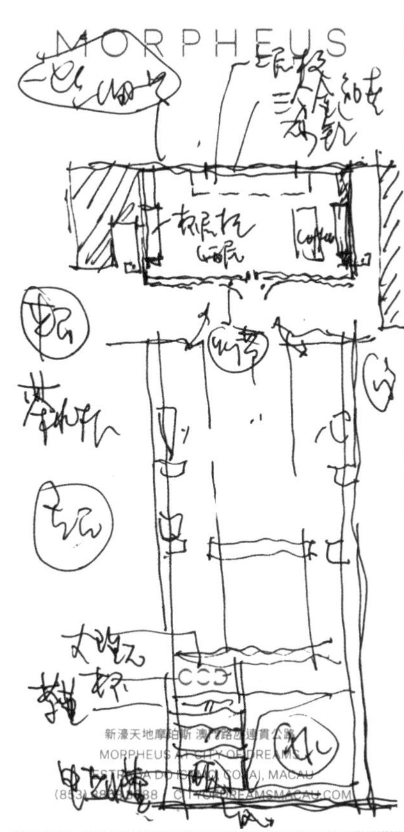

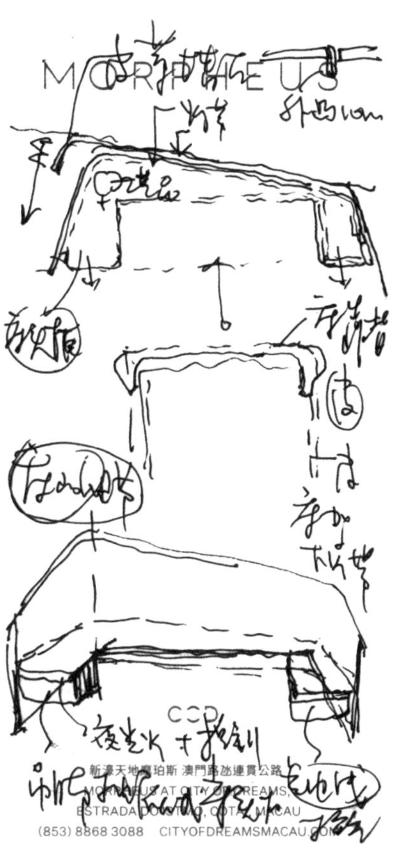

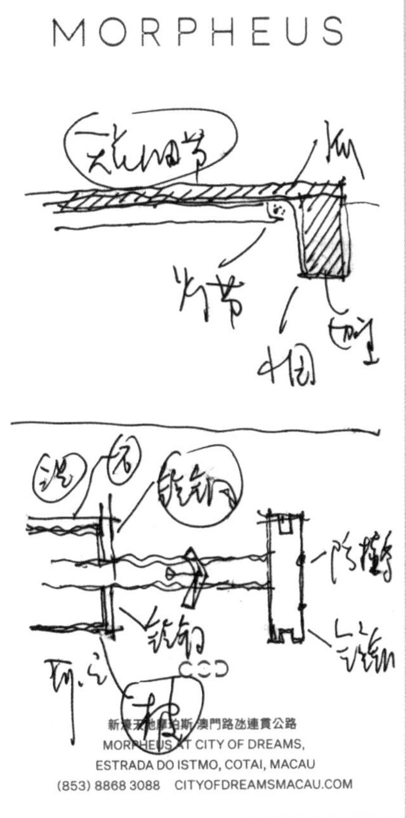

向扎哈大神尊敬·澳门新濠天地
Morpheus, Macau, China

新濠天地摩珀斯 澳門路氹連貫公路 MORPHEUS AT CITY OF DREAMS, ESTRADA DO ISTMO, COTAI, MACAU (853)8868 3088 CITYOFDREAMSMACAU.COM

新濠天地摩珀斯 澳門路氹連貫公路 MORPHEUS AT CITY OF DREAMS, ESTRADA DO ISTMO, COTAI, MACAU (853)8868 3088 CITYOFDREAMSMACAU.COM

Distance Around
身边的距离

51岁（无忧的年纪）生日选择入住澳门新开的酒店——扎哈·哈迪德大神（大婶）的“遗作”：摩珀斯酒店。

At the age of 51 (the worriless age), I chose the new hotel Morpheus— ZAHA's final work in Macao.

当然还有一个原因，室内设计师是我们一个项目的主创设计师Peter，一个挺“猛”的帅哥，也算是来看看他的大作吧。

Of course there was also another reason. The interior designer Peter was the chief designer in one of our projects. What a powerful handsome guy! I took the chance to have a look at his work.

房间放着一本这个酒店“创造产生”过程的册子。牛！从想法到结构、施工、加工等等，都是穷尽心思和不可思议，室内的设计也是很有延续性和独特性（酒店大堂及餐厅可能不是Peter先生的作品，是扎哈·哈迪德团队的一体化设计）。特别是我们入住的端头房型——“三尖八角”（不规则），真的可以说“搞死人”啦！

There was a book which introduced how to design and build the hotel in the room.Amazing! From concept to structure, construction, processing and so on. All were the best and unimaginable. The interior design had continuity and uniqueness(The lobby and dining hall might not be designed by Mr. Peter, it was the integrated design by ZAHA's work group). Especially the guestroom at the end which we stayed in was irregular. It really made the designer mad!

平面布局在建筑基础上更是发挥了优化的功力，更加在不知不觉中化解了原来的瑕疵（毛病）。

The plane layout on basic construction showed the optimize ability and also solved the flaws unknowingly.

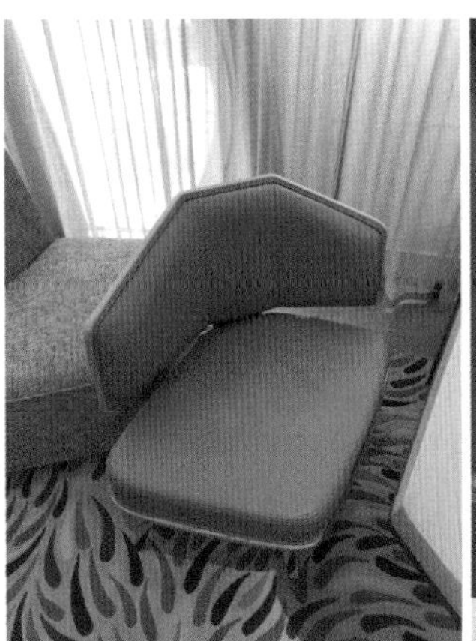
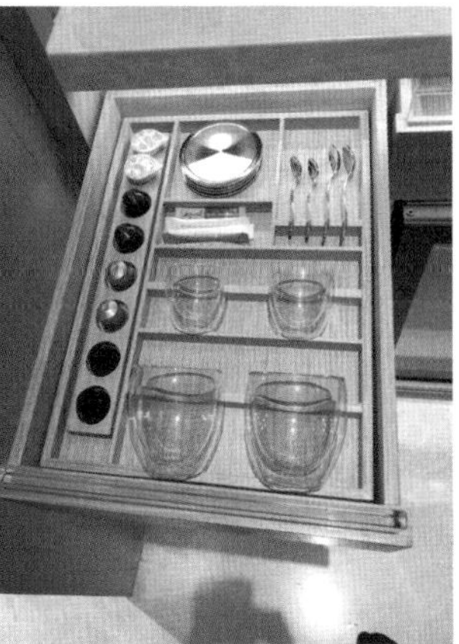

极具匠心的家具·澳门新濠天地
Furniture with great ingenuity · Morpheus, Macau, China

顶棚与立面简约，大尺度的设计也是室内设计师的老招式（与西安合作的项目如出一辙），当然成本高得惊人。但家具的作用在这个室内设计中，我认为起到了“化腐朽为神奇”的效果(不为过)。

The ceiling and facade were simple. The large-scale interior design was also the old style (the cooperative project with Xi'an was the same), of course the cost was breathtaking. But I thought the furniture played an important role in the interior design, just like the old saying "turn decayed into magic"(not too much).

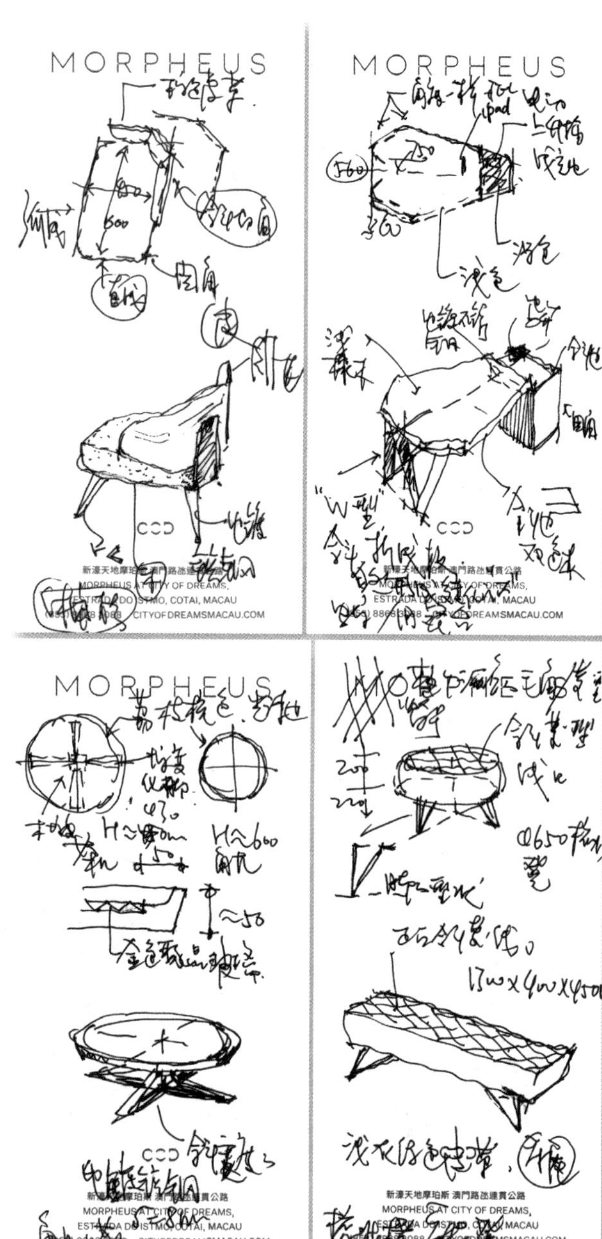

飞镖形的大沙发，与近菱形的书写台、橙色的扶手皮椅算是第一个主空间：客厅的主角，而客房的变化高度的弧形床背、床靠相结合，令人惊叹设计师的功力。

The dart-shaped sofa, the diamond-shaped writing table and the orange leather armchair made up the first main space: the leading part in the sitting room. The different height of the guestroom made combination of both the backboard and headboard of the curved bed. The ability of the designer was surprising!

细节家具如床头凳、梳妆凳等等，带灯光的茶水柜门也是小亮点，多层次的灯光系统让人入住的氛围相当亲切。

The details of furniture like the bed stool, the dressing stool, the door of the tea cabinet with lights was also the high light. The multi-level light system made you feel at home.

认认真真地画了其中的几件家具，感到距离真的在我们身边，值得学习。

I drew some pieces of the furniture carefully and really felt the distance around us. It was worth learning.

入住，正是学习的方式与渠道。

Staying in it is the best way to learn.

不枉住这，《住哪？》都是幸福的事情。

It deserved my staying here. Where to Stay? is also a happy experience.

新濠天地摩珀斯 澳門路氹連貫公路 MORPHEUS AT CITY OF DREAMS, ESTRADA DO ISTMO, COTAI, MACAU (853)8868 3088 CITYOFDREAMSMACAU.COM

Curve + Triangle
曲线+三角形

因为入住的是“摩珀斯”的洞口边上的房间，不规则的平面，真是难为室内设计师，几乎没有一个横平竖直的空间，室内设计师在建筑师扎哈·哈迪德（团队）的基础上，用“小曲线”+“三角形”很好地“化解”了这个刁钻的房型。

Because of staying in a guestroom which was on the side of the hole, the irregular floor plan was really difficult for the interior designer. It hardly had a horizontal and vertical space. Based on the ZAHA's architect work group, the interior designers dealt with the problem of this irregular shape tactfully with a tiny curve and triangle.

端头，一个大弧面的采光，分隔客厅区与睡眠区，以“回”字路线处理各个空间的分、连关系。入口前区连一个小客厕，天然大理石“挖出”一个圆洗手盆，土豪，嵌墙式的对讲式电话（挺无用的），外抛的斜线“腾出”一个衣服间、行李间，巧妙地让客厅的飞镖形沙发有了“正向”看电视的角度，“三角”的写字台，菱形的橙色椅子，设计师的心思尽现，沙发大的靠背将杂志托盘自然而然地设在其上，斜角“留出”一点空间优雅地放了大大的落地灯。 干净、霸气的到顶无缝的木饰石墙，高成本而尽显低调。

The room at the end had natural lighting through a big cambered surface. It was divided into the sitting room area and sleeping area.Each space was connected and separated by a shape of “回”. There was a small washroom connecting the front area at the entrance. A round marble lavabo was luxury. The in-wall interphone was useless. Because of an outward oblique line, it formed a cloakroom and a luggage room.The artful design made the guests sit on the dart-shaped sofa to face the TV set. Both the triangular table and the orange diamond-shaped chair could embody out the designers inspiration. The magazine tray was naturally put on the large back of the sofa. A big floor light was gracefully put at the corner. The clean and domineering wooden stone wall was from top to the end without any space. The high cost showed the low-key luxury style.

夹丝玻璃，直与圆弧结合，让灯光可温柔地透到客厅。超比例的大滑门，让进入卧室和步入大大的洗手间都有“手感”和仪式感，最绝的是以斜三角的平面及空间组合弧形床背板、床头靠及床头柜，也更“自然而然”地处理房间的多条不可分解的“轴线”，慢慢地一点点地“纠正”。啊！高手，连地毯都是没有规律的。

Wired glass combined straight line and the cambered surface to let the light into the sitting room gently.The oversize sliding door created the sense of ritual when guests entered the bedroom and the big washroom. The most perfect part was that the plane and space of the oblique triangle made up the backboard of the curved bed, the headboard of the bed and also the night table. Some indecomposable axes in the room were naturally corrected little by little. Amazing! Zaha is an absolute master! Even the blankets were also irregular.

“无章法”也许才是最难得！

No rules might be the most difficult!

洗手间的订制浴缸（唯一住过可以看到电视的浴缸，因为安装得够大和够低），双云石洗手台和大大的真皮梳妆椅，外伸式的小创新梳妆镜，倍感人性，大的如厕位，可以供两人同时使用。

The customized bathtub in the washroom was so large and low that you could watch TV while having a bath(It was the only hotel that I had stayed in).There were double marble lavaboes and a big leather dressing stool in it. An outward innovative toilet glass felt humanized. The toilet room was big enough for two persons.

不简单的一个客房，真的有趣。空间上更是以斜线+微弧线解决了不规则的问题。

The uneasy guestroom was very interesting. What was more, the irregular space was solved by oblique lines and arcs.

无招胜有招，令我佩服的一个案例。

No rules are better than rules. This is the example I admire most.

魔幻的空间 · 澳门新濠天地

The magical space · Morpheus Macau, China

NICCOLO
CHANGSHA

长沙尼依格罗酒店

62 ★★★★★

NICCOLO CHANGSHA CHINA

Address : 4/F, building 1, Guojin center, 188 Jiefang West Road, Furong District, Changsha, China
中国湖南省长沙市
芙蓉区解放西路188号
国金中心1号楼4层

Telephone : +86 731 8505 2137

冷酷的酒店・长沙尼依格罗酒店

Niccolo, Changsha, China

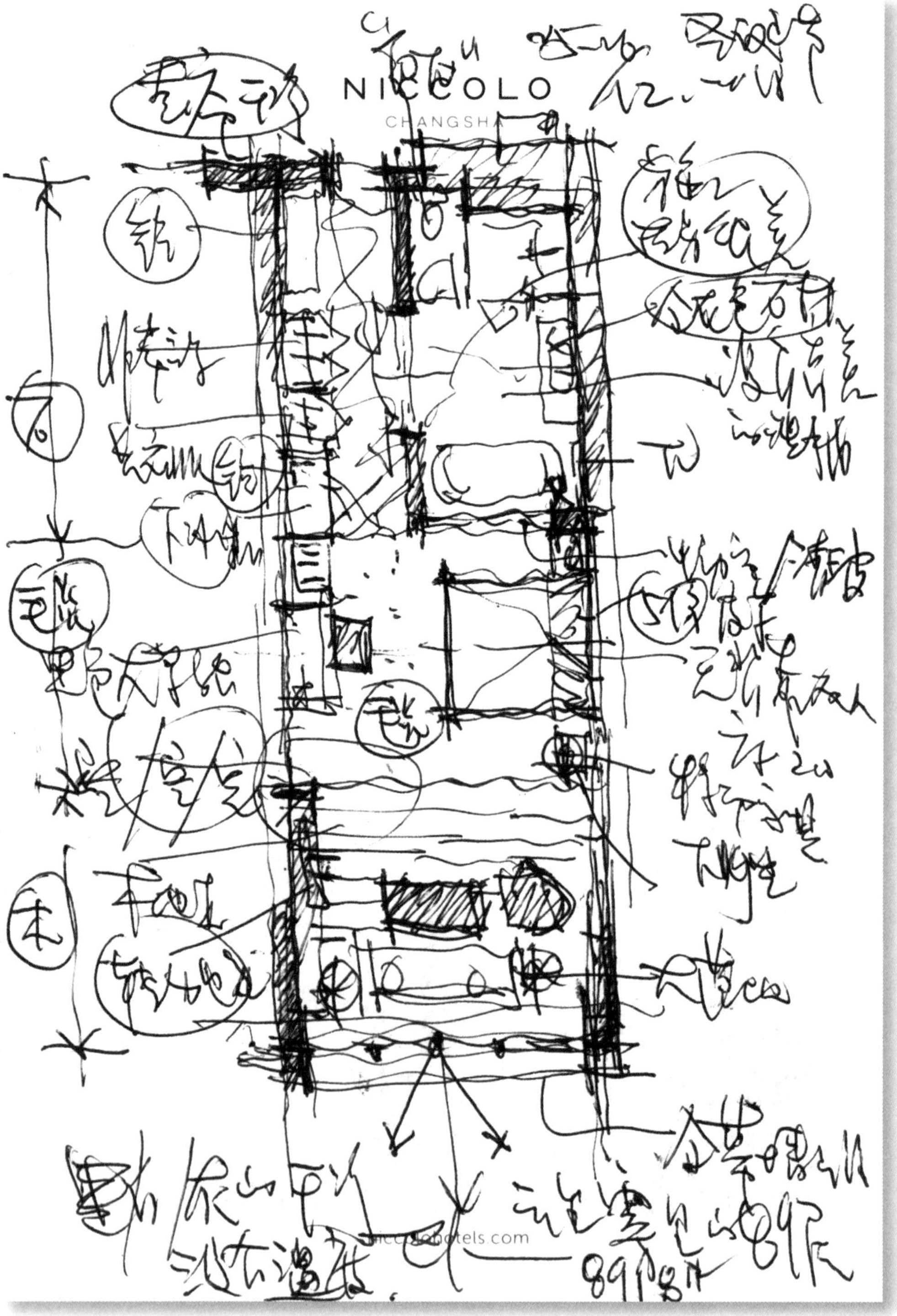

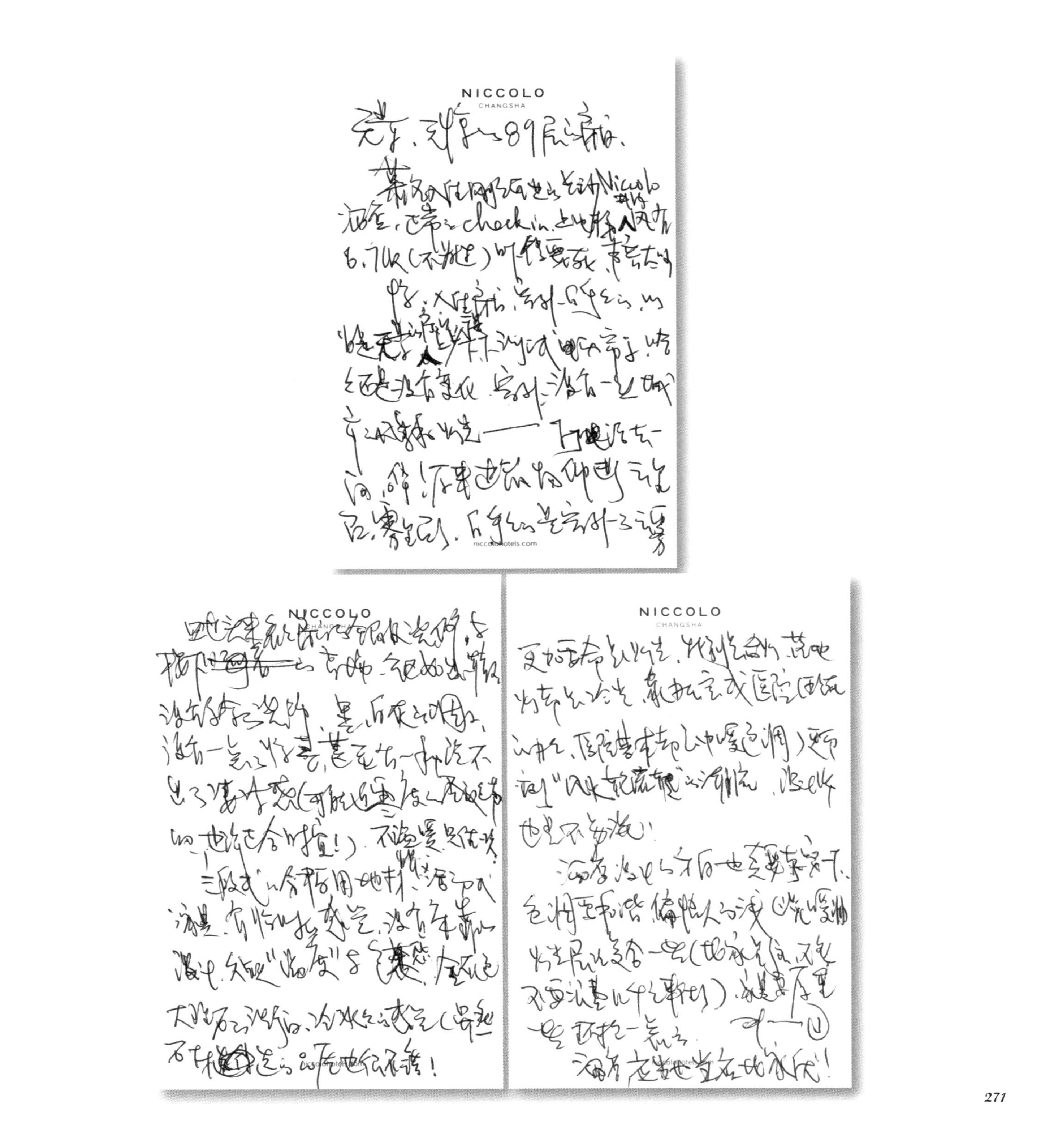

NICCOLO
CHANGSHA

niccolohotels.com

NICCOLO
CHANGSHA

niccolohotels.com

NICCOLO
CHANGSHA

niccolohotels.com

No View, No Surprise

无景，无惊（喜）

大雾无景的房间 · 长沙尼依格罗酒店
The room without view but fog · Niccolo, Changsha, China

慕名入住刚刚开业的长沙尼依格罗酒店，正常地check in，上电梯，井内风有6、7级（不为过），吓得要“死”，声音太吵。

I was attracted to stay in the new Noccolo Hotel in Changsha. Checked in as usual and took the elevator. The wind was up to six or seven degrees in it and made too much noise. I was scared.

惊，入住房间，室外白乎乎的，以为是无景的房间，于是上上下下测试电动帘子，哈哈！还是没有变化。室外没有一丝城市的风景和灯光。打电话去一问，哗！ 原来建筑物伸进了云里面、雾里面了。白乎乎的是窗外面的云雾。

I felt surprised when entering the room. It was white outside. I thought the room had no views. Then I tested the electric curtain up and down. How interesting! Nothing changed. There was no view of the city or lights outside. I called the reception. In fact, the building was so tall that it reached cloud and fog. The white was cloud and fog outside.

回过头来看看房间的布局与装修，与楼下的商场、公共区如出一辙。没有多余的装饰，黑、白、灰的调子，没有一点点惊喜，甚至有一种说不出来的凄凉感（可能赶上近零度的圣诞节吧，也许正合时宜！），不温暖，只住一次。

Turned around and looked at the layout and decoration of the guestroom, they were the same as the shop downstairs and the public area. No spare decorations, black, white and grey color had no surprise but a little sad.(It was nearly zero degree centigrade, maybe it was the perfect temperature for Christmas.) I couldn't feel cozy and didn't want to stay here again.

三段式的分隔用地材为界，活动式家具，有临时的感觉，没有床靠的设计，欠缺“温度”与包裹感。 全灰色大理石的洗手间，给人冷冰冰的感觉（虽然选的石材品质也很不错！）。

The guestroom was divided into three parts by flooring. Movable furniture showed temporary feeling. Without design, the bed back lacked warmth and sense of parcel. The grey marble washroom made me felt cool. (Although the quality of the marble was quite good.)

更加要命是灯光，特别是台灯、落地灯都是冷光，像办公室或医院（反而现在的办公、医院基本都是中暖色调），更印证了“风水轮流转”的潮流，设计师也是不易混！

The most terrible was light, especially the desk lamp and floor lamp were both cold light just like that in offices or hospitals (But nowadays the lights there were warm colors). It was a perfect example of changeable fashion. It was not easy to be a good designer.

酒店设计的方向也真要考察一下，色调要和谐，偏怡人的浅（淡暖调），灯光层次复合一些（比家里强，不然也不用浪费几千元来住了）。家具要厚重一些，环抱一点的。

The direction of hotel design should be thought over. The color should be harmonious, pleasant light color(Warm and light color). Lighting levels should be complex (much stronger than that at home, otherwise we won't spend thousands of yuan staying here) .The furniture should make you feel a little heavy and surrounded.

酒店，应当也当应比家优！

Hotels, should be and must be better than home!

大雾无景的房间・长沙尼依格罗酒店

The room without view but fog · Niccolo, Changsha, China

63 西安W酒店 W HOTELS XI'AN CHINA

★★★★★

Address : No.333 Qujiangchi East Road, Yanta District, Xi'an, Shanxi, China
中国陕西省西安市雁塔区曲江池东路333号

Telephone : +86 29 8579 6082

Http : //www.m.whotelxian.cn

小玩意儿・西安W酒店
A gadget · W Hotels, Xi'an, China

NOTE TO SELF
自写自画

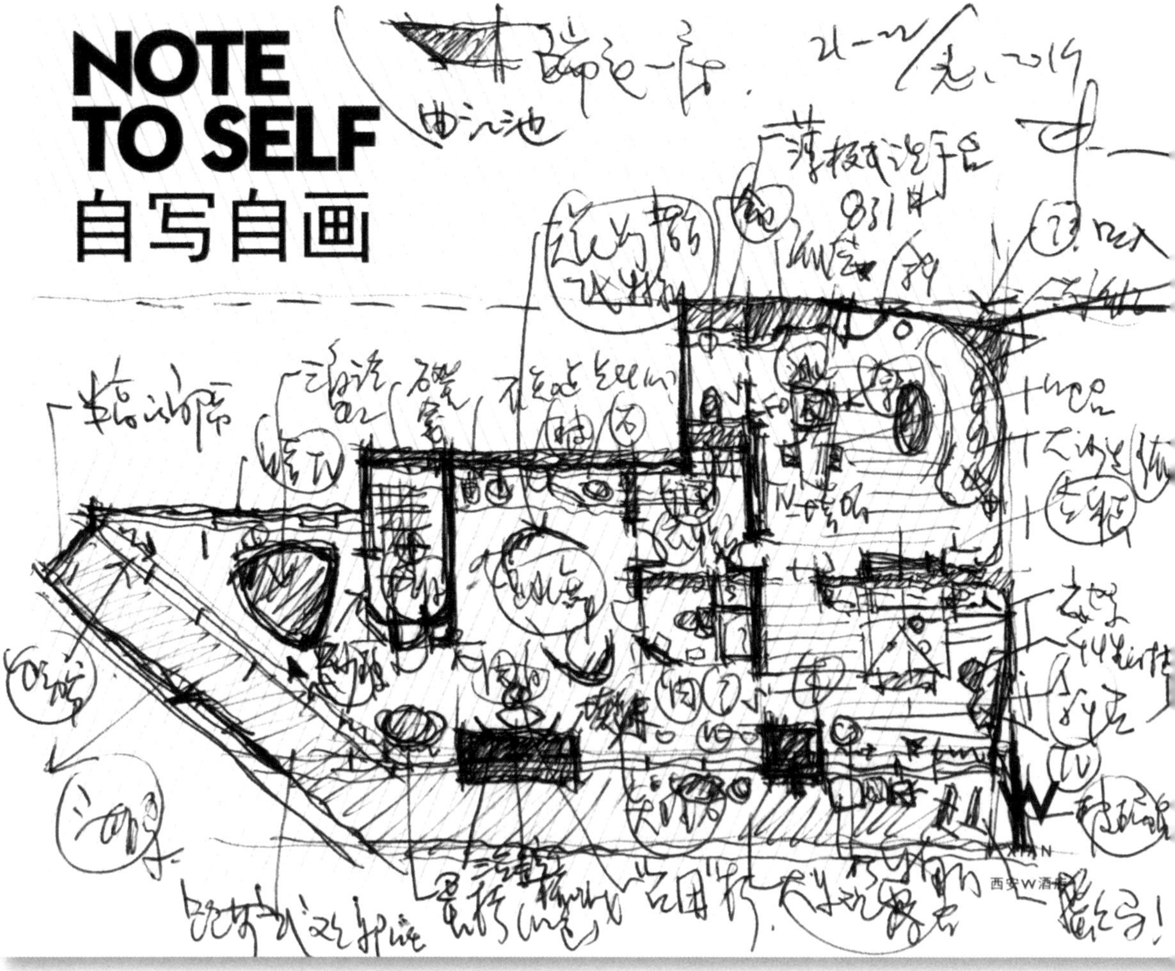

W XIAN 西安W酒店

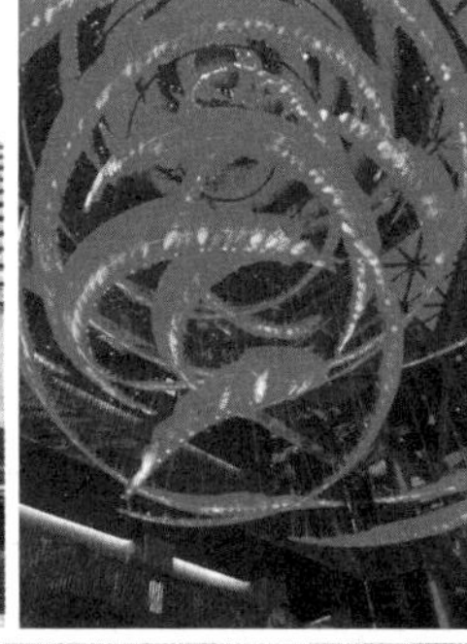

一千多万的吊灯・西安W酒店

A chandelier worth more than 10 million yuan · W Hotels, Xi'an, China

NOTE TO SELF
自写自画

[illegible]

W
西安W酒店

NOTE TO SELF
自写自画

[illegible]

W
XIAN
西安W酒店

NOTE TO SELF
自写自画

[illegible]

W
XIAN
西安W酒店

Do You Want a Home Like " W "?

像"W"的家，爱吗？

因为太"热"了，这是我们的客户开发的第一家酒店，就"一炮而红"，成为全中国最"火"的酒店之一，所以借来西安汇报方案之机，入住"W"，体验一下，什么是最炫之"疯"！

Because it was too hot, "W" Hotel, which was developed by our client, made a sudden rise and became one of the most popular hotels in China. We took the opportunity to stay in when we came to Xi'an to report our project. I wanted to experience what was the craziest!

AB concept的再续华丽、艳俗之风，从大堂到餐饮再到客房层，确是有其独特的地方，当然业主的"任性"和不惜重金为设计师的为所欲为提供助力。同事住标准双床房，我住端头，一个套房。哇，确实没有想到这么大！

AB concept continued its magnificent and gaudy style. It was really unique from lobby to dining hall and also the guestrooms floor. Of course the wayward owner provided a large sum of money for the designers, so they could do whatever they desired. My colleague stayed in a standard twin beds room while I stayed in a suite at the end. OMG! I had never thought that it was so large!

微微斜向的大门入口，几处不规则的圆弧线组合客厅，成为自然而然的"包裹"，大型岛式的小吧台，充满娱乐感，可以三五知己聊聊，吃吃，喝喝，看看电影。（音响太猛了！！！）

The entrance was a little diagonal. The living hall was naturally wrapped by a few irregular curved lines. A large island bar was full of entertainment. Several friends enjoyed chatting, eating, drinking and watching movies here. (The sound was too loud.)

环形回旋的设计，是"大房子"才能有的想法，客厅有门与衣帽间连接，而入卧室的另一条路线，连起了不可想象的厕、洗、淋、泡的空间；可能也是端头"三尖八角"之缘故，洗手间的湿区格外"不和谐"，非常长的流线，独立的大如厕间，包裹奢华的绿色纹理大理石，无敌的景观，真是勾起了"露体"的欲望；过大的通道，近3米x5米，让设计师不知如何是好（我猜的），空空的梳洗区，高低式的分设洗手盆，大镜不规则，与整体手法一致。顶棚更是"点到即止"的蓝光，如夜总会。面团式的梳妆椅与休闲椅设计，奇怪的红色，过长过深的淋浴间。尽头是一个三角大浴缸，一泡去忧愁……

The guestroom was so big that it could be circle round. The door of the sitting room was connected to the cloakroom while another way was connected to an unimaginable space including a toilet, washing, shower and bath. Maybe the guestroom at the end was irregular, so the wet area in the washroom didn't go together. A very long streamline, the independent toilet wrapped by luxurious green marble as well as the nonpareil landscape arouses my desire of nudity; the designers confused about such a large passage which was nearly three meters wide and five meters long. The ceiling was "just the perfect" blue light, like night clubs. The dough-shaped dressing chair and the leisure chair were in some kind of strange red color. At the end of the guestroom, there was a big triangle bathtub which could relieve worry when bathing in it.

通长超10米的外阳台，有一个大的娱乐搞笑的一大木摇摇马仔，有不甚“高档” 的户外休闲椅，当然景观（曲江湖）无敌是关键。

On the over ten-meter corridor, there was a big funny rocking wooden horse and some cheap outdoor leisure chairs. Most importantly, the nonpareil landscape was the attraction.

多彩，如果你自己的家像这个样子，你接受吗？——客厅为灰色（包括木饰面墙身都是银灰色现纹漆），超大的红色调的立体装饰画；卧室设大大的鲜艳蓝色背幅，亦清新。床上的小笼包也可爱。还有洗手间区两张红色椅子成为主角，随时随地让你有置身“夜场”的感觉和有着“异想天开”的“堕落”（坠落）感。 也难怪一开业就生意火爆，生活需要增加“色彩”。W正好满足了你我的要求，也许，艳丽还会流行一段时间。

Colorful, but can you accept your home like it? The sitting room was grey color(including the wooden wall with grey silver painting), and the huge bright red solid decorative painting. In the bedroom, the light blue background was fresh. "Xiao Long Bao" on the bed was lovely as well. Two red chairs in the restroom were certainly the leading roles. They were ready to make you feel at the nightclub and fantastic corruption. No wonder it was popular as soon as it opened. Life needs colors and excitement. W Hotel exactly satisfies you and me. Maybe gaudy style will last for some time.

炫而豪・西安W酒店
Dazzle and luxury · W Hotels Xi'an, China

GRAND HYATT CHANGSHA

长沙君悦酒店

★★★★★

GRAND HYATT CHANGSHA CHINA

Address : No.36 Middle Xiangjiang Road, Tianxin District, Changsha,Hunan, China
中国湖南省长沙市
天心区湘江中路36号

Telephone : +86 731 8823 1234

E-mail : Changsha.grand@hyatt.com

Http : //www.hyatt.com

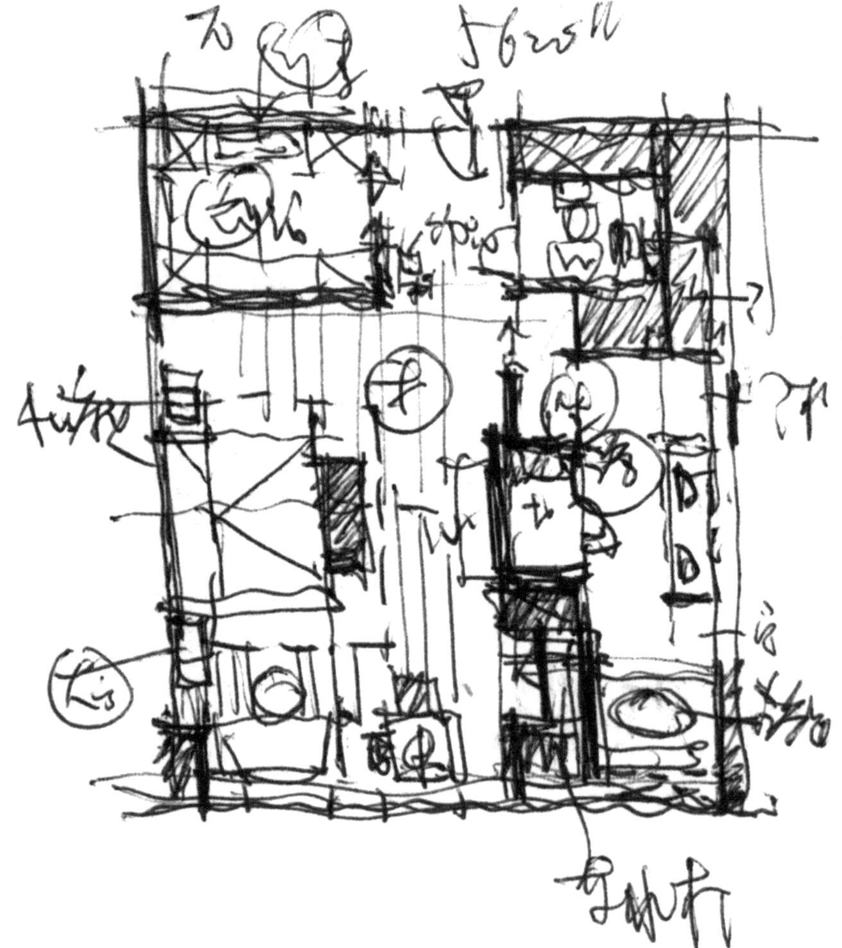

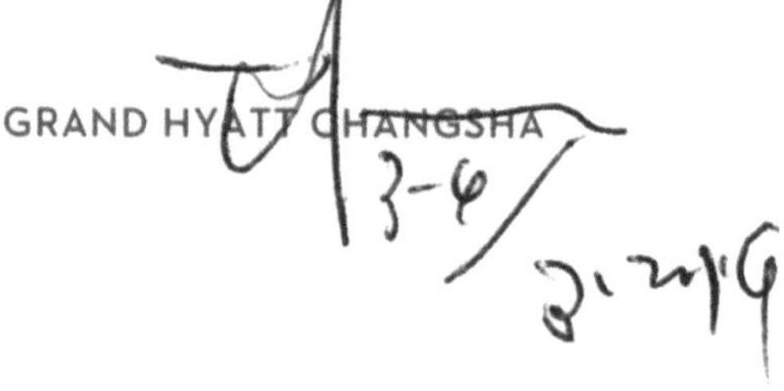

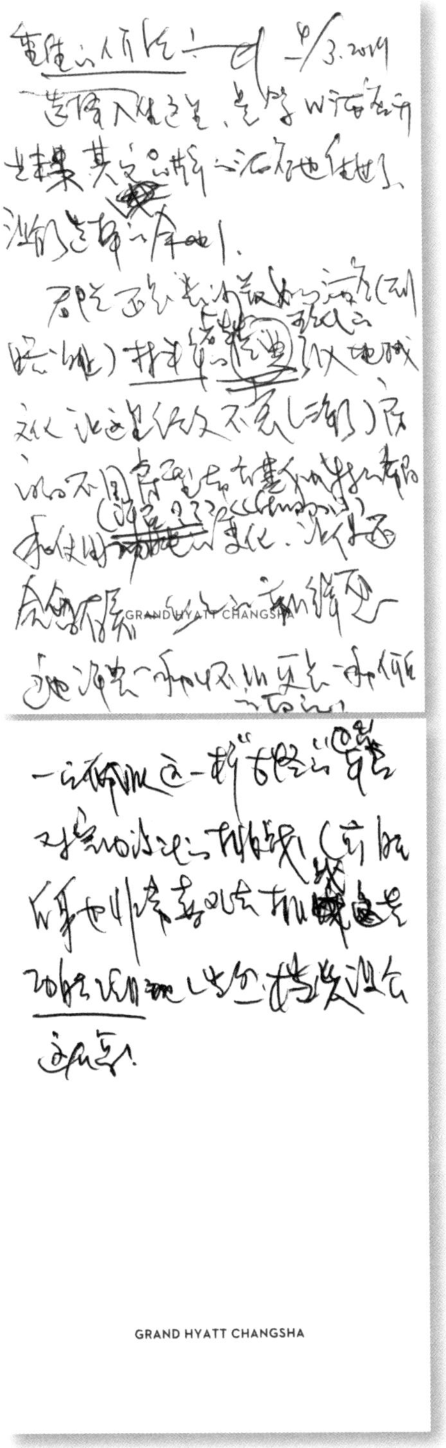

The Value of Staying Again

重住的价值

选择入住这里，是因为等W酒店开业未果，其他品牌的酒店也住过了，没有了选择的余地。

Because W Hotel hadn't opened yet, I had to stay in Grand Hyatt again. There was no choice because I had stayed in other brands.

君悦还是长沙最好的酒店（到目前为止），是林丰年老先生经典而现代的设计。引入地域文化，让这里经久不衰（三年了）。不同房型的房间都有独特的布局和使用功能的变化（详见《住哪？3》），让你还会留有着一丝丝的新鲜感，这也许是一种怀旧，更是一种价值的印记！佩服这样“古怪的”建筑平面对室内设计的挑战（而自己自身也非常喜欢去挑战这些功能项目），当然，设计档次没有这么高！

So far,Grand Hyatt was the best hotel in Changsha. It was classical and modern by Mr. Lin Fengnian. Because the local culture was introduced to the hotel, it had been popular for long (already three years). Different types of guestrooms had their unique layout and changeable function(See Where to Stay?3). It made you remain a little fresh. It was nostalgia and a mark of value as well. On the other hand, I admired that the strange building plane was a big challenge for the interior design(And I like to challenge these functional items). Of course, the level of our projects isn't so high!

巴西之旅（2019年3月9~19日）

The Trip to Brazil

INTERCONTINENTAL®
SÃO PAULO

Alameda Santos, 1123
São Paulo SP Brasil CEP 01419-001
Tel.: 55 11 3179-2600 Fax: 55 11 3179-2666
Central: 0800 11 8778
reservas@ihgbrasil.com www.intercontinental.com/saopaulo

Esse hotel pertence à IHC São Paulo Hotelaria Ltda. e é operado pelo InterContinental Hotels Group

This property is owned by IHC São Paulo Hotelaria Ltda. and operated by InterContinental Hotels Group

65 里约热内卢JW万豪酒店

★★★★

JW MARRIOTT RIO DE JANEIRO SAO PAULO BRAZIL

Address : Av. Atlantica, 2600- Copacabana, Rio de Janeiro-RJ, 22041-001, Brazil
Telephone : +55 21 2545-6500
Fax : +55 11 3179 2666
E-mail : reservas@ihgbrasil.com
Http : // www.intercontinental.com/saopaulo

66

卡利南喝普鲁斯尊贵酒店

★★★★★

CULLINAN HPLUS PREMIUM BRAZIL

Address : SHN Q. 4 BLE-Asa Norte, Brasilia- DF, 70704-050,Brazil

Telephone : +55 61 3426 5000

教育和卫生总部外观・卡利南喝普鲁斯尊贵酒店

The appearance of Education and Health Headquarters · Cullinan Hplus premium

hplus.com.br HplusHotelaria Hplus

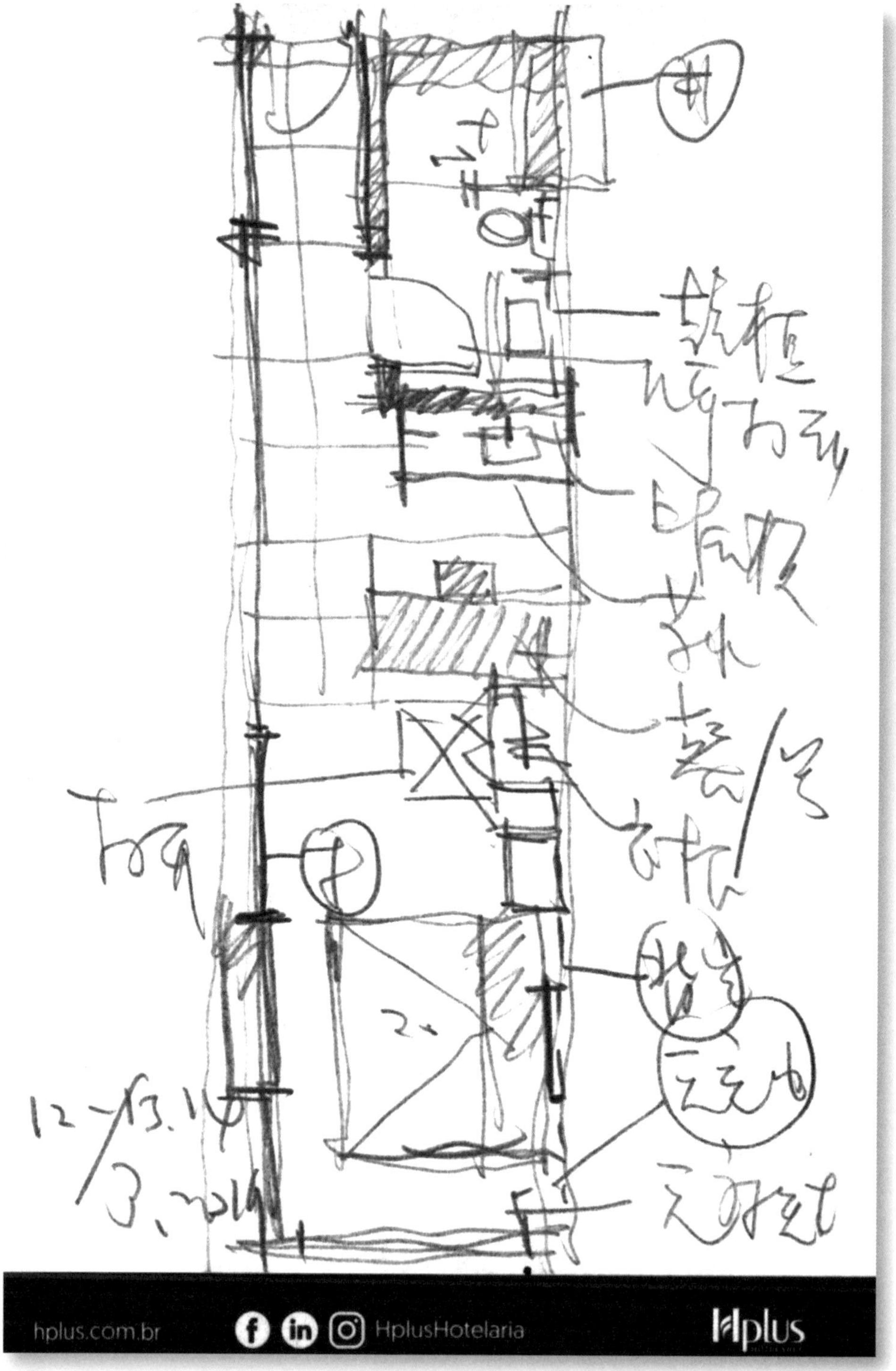

Museu de Arte Moderna do Rio de Janeiro

ACERVO / HORIZONTES
ILUSIONISTA / CONSTELAÇÕES

12-03-19 15:32 R$14,00 VISITANTE
prédio principal

não é permitido comer beber ou fumar dentro dos salões de exposições
o ingresso dá direito a apenas uma entrada

02297195

INTERCONTINENTAL
SÃO PAULO

67

圣保罗洲际酒店

★★★★

INTER CONTINETAL SAO PAULO BRAZIL

Address : Alameda Santos, 1123, Aden Paulista, Sao Paulo
Telephone : +55 11 3179 2600
Http : // www.intercontinental.com/saopaulo

INTERCONTINENTAL
SÃO PAULO

Alameda Santos, 1123
São Paulo SP Brasil CEP 01419-001
Tel.: 55 11 3179-2600 Fax: 55 11 3179-2666
Central: 0800 11 8778
reservas@ihgbrasil.com www.intercontinental.com/saopaulo

Esse hotel pertence à IHC São Paulo Hotelaria Ltda. e é operado pelo InterContinental Hotels Group

This property is owned by IHC São Paulo Hotelaria Ltda. and operated by InterContinental Hotels Group

INTERCONTINENTAL
SÃO PAULO

Alameda Santos, 1123
São Paulo SP Brasil CEP 01419-001
Tel.: 55 11 3179-2600 Fax: 55 11 3179-2666
Central: 0800 11 8778
reservas@ihgbrasil.com www.intercontinental.com/saopaulo

Esse hotel pertence à IHC São Paulo Hotelaria Ltda. e é operado pelo InterContinental Hotels Group

This property is owned by IHC São Paulo Hotelaria Ltda. and operated by InterContinental Hotels Group

INTERCONTINENTAL
SÃO PAULO

Alameda Santos, 1123
São Paulo SP Brasil CEP 01419-001
Tel.: 55 11 3179-2600 Fax: 55 11 3179-2666
Central: 0800 11 8778
reservas@ihgbrasil.com www.intercontinental.com/saopaulo

Esse hotel pertence à IHC São Paulo Hotelaria Ltda. e é operado pelo InterContinental Hotels Group

This property is owned by IHC São Paulo Hotelaria Ltda. and operated by InterContinental Hotels Group

圣保罗洲际酒店
Inter Continetal, Sao Paulo,Brazil

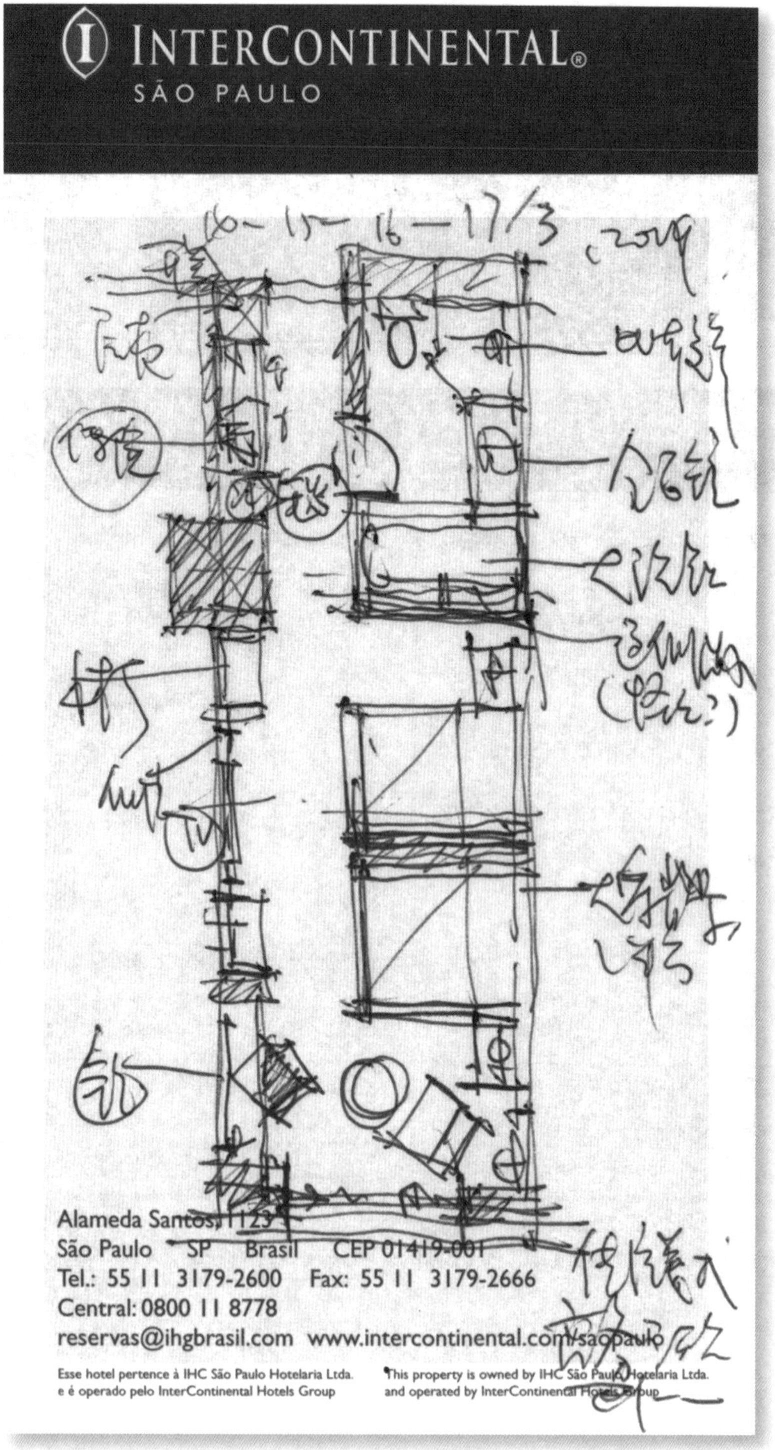

INTERCONTINENTAL®
SÃO PAULO

Alameda Santos, 1123
São Paulo SP Brasil CEP 01419-001
Tel.: 55 11 3179-2600 Fax: 55 11 3179-2666
Central: 0800 11 8778
reservas@ihgbrasil.com www.intercontinental.com/saopaulo

Esse hotel pertence à IHC São Paulo Hotelaria Ltda. e é operado pelo InterContinental Hotels Group

This property is owned by IHC São Paulo Hotelaria Ltda. and operated by InterContinental Hotels Group

Runaway Temperature

失控的温度

2019年3月16日

March 16, 2019

最后一站：巴西圣保罗，大城市，入住的是大品牌的洲际，想必组织方也想以“平稳”结束整个行程(当然也听到同行的团友对这一贯的商务酒店的吐槽！)，可以理解！

The last stop of this trip was Sao Paulo, a big city in Brazil. We stayed in Inter Continental, the big brand. I thought the organizer wanted to end the trip steadily. (Of course, our group members complained about this kind of business hotels.) I understood their complaint!

标准化的平面，不过不失，过度的床品配饰，红白搭配，不高档而易陈旧（红色不易保持）。天气热，开空调，理所当然。而过大的噪声，不可忍受，被迫关了空调，心静自然凉！

The standard plane was neither good nor bad. Excessive bed accessories in red and white were not fancy but easy to get old (red color was not easy to keep clean). The weather was so hot that of course I turned on the air conditioner. But the noise was too much for me to bear. I had to turn it off. The peace nature made me cool down.

而洗手间（四件配套）颇窄，真的不明白，外国人怎么可以塞进“淋浴间”去冲凉（大约800x830，净空，斜切角）。最要命的是热水时冷时热，大约30秒一变化，让人没法忍受，可能以前的老酒店不是双供热水系统（专业分析一下），但这么频繁的温度变化，确实少见。我也难以想象“这种结果”如何形成，也许以后的《住哪？》可以就问题而提出并解决，就更有研究和操作的意义。

The washroom (with four pieces) was quite narrow. I was really hard to understand how the strong foreigners had a shower in such a small room (About 800×830, clear space, angle of chamfer).What's more, the hot water was sometimes hot but sometimes cold, it changed every thirty seconds. It was hard to stand for us. Perhaps these old hotels didn't have double heating system (according to the specialty analysis), but the frequent changing temperature like this was rare.I could hardly imagine how to make it. Maybe in the following Where to stay?, it was more meaningful to study and practise if I could find out the ways to solve the problems.

房间温度（水温及空调）要做得好（完美），真的不容易，当然住客的意见（投诉）也是很重要的，你，要敏感一些啊!

It was not easy to keep the room temperature (including water temperature and the air conditioner) good (perfect). The opinions (or complaints) of guests were also important. So you should be more sensitive!

海口香格里拉大酒店
Shangri-La hotel
HAIKOU

68 海口香格里拉大酒店

★★★★★

SHANGRI-LA HOTEL HAIKOU CHINA

Address : 256 Binhai Road, Xiuying District, Haikou, Hainan, China
中国海南省海口市
秀英区滨海大道256号

Telephone : +86 898 6870 7799

Fax : +86 898 6872 2111

Http : //www.shangri-la.com

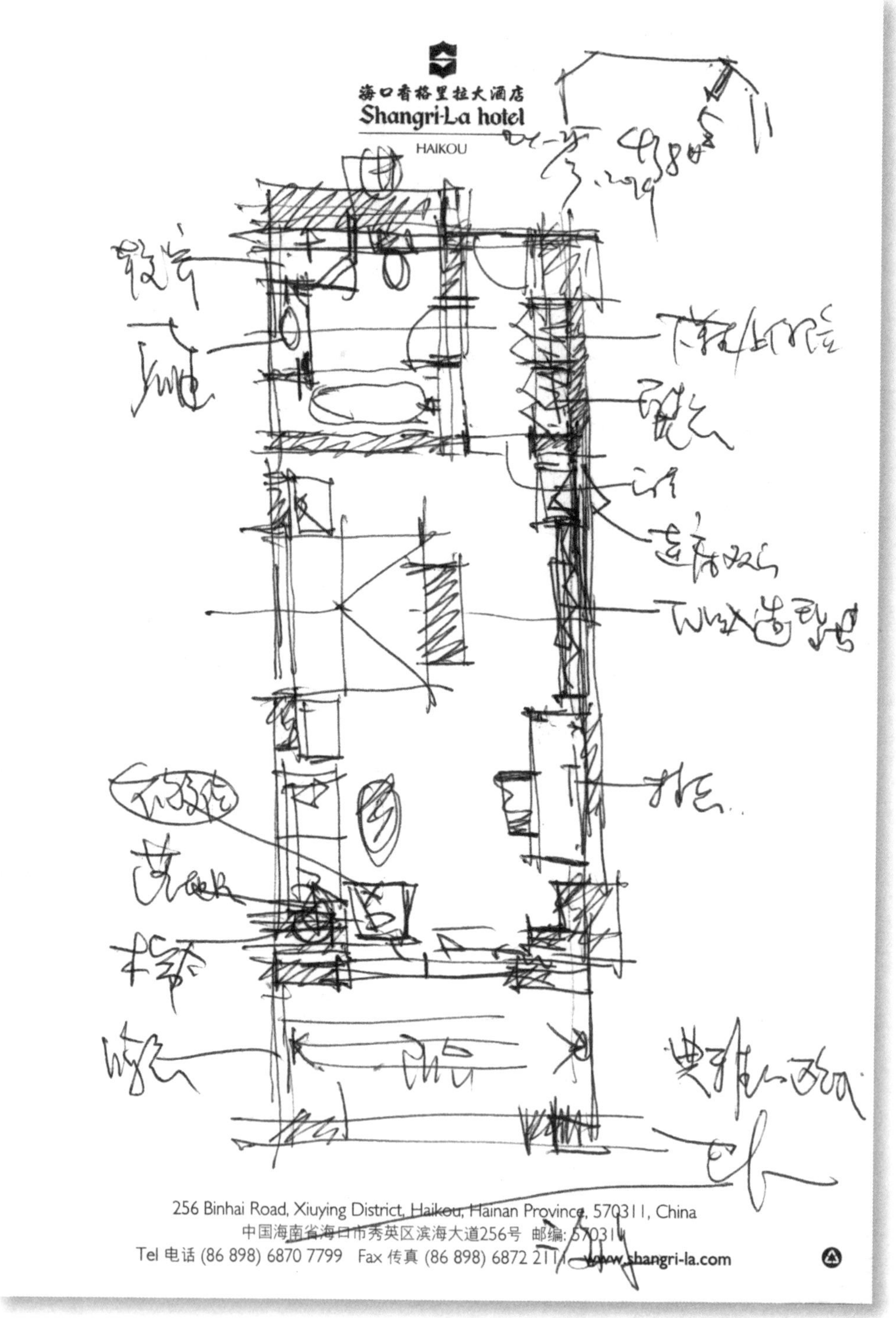

海口香格里拉大酒店
Shangri-La hotel
HAIKOU

256 Binhai Road, Xiuying District, Haikou, Hainan Province, 570311, China
中国海南省海口市秀英区滨海大道256号 邮编: 570311
Tel 电话 (86 898) 6870 7799 Fax 传真 (86 898) 6872 2111 www.shangri-la.com

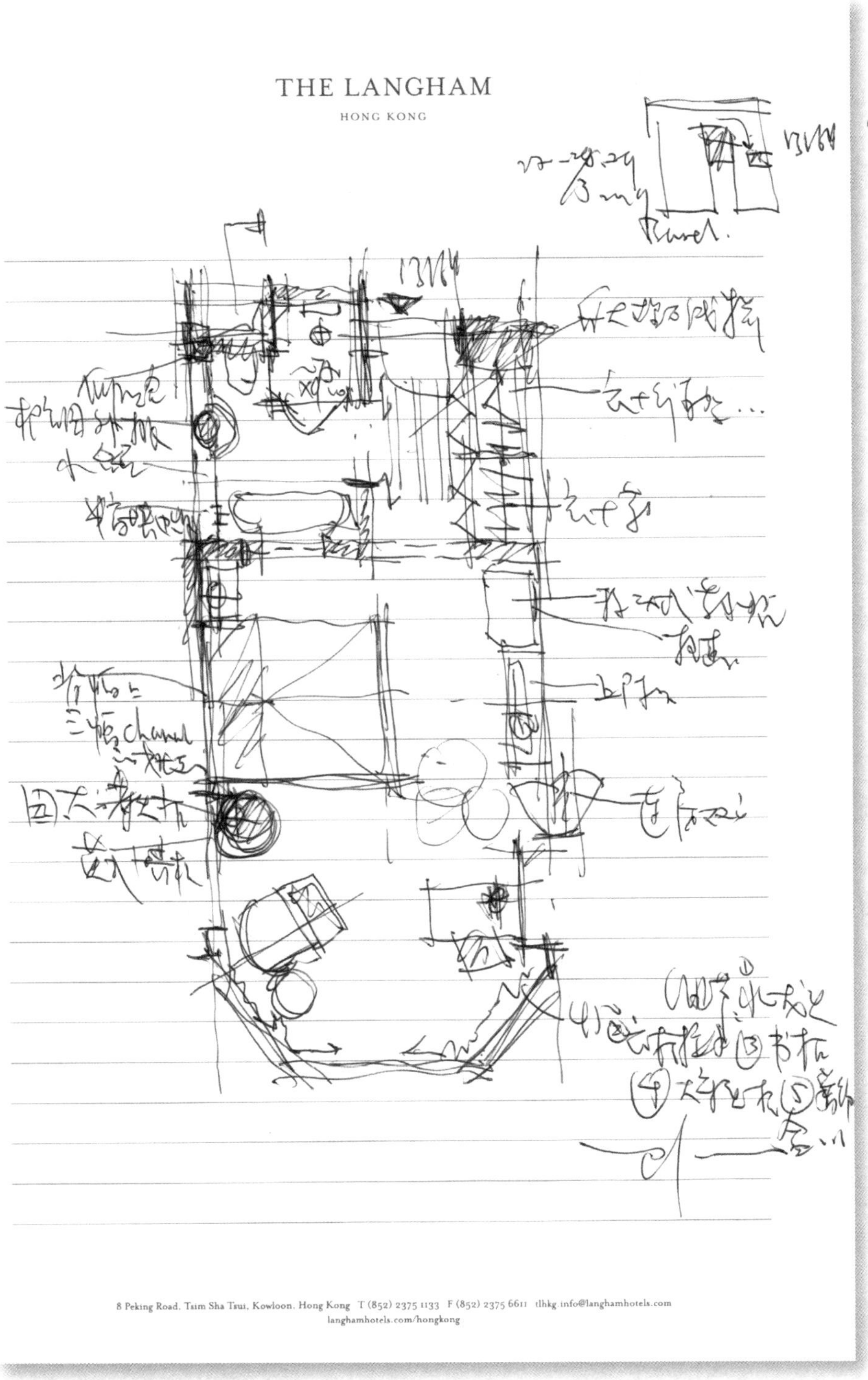

THE LANGHAM

HONG KONG

8 Peking Road, Tsim Sha Tsui, Kowloon, Hong Kong T (852) 2375 1133 F (852) 2375 6611 tlhkg.info@langhamhotels.com
langhamhotels.com/hongkong

THE LANGHAM

HONG KONG

香港朗廷酒店

★★★★★

The LANGHAM HONG KONG CHINA

Address : 8 Peking Road, Tsim Sha Tsui, Kowloon, Hong Kong, China
中国香港特别行政区尖沙咀北京道8号

Telephone : +852 2375 1133

Fax : +852 2375 6611

Http : //www.langhamhotels.com/hongkong

THE LANGHAM

HONG KONG

IN RESIDENCE

8 Peking Road, Tsim Sha Tsui, Kowloon, Hong Kong
T (852) 2375 1133 F (852) 2375 6611 tlhkg.info@langhamhotels.com
langhamhotels.com/hongkong

THE LANGHAM

HONG KONG

IN RESIDENCE

8 Peking Road, Tsim Sha Tsui, Kowloon, Hong Kong
T (852) 2375 1133 F (852) 2375 6611 tlhkg.info@langhamhotels.com
langhamhotels.com/hongkong

THE LANGHAM

HONG KONG

IN RESIDENCE

8 Peking Road, Tsim Sha Tsui, Kowloon, Hong Kong
T (852) 2375 1133 F (852) 2375 6611 tlhkg.info@langhamhotels.com
langhamhotels.com/hongkong

THE LANGHAM

HONG KONG

IN RESIDENCE

8 Peking Road, Tsim Sha Tsui, Kowloon, Hong Kong
T (852) 2375 1133 F (852) 2375 6611 tlhkg.info@langhamhotels.com
langhamhotels.com/hongkong

Old Hotels, How to Win the Market?

老酒店，凭什么？（取胜市场）

香港巴塞尔展览，组织方安排入住有相当历史的位于尖沙咀的朗廷酒店（1989年开业，2010装修），本应不会有惊喜。

During Hong Kong Basel Exhibition, the organizer arranged us to stay in the Langham Hotel with long history(which opened in 1989, and was refurbished in 2010). I didn't hope any surprises.

入住，大堂古典豪华，当然也生意兴隆，因为地段好，邻近购物圣地：海港城，广东道，1881号。

Checked in, I found the lobby was classic and deluxe. The business was booming because of the good location which was near to the shopping paradise: Harbor City, at 1881Canton Road.

香港朗廷酒店

The Langham, Hong Kong, China

老酒店还有什么好看的呢？凭什么来留住到访的客人呢？一向的朗廷，英伦风格，房间严谨精美：白色墙面（手扫漆），基本对称；严谨的布局给人第一感觉就相当“绅士”，不算很大的客房，循例(天书式的) 洗手间四件套。但是细节可谓做到极致，偏心的外凸式洗手台，大理石圆角一气呵成，关键是尺度，开关几乎贴墙，分毫不差；挂墙式马桶左右不差，而不过分；小的淋浴间亦是精致，双层的小转角层板极尽心思，浴缸稍短。开关墙面布局合适，左右双门的处理解决了走廊的左右关系，亦作为全身镜。还有其善用房间的对比尺度，家具的造型选式、选色相当微妙，白墙，深色家具，小写字台，大的圆的床头柜（其中的一边）可与休闲区组合成聊天区域。难得之处，这么紧凑的平面层居然茶水柜独立而舒适，还配有书柜，真的“五脏”俱全。旧酒店虽侧拼不过时尚流行的“样子”，但其细节和体验还是相当有深度和其独特的吸引力。我相信这是留客的重要手段。

What was the charming of the old hotel? Why attracted the guests? The Langham Hotel was always in British style. The guestrooms were rigorous and exquisite with white wall (hand painting). The structure was generally symmetrical. You would feel gentlemen style at first glance by the rigorous layout. The living room was not very big. The bathroom follows the common pattern with four sets. But the details were really freaky. the eccentric outer sink with marvel round corner was done in one take. The scale was the key. The switches were embedded into the wall without any space; the wall-hanging toilets was put in the middle precisely. The small washroom was also exquisite. The perfect part was the small double corner shelf. It was a pity that the bathtub was a little short. The wall with switches was arranged reasonable. Double doors which were on both sides dealt with the relationship of the corridor. They were also used as full-length mirrors. The details which impressed me were the good use of contrast size of the guestroom. The style and color of furniture were very subtle including white wall, the dark- color furniture and a small table. The chat area was made up of one side of the big round night table and the leisure space. It was rare that the tea cabinet was independent and comfortable in such a compact room. There was also a bookshelf to show off. It was really fully-equipped. Although this kind of old hotels was not as fashionable as the trendy ones, the details and the experience in them still had unique attractions. I believe it was the most important way to retain the guests.

老酒店，有情怀，仅仅地段就能加分，而且有品位的细节，相信增色不少，值得！

The old hotel is full of feelings, not only the location but also the tasty details can make a deep impression on guests, it is worth staying!

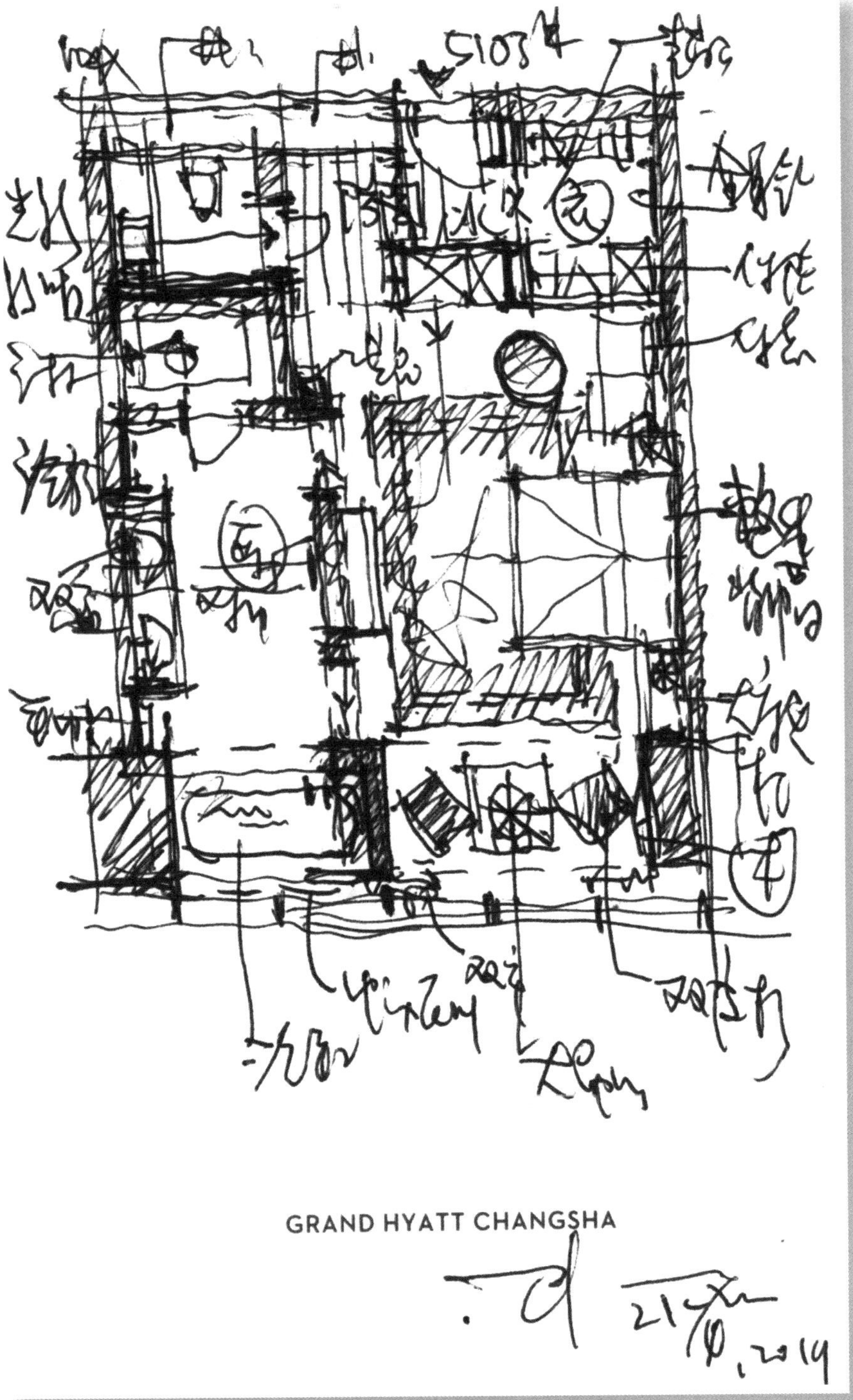

GRAND HYATT CHANGSHA

长沙君悦酒店

★★★★★

GRAND HYATT CHANGSHA CHINA

Address : No.36 Middle Xiangjiang Road, Tianxin District,Changsha, Hunan,China
中国湖南省长沙市
天心区湘江中路36号

Telephone : +86 731 8823 1234

Http : //www.hyatt.com

E-mail : Changsha.grand@hyatt.com

GRAND HYATT CHANGSHA

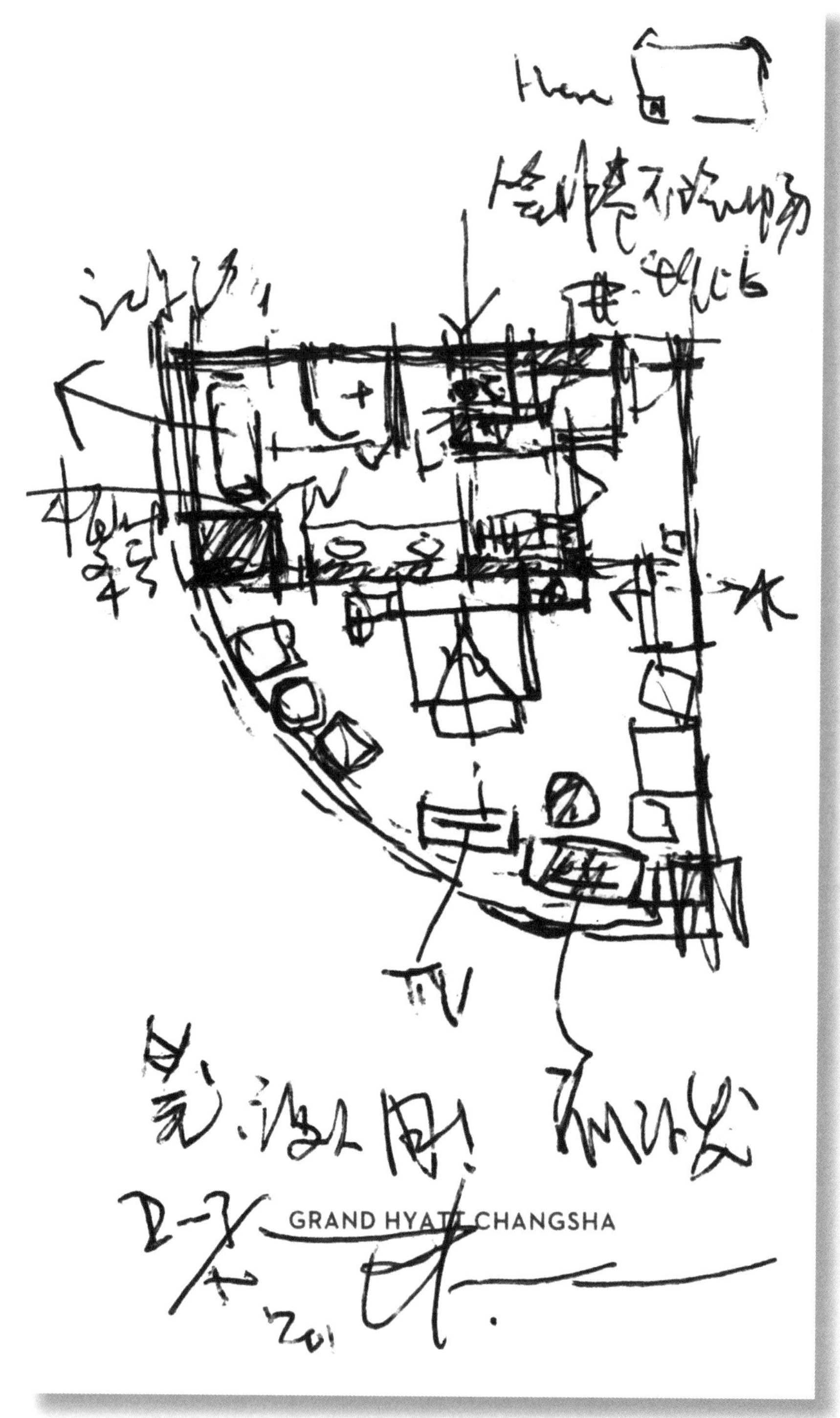
TV
GRAND HYATT CHANGSHA

[illegible]：

来长沙看了热闹，——一个1988[illegible]

[illegible]

[illegible]：“一个[illegible]

[illegible]”，[illegible]

[illegible]

[illegible]

[illegible]

[illegible]

[illegible]

[illegible]！

GRAND HYATT CHANGSHA

[illegible]

[illegible]

[illegible]

[illegible]

[illegible]

[illegible]

[illegible]

[illegible]

[illegible]

[illegible]

GRAND HYATT CHANGSHA

[illegible]

[illegible]

"Highlights" and Regrets

"亮点"与遗憾

来长沙凑个热闹——参观一个1985年出生的艺术家董晓亮小帅哥的展览，有策划："一个期待或者新的奇迹"，选址新潮的时尚网红地点：谢子龙影像艺术馆。

Gather in Changsha to join the fun—I came to visit an exhibition of Dong Xiaoliang, a handsome artist who was born in 1985. The exhibition was planned, "An Expecting or New Miracle" in Xie Zilong Image Gallery which was the latest fashion internet popular spot.

这哥们的作品怎么定义，不懂。我以为只能是合成影像艺术品（不过艺术无边界，也无固有的章法），而且喜欢炒新：新人，新技法，新藏品，便宜！

I didn't know how to define this guy's works. I thought they were just artworks of composite image(But art has no boundaries or regular patterns). Much hype was popular, new people, new techniques and new collections, they were cheap!

还是住在江边的君悦，不同的房型。细看，细拍，细画，真的还是有发现老先生林丰年（LTW公司）的亮点。

I still stayed in the Grand Hyatt by the river but in different types of guestroom. Watched carefully, took pictures and drew carefully, I really found the highlights of Mr. Lin Fengnian (LTW Company).

如1. 回风口在靠窗边的休闲、写字区；2. 巧妙的入口迂回令空间变化而隐私；3. 大尺度的门（滑门）装饰，有档次(当然有投入的保障)，"百看不厌"。

For example, 1. The air outlet was near the leisure and writing area by the window; 2. A clever circuitous entrance made the space changeable and private. 3. A large sliding door was used as decoration, with style (of course the investment was guarantee). Never be bored of seeing.

当然也会有遗憾：1. 洗手间的门是否选用趟门（有声），视线反光，声的干扰无法解决；2. 好几个插座已经"半坏"了（难用，像小兔一样跳来跳去）。

There were still some regrets, 1. Whether chose sliding door for restroom(it made a lot of noise), reflected light. There was no way to solve the sound interference; 2. A few sockets had almost been broken(They were difficult to use or jumped like rabbits).

做一间耐得住时间考验的酒店，不易（当然五年左右会更新）。

It is not easy to build a hotel which can stand the test of time(Of course it should be update around five years.)

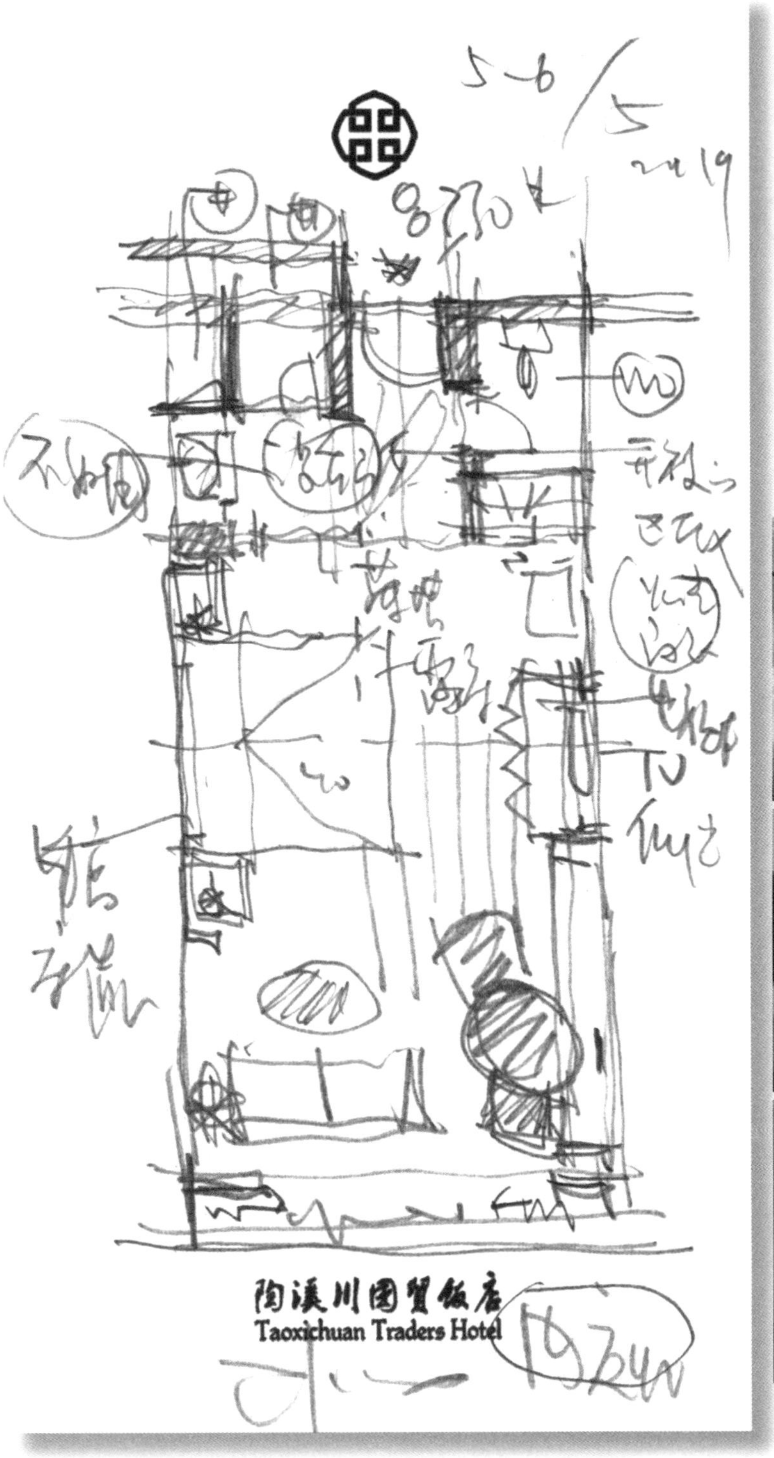

陶溪川国贸饭店
Taoxichuan Traders Hotel

陶溪川国贸饭店

71

TAOXICHUAN TRADERS HOTEL CHINA

Address : No.150 Xinchang West Road, Zhushan District,Jingdezhen, Jiangxi, China
中国江西省景德镇市
珠山区新厂西路150号

Telephone : +0798 873 6666

陶溪川国贸饭店
Taoxichuan Traders Hotel, China

陶溪川国贸饭店
Taoxichuan Traders Hotel

陶溪川国贸饭店
Taoxichuan Traders Hotel

陶溪川
Taoxichuan

Am I in the City of Ceramics?
可以来点陶都Feel吗?

这里418元/天，前一天的民宿480元，有对比，才知道谁好谁不好。

It cost 418 yuan one day. The B&B which I stayed cost 480 yuan the day before yesterday. It showed which was better by comparing them.

过节来凑热闹，景德镇“居然”有这么好的文创项目——陶溪川。而这间国贸酒店就在其中，最后一晚住这里，方便逛逛集市，感受这里的白天和晚上。的确非常值得学习，一个俄式建筑改造的创意社区（虽然未完完全全地招商结束）。

I came here to join the fun during the festival. To my surprise, there was a Cultural Creativity Project in Jingdezhen —Taoxichuan while this Traders Hotel was in it. I stayed here for the last night. It was convenient to experience the street fair, enjoyed the day and night here. It was really worth studying, this cultural creativity community was transformed by a Russian architecture(Although the investments hadn't finished yet).

这间酒店也是红砖+浅米色的阳台建筑造型。细节、光影、比例、布局可圈可点。每层有近60间客房，商务会议型，倒是房间有不少的毛病：

The architectural style of this hotel was red bricks and light beige balcony. Details, light, proportion and plan were perfect. As a business conference hotel, there were nearly 60 guestrooms on each floor, but still some defects in the room.

陶溪川
Taoxichuan

追求新颖而将厕、浴分离，开门见床，不知入住的其他顾客有什么意见，我感觉不好。

In order to pursue new and unique, toilet and bathroom were set apart. The bed was seen while opening the door. I didn't know how other guests felt, but I felt terrible.

开放式的洗脸区，最大的问题是灯光对睡眠区的影响，没有设滑门。淋浴间的水易溢满（果然中招了）。

The most problem was that the light in the open face-washing area affected the sleeping area because there was no sliding door here. Water overflowed easily. (I was really caught!)

灯光控制相对乱，没有“陶都”的特色。

The control of lights was quite confusion and lacked specialty of "City of Ceramics".

当然当下也没有配我需要的签字笔，只能用铅笔了。

Of course there was no pen I needed at present. I had to use pencil instead.

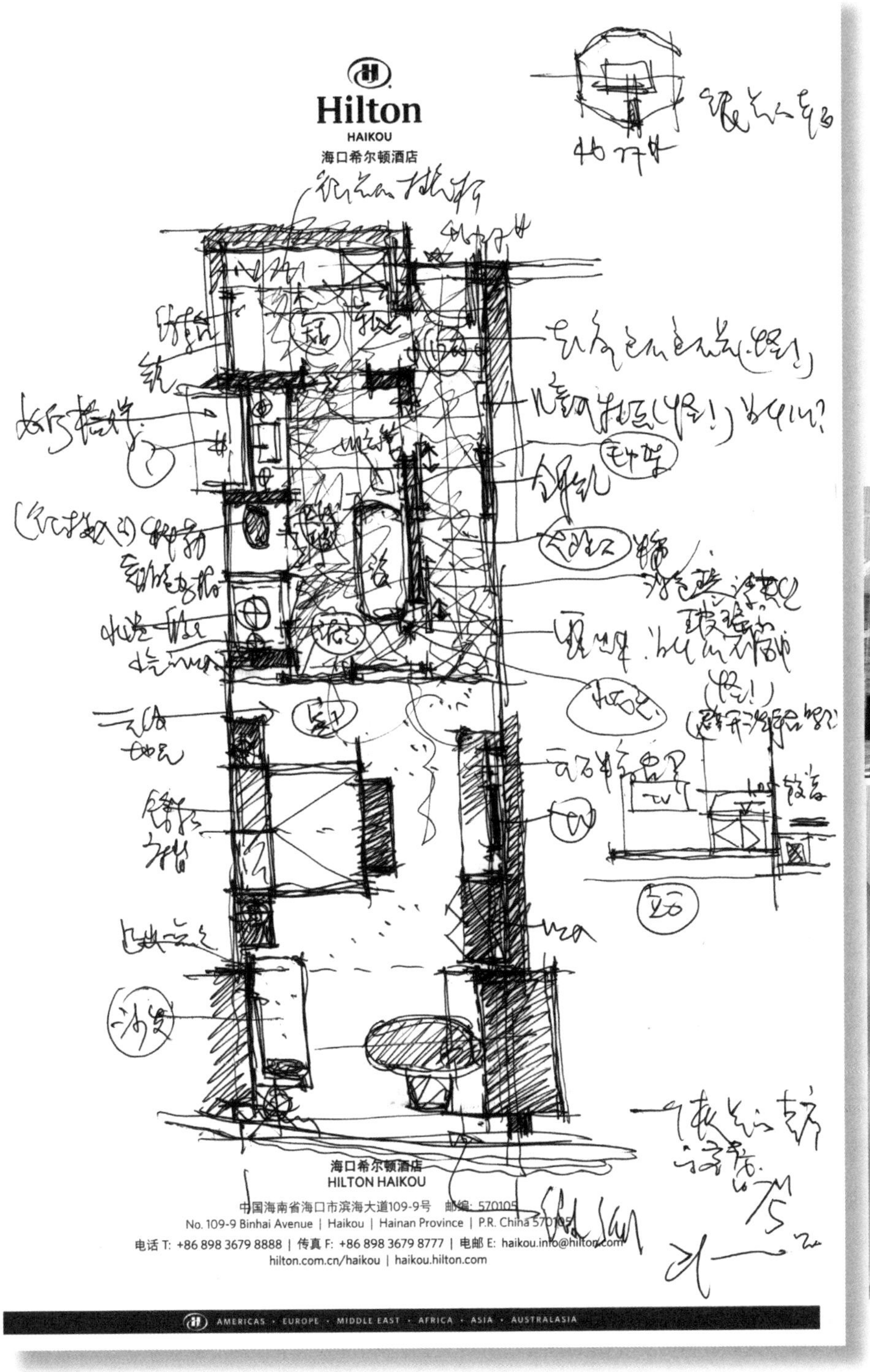

海口希尔顿酒店

72

HILTON HAIKOU CHINA

Address : No.109-9 Binhai Avenue, Haikou, Hainan, China
中国海南省海口市
滨海大道109-9号
Telephone : +86 898 3679 8888
Fax : +86 898 3679 8777
Http : //www.hilton.com.cn/haikou
//www. haikou.hilton.com
E-mail :haikou.info@hilton.com

Hilton
HAIKOU
海口希尔顿酒店

海口希尔顿酒店
HILTON HAIKOU
中国海南省海口市滨海大道109-9号 邮编：570105
No. 109-9 Binhai Avenue | Haikou | Hainan Province | P.R. China 570105
电话 T: +86 898 3679 8888 | 传真 F: +86 898 3679 8777 | 电邮 E: haikou.info@hilton.com
hilton.com.cn/haikou | haikou.hilton.com

AMERICAS · EUROPE · MIDDLE EAST · AFRICA · ASIA · AUSTRALASIA

海口希尔顿酒店
Hilton, Haikou, China

第一版很好，修了也许就明快

可能是因为受第二版的影响（神话故事、中国房子与材料的关系）推出西方房子，将情感推到一个宏观的问题，反而没有意识到了！

第一版还[illegible]加上我的没有与本质（也许也是本书的[illegible]的指导吧！）太多的描述，更加说明和完全感觉加强，情不自禁地忘记了，那会思路也会变为[illegible]的建筑和各种形式（如我未注意的），若能还原本意，是感受一下没有[illegible]的重要性。

其实我写的是想象的"界定"，如何住在不同地方的区域，那是[illegible]、[illegible]、[illegible]（用[illegible]），[illegible]，那就是一条[illegible]的线。不能放弃[illegible]（北京）《[illegible]》《[illegible]》[illegible]的[illegible]，[illegible]一下[illegible]的合适的区域，[illegible]，但这都不是[illegible]。

也许[illegible]"[illegible]"[illegible]。

[illegible]

Nothing Strange

见怪不怪了

可能是因为总平面的原因，中间鼓起来，中间的房间特别长。推门进入房间，哗，非常非常长的一个走廊啊！很久没有遇到了！

Perhaps because of the total plane, the middle part rose up so the guestroom was quite long. When entering the room, Wow, what a long corridor! I had never seen such a long corridor for long time.

第一眼还是挺不舒服的！加上较为深色的木饰面（也许也是希尔顿品牌手册的指导吧！），大的对比度，更加让这种窄长的感觉加强，情不自禁地这里看看，那里翻翻，也思考着怎样可以让这种感觉减轻（如果我来主笔的话），最后还是拿起笔感受一下设计师的体验。

I felt uncomfortable at the first sight. The dark wooden wall (maybe it followed the Brand book of Hilton) showed a big contrast and made the room much longer and more narrow. I couldn't help having a look here and there, and also thinking about how to reduce this feeling(If I am the designer). Finally, I took up the pen to experience the designer's feeling.

真的不容易，关键是要“界定”，如果传统地界定不同的功能区域，那只能是干区：衣帽间、睡眠区、休闲区、写字/用餐区，湿区：洗、厕、浴，那就是这个一条长走道的效果。不然就像柏悦(北京)（见《住哪？3》P118）、君悦（广州）（见《住哪？1》P32），以及广州康莱德的穿越式：打散洗手间，故意让客人迂回一下，穿过洗手间的恰当区域，解决一条长走廊的问题，但这样不一定会满足品牌的要求。

It was really not easy. The key was to "Define". According to the tradition, it was divided into two parts, dry area including coatroom, sleeping, leisure and writing or dining area; wet area including washing, toilet and bathroom. All these made the long path. Otherwise, it would be just a passage like Park Hyatt (Beijing) (See Where to stay? 3, Page 118), Grand Hyatt (Guangzhou) (See Where to stay? 1, Page 32), as well as Guangzhou Conrad, set the bathroom apart to make a circuitous route. Solved the long corridor by going through the right area of restroom, but it might not satisfy the requirement of the band.

也许设计就是这样，“顺得哥情失嫂意”，见多了，就不怪了。

Maybe this is design. It can't please everyone. You won't feel strange if you see it more.

海口希尔顿酒店

Hilton, Haikou, China

Hilton

HAIKOU

海口希尔顿酒店

收件人 / TO:	发件人 / FROM:
电话 / TEL:	电话 / TEL:
传真 / FAX:	传真 / FAX:
抄送 / CC:	
地址 / LOCATION:	主题 / SUBJECT:
日期 / DATE:	页数 / PAGES:

备注 / NOTES

17.5.28 2019

[illegible]

海口希尔顿酒店
HILTON HAIKOU
中国海[illegible]省[illegible]市滨海大道109-9[illegible] 邮编：570105
No.109-9 [illegible]hai Avenue, Haikou, [illegible] Province | P.R. China 570105
电话 T: +86 898 3679 8888 | 传真 F: +86 898 3679 8777 | 电邮 E: haikou.info@hilton.com
hilton.com.cn/haikou | haikou.hilton.com

[illegible]

I am When I Writing

我写故我在

是否其他人都会有这样的迷惘和停顿，做一件事，做着做着会突然问：这件事有意义吗?

Do other people feel confused and want to stop at some moment like me? When I do something for long, I will ask myself suddenly whether it has meaning or not.

资讯年代，手机就是宠儿， 纸质产品成“恐龙”，公司取消了坚持了十几年的纸质季刊（作品集），也取消了每年的书籍及杂志的订阅，那纸质书还有意义吗?

The information age, mobile phones are the minions Paper products become "dinosaurs". Our company has cancelled our quarterly magazine (our works collection) which I have persisted for more than ten years and also cancelled magazines and books subscriptions. So are paper books meaningful?

偶尔收了一套“莎士比亚全集”，虽然内容经典，但排版方式新颖，包装精美，等等。也许我是传统的人（已过51之年，52了）。墨香（淡），手感（抚摸），温度（心情），声音（翻页）……爱不释手（机）。

I got a set of Complete Works of William Shakespeare occasionally. Though the content is classic, the typesetting is innovative and the package design is beautiful and so on. Maybe I am a traditional person (fifty-one years has passed and I am fifty-two years old now). I like the faint smell of ink, the good feeling (touch), the temperature (mood), the sound (turning pages) …I can't put it down (maybe "it" refers to mobile phones).

《住哪? 》只是杂文，记录我的到处吃吃喝喝、走走看看、住住睡睡的琐碎事情，一不经典，二不惊艳，更加不是一个什么名人的手笔，也许只剩下坚持、持续和持久了。30年十本《住哪? 》（计划进行中），也许是“时间”让它有了意义，有了传递的寄托和期望。

Where to stay? is only an essay which records my traveling and some trivial matters. It is neither classical nor amazing, even not a famous person's works, maybe only insistent, continuation and lastingness left. My plan is to spend thirty years writing ten books of Where to stay? (My plan is under way).I believe time will make my books meaningful and also gives sustenance and expectation of passing on.

那就继续吧，管它有什么意义，有没有意义，可以自娱自乐也是不错的，最好也把我的快乐，传递一点点给你。

I will go on. No matter what or whether it is meaningful or not. Enjoy myself. What's more, share my happiness with you.

亦间酒店 · 一座城

73

A HOTEL A CITY
CHANGSHA CHINA

Address : No.171 South Station Road,Yuhua District, Changsha, Hunan, China
中国湖南省长沙市
雨花区车站南路171号
Telephone : +0731 85900666

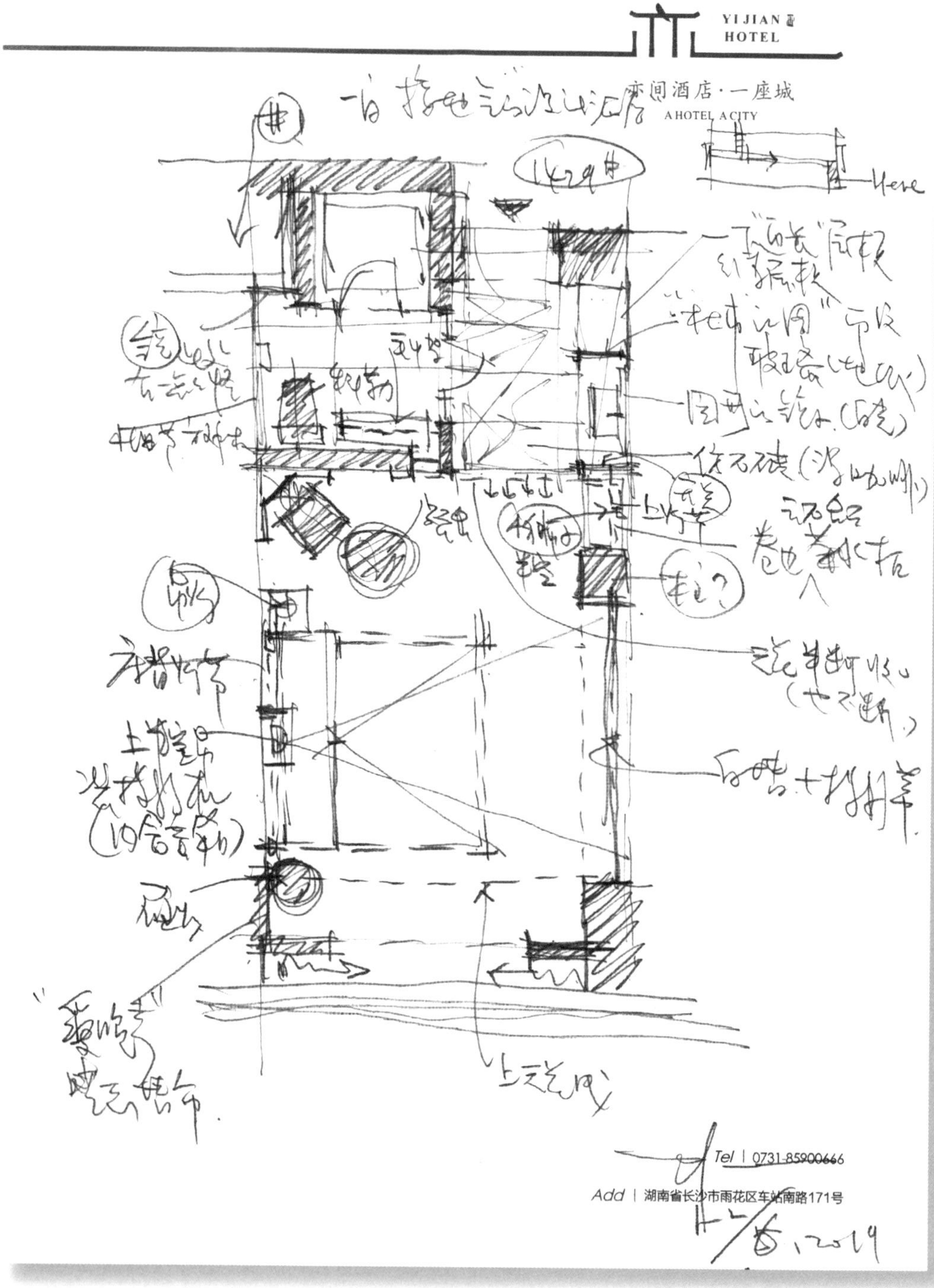

YIJIAN HOTEL

亦间酒店 · 一座城
A HOTEL A CITY

Tel | 0731-85900666
Add | 湖南省长沙市雨花区车站南路171号

亦间酒店 · 一座城
A Hotel A City, Changsha, China

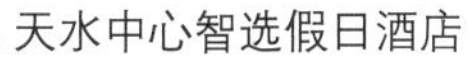

74 天水中心智选假日酒店

HOLIDAY INN GANSU CHINA

Address : No.78, Minzhu East Road, Qinzhou District, Tianshui, Gansu, China
中国甘肃省天水市
秦州区民主东路78号
Telephone : +86 0938 8237 777
E-mail : THQCCRSVN@HIEX8.com

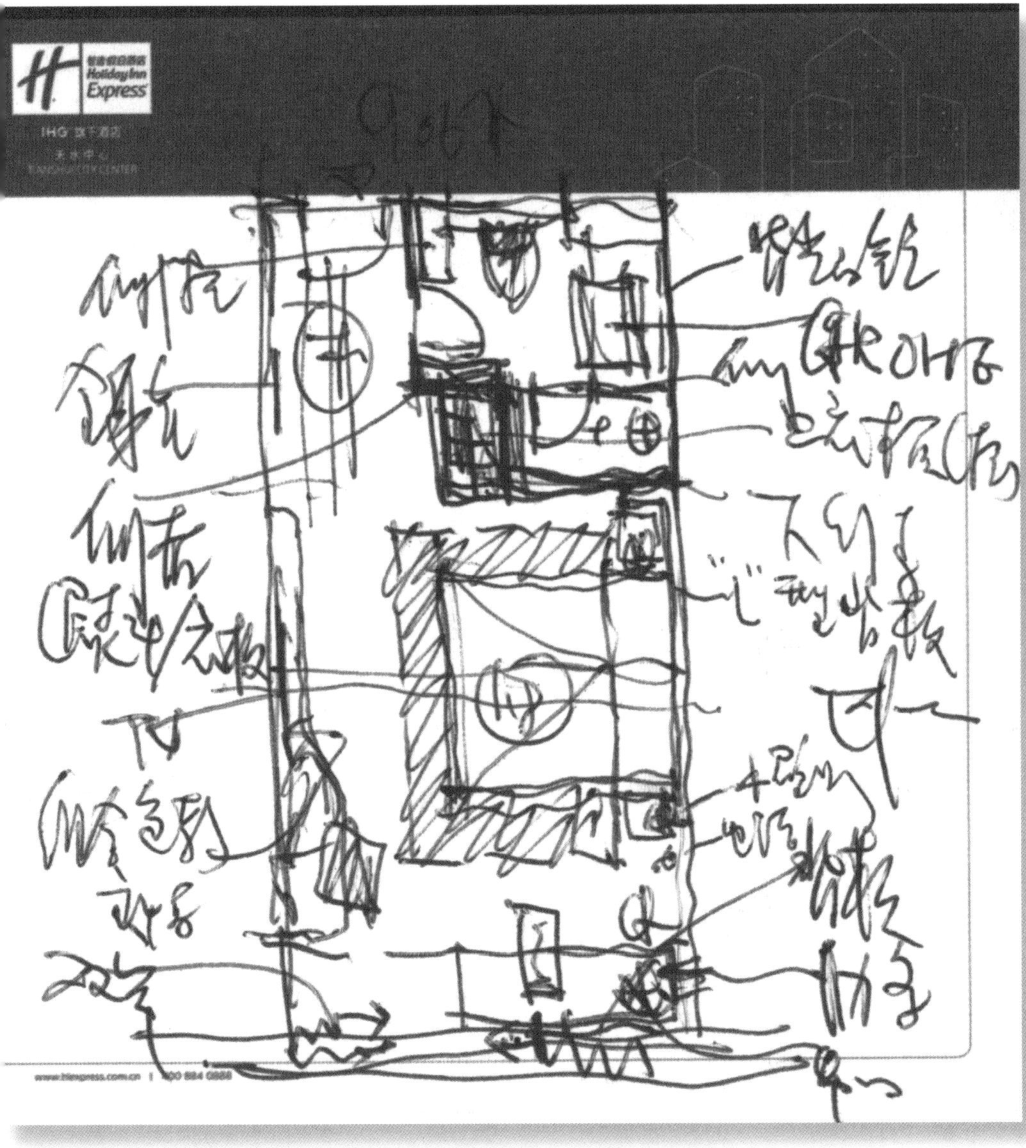

天水中心智选假日酒店
Holiday Inn Ganshu, China

75

龙潭湖宾馆

★★★★★

LONG TAN HU HOTEL
JIANGXI CHINA

Address : No.6 Jiyang Middle Road, Shangrao, Jiangxi, China
中国江西省上饶市
吉阳中路6号
Telephone : +0793-7936666/7938888
Fax :+0793-7936688

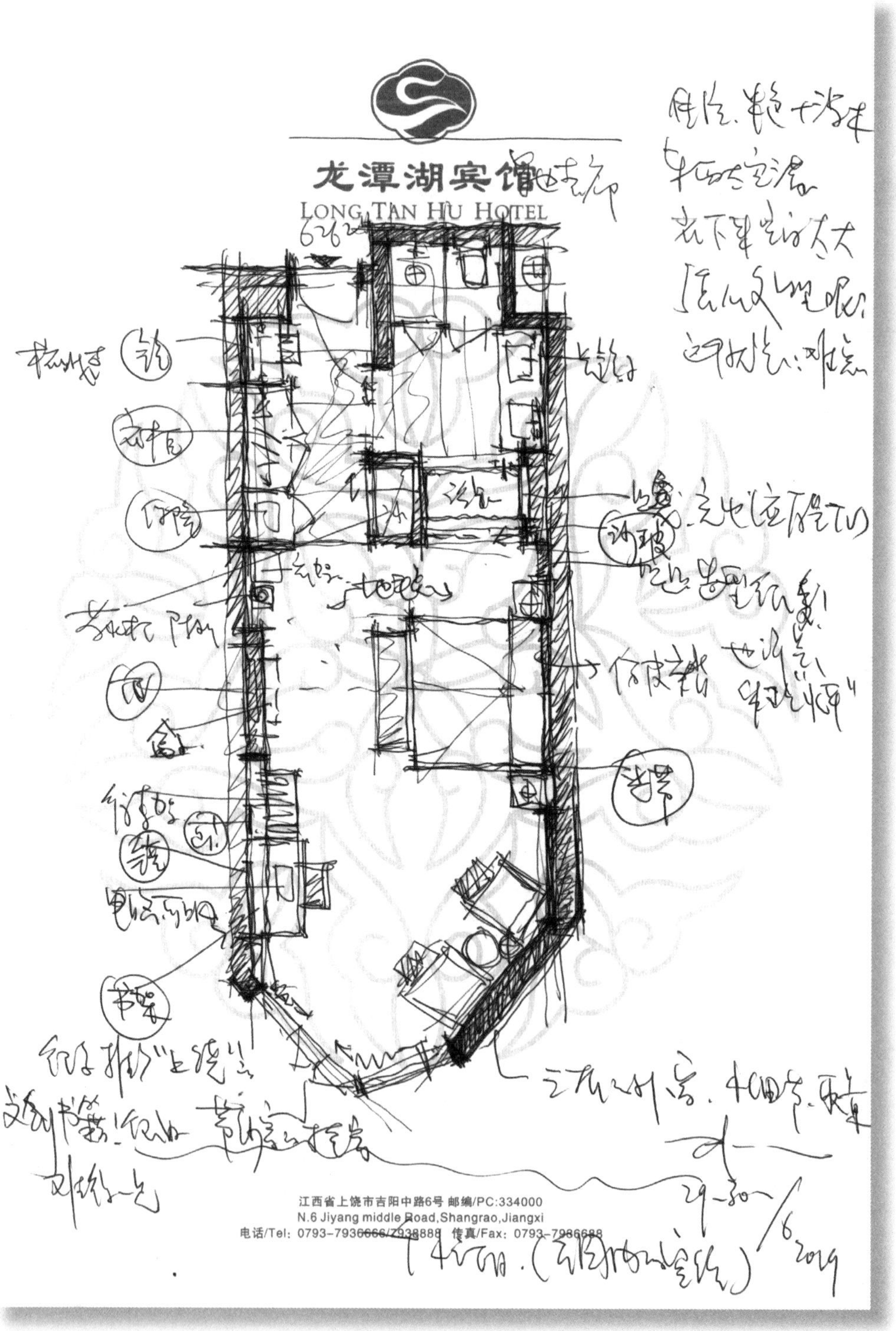

龙潭湖宾馆
LONG TAN HU HOTEL

江西省上饶市吉阳中路6号 邮编/PC:334000
N.6 Jiyang middle Road,Shangrao,Jiangxi
电话/Tel：0793-7936666/7938888 传真/Fax：0793-7936688

龙潭湖宾馆
LONG TAN HU HOTEL
| 江西 · 上饶 |

30/6.2019

没到人家那，如果你来的，会怎么样？

[illegible]

江西省上饶市吉阳中路6号 邮编/PC:334000
6 Jiyang middle Road,Shangrao,Jiangxi
电话/Tel: 0793-8366666/7838888 传真/Fax: 0793-7836688

龙潭湖宾馆
Long Tan Hu Hotel, Jiangxi, China

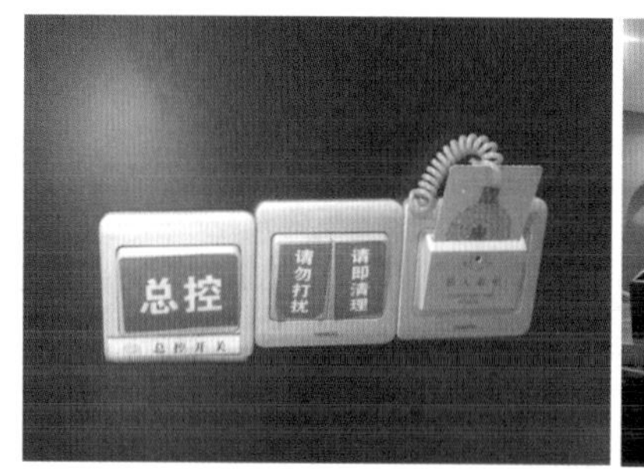

What If You are the Designer?

换你，设计会更好吗？

第一次到达这个地方，风景人文，吃吃喝喝，当然最后肯定是“住”。上饶是一个好地方，入住这间龙潭湖宾馆也是“有备而来”，面湖多树的园林式会议酒店，果然！

It was my first coming here. I walked around, enjoyed the local culture and food. Of course I stayed here at last. Shangrao was a good place. I was well prepared to stay in the Long Tan Hu Hotel which was a garden conference hotel with lots of trees and faced a lake. Sure enough!

房间很大，很传统的调子：米色基调加深色的木，但，比我想象中的“猛”；细节多，功能强，齐全，但是感觉“多余”中显老到，一侧梳妆台（少见），长的衣柜（可以住一个月啊！），右侧洗手间，大，四件，凹入式的茶水柜相当耗钱！

The large guestroom was quite traditional, beige style with dark wood. But, it was much better than I imagined: more details, powerful and fully functions. I thought it was so experienced though a little redundant. The dresser was on one side (seldom to see). The long wardrobe was big enough for the guests to stay for a month. The big restroom with four sets was on the right. The embedded tea cabinet cost a lot of money.

睡眠区挺宽大，复杂多变的床背幅，床头柜反而小小的，休闲区面湖，很不错。变化的外窗折线式亦让风景尽多地纳入眼帘。写字台，牛！有上网电脑（很老人家啊！），有一个书架，非常多的书籍，不常见，很大部分是关于“上饶”的：风景、美食、人文、历史……真的可以“涨知识”，也更加适合长住，解无聊，是很细心的软推广。大空间让你来做设计，看别人的挑刺容易，我来操刀，会更加好吗？值得思考！

The sleeping area was wide and the back of the bed was complex and changeable while the night tables were small. The leisure area faced the lake. It felt great. I enjoyed as many views as possible by the broken line window. The writing table was so cool! There was a desktop computer for surfing the net(What an old style!). The bookshelf with plenty of books was rare. Most of them were about scenery, delicious food, culture and history in Shangrao. It really made me learn a lot. I thought it was more suitable to stay here for long. I wouldn't feel bored. What a careful soft promotion! How to design such a large space? It's easy to find faults in other people's work. Will it be better if I design it? It is worth thinking over.

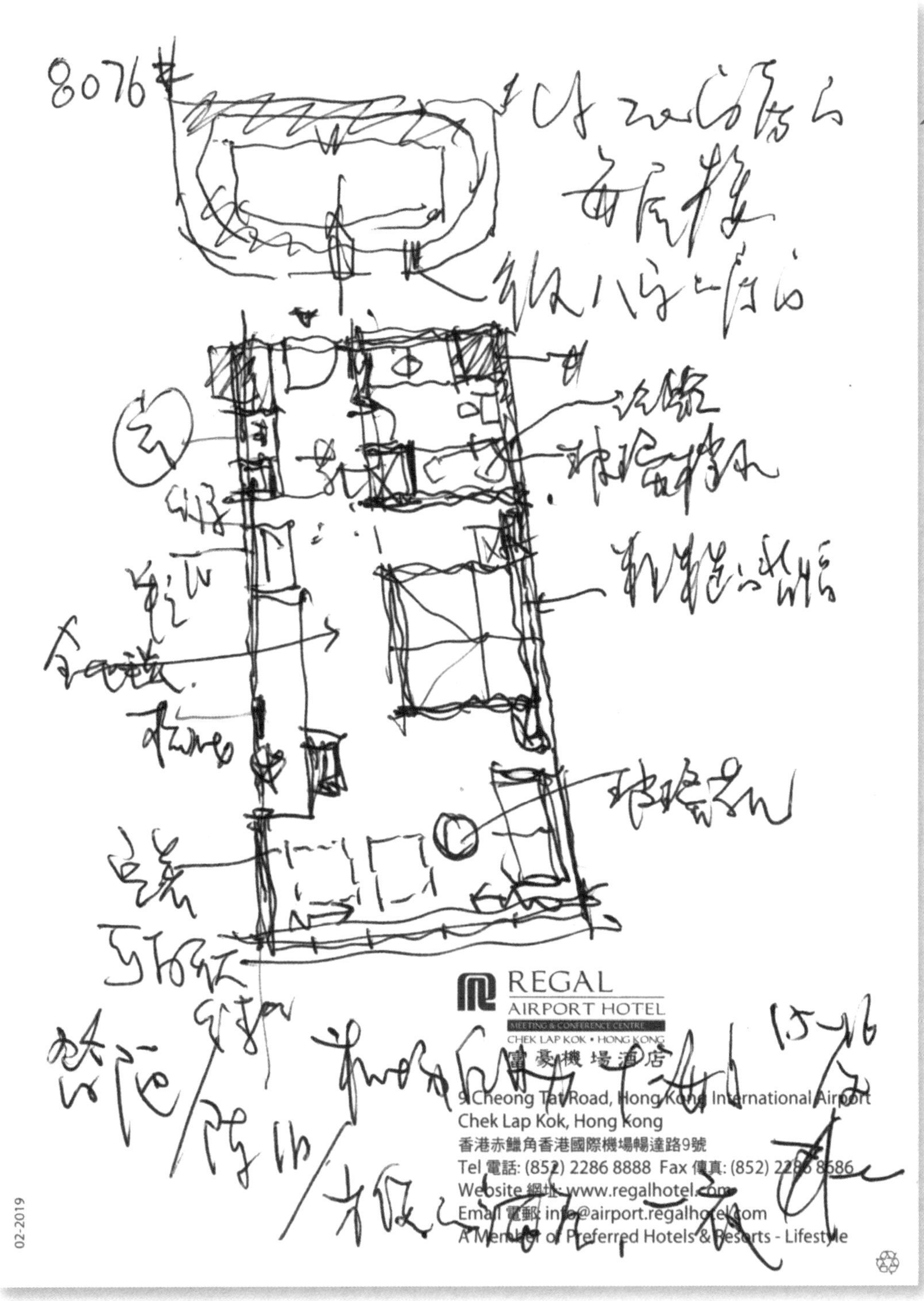

76 富豪机场酒店

★★★★★

REGAL AIRPORT HOTEL CHEK LAPKOK HONG KONG CHINA

Address : 9 Cheong Tat Road, Hong Kong International Airport Chek Lap Kok, Hong Kong,China
中国香港赤鱲角香港国际机场畅达路9号

Telephone : +852 2286 8888

Http : //www.regalhotel.com

E-mail : info@airport.regalhotel.com

8076

REGAL
AIRPORT HOTEL
MEETING & CONFERENCE CENTRE
CHEK LAP KOK • HONG KONG
富豪機場酒店

9 Cheong Tat Road, Hong Kong International Airport
Chek Lap Kok, Hong Kong
香港赤鱲角香港國際機場暢達路9號
Tel 電話: (852) 2286 8888 Fax 傳真: (852) 2286 8686
Website 網址: www.regalhotel.com
Email 電郵: info@airport.regalhotel.com
A Member of Preferred Hotels & Resorts - Lifestyle

02-2019

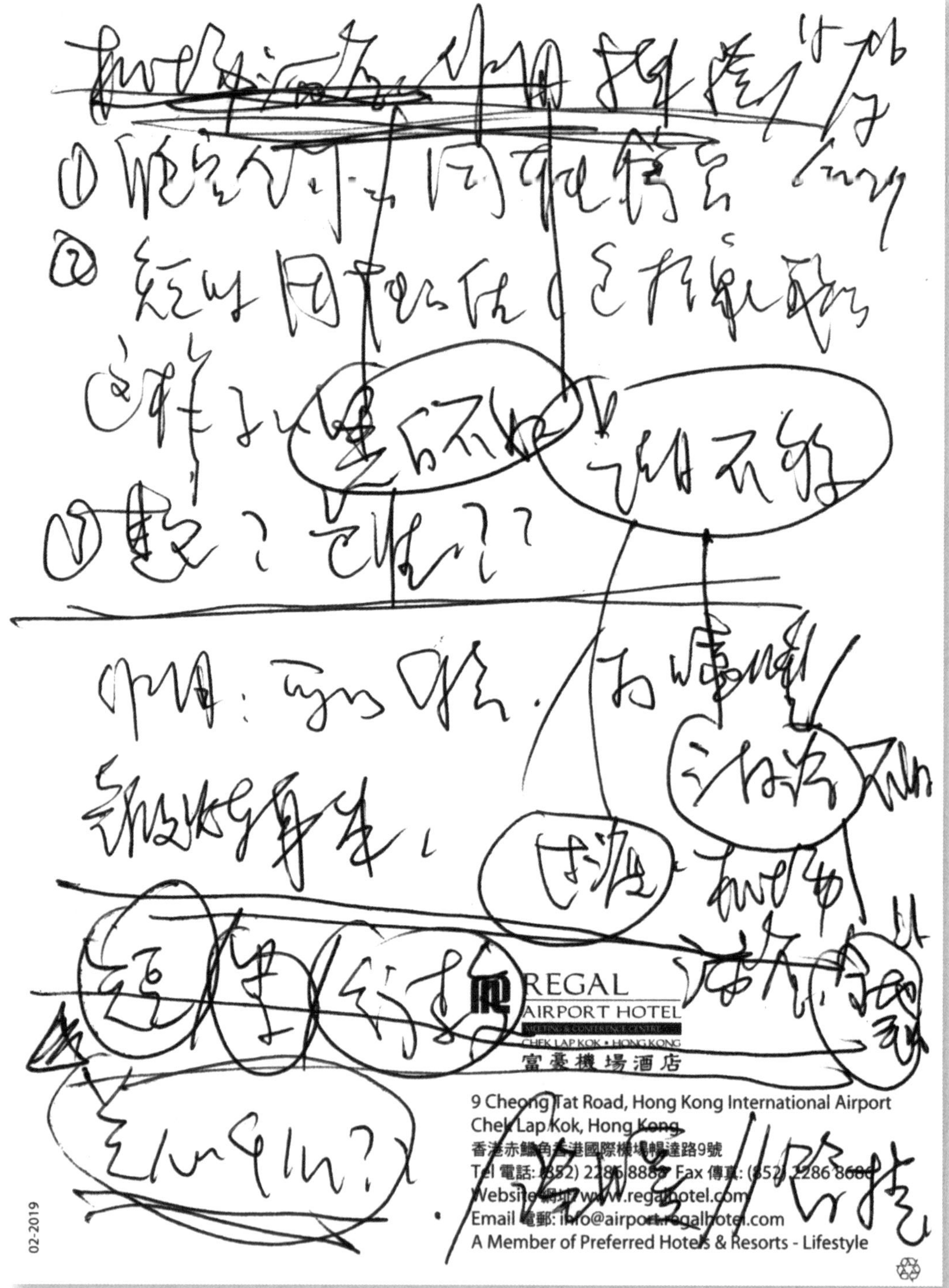

REGAL
AIRPORT HOTEL
MEETING & CONFERENCE CENTRE
CHEK LAP KOK • HONG KONG
富豪機場酒店
9 Cheong Tat Road, Hong Kong International Airport
Chek Lap Kok, Hong Kong
香港赤鱲角香港國際機場暢達路9號
Tel 電話: (852) 2286 8888 Fax 傳真: (852) 2286 8686
Website 網址: www.regalhotel.com
Email 電郵: info@airport.regalhotel.com
A Member of Preferred Hotels & Resorts - Lifestyle
02-2019

2019年7月15、16日
July 15-16,2019

1. 航空公司的周转停留。
2. 短期周转时住（包括像我们这样），黑白不好，题目不够。
3. 其他？是谁？？

作用：可以休息/打瞌睡/锻炼身体。过渡，机场，酒店，全地毡，
　　短、累、行李多。
　　关心什么？？温暖？简捷？淋浴不好。

1. Transfer and stay of the airline.
2. A short stay (like us), black and white are not good, the title is not enough.
3. What else? Who else?
Function: Rest, power snap, exercise.
The transition, the airport, the hotel and all the carpets.
Short, tired, too many languages.
What is concerned? Warmth?Convenience?The shower is not good.

美国东海岸游（2019年7月16~25日）

The Trip to the East of America

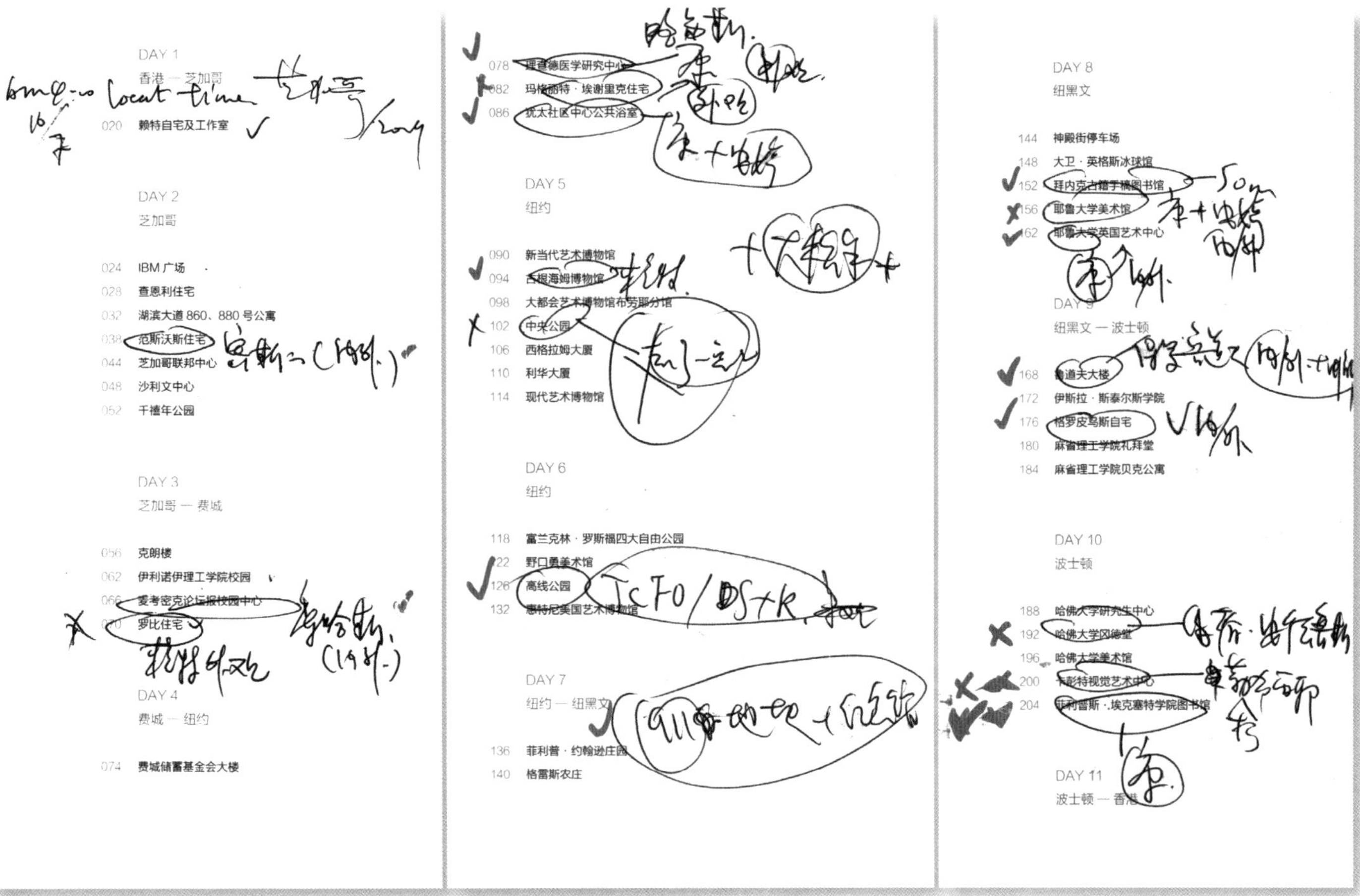

DAY 1
香港 — 芝加哥

020 赖特自宅及工作室

DAY 2
芝加哥

024 IBM 广场
028 查恩利住宅
032 湖滨大道 860、880 号公寓
038 范斯沃斯住宅
044 芝加哥联邦中心
048 沙利文中心
052 千禧年公园

DAY 3
芝加哥 — 费城

056 克朗楼
062 伊利诺伊理工学院校园
066 麦考密克论坛报校园中心
070 罗比住宅

DAY 4
费城 — 纽约

074 费城储蓄基金会大楼
078 理查德医学研究中心
082 玛格丽特·埃谢里克住宅
086 犹太社区中心公共浴室

DAY 5
纽约

090 新当代艺术博物馆
094 古根海姆博物馆
098 大都会艺术博物馆布劳耶分馆
102 中央公园
106 西格拉姆大厦
110 利华大厦
114 现代艺术博物馆

DAY 6
纽约

118 富兰克林·罗斯福四大自由公园
122 野口勇美术馆
126 高线公园
132 惠特尼美国艺术博物馆

DAY 7
纽约 — 纽黑文

136 菲利普·约翰逊庄园
140 格雷斯农庄

DAY 8
纽黑文

144 神殿街停车场
148 大卫·英格斯冰球馆
152 拜内克古籍手稿图书馆
156 耶鲁大学美术馆
162 耶鲁大学英国艺术中心

DAY 9
纽黑文 — 波士顿

168 鲁道夫大楼
172 伊斯拉·斯泰尔斯学院
176 格罗皮乌斯自宅
180 麻省理工学院礼拜堂
184 麻省理工学院贝克公寓

DAY 10
波士顿

188 哈佛大学研究生中心
192 哈佛大学冈德楼
196 哈佛大学美术馆
200 卡彭特视觉艺术中心
204 菲利普斯·埃克塞特学院图书馆

DAY 11
波士顿 — 香港

77 芝加哥壮丽大道洲际酒店

★★★★★

INTER CONTINENTAL CHICAGO MAGNIFICENT MILE AMERICA

Address : 505 N. Michigan Avenue, Chicago, IL 60611

Telephone : +312 944 4100

Http : // www.ihg.com.cn

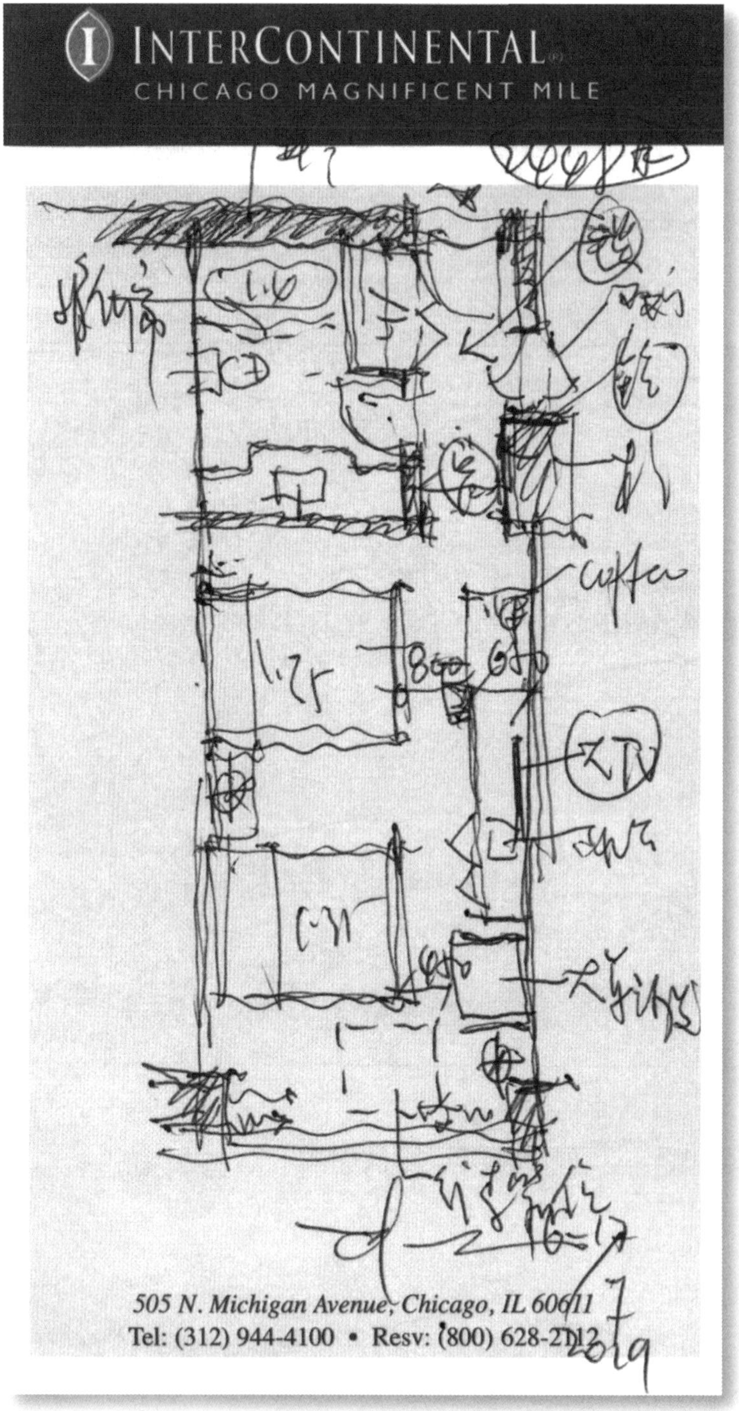

INTERCONTINENTAL

CHICAGO MAGNIFICENT MILE

505 N. Michigan Avenue, Chicago, IL 60611

Tel: (312) 944-4100 • Resv: (800) 628-2112

11/7，2014 [illegible]（[illegible]经理）

[illegible]斯[illegible]）

第一[illegible] + 第二[illegible]（[illegible]

[illegible]入[illegible]……）

---

① [illegible]（上海[illegible]）[illegible]

[illegible]

[illegible]

⑤ [illegible]

② 11/7 [illegible]

[illegible]也OK

③ [illegible]第三、四

[illegible]

[illegible]

④ [illegible]

十 [illegible]

July 17, 2019 On the Bus (On the Way to Visit Ludwig Mies Van der Rohe's Farnsworth House)

2019年7月17日在车上（去看密斯·凡·德罗的范斯沃斯住宅的路上）

第一天飞机上+第二天到达(下午2:00左右到达芝加哥，还是要近1.5小时入境……)

The first day, we were on the airplane, and then arrived the next day. (Arrived at Chicago at around 2 p.m.It took us one and a half hours to go through the customs.)

密斯的湖滨公寓

Mies'Lake Shore Drive Apartment

1. 提前一天乘高铁达香港，方便。在“圆方”吃了小南天，很难吃（上海快餐），机场快线倒是方便入住机场的富豪酒店。匆匆地在香港机场完成5公里的徒步打卡。

1. I got to Hong Kong by high speed train one day in advance, it was convenient. I had my supper at Xiao Nantian in the Elements Mall. It was like Shanghai food and tasted terrible. It was convenient to take the airport Express to the Regal Hotel. I finished the five-kilometer-hiking in my spare time at the airport.

密斯的湖滨公寓可谓是现代建筑的“活化石”

Mies' Lake Shore Drive Apartments are the living fossil of modern architecture.

2. 7月16日上午集合，11:35分的飞机，大约12:30分起飞，在飞机上看看电影，吃吃喝喝，也ok。

2. We gathered on the morning of July 16. The plane was scheduled for 11:35, but took off at 12:30. I watched movies, ate and drank a lot. It was ok!

3. 入境，一如既往的慢，应当是第三、四次落地芝加哥了（两次考察科勒，一次探望家人），拿行李顺利。

3. Immigration procedure was slow as usual. It should be my third or fourth time landing in Chicago(two visits to Kohler, another visit with my family). I got my baggage easily.

4. 下午去看了赖特的罗比住宅、湖滨公寓（密斯）+入住，吃饭，逛逛，入夜倒时差。

4. We visited Robie House which is designed by Frank Lloyd Wright and Lake Shore Drive Apartments by Ludwig Mies Van der Rohe. After visiting, we checked in, had supper and walked around and beat jet lag at night.

费城洛伊斯酒店

78 ★★★★

LOEWS HOTELS PHILADELPHIA AMERICA

Address : 1200 Market Street, Philadelphia

Telephone : +215 627 1200

Http : // www.loewshotels.com

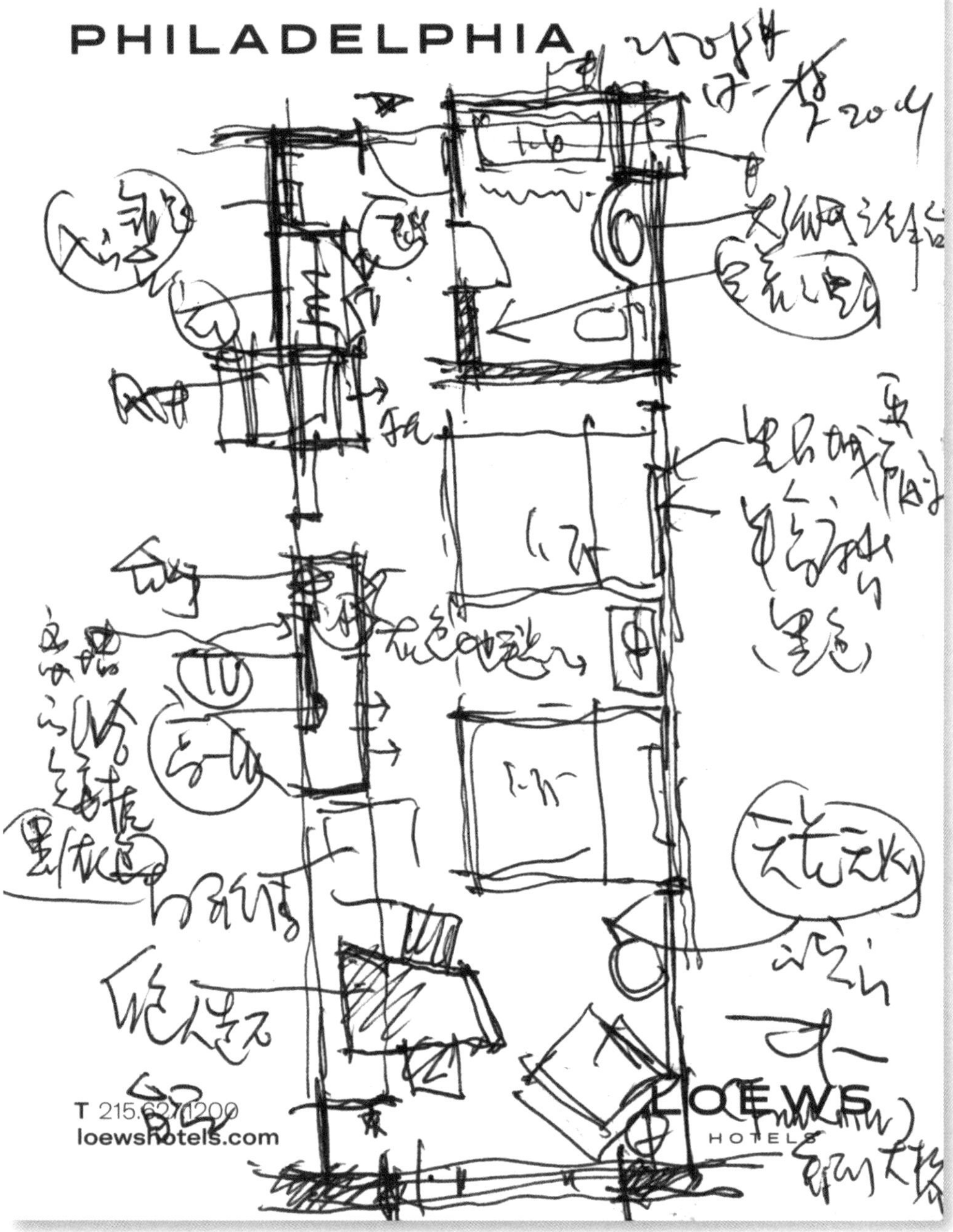

PHILADELPHIA

T 215.627.1200

loewshotels.com

LOEWS

HOTELS

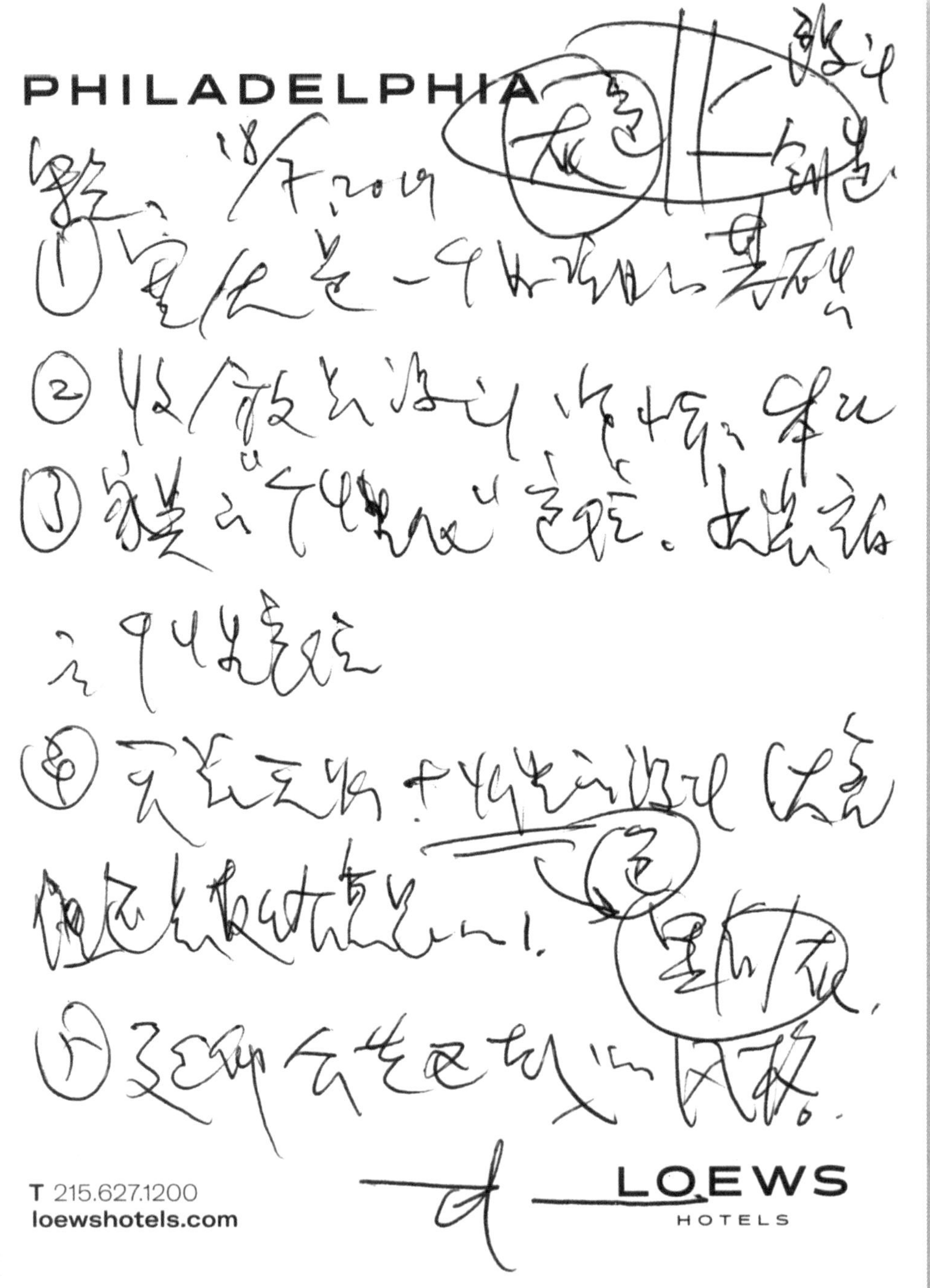
PHILADELPHIA
T 215.627.1200
loewshotels.com
LOEWS
HOTELS

餐厅不错·费城洛伊斯酒店
A nice dining room · Loews Hotels, Philadelphia

2019年7月17、18日
July 17-18,2019

1. 宽/大是一个项目的基础。
2. 收/放是设计师水平的体现
3. 家具的个性化“定位”。就是空间的个性定位。
4. 顶棚无灯+灯光（白色）的设计（太亮，但还是挺有意思的）。
5. 延伸公共区域的风格（黑/白/灰）。

1. Width is the basis of a project.
2. Designing easily and freely shows the designer's ability.
3. The position of the personalized furniture is just the position of the personalized space.
4. The ceiling without lights and white light (too bright but quite interesting).
5. It extends the style of the public area (black, white and grey).

黑白房间·费城洛伊斯酒店
The black and white guestroom · Loews Hotels, Philadelphia

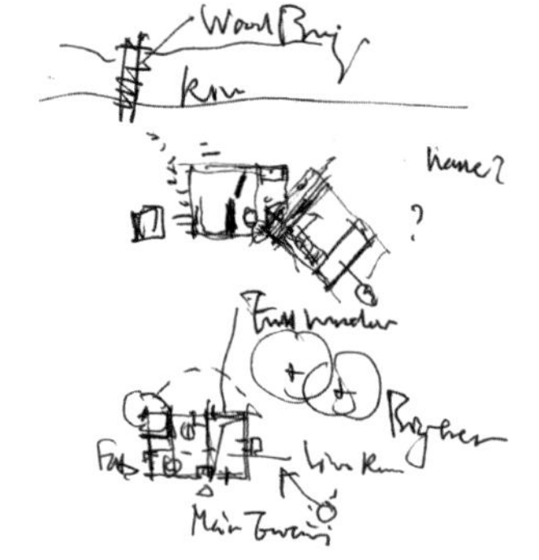

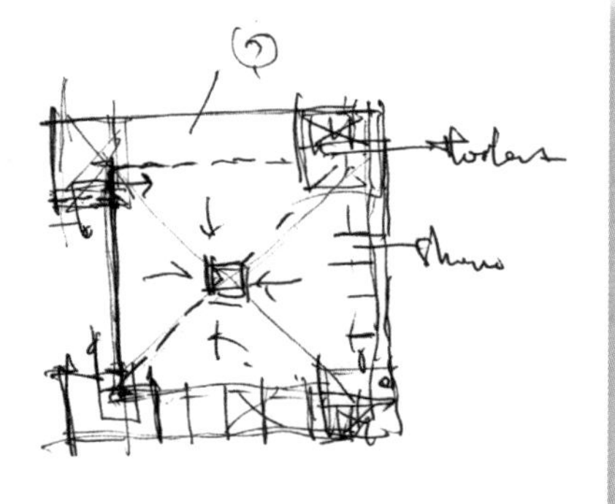

On the Way to Big Apple New York, July 18, 2019

2019年7月18日快到Big Apple New York的路上

倒叙。车上各自介绍自己，似乎说得不少，也许是这几天看得太多了吧！

Backward.We introduced to each other and seemed to speak a lot. Maybe we saw too much during these days.

7月18日，看了几个路易斯·康的作品。

1. 玛格丽特住宅，严谨的外观设计与实践，经典，紧凑，有独特而近乎“变态”的对光的“崇拜”，也许设计的本质就是对怎么用光、用景、用材的研究；相反，在当时，这些小建筑对技术的依赖反而不大。

July 18, We saw several works of Louis Kahn's.

1. Margaret's House, rigorous appearance design and practice were classical and compact, with the unique worship of bright, which was nearly insane. Maybe the nature of design was the study of using light, scenery or material. Instead, these small buildings never relied on technology too much at that time.

2. 早上看到宾夕法尼亚大学理查德医学研究中心的大楼（不错的样子），看到了其新技术（预应力混凝土构件运用的独到之处，“外表皮”）——门窗的设计令人叹为观止，比例、形式、构思……

2. In the morning, we went to University of Pennsylvania and visited the Richards Medical Library (the building looked perfect). We saw the new technology in it(the use of prestressed concrete was unique, "the outer skin")—the design of doors and windows were stunning, the proportion, the shape , the conception and so on.

3. 下着毛毛细雨去看了“犹太社区中心公共浴堂”，牛！四个十字+交叉的方形巧妙地产生虚实、错位、高低、通闭的对比，放在现在的公共配套建筑领域还是相当有独特的领先性。

3. It was drizzling and we went to a public bath hall in the Jewish Community Center, Amazing! There were four crosses and the cross-shaped squares. They created a contrast — solid and empty, different positions, heights and transparency. Compared with the modern public supporting buildings, it also had its leading advantages.

路易斯·康的手稿还是挺有趣的（只可惜他最终只能卖图纸手稿还债）。坚持就是胜利，脑+机合一的时代，我们怎样才能幸福地做设计呢?

Louis Kahn's manuscripts were also interesting (It is a pity that he had to sell his drawings to pay off his debt). Persistence is success! It's the era of brain and machine, how can we design happily?

玛格丽特住宅
Margaret's House

精品住宅的研究对象
It is the study object of quality residence.

理查德医学研究中心
The Richards Medical Library

惊叹当年如此成熟的装配式建筑，这么复杂丰富的立面变化。
We were surprised at such a mature prefabricated building of those years. What a complex and rich façade change!

非常应景，下着毛毛雨，对自然的尊重和若有若无的空间效果放在现在还是值得好好研究的。

It was drizzling in time. The respect for the nature and the effect of the space which seemed to be fading in and out were worth studying.

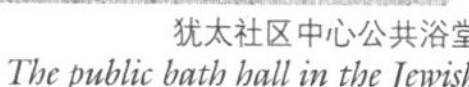

犹太社区中心公共浴堂

The public bath hall in the Jewish

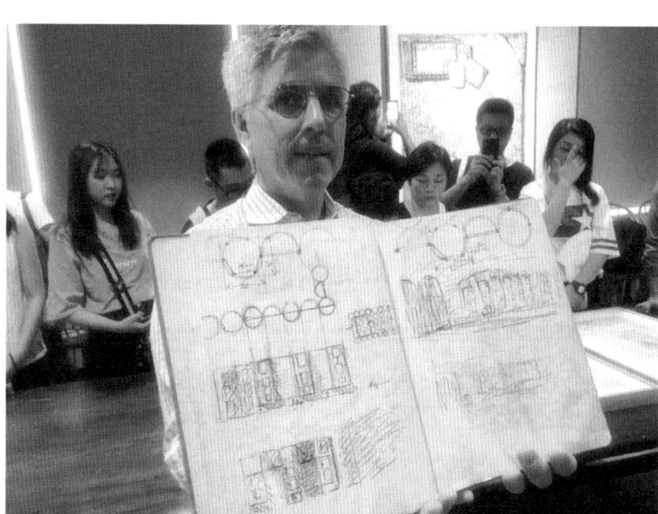

一本一本的路易斯·康的手稿，大师啊！

So many Louis Kahn's manuscripts! He is really an Architect.

路易斯·康的手稿

Louis Kahn's manuscripts

OMNI HOTELS & RESORTS

新港耶鲁奥姆尼酒店

79

★★★★★

OMNI HOTELS & RESORTS NEW HAVEN AMERICA

Address : 155 Temple Street, New Haven, America

Telephone : +888 444 6664

Http : // www.OMNIHOTELS.COM

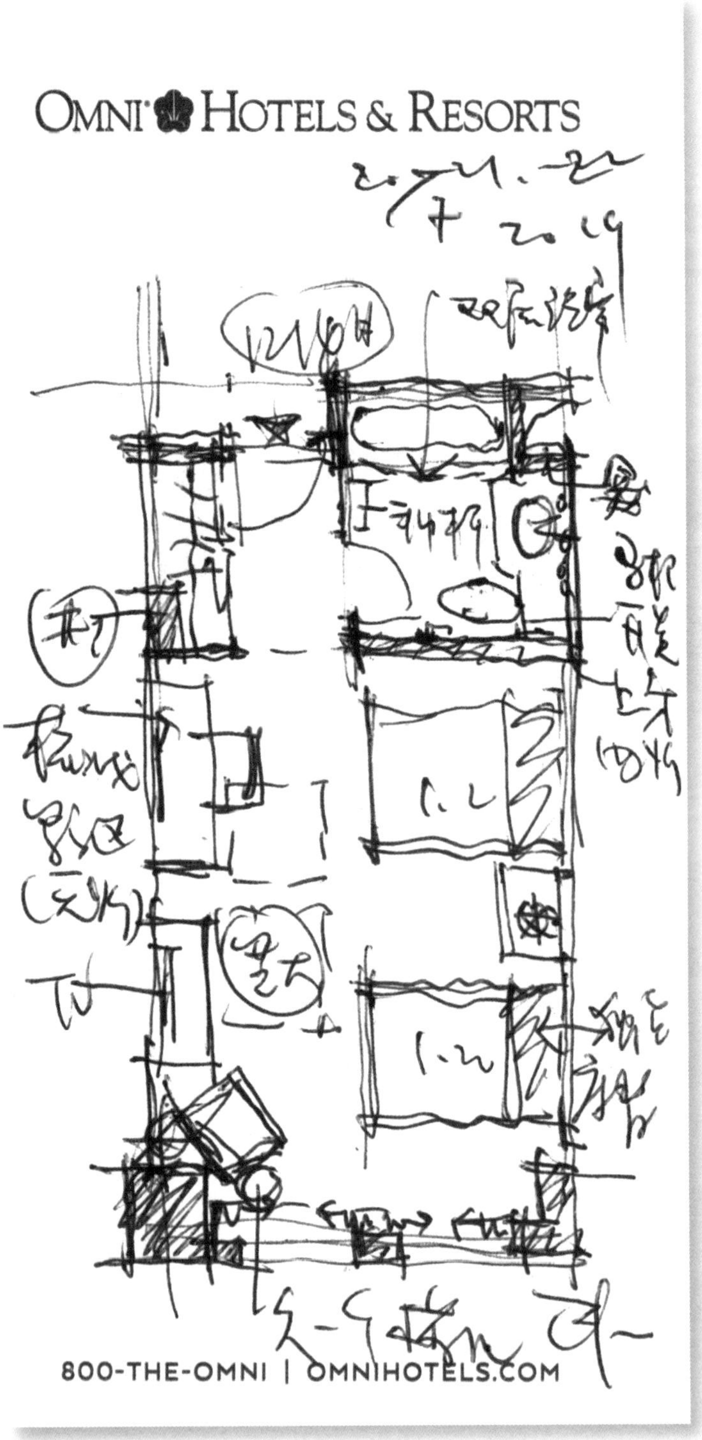

OMNI HOTELS & RESORTS

800-THE-OMNI | OMNIHOTELS.COM

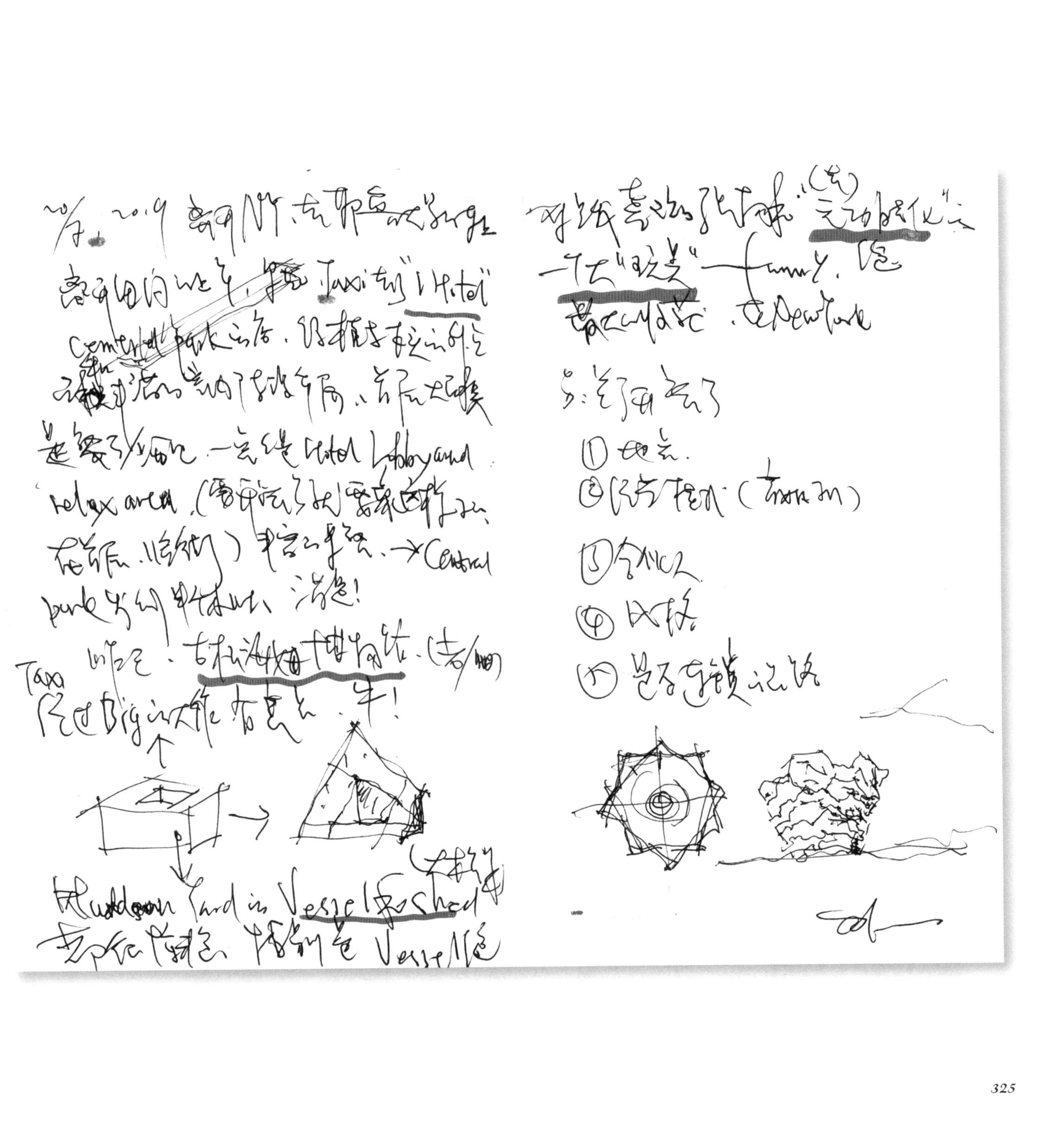

July 20, 2019 Leave New York, on the Bus to Yale University

2019年7月20日离开纽约，去耶鲁大学的车上

离开纽约的上午，早起，乘出租车去了“1 Hotel”中央公园店，绿植与木头的外立面和丰满的室内陈设布局，首层大面积是餐厅/酒吧，一点点是酒店大堂和休闲区（要开餐厅就要像这样在首层，临街）。用过丰富的早餐，在中央公园步行半个小时，满足！

On the morning of leaving New York, I got up early and took a taxi to “1 Hotel”which was in the Central Park. The facade was decorated with green plants and wood. The interior furnishings were abundant and rich. The large part of the first floor is a dining hall or a bar while the rest was a Hotel Lobby and relaxing area. (The restaurant should open on the first floor like this or on the street.) After having a rich breakfast, I had a walk in the Central Park for half an hour. Contented!

昨天，参观了古根海姆博物馆（老/旧），乘出租车经过BIG的大作，有意思，牛！

哈德逊园区的Vessel（大松果）和Shed都很惊艳，特别是Vessel，绝对是我喜欢的张扬和“去（无）功能化”的一个大“玩具”——有意思，绝！

最大的收获，在纽约。

The day before, we visited the old Guggenheim Museum and saw BIG’s masterpiece on the taxi. How interesting! Cool!

Both Vessel and the Shed in Hudson Yard were very amazing, especially the Vessel. It was my absolute favourite of aggressiveness and a large toy with no function—funny , absolutely unique!

I learned most in New York.

印象深，地段好，早餐靓（漂亮）
Deep impression, Good location and flavorful breakfast.

在1hotel吃早餐
Having breakfast in 1 Hotel

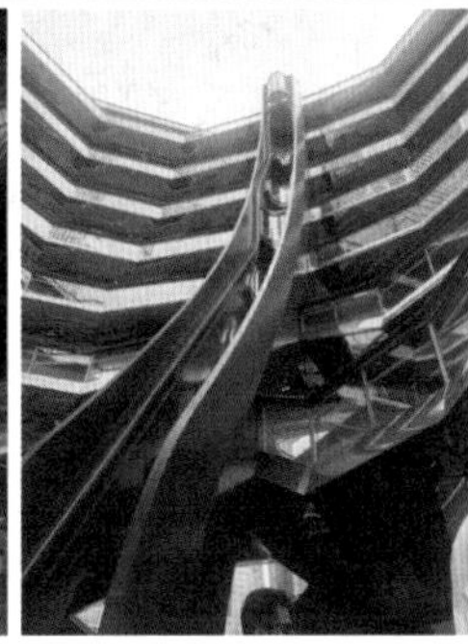

大松果Vessel
这一次美国东部游，“无用”的装置是我最喜欢的一个地方。
The useless equipment I have seen was my favoriteduring this trip to the East.

古根海姆博物馆
之前经过纽约几次，没入内参观，这一次终于如愿以偿了。
I've been to NYC for several times, but never gone inside for a visit. This time, I finally fulfilled my dream.

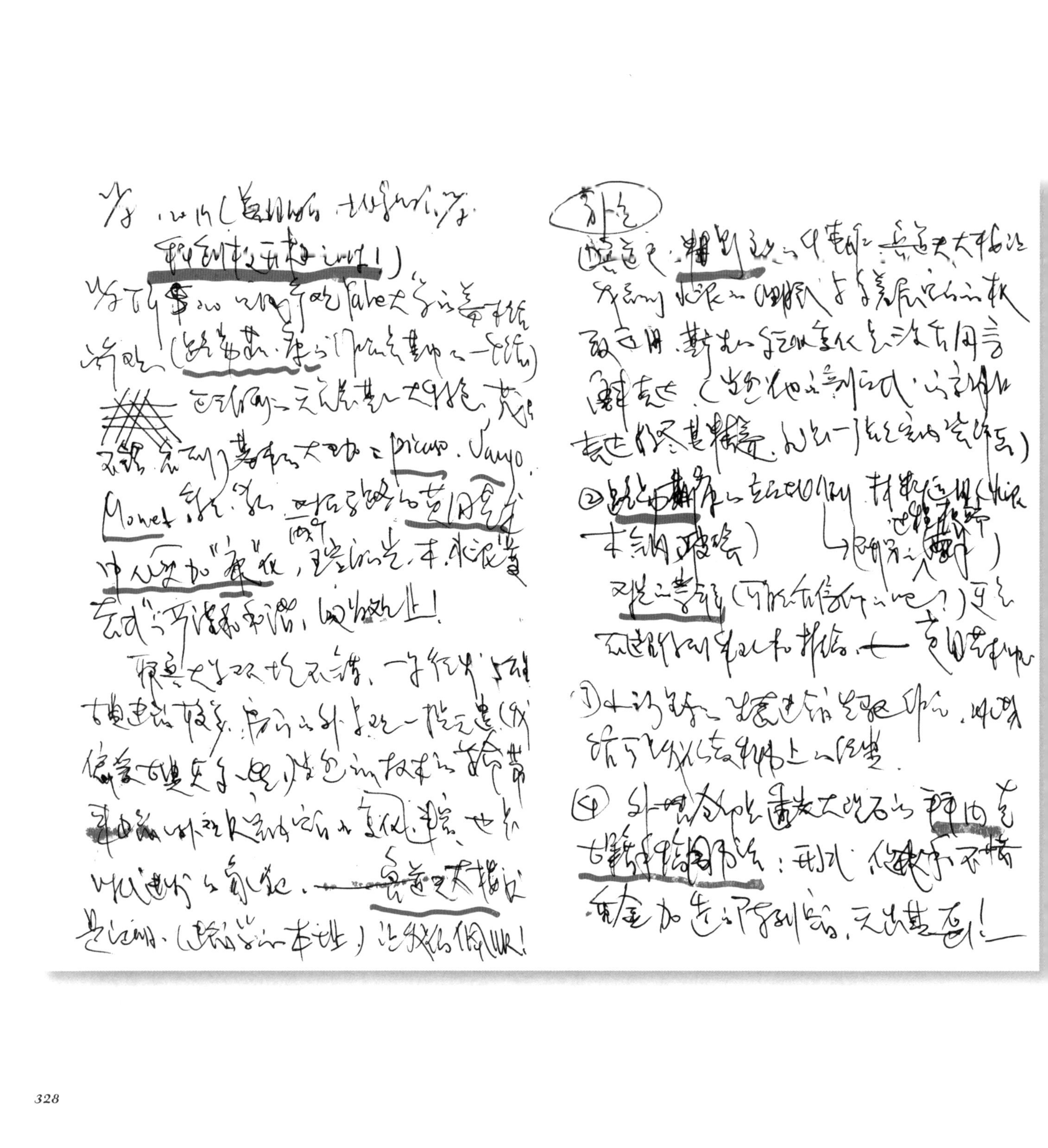

July 21, 2019 in American, July 22, 2019 in Beijing, the Opening of Sci-Tech Innovation Board

美国时间2019年7月21日，北京时间2019年7月22日，科创板开板之时！

耶鲁大学美术馆

这一次几乎是“康”之旅，有垄断市场的感觉，很多地标性的建筑，都出自他的手笔。

It was really a trip of Louis Kahn. A lot of landmarks were designed by Kahn and cornered the market.

7月21日下午5:00完成耶鲁大学美术馆的参观（路易斯·康的作品是其中的一个馆）。

正三角形的顶棚是其一大特色，藏品不错，看到了毕加索、梵·高、莫奈等大咖的作品，等等。对面马路的英国艺术中心更加“康化”，两个主空间的光、木、水泥“变态式”的严谨和和谐，叹为观止！

On the afternoon of July 21, we finished visiting the Art Gallery of Yale University at five. (One of the building s was Louis Kahn's work.)

One of its features was the equilateral triangular ceiling. The collections were wonderful including Picasso, Van Gogh, and Monet's works, and also many other masters'. Yale Center for British Art across the road was more Kahn's style. The bright, wood and concrete in the two main spaces were so rigorous and harmonious that it made me stunned.

耶鲁大学环境不错，一早徒步5公里。古典建筑较多，房间外的景观一览无遗（我偏爱古典更多一些），当然新技术的革命带来建筑的外观及室内空间的丰富变化，也是时代进步的象征——鲁道夫大楼就是证明(建筑学院的本址)，让我们佩服！

The environment of Yale was comfortable. I walked for 5 kilometers in the early morning. There were more classical buildings. I had a broad view of all the scenes outside (I preferred the classic). It was sure that the revolution of the new technology brought the appearance and interior design more changes. It was also the symbol of the progress of the times—Rudolph Hall was here to prove it(The old Yale School of Architecture), made me admire!

拜内克古籍手稿图书馆
Bynek ancient manuscript Library

精细，半透光的大理石与粗犷的混凝土的碰撞，真能想啊！
So fine work! The translucent marble collides with the rough concrete. How creative!

补充：

1. 鲁道夫大楼是粗野主义的代表作，让我们看到了水泥的细腻与多层空间的极致运用，其产生的多元化变化是无法用言语来表达的（当然他的剖面式的效果图表达得尽其精美，就是一张张室内装饰画）。

PS:

Rudolph Hall was the masterpiece of Brutalism. It showed us the fine concrete and the use of multi-level space were well executed. It couldn't express the changes which were made by the multi-level space. (His section sketches were so perfect that they were just like interior decorative pictures.)

路易斯·康的立面比例（极端研究的结果），材料运用（水泥、木、钢、玻璃），对光的崇拜（可能有信仰的吧？）更多在这里得到体现和推高——英国艺术中心。

As a result of his extreme study, Louis Kahn's façade proportion, the use of material (concrete, water, steel and glass) as well as the worship of bright (maybe his belief) were shown and improved here—Yale Center for British Art.

2. 小沙里宁的生态建筑先驱作品——冰球馆可是我们教科书上的经典。

Eero Saarinen's work was the pioneering design in the ecological architecture, especially David S. Ingalls Rink (the Whale) was the classic in our textbooks.

3. 外墙全部是透光大理石的拜内克古籍手稿图书馆：形式、仪式感不错，重金打造的陈列馆，无出其右！

The outer wall of the Beinecke Rare Book and Manuscript Library were all transparent marble full of sense of ritual. The show room made with capital, which was top the list!

波士顿剑桥凯悦酒店

80

★★★★

HYATT REGENCY CAMBRIDGE AMERICA

Address : 575 Memorial Drive, MA 02139

Telephone : +(617)568 1234

Http : //www.hyatt.com

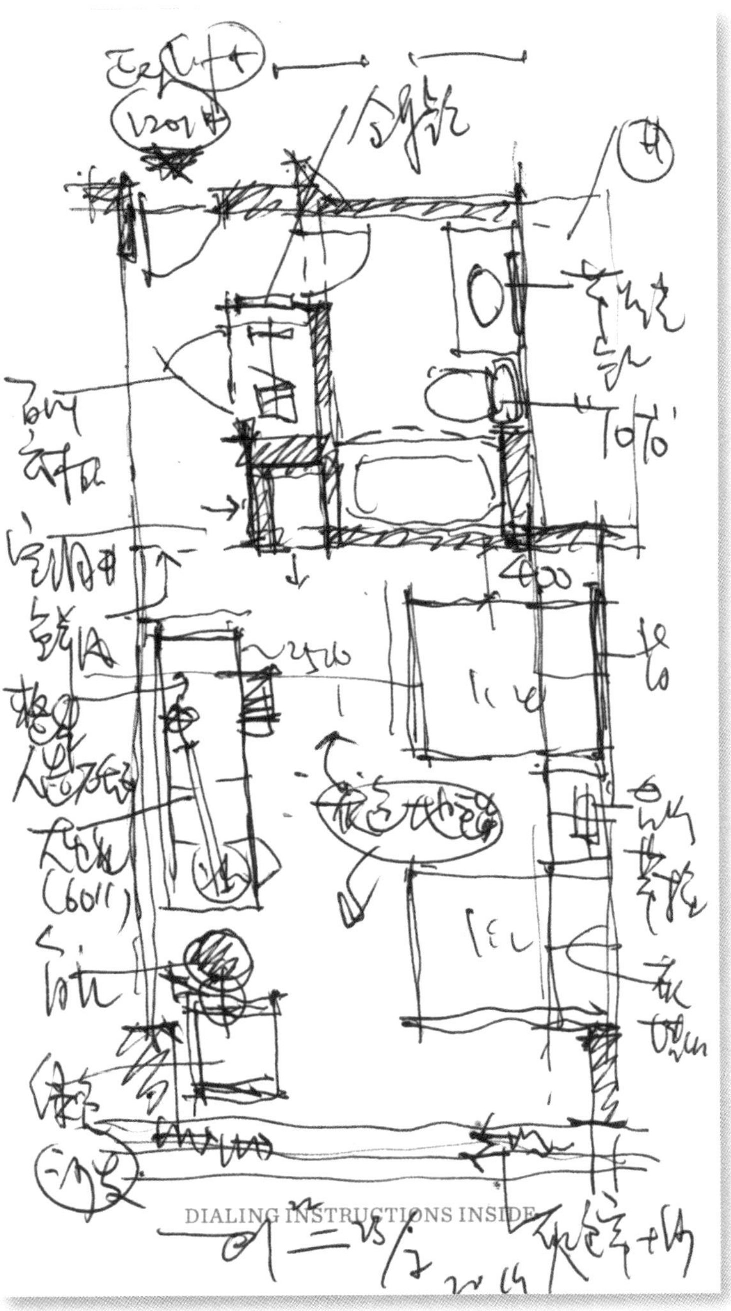

DIALING INSTRUCTIONS INSIDE

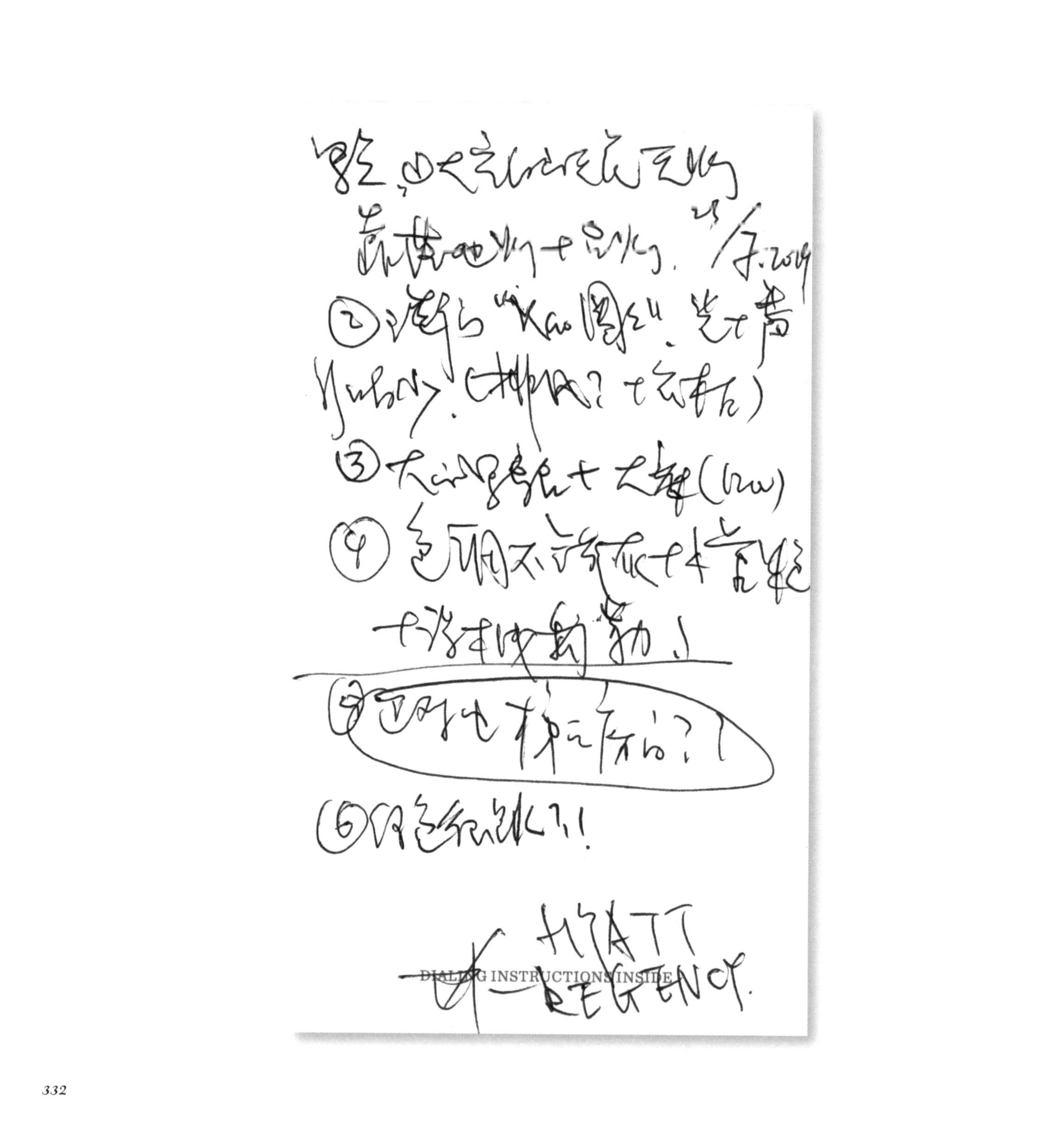

HYATT
REGENCY.

DIALING INSTRUCTIONS INSIDE

2019年7月23日
July 23,2019

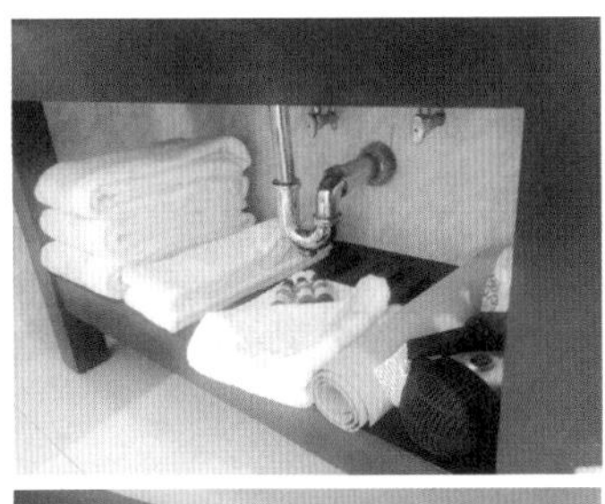

写点：

1. 大空间的顶棚无灯，靠落地灯+台灯
2. 洗手间“绕圈圈”，光+声，影响小（排风+衣柜）
3. 大的写字台+大床（1200）
4. 色调不错，灰+小量米色+浮木浅勾勒！
5. 正对电梯的房间？！
6. 绿色很跳？！

Highlights:

1. *A large space without ceiling light, only the floor lamp and table lamp.*
2. *Zigzag to washroom, little effect from light and sound (exhaust air and wardrobe)*
3. *A big table and a king-sized bed (1200 wide)*
4. *The color is good, grey and a little beige, and is drawn the outline by drift wood.*
5. *The guestroom faces the elevator.*
6. *Green is sharp.*

波士顿剑桥凯悦酒店
Hyatt Regency, Cambridge, America

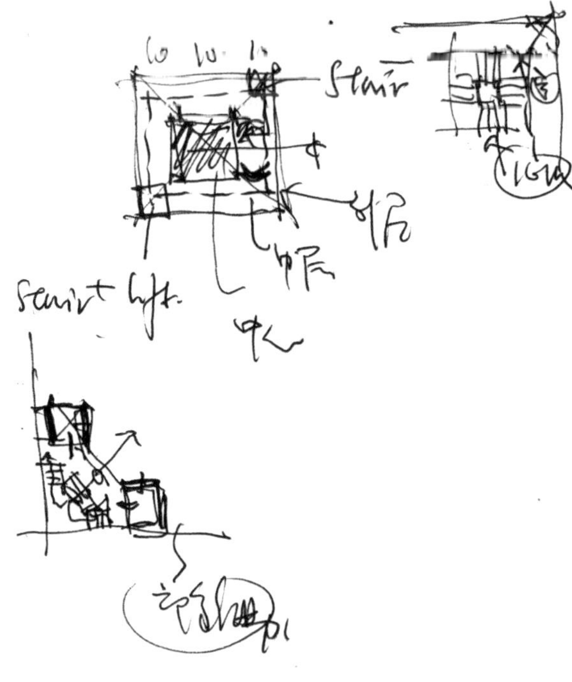

Phillips Exeter学院图书馆

精彩的中空大厅，连系着丰富的空间变化，耐用耐看。

A wonderful hollow hall isconnected with the rich space change. It is durable and will look wonderful for long.

July 23, 2019 The Last Programme

2019年7月23日介绍最后一个项目

路易斯·康的Phillips Exeter 学院图书馆：惊艳与极端佩服（1972年）。

Phillips Exeter Academy Library was one of Louis Kahn's best works. I felt surprised and admired. (It was made in 1972.)

外观的严谨、成熟的比例，与材料的结合：红砖，水泥，木，少量钢。

Its appearance was rigorous, the proportion and the material were mature, red bricks, concrete, wood and a little steel.

室内空间：极端的“内外不一致”，惊艳的共享与丰富，不断的不同凡响的享受（机电/灯光/家具……），非常舒服与温暖。

Interior space: the external and the interior were extremely different, stunning, sharing and abundant. It was constant unreasonable and unusual enjoyment (electromechanics, lights and furniture and so on).

学习之旅，继续学习！

A study trip, keep studying!

深圳柏悦酒店
PARK HYATT SHENZHEN

81 深圳柏悦酒店

★★★★★

PARK HYATT SHENZHEN CHINA

Address : 5023 Yi Tian Road, Futian District, Shenzhen Guangdong Province, China
中国广东省深圳市
福田区益田路5023号
Telephone : +86 755 8829 1234
Fax : +86 755 8830 1234
Http : //www.www.parkhyattshenzhen.com

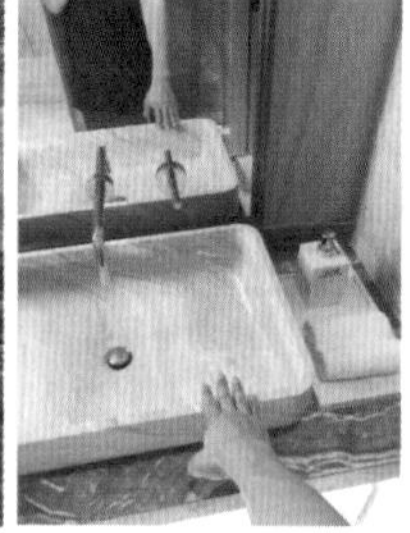

深圳柏悦酒店
Park Hyatt, Shenzhen,China

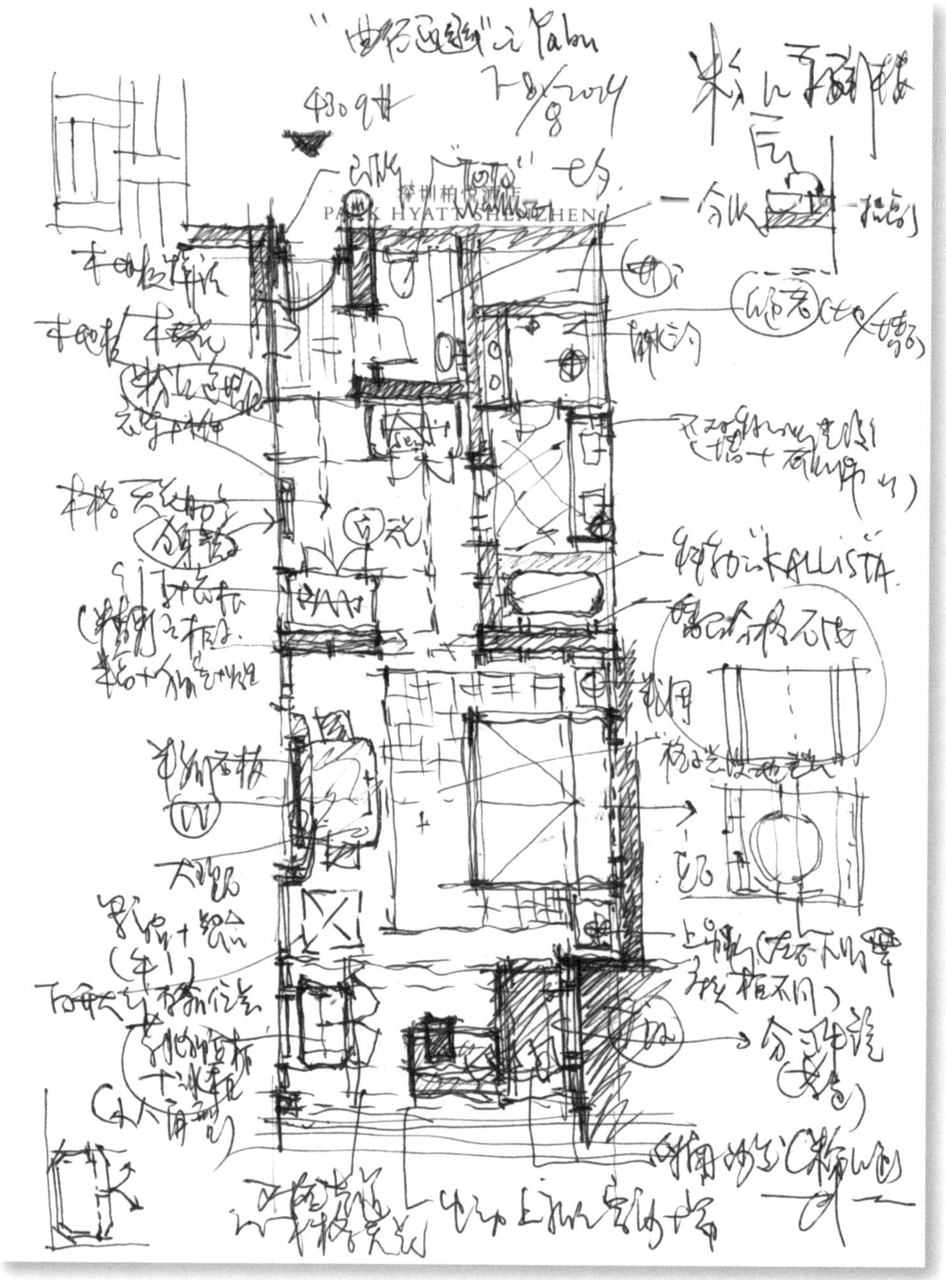

深圳柏悦酒店
PARK HYATT SHENZHEN

深圳柏悦酒店
Park Hyatt, Shenzhen,China

A Room Worth Writing

可写性较多的一间房

网红打卡点：新开业一个多月的深圳柏悦酒店，Yabu的又一名作。之前住过杭州的柏悦、北京的华尔道夫等都是出自这间公司。

Internet celebrity spot: Park Hyatt Shenzhen was Yabu's another famous work which has been open for a month. I stayed in Park Hyatt Hangzhou, Waldorf Astoria Beijing and so forth which were designed by this company before.

这一次的房间布局有点北京华尔道夫的味道：狭长的平面，曲径通幽的方法"消耗"空间与融入各种功能，当然优点在使用中渐渐显露：当下流行的厕浴分离，近门的洗手间，尺度比一般的大，双衣柜的设置，颇有新意，也自然而然地让过渡空间功能化，入口设置挂衣架和包包、房卡放置处，细腻，有装饰感、仪式感；洗浴区尺度也是很"土豪"，设计与布局却相当标签化：IP；对床的大理石的异形写字+餐饮区，有张力，有画面感，高档大气；近窗的休闲区，茶水柜成为主角，窗外的远眺风景亦是抢眼，转角沙发巧妙地弱化了大柱位的唐突。当然，家具的设计之妙也是大师的一大优势，对中式文化的"一知半解"恰到好处地显现出当下流行的所谓的"新"。"藻井式"中式的顶棚，木漆结合，墙、地、天的和谐统一，对应地也让这个"中式"味道时髦化地浓重起来。粉色渐变的衣柜内漆颇有创意，一幅中式月亮门的床背，轻松有趣。"半懂"正是大师的高明之处。

The plan of the guestroom was a little Beijing Waldorf Style, the long and narrow plane; a winding path blended different functions in the space. Of course, the advantages were shown while using. The trendy design was separated toilet and bath. The restroom by the door was a little larger than regular. The creative double-door wardrobe made the transition space functional naturally. The coat hanger and a card bag at the entrance were detailed with sense of decoration and ceremony.The size of the bathroom was also luxurious while the design and layout were quite tagged, IP. The marble irregular writing and dining area facing the bed was not only tension, but also had sense of pictures. It was also high-grade and elegant.The tea cabinet became the leading role in the leisure area by the window. The distant views outside the window caught your eyes. The large abrupt column was weaken by the corner sofa skillfully. Of course, the perfect design of furniture was one of the Master's advantages. The design with a smattering of Chinese culture just showed the latest "New" style. The caisson ceiling was made of wood and paint. The wall, floor and ceiling were harmony and unity. The corresponding made the Chinese style smarter.The wardrobe with pink-gradient paint inside was quite creative. The bed back which looked like a Chinese Moon door was funny and relaxing. Knowing a little was the Master's wisdom.

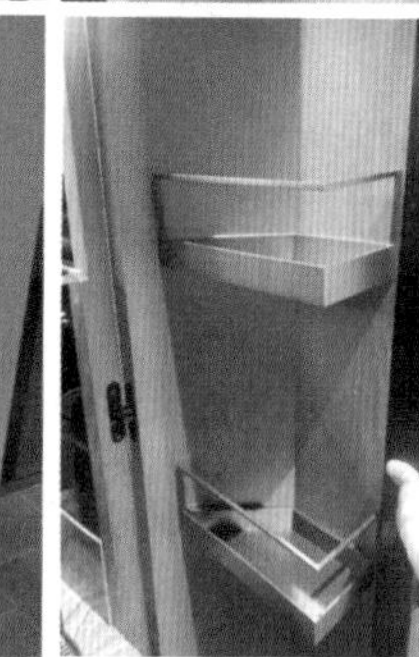

深圳柏

Park Hyatt, Shenzhen,

82 上海虹桥英迪格酒店

★★★★★

HOTEL INDIGO SHANGHAI HONGQIAO CHINA

Address : No.43 Lane 188, Yonghong Road, Minhang District, Shanghai, China
中国上海市
闵行区甬虹路188弄43号

Telephone : +86 21 3323 3666

Fax : +86 21 3323 3555

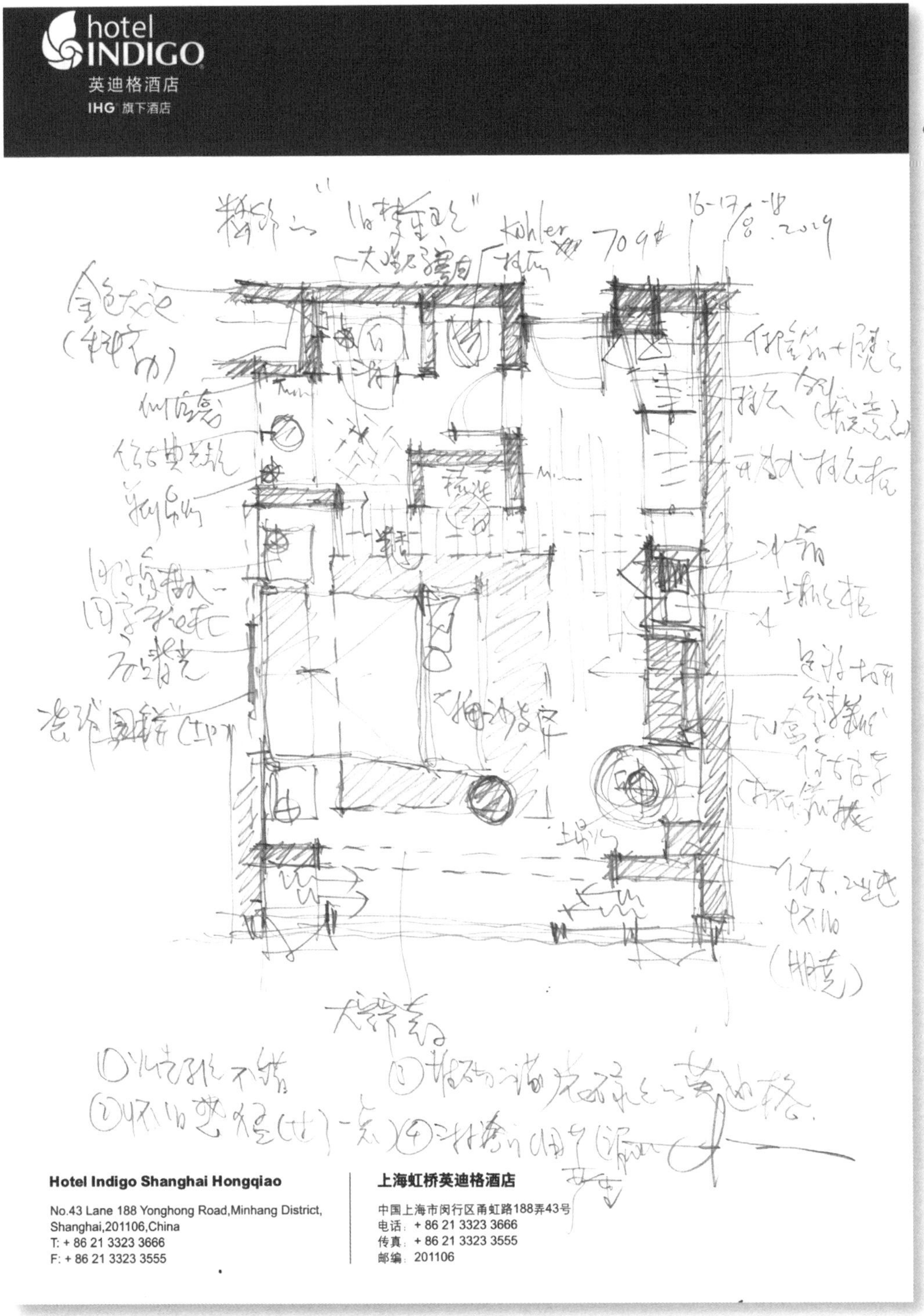

Hotel Indigo Shanghai Hongqiao

No.43 Lane 188 Yonghong Road,Minhang District,
Shanghai,201106,China
T: + 86 21 3323 3666
F: + 86 21 3323 3555

上海虹桥英迪格酒店

中国上海市闵行区甬虹路188弄43号
电话：+ 86 21 3323 3666
传真：+ 86 21 3323 3555
邮编：201106

上海虹桥英迪格酒店
Hotel Indigo,Shanghai Hongqiao,China

[illegible]

Hotel Indigo Shanghai Hongqiao
No.43 Lane 188 Yonghong Road,Minhang District,
Shanghai,201106,China
T: + 86 21 3323 3666
F: + 86 21 3323 3555

上海虹桥英迪格酒店
中国上海市闵行区甬虹路188弄43号
电话：+ 86 21 3323 3666
传真：+ 86 21 3323 3555
邮编：201106

Be Picky

挑“毛病”

可能已经做好了心理准备，“英迪格”酒店品牌也住了几间：上海外滩的，香港的（湾仔），还有……忘记了！

I have stayed in a few hotels of Indigo, including the ones in the Shanghai Bund, in Hong Kong Wanchai and so forth. I forget some others. So I know this brand very well.

“花花绿绿”“堆砌”成了这个品牌的代名词，也符合某一领域或年龄层的人群的“审美要求”（也许W酒店的张扬也有这个“问题”吧）。

The fancy packing becomes the symbol of this brand. It just meets some people's aesthetic. (Maybe the flamboyant W hotel has the same style.)

生意还挺好的，近机场，大商业圈。“潮”是关键，倒是进入房间马上“衰老了”。所谓的“新海派”大绿色，重装饰，相当厚重的“钉钉”墙面和仿古家具，白+金色的浴室倒是唯一的“小清新”。当然，拍照很上镜，也算成功。

The business was very good because it was near the airport and a large business circle. "Fashion" was the most important. But it turned into old style at once while entering the guestroom. Green was said to be the "New Shanghai style". The heavy adornment was like the wall with quite heavy nails and antique furniture. The white and golden restroom looked a little fresh. Of course, it was sharp to take photos. What a success!

小创新点：1. 斜向的入口衣柜组合；2. 灯光系统多样而易控；3. 小休闲沙发区，亲切舒适（刚好晚上有朋友来访，与小茶水柜还是有不错的联动）。

Several innovations: 1. The wardrobe combination at the angled door; 2. The various light system was easy to control; 3. The sofa in leisure area was comfortable. (A friend came over at night; It interacted well with the small tea cabinet.)

找找毛病吧（住多了的职业病）：1. 洗手间的平台挺不规则的，因为嵌入了梳妆台，不知是否可以取消或优化；2. 淋浴间门的位置是否应尽量远离淋浴头？防水条失效，门关不严，水外溢非常严重，不可接受；3. 环形通道是否有“必要性”？洗手间近床的门是半透明的，光与噪声都会影响睡眠。也许趣味多了，布局也多变，有时候在想，什么是好：是舒服，还是吸引？当下，也许后者为大！

Try to find faults(because of my occupational habit):1. The toilet table was quite irregular because the dresser embedded in it, I wondered if it could be cancelled or optimized; 2. Was it better if the door of the bathroom was far away from the shower head? It was unacceptable that water overflowed seriously. The door was hard to close entirely because of the broken waterproof bar; 3. Was the circle path necessary? The door near the restroom was translucent so that light and sound would affect sleep. Maybe it was funnier, and the layout was changeable. Sometimes I wondered which was better, comfort or attraction? Nowadays attraction may be the first choice.

全季酒店

上海浦东全季酒店

83 ★★★★★

ALL SEASONS SHANGHAI CHINA

Address : No. 33 Henan South Road, Huangpu District, Shanghai, China
中国上海市黄埔区
河南南路33号
Telephone : +86 21 6191 9978

上海浦东全季酒店
All Seasons,Shanghai,China

静下心来，拾一诗，赏一画

竹庵 | 山中何所有 | 艺巢ART 授权 |

全季酒店

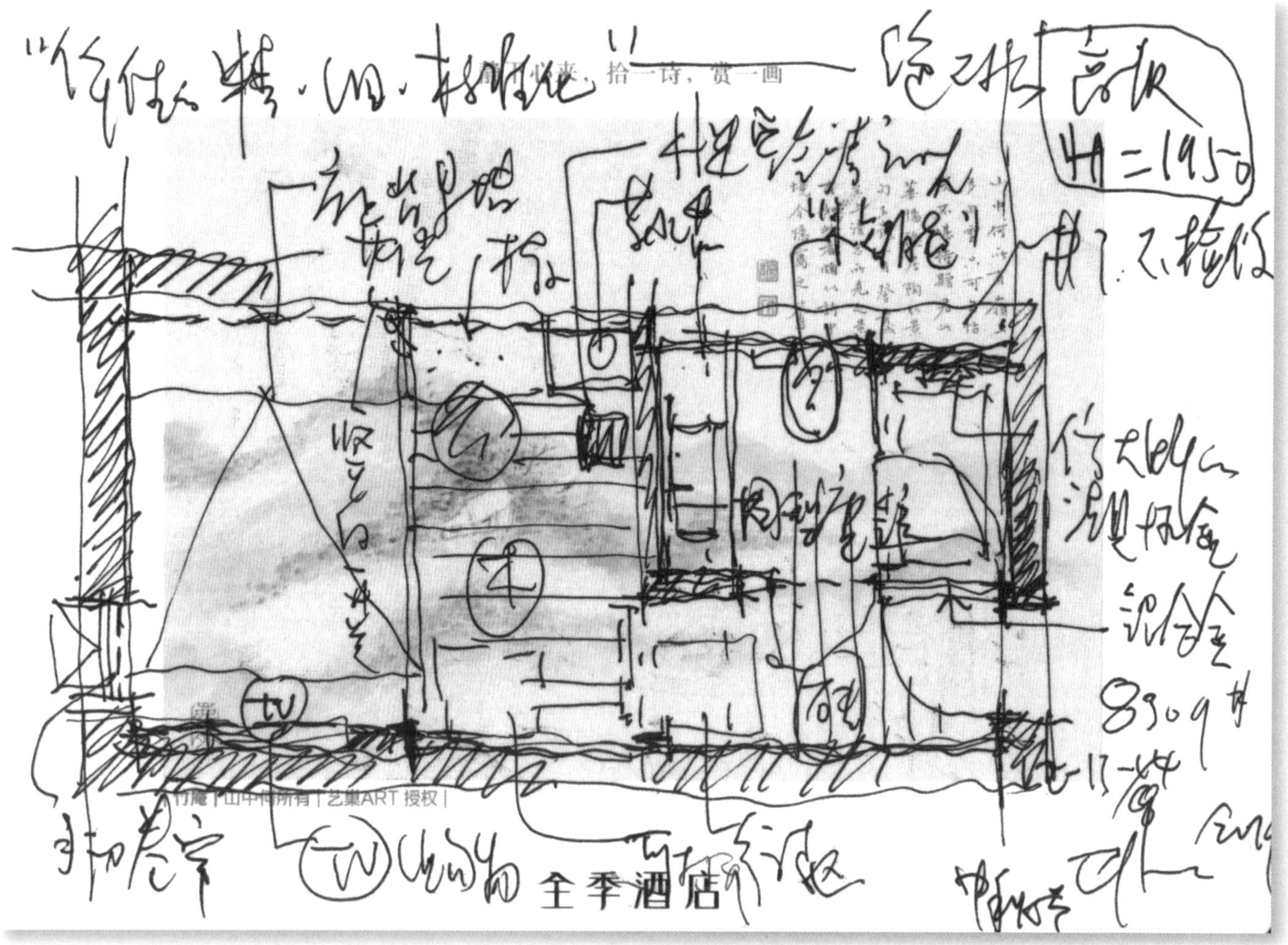

静下心来，拾一诗，赏一画

[illegible]

全季酒店

静下心来，拾一诗，赏一画

[illegible]

全季酒店

上海浦东全季酒店

All Seasons,Shanghai,China

Save Money to the Extreme II

能"悭"则"悭"

参观家具展，入住在浦东展会边上的这一家"全季"酒店。哔！第一次有这么绝的酒店：外墙三层，内部六层，一割为二。层高（净空）绝对的反人类：1950mm，难得的体验，可惜已交了两个晚上的钱，被迫将就！

I stayed in All Seasons Hotel near Putong Expro when attending the Shanghai Furniture Fair. Impressive! What a crazy hotel! There were three storeys outside while six storeys inside, divided into two. The low headroom was absolutely disgusting: only 1950mm. Such an unforgettable experience! I had to stay there because I paid for two nights.

大堂有"书吧"。就是卖书的地方，没有一本可以免费看的！牛！ 倒是房间内放着老板的书：季琦帅哥的手记（这位确是生意天才），我很耐心地看了书，佩服，绝对的成功的投资人和酒店、商旅的行业泰斗！每一处的营销，到位！

There was a "Book Bar" in the lobby. It was just for selling books, no one for free. Cool! I found a book written by the handsome boss Jiqi in the guestroom. (He was really a business genius). I read the whole book patiently and admired him. He was an absolutely successful investor as well as a head of the hotels and travels. Business could be seen everywhere.

房间也叹为观止。

Guestrooms were also stunning.

1.尺度严谨而有"钻研"，靠窗的床，1.5m，房间尽宽约2400mm， 不要忘记了， 层高1950（我猜， 一年回本！ ）。

The size was strict but studied. The bed beside the widow was 1.5 meters wide. The room was exactly 2400mm wide. Don't forget that the headroom was only 1950mm. (I guessed its investment could be recovered in one year.)

2.洗手间使用高档大牌的洁具（电动马桶），五金龙头（科勒，几千元的），特别是电热毛巾架，心思棒棒哒！

The upscale sanitary appliances were used here, such as electric toilet, hardware faucet (Kohler, cost several thousand yuan), especially the electric towel rack was the most thoughtful.

3."小客厅"配备平实的凳子（不能说是沙发）+小茶几，小茶台配电热水壶，茶品（托+杯）都可以卖（打开电视就可以下单，佩服），倒是空调和小米的空气清新器贴心。

There were a plain stool (not a sofa) in the mini living room and a small tea table with electric teapot. The tea set (including tray and cups) could be sold. (You might order just by turning on TV, I admired!)The air conditioner and Mi air fresher were so considerate.

4.灯光控制也"简单粗暴"，总而言之，还是非常有思考的一个品牌！

The control of lights was "simple and rude". In a word, it was a thoughtful brand!

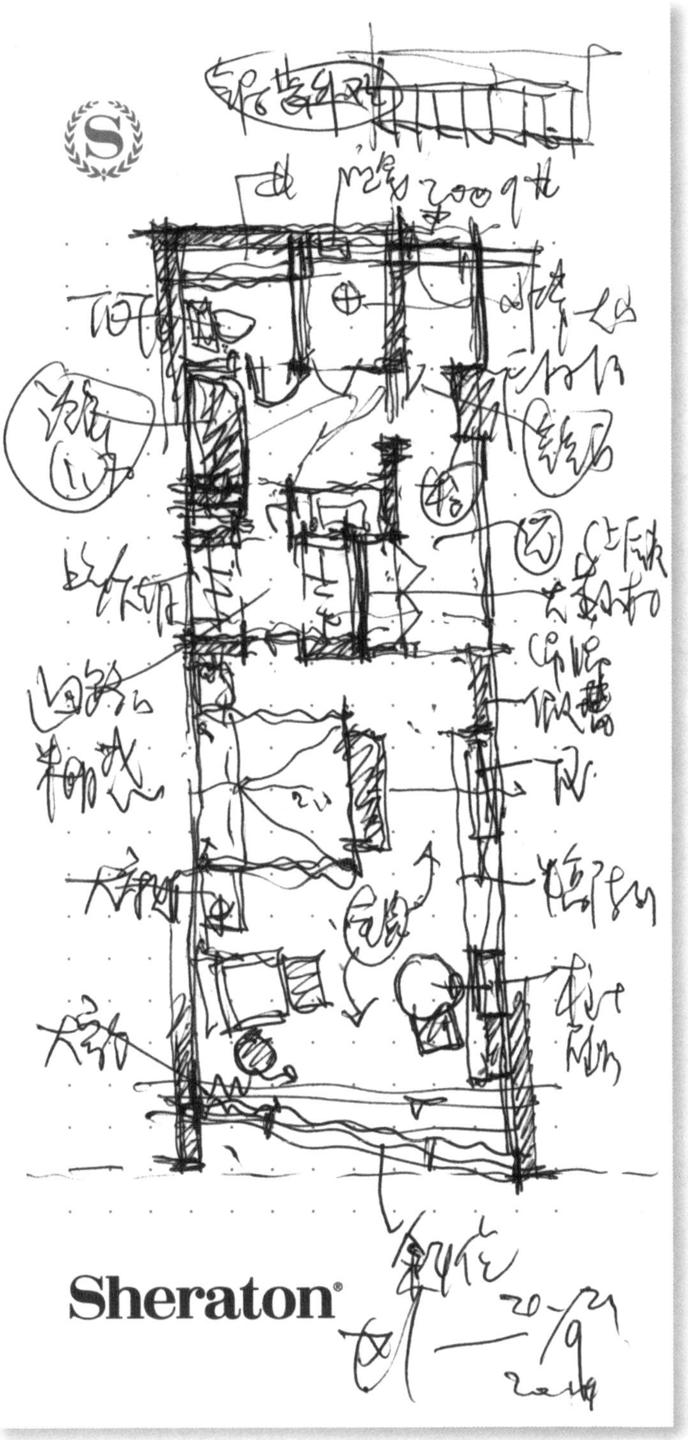

84

上海崇明由由喜来登酒店

★★★★★

SHERATON SHANGHAI CHONGMING HOTEL CHINA

Address : No. 2888 Lanhai Road, Chongming District, Shanghai, China

中国上海市
崇明区揽海路2888号

Telephone : +86 21 3936 1888

Sheraton®

[illegible]

Sheraton

[illegible]

Sheraton

[illegible]

Perfect Arrangement
完美空间

对于传统的品牌喜来登来说，度假酒店可谓驾轻就熟（之前也写过类似的酒店），度假/会议型的酒店，房间面积较大，分配就显重要。 入住位于第三大岛的崇明喜来登。房间面积基本是55:45左右（睡眠区与湿区的分配比），反而开间不是很宽（层高约4.3m）。对于选用单边走廊，合一的洗手间、衣储空间，就会露拙：1. 走道有一条柱（还对着洗手间的门，刻意的吧）；2. 洗手间的平面布局难以流畅和舒适。浴缸选用圆角式产品，两边夹墙，感受比较一般，倒是多出大大的淋浴间，很棒，徒步回来，可以静静地坐下来，慢慢沐浴。

As a traditional brand—Sheraton Hotel, it handles the resort hotel rather well(I wrote about this kind of hotels before). Because the guestrooms of resort hotels or conference hotels were a little big, the space allocation was quite important. Staying in the third largest island in China—Sheraton Shanghai Chongming Hotel. The room was basically around 55:45 in area (sleeping and wet area), while it was not wide (about 4.3metres). The one-side corridor, a combination of bathroom and wardrobe showed the weakness: 1. There was a pillar (facing the door of the bathroom, it was intentional, wasn't it?); 2. The plane of the bathroom was hard to be smooth and comfortable. The bathtub with rounded corners between two walls Let me feel not so much and the extra big shower was awesome. Came back from hiking, I sat down quietly and enjoyed shower slowly.

睡眠区把握老到，不惊艳，合适；“浅薄”的电视主幅，配合写字台（较小）。 大床背摒弃了流行的背藏光，大大的床头柜可以摆相当多的东西，方便实用。

Sleeping area was experienced, but nothing surprising just fit everything, a thin TV set matched the small table. The back of the bed wasn't the popular style which hid light. The night table was big enough to put many things, convenient and practical.

休闲椅我认为选择可以移动的更加合适。窗外，风景不错。 小茶几设计得有细节（有档边，不容易掉东西）。整休来说，收货啦！（广州话——合格了！）

I thought the movable leisure chair was more suitable. The view outside the window was pleasant. The design of the small end table had specifics so that it was not easy to drop things. Overall, I accepted it.

上海崇明由由喜来登酒店
Sheraton Shanghai Chongming Hotel,China

[illegible]

大佳中国第三大岛——上海崇明岛，原来和我一样，[illegible]

[illegible] 大佳喜来登酒店一个好有意思的名字：由由）[illegible]

[illegible]

不错。

[illegible]

Sheraton

[illegible] Mark，[illegible]

[illegible]

Sheraton

[illegible]/9.2011

Point Out Something Fresh

“点”出新意

入住中国第三大岛——上海崇明岛。原来和我一样，这里很少人来过（指身边的人）。入住喜来登酒店（一个好有趣的名字：由由）。阳光大堂，全玻璃锯齿顶棚，室内室外一体化，度假型（会议型）的酒店，房间多多的，档次也可以（和业主的投入资金有关），感觉不错。

Shanghai Chongming Island is the third largest island in China. I find that few people have visited here like me the people around me . I stayed in Sheraton Shanghai Chongming Hotel which had a funny name—Youyou. The lobby was full of sunshine. The zigzag rooftop was made of glass. The exterior and interior were unity. There were plenty of guestrooms in resort hotels or conference hotels, as well as its high level(related to the investment of the proprietor). I felt great!

循例地找找信纸（笺），写写画画平面图。因为临近退房了，只十分钟，心急，居然看到信笺上有纵横的一点一点的标记，那就容易多了。于是，十分钟完成了。顺手数了一小撮信笺：7张，不知道是原来的数量还是未消耗完的，有没有特定的（或者8张）。

As usual, I tried to find some paper to write and draw the plan. There were only ten minutes left to check out. Hurriedly, I found that the paper was full of marks —vertical and horizontal points. It was so easy that I only spent ten minutes finishing my drawing. By the way, I counted a handle of paper, seven pieces. I wondered if this handle of paper hadn't finished or it had the specific number (it might be eight pieces).

点点的，少见，比画线的优雅而含蓄，也算是一个“标签化”。对于我们要用纸的人，还是一种温暖的体贴。

Paper with points was rare, more graceful than the one with lines. It was also a tag. People who needed paper would feel a warm consideration just like me.

店住多了，能找到一点点新意的温暖感，还是难得的！

The more hotels I stayed in, the harder I could find something new and warm. So it was great!

上海崇明由由喜来登酒店
Sheraton Shanghai Chongming Hotel,China

85 广州瑰丽酒店

★★★★★

ROSEWOOD GUANGZHOU CHINA

Address : 95th floor, Guangzhou Zhoudafu Financial Center, No.6 Zhujiang East Road, Tianhe District, Guangzhou,China

广州市天河区珠江东路6号

广州周大福金融中心95层

Telephone : +86 20 8852 8888

Fax : +86 20 8852 6880

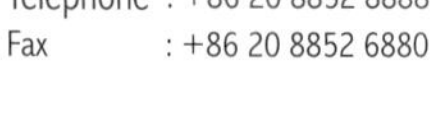

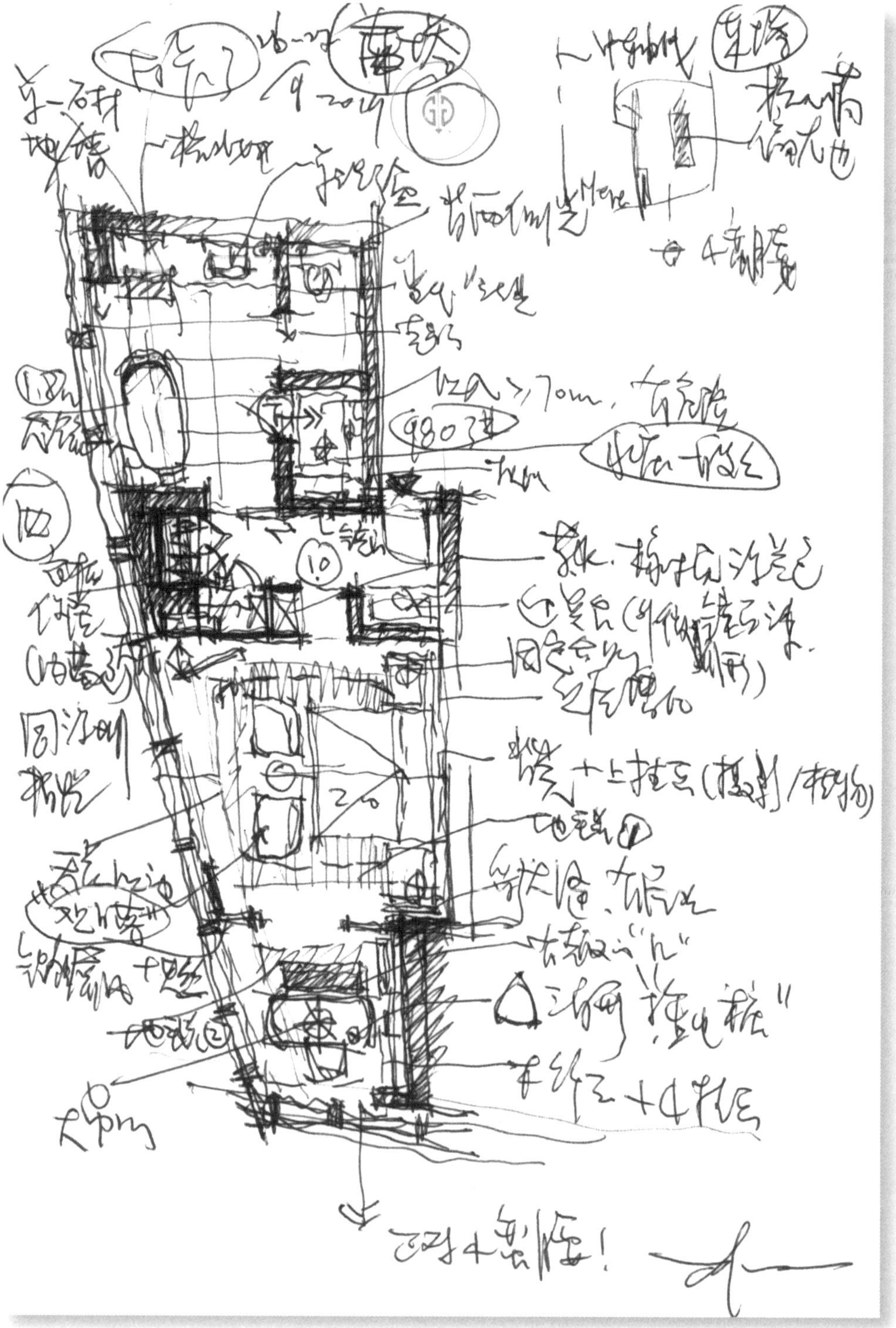

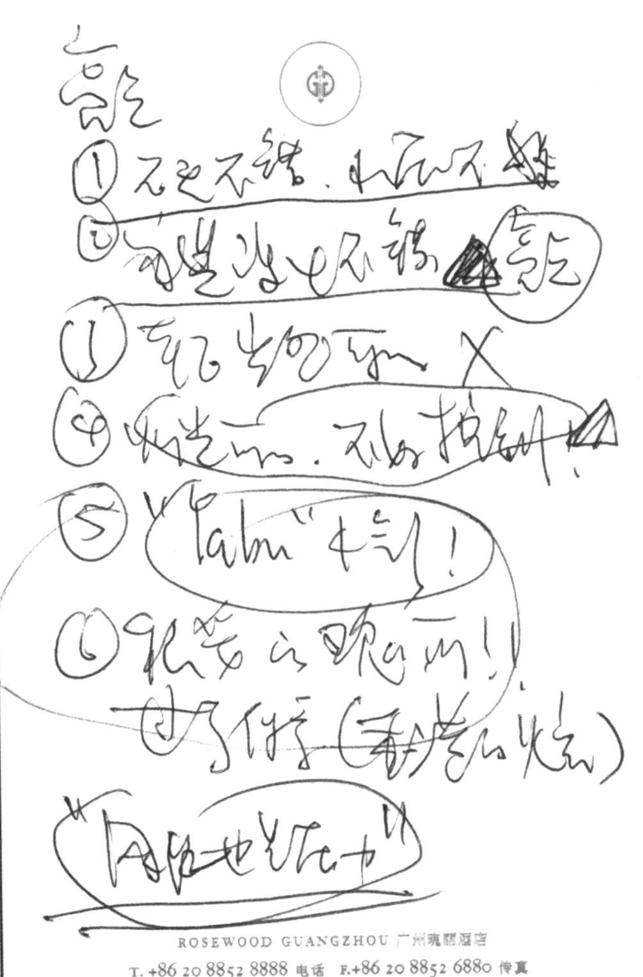

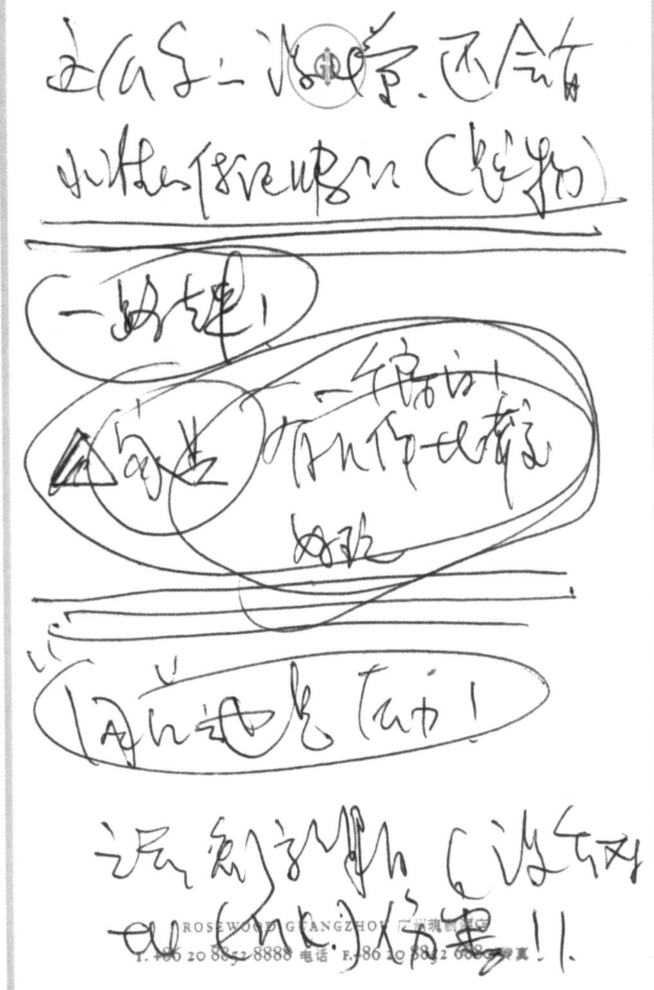

广州瑰丽酒店
Rosewood,Guangzhou,China

Highlights
亮点

1.石头不错，水压不够。
2.家具设计不错（亮点），
一个房间有几件比较好玩的。
3.平面当然可以。
4.灯光可以，不好控制。
5.Yabu小气了！
6.稳妥的瑰丽！过于保守(香港的炫)。
“网红也是压力”。
这么多的设计量, 还会有水准的保证吗?
7.“网红”也是压力！
之前看了效果图（没有对比香港），伤害！！

1.The stones were quite good, while the water pressure was not enough.
2.The design of furniture was perfect (the highlights).
Furniture: there were some interesting pieces in a guestroom.
3.The plane of the hotel was of course all right.
4.The lights were good but not easy to control.
5.Yabu was stingy this time!
6.Rosewood was pretty good! But too conservative.(the one in Hong Kong was cool.) An Internet celebrity was also under stress.
There were a large number of designs, how to guarantee the standard?
7.An Internet celebrity was also under stress.
I saw the design sketch before(no comparison with the one in Hong Kong), harm!

广州瑰丽酒店
Rosewood,Guangzhou,China

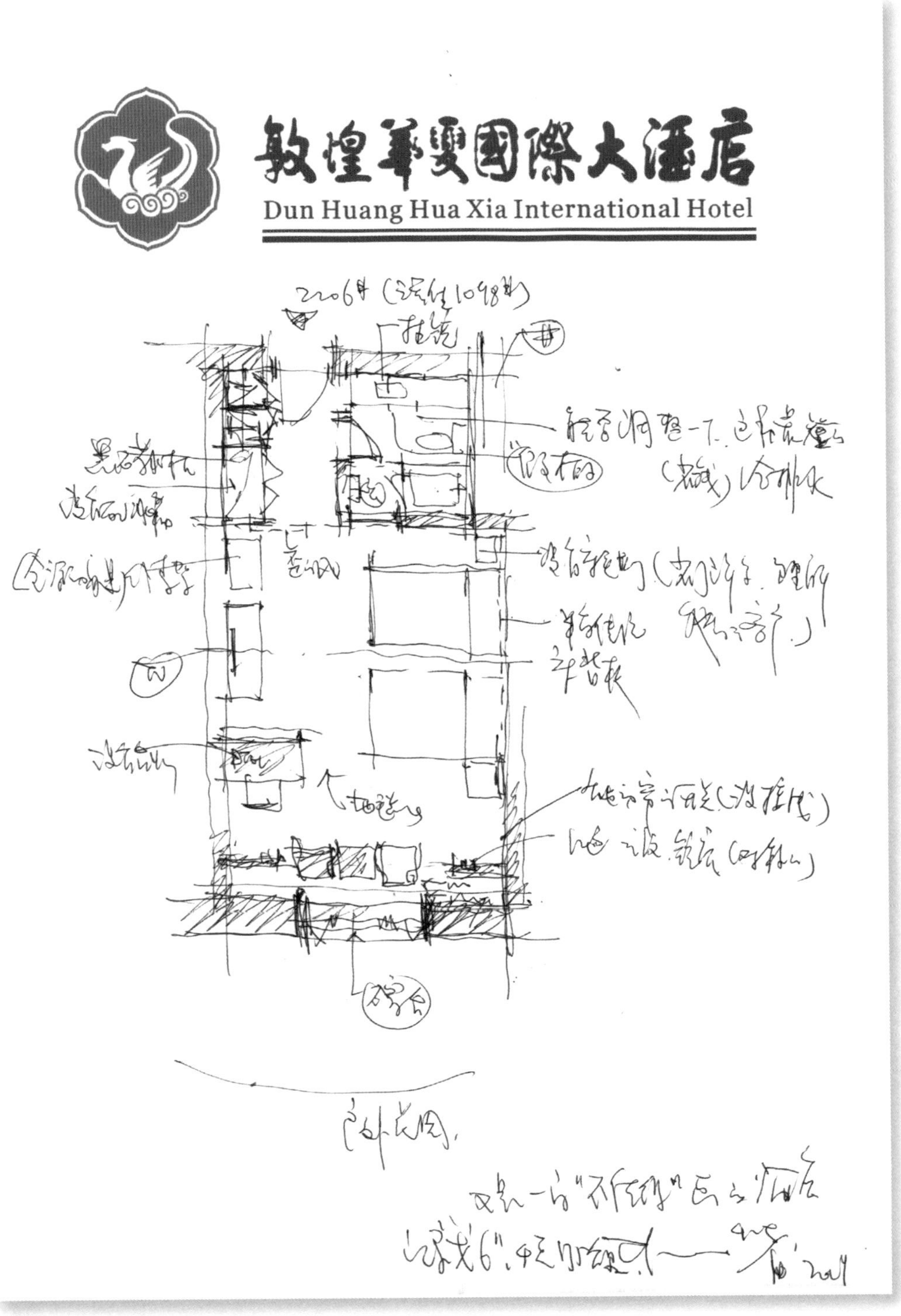

敦煌華夏國際大酒店

Dun Huang Hua Xia International Hotel

86

敦煌华夏国际大酒店

★★★★

DUN HUANG HUA XIA INTERNATIONAL HOTEL GANSU CHINA

Address : Middle of Dunyue Road, Dunhuang, Gansu,China (South of Dunhuang Middle School)
中国甘肃省敦煌市
敦月公路中段（敦煌中学南侧）

Telephone : +86 937 888 7111

Fax : +86 937 888 7188

敦煌華夏國際大酒店
Dun Huang Hua Xia International Hotel

地址：甘肃省敦煌市敦月公路中段（敦煌中学南侧）
Add:Middle of Dunyue Road, Dunhuang,Gansu(South of Dunhuang Middle school)
电话总机 Tel: 0937-8887111 传真 Fax: 0937-8887188 邮编 P.C: 736200

敦煌華夏國際大酒店
Dun Huang Hua Xia International Hotel

地址：甘肃省敦煌市敦月公路中段（敦煌中学南侧）
Add:Middle of Dunyue Road, Dunhuang,Gansu(South of Dunhuang Middle school)
电话总机 Tel: 0937-8887111 传真 Fax: 0937-8887188 邮编 P.C: 736200

敦煌華夏國際大酒店
Dun Huang Hua Xia International Hotel

地址：甘肃省敦煌市敦月公路中段（敦煌中学南侧）
Add:Middle of Dunyue Road, Dunhuang,Gansu(South of Dunhuang Middle school)
电话总机 Tel: 0937-8887111 传真 Fax: 0937-8887188 邮编 P.C: 736200

Have You Ever Used the Pens in Hotels?

酒店的笔，你用过吗？

看到了传统的“羽毛笔”，是一头可以写字，另一头可以裁纸的“稀缺品”。怎么说呢？ 新一次戈壁活动入住的这间酒店：华夏国际，迷宫一样的房间区，经常迷路。对于戈友来说——晕。幸亏服务一流，行李早早就放在房间了，不然要拖着行李在地毯上走这么“一长路”，结果很可能戈壁四天（135公里）没有受伤，最后一天，在这里中招（入住还是自己拉行李的话）！ 啊啊！

I saw the traditional quill pen which one head was for writing, the other was for cutting paper. It is really a rarity now. But what do you think of it? I often got lost in this hotel, Dun Huang Hua Xia International Hotel, where we stayed in the Sixth Gebi Hiking Activity. The zone of the guestrooms was just like a maze for us. Thanks to the first-class service, our luggages were taken to our rooms. Otherwise we might hurt ourselves walking such a long way on the carpet with the heavy bags on the last day (We didn't hurt during the four-day hiking, totally 135 kilometers), if we carried the luggages by ourselves. OMG!

会议式度假酒店，没有什么特色，丰富、齐全的功能，接待能力一流（可能有近千间客房吧），20000元以内一间房间的装修费用吧（四星级左右），勉勉强强记录一下就好了！ 作为“非正面”的教材（如果我们设计，可以怎样去引导甲方呢？这个也是我们在类似项目要考虑的），这一类项目最不容易设计了，包括选材（设备及用品）、选色（大众）、布局（入住的都是大件行李，匆匆忙忙的，不考虑使用者）。设计师的“一厢情愿”，有时候是经不起时间的考验的，你说呢？ 就像选用这支笔一样——现在写字的人都少了，还有裁纸的必要吗？一不小心，弄伤自己，惨！

A conference-resort hotel has no features but is fully functional and a huge receiving capacity (nearly one thousand guestrooms). The decoration of each room costs less than 20,000 yuan (almost the four-star standard). I just wanted to record it as a negative example. (If we have a chance to design it, how should we guide Party A? We should also think about the projects like that.) It's not easy to design this kind of projects, including materials (facilities and supplies), colors (common), hotel layouts (the most guests are those who stay with big luggages in a hurry) . The designers' own wishful idea sometimes can't stand the test of time. What do you think of it? Just like the traditional quill pen—fewer and fewer people write with pens now, does it still need to cut paper? You might cut yourself accidently. So awful!

敦煌华夏国际大酒店
Dun Huang Hua Xia International Hotel,China

西班牙游记（2019年10月17～28日）

The Trip to Spain

去哪?
22年，
每一年的员工旅游或拓展活动，
一直坚持着，简单成自然。
Where to go?
We have insisted on organizing
ourcompany trip or outreach activities
for twenty-two years.
It becomes natural and simple.

我们，
有资格，更应当让每一次远行留下一笔
财富。
We are qualified
to leave a legacy for every travel.

以前忽略了，
介绍路线、建筑、风景名胜。
留白，写写画画的记录，
每时每刻。
We ignored before
to introduce the routes,the buildings
and the places of interest.
Leave a note for you to write, to draw
at anytime ,anywhere.

当然，
有你的名字，让时间暂停，
走走、看看、吃吃、喝喝、玩玩、睡睡。
Of course,
With your name, stop the time,
to walk, watch, eat, drink, play and sleep.

一本经得起你收藏的书，
每一天，
开心。
The book is worth collecting by you,
and happy every day.

2019年10月12日启用
October 12, 2019

不要问意义在哪，迈开脚步就好。
不要担心饮饮食食，放开肚皮就好。
不要担心天气如何，天晴抬头。
下雨担“姐”（伞）。
不要担心工作没人做。
我们有同事在。
有你们真好！
Don't ask what the meaning is, just start to walk.
Don't worry about what you eat, just enjoy.
Don't worry about what the weather is like.
Just look up when it is fine.
Hold up an umbrella when it rains.
Don't worry about your work.
Our colleagues are always there.
So good to be with you!

随·笔 essays

18/10 2019 pm 10:00

[illegible]

11

12

这几年习惯了有雪茄陪伴我出行。但，第一次体验到“塞不进人”的机场吸烟区：多哈机场，有香蕉泰迪熊的国际机场。

I've got used to travelling with cigars these years. However, it was my first experience that the smoking area was overcrowded in the Doha International Airport which has the yellow Teddy Bear.

October 18, 2019, Local Time 10:00P.M.

2019年10月18日,当地时间10:00P.M.

算是第二天了吧。10月17日半夜到白云新机场T1，半夜1:30（10月18日）飞；经停多哈3小时（9+3+7），到达热闹的城市（闹革命）的西班牙巴塞罗那，“越穷越出屎”，近20%的失业率，懒人越来越懒，穷人越来越穷。

It was just about the second day. We arrived at the Baiyun airport on the midnight of October 17. The plane took off at 1:30 a.m on October 18 with a stopover at Doha for three hours(9+3+7). After nineteen hours, we finally arrived at Barcelona which was busy revolting. The poorer the people were, the more trouble they made. Here the unemployment rate was almost 20%. The lazy people became lazier and lazier, the poor became poorer and poorer.

到了多哈机场，人山人海，也有趣，我只负责大家间歇性的补充：在多哈机场中转喝咖啡，在巴塞罗那吃吃雪糕，在兰布拉的步行街上厕所、火腿+水果的“牙缝”餐。

Arriving at the Doha airport, there were people mountain people sea. How funny! I treated my companions to some snacks on the way, drink coffee at Doha airport, taste ice-creams in Barcelona, go to toilet in the walking street,and have some hams and fruits.

也看了一下游行聚会，还不如正儿八经地去“闹”好经济，惠及民生，来的实际，愿平安和平。

We also watched the demonstration.It was more practical to develop economic to improve people's livelihood. Wish people peace and safety.

中餐，还可以。回到酒店。

Lunch was not bad and then we returned to the hotel.

万豪旗下的一个低档酒店，干干净净，还不错，房间也多，一点没有配套，也整齐，可谓正好给“团队”的。基本没有一盏灯，可以一切都省下来，没手纸，无多余的装饰，无顶棚（顶棚灯）。

We stayed in a low-class hotel under Marriott International, Inc. It was clean and tidy. There were plenty of guestrooms without any facilities. It was just for tour groups. There was not a light on the ceiling or any toilet paper. It might save a lot of money. No extra decoration, no ceiling (no lights on the ceiling).

这间酒店非常明确自己的定位，相信是一个不错的投资项目，值得学习！

The position of the hotel was very clear. I believed it was a good investment project and worth studying.

万豪AC酒店

87

★★★★

AC HOTELS SANT CUGAT BARCELONA SPAIN

Address : Plaza Xavier Cugat,
Sant Cugat del Valles,
Barcelona, Spain 08174

Telephone : (+34)935894141

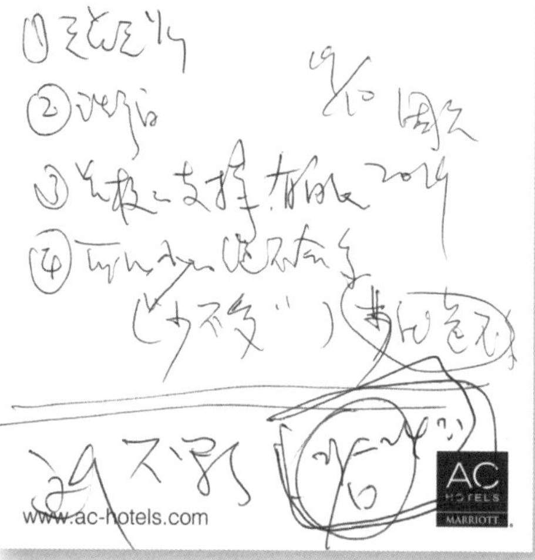

万豪AC酒店

AC Hotel Sant Cugat Barcelona

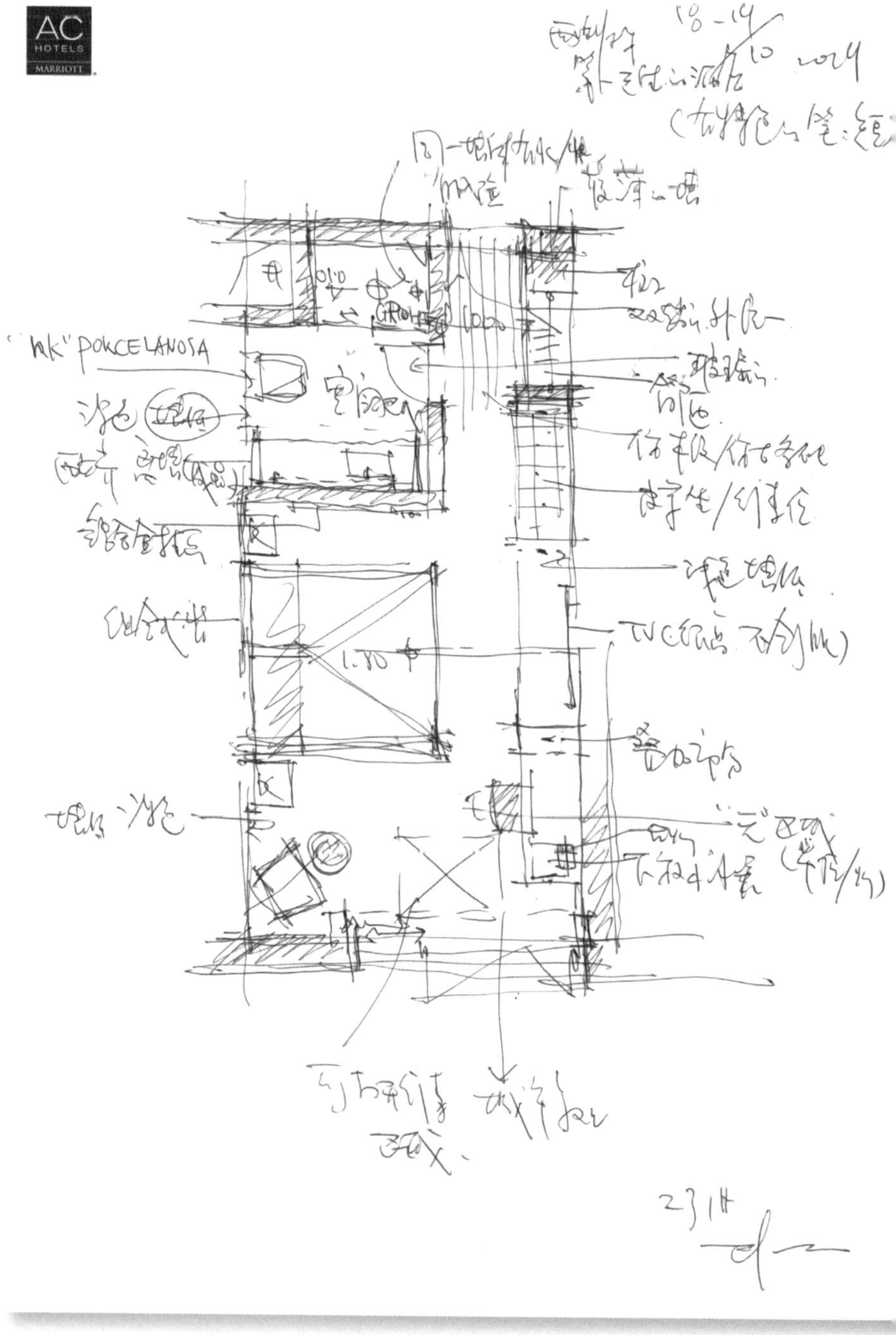

Write Back the Diary of the Third Day on the Bus

早上车上补第三天的日记

第三天信息量太多了，以至于发了两条微信。上午去了巴塞罗那俱乐部与奎尔公园。

There was so much information on the third day that I sent two Wechat messages. In the morning we visited Barcelona Club and Park Guell.

第一次亲见高迪大咖的惊世之作。好甲方+高迪的2间别墅+公园。一个为所欲为的甲方和一个“疯狂”的“天才”，现场叠砖取材。没有“规则”的肆意创意，建筑作雕塑，雕塑如建筑。我认为扎哈也望尘莫及（也是天才）。

It was my first time to see Gaudi's world-shocking masterpiece. A good Party A, Gaudi's two pink villas and Park Guell. The Party A who got his own way and a crazy genius made the wonderful buildings. They were made with local material to pile up bricks at the site. Gaudi's creativity was so free that the building was like a sculpture, and the sculpture was like a building. I considered that Zaha Hadid, who was also a genius, but far behind him.

40多亿的欧元市值，7亿~8亿欧元的年收入，巴沙俱乐部，牛！参观也棒棒哒。一个多小时，连铲起的草皮也可分小小块“卖”，无所不能而又贴心关怀！

The Market value of Basha Club was more than four billion and annual income was about seven or eight hundred million. It was exaggerating! Our great visit took over one hour. The small pieces of grass which was scooped up were also sold. It could satisfy everyone and was quite considerate.

下午的精彩高迪，高迪，高迪：圣家（族）大教堂（可惜旅游公司失误：没有门票了！）。新旧不和谐，不予评论。

In the afternoon, what impressed me most was Gaudi, Gaudi and also Gaudi, Basilica Temple Expiatori De La Sagrada Familia. (What a pity that we didn't buy tickets because of travel agency's mistake!) The old and new felt disharmony. No comments.

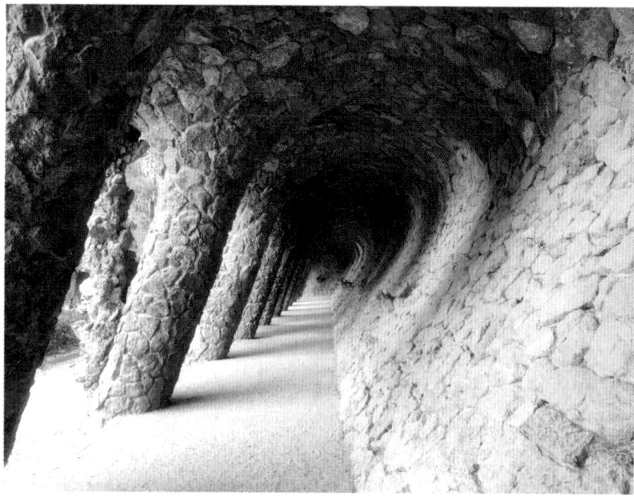

特色的风景、建筑常常会激发我们拍摄的创意。你也可以来凑热闹，另一个角度看自己，看我们。

The unique scenery and buildings always inspire our shooting creativity. Come to join us for fun. Look at us and yourself from another angle.

米拉之家——没有一样的两块石头，高投入、高技术（现代的数码化也难以实施）。幸运去了巴特罗住宅，中文+AR导游器，好牛！真是一个天才：建筑、室内、家具、饰品……一切都是高迪所设计，佩服。用光、用风、用景，真是领先我们100多年。

Casa Mila had high investment and high technology but not same stones. (It was hard to implement with modern digitalization.)Luckily, I went to visit Batllo with Chinese and AR guide device, cool! What a genius! The buildings, interior, furniture and accessories were all designed by Gaudi. I respected him. He was very skilled at using light, wind and scenery, really more than 100 years ahead of us.

当然买了一些纪念品，喜欢的，不能浪费每一次的旅行。

Of course I bought some favourite souvenirs. I wouldn't waste every travel.

高迪圣家族大教堂
Cathedral of the Sagrada Familia in Gaudi

O.'s File (区生词典)

西班牙巴塞罗那的米拉之家建于1906~1912年，是高迪设计的最后一个私人住宅，整栋建筑如波浪汹涌的海面，极富动感。屋顶是奇形怪状突出物做成的烟囱和通风管道，米拉之家被认为是所有现代建筑中最具代表性的，也是最具创造性的建筑，是20世纪世界上最重要的建筑之一。该建筑无一处是直角，这也是高迪作品的最大特色。（百度百科）

Barcelona Casa Mila was built from 1906 to 1912. It is the last private house designed by Gaudi. The whole building is like rough sea waves, full of dynamic. The top has chimneys and ventilation pipes which were made of strange-shaped protrusions. Casa Mila is considered to be the most representative of all modern buildings, as well as the most creative architecture. It is one of the world's most important buildings in the 20th century. The building has not any right angles, which is the greatest feature of Gaudi's works. (From Baidu Encyclopedia)

高迪的每一个作品都令人叹为观止，能进去现场体验一次，就是不枉此行了。AR的中文讲解更是让我“汁都捞埋”（广州话，意思是一点都没有漏掉）。

Every work of Gaudi's is stunning.I deserved the trip having a live experience. The narration in Chinese of AR was so good that I enjoyed it to the last bit.

Mercure
HOTELS
瓦莱斯美爵酒店

88

★★★★

MERCURE ATENEA AVENTURA HOTEL VILA-SECA TARRAGONA SPAIN

Address : Avda. Ramon D'olzina NO.52 Vilaseca, Spain 43480

Telephone : (+34)977396278

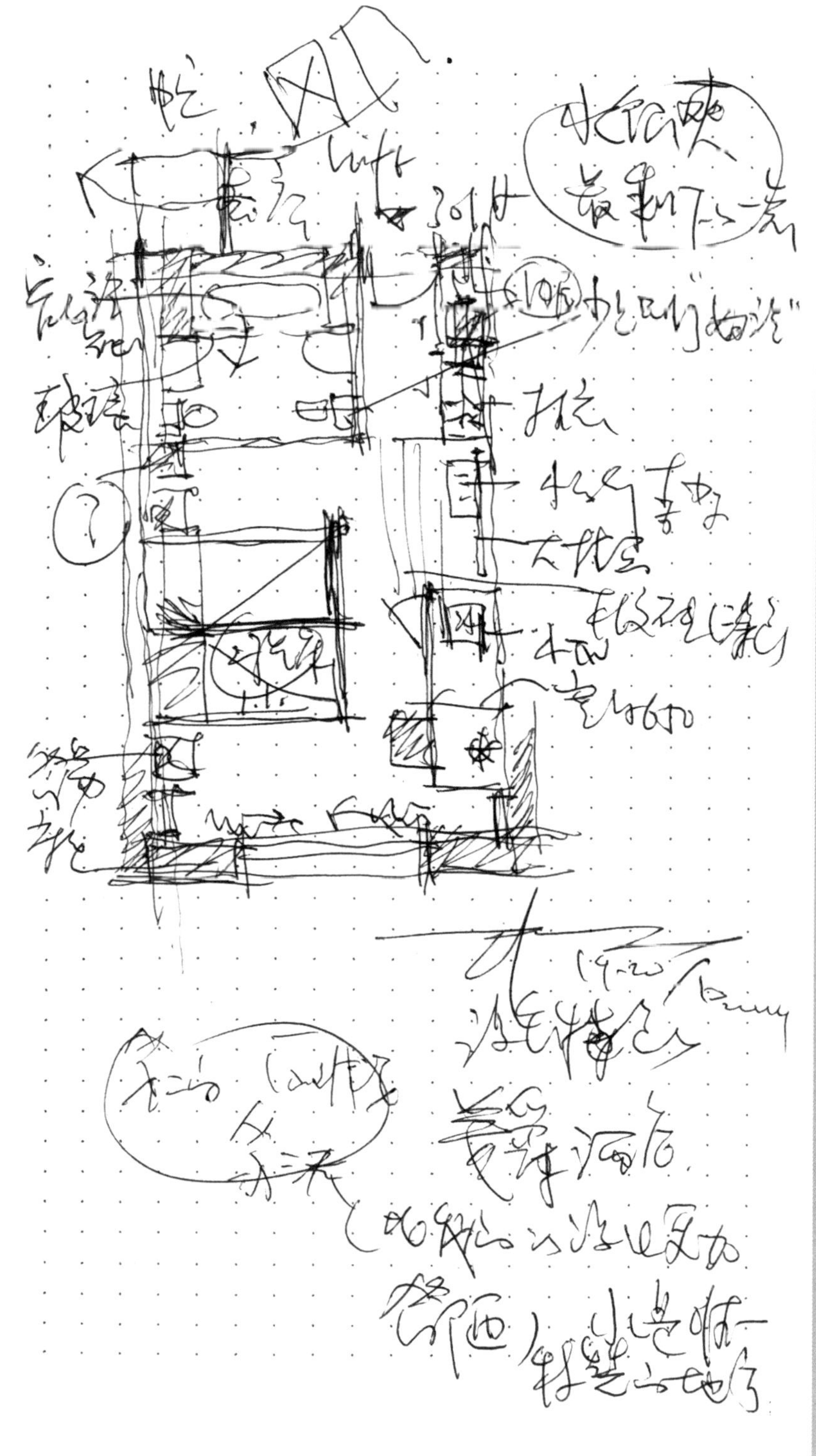

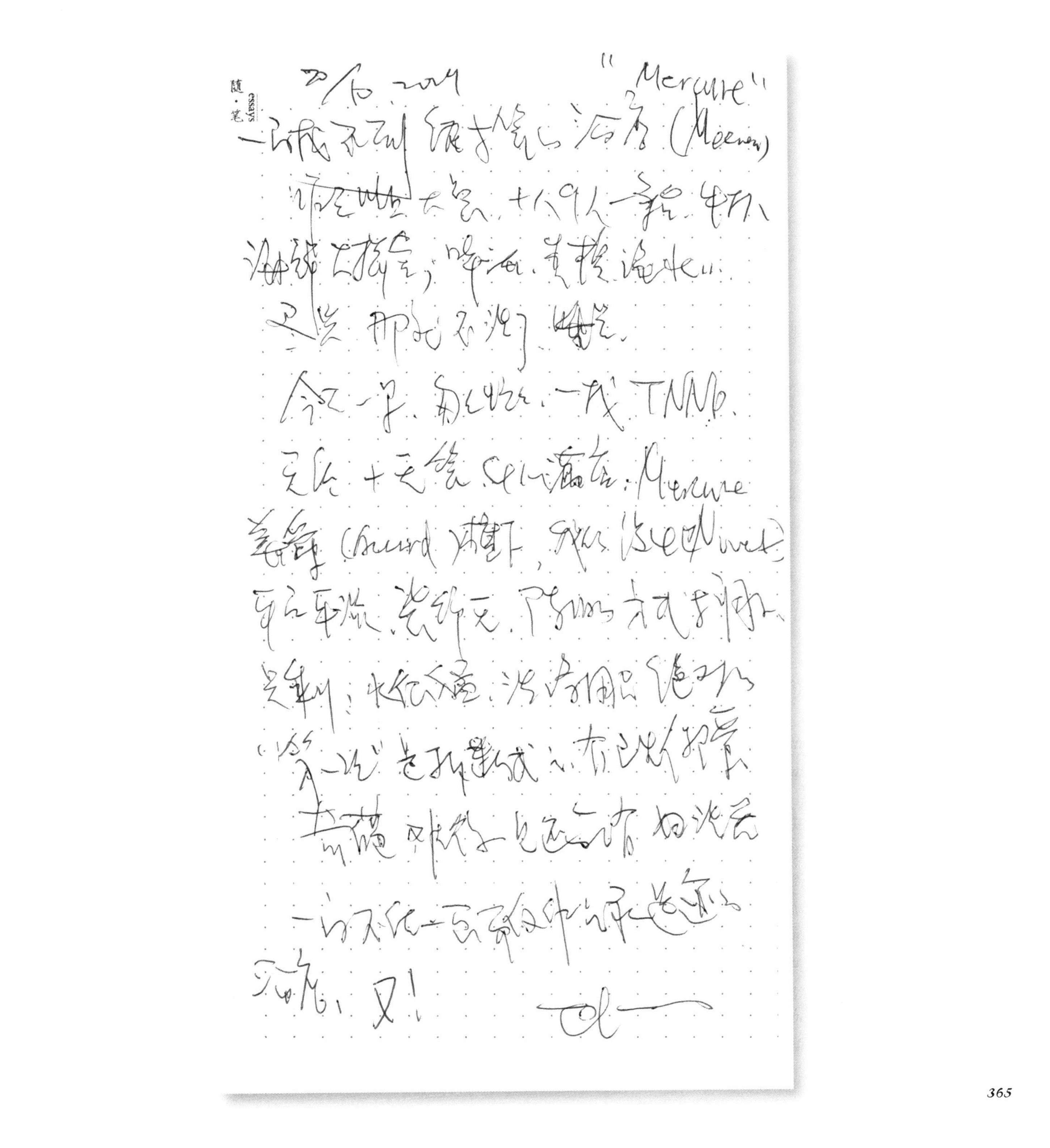

Mercure—A Hotel without Paper or Pens
一间找不到纸与笔的酒店——美爵

2019年10月20日
October 20, 2019

昨天晚上大餐，18个人一条台，牛扒海鲜大拼盘，啤酒，香槟，汽水，尽兴，那就不洗了，睡觉。

We had a big dinner last night. We eighteen people sat at a long table and enjoyed steak, seafood, beer, champagne and soft drinks. We had a great time, and then slept without a shower.

今天一早，匆匆忙忙，一找，无纸+无笔，什么酒店？美爵（Accord麾下，和我设计过的Novotel诺富特一样），平面平淡，装饰无，陈旧的方式与调子，只剩：水很猛，洗涤用品绝对的“第一次”，是拆断式的，有卫生保障，奇葩，难得一见，还设有妇洗器。一间不值得一画而仅作记录足迹的酒店，又！

This morning, I hurriedly found that there were no pens or paper in the room.What a hotel! Mercure was under Accord, the same as Novotel which I had designed. The plane was ordinary without any decoration, also old style. There was only strong water left. Washing products were absolutely for the first time because of the once broken form.They were sure clean. It was rare to see the women wash device here. I didn't think this hotel was worth drawing, just recorded it as my footprint. Another again!

瓦莱斯美爵酒店
Mercure Atenea Aventura Hotel, Vila-Seca Tarragona,Spain

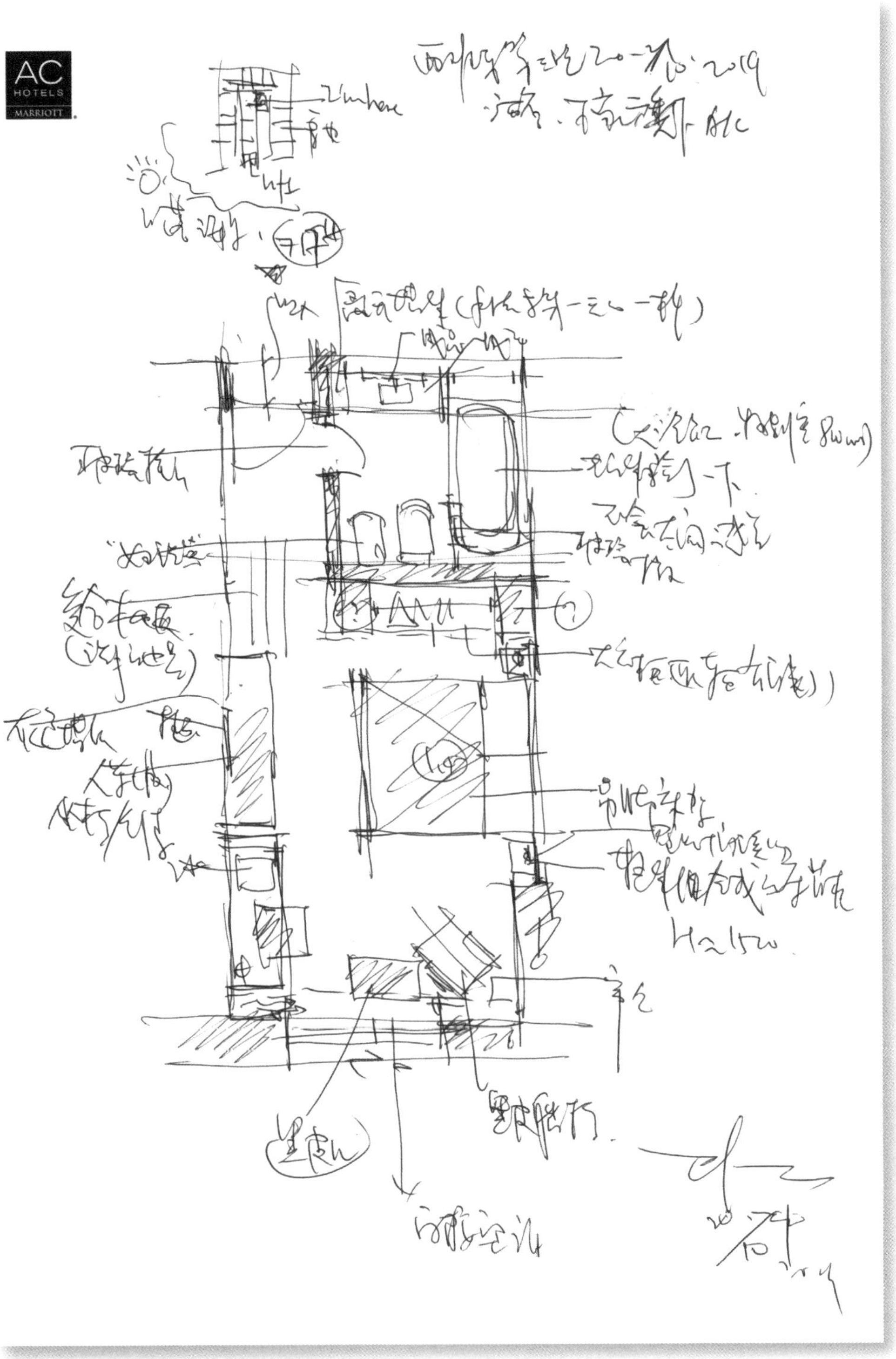

万豪AC酒店

89

★★★★

AC HOTELS ALICANTE SPAIN

Address : Avda. de Elche, 3, Alicante, Spain 03008

Telephone : (+34)965120178

万豪AC酒店

AC Hotel, Alicante, Spain

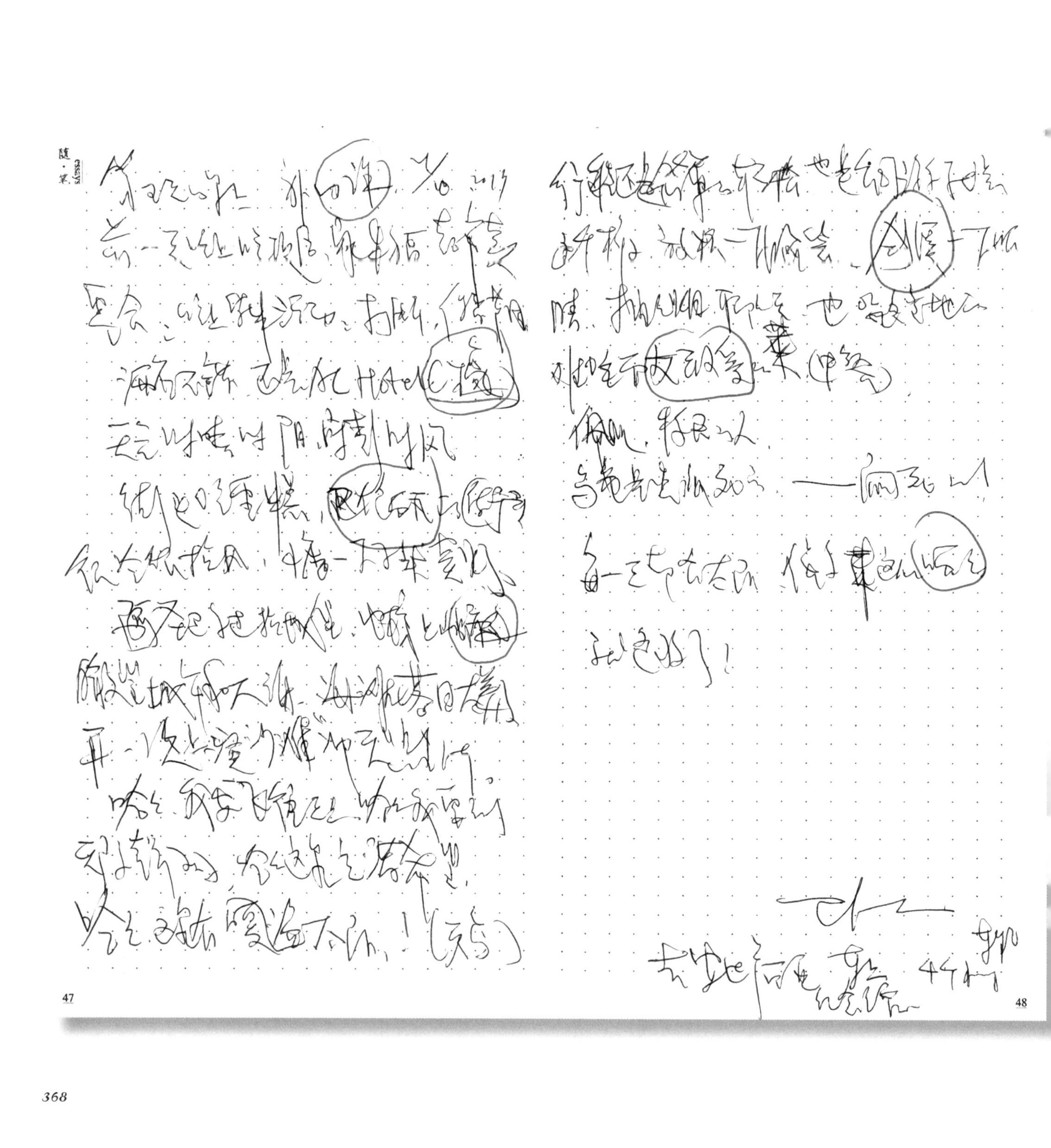

47

48

On the Morning of the Fifth Day, Write Back the Diary
第五天的早上，补“功课”

去安达卢西亚纪念馆，车程4个小时
2019年10月21日
On the way to Andalucia's Museum Of Memory, a four-hour drive
October 21, 2019

前一天晚上吃饭后，在家乐福超市买零食，晚上集体活动：打牌，保留节目。

酒店不错，还是AC Hotel（万豪）。

天气时晴时阴，时静时风。

街头吃雪糕，在瓦伦西瓦的街头很冷、很拉风，小店一下子就卖火了。

圣巴尔巴拉城堡，乘电梯上山，瞰望城市和大海，海滩落日太美了！再一次上演沙滩“冲天”照片。

哈哈（原歌词：啊哈），我要飞往天上，哈哈，我要到天际翱翔，哈哈，这里充满希望，哈哈，这里有温暖太阳！——《天鸟》

We went to Carrefour to buy some snacks after dinner the day before yesterday. And then we played cards which was our reservation program in the evening.

We still stayed in AC Hotel , of course it was perfect.

Sometimes the weather was fine and cloudy, sometimes quiet and windy.

We ate ice cream in such cold weather in the street of Valencia. Cool! The little store became a hit at once.

In Castillo de Santa Barbara, we climbed up the mountain escalator and had a bird's eye view of the city and sea. What a beautiful sunset on the beach! We took a lot of photos of jumping again.

"Haha!I want to fly to the sky; Haha! Iwant to fly high in the sky; Haha! There is full of hope; Haha! The sun is warm here".
(From Birds)

行程还是简单轻松的，公司游就是这个样子，放松一下脑袋，舒缓一下眼睛，抽抽烟，聊聊天，也吃吃当地的难吃而又至爱的菜（中餐）。佩服，移民的人。

乌龟是怎样死的——憋死的！

每一天都有太阳，偶尔来这里晒晒就足够了！

The travel was simple and relaxing. Company travel was a chance for us to relax ourselves, relieve our eyes, smoke and chat together. Of course we also tasted some terrible but our favorite Chinese food. I admired the immigrants.

How do tortoises die? —They are bored to die.

The sun rises every day. Coming to enjoy the sunshine occasionally is enough.

万豪AC酒店
AC Hotel Alicante

晚间的娱乐成为国外游的保留节目，聊聊天，放松放松。

The evening entertainment becomes our reservation program during travelling in foreign countries. We Chatted together and relaxed ourselves.

随·笔 essays

[illegible]

55

The Sixth Day+ the Morning Write Back the Diary of the Day Before (5th+6th)

第六天+上午，补上一天的日记（第五天+第六天）

游完了“阿尔汉布拉宫”，叹为观止，真的有其特色：依山而建，外粗犷，内细腻，十分典型的伊斯兰风格，光影，围合。天、地、墙，真的不枉此行。可惜逛到中午快离开时太阳才真正地出来了！也是给面子的。

After visiting the Alhambra Palace, we all thought it was an amazing artwork with its own features:it was built into a hillside, the rugged look but delicate interior, with an typical Islamic Style which the light and shadow were made into an enclosed space. Looking at the sky, the ground and the wall, we had no regrets really. It was a pity that the sun came out at noon while we were leaving. Nice enough!

昨天看了一个好建筑，4个小时车程到达安达卢西亚纪念馆和银行，光与石。

圆形中空共享，纪念馆中庭阳光下，拍照的好地方。周一闭馆，特地开放参观，牛！

We visited Andalucia's Museum of Memory which was a good building as well as a bank yesterday after a four-hour drive. How impressive the light and the stones were!

The round atrium was a public area. It was a good place to take photos in the atrium of the museum in the sun. The museum was closed on Mondays, but open for us to visit. How great!

米哈斯小镇的暴走，阳光下非常的有趣，拍照、摆拍，绿植、鲜花、小园，太阳下的下午茶、体验海鲜的小吃。

It was very funny to aimlessly walking in the Mijas Town in the sun. We took photos with a variety of poses, a lot of plants and flowers in a garden. We had sea food snacks for afternoon tea in the sun.

真正地吃吃、喝喝、走走、看看……

We really came here to eat, drink, walk and watch…

***O.'s File* (区生词典)**

阿尔汉布拉宫是西班牙的著名宫殿，是中世纪摩尔人在西班牙建立的格拉纳达埃米尔国的王宫，有“宫殿之城”和“世界奇迹”之称，是伊斯兰教艺术在西班牙的瑰宝。（百度百科）

Alhambra Palace is the famous palace in Spain which was built by the medieval Moors, it was the palace of the Emirate of Granada at that time. Alhambra is known as "the city of the palace" and "a wonder of the world", as well as the treasure of Islamic art in Spain.

(From Baidu Encyclopedia)

阿尔罕布拉宫，值得长呆的地方

Alhambra Palace, worth staying for long

是值得慢慢走、细细品的一个经典，一首吉他名曲更是让其名扬远播。我们走走停停，好几个小时，尽兴的一天。

Alhambra Palace is a classic which is worth walking in it and enjoying it slowly. She is famous far and wide for wonderful guitar music. We walked and stop for a few hours. What an enjoyable day!

米哈斯小镇

The Mijas Town

午后的米哈斯，每一处都像明信片一样。

Afternoon came, each view in Mijas was like a postcard.

ABADES酒店

90 ★★★★★

ABADES NEVADA PALACE HOTEL GRANADA SPAIN

Address : Calle de la Sultana, 3
Granada, Spain 18008
Telephone : (+34)902 222 570

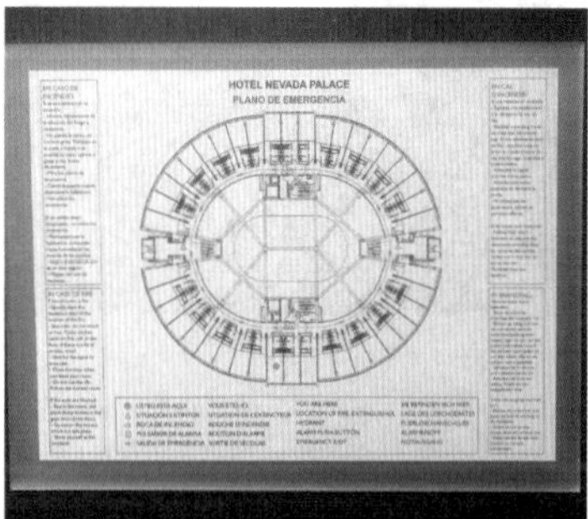

Abades Nevada Palace Hotel Granada

ABADES HOTELES

www.abadeshoteles.com

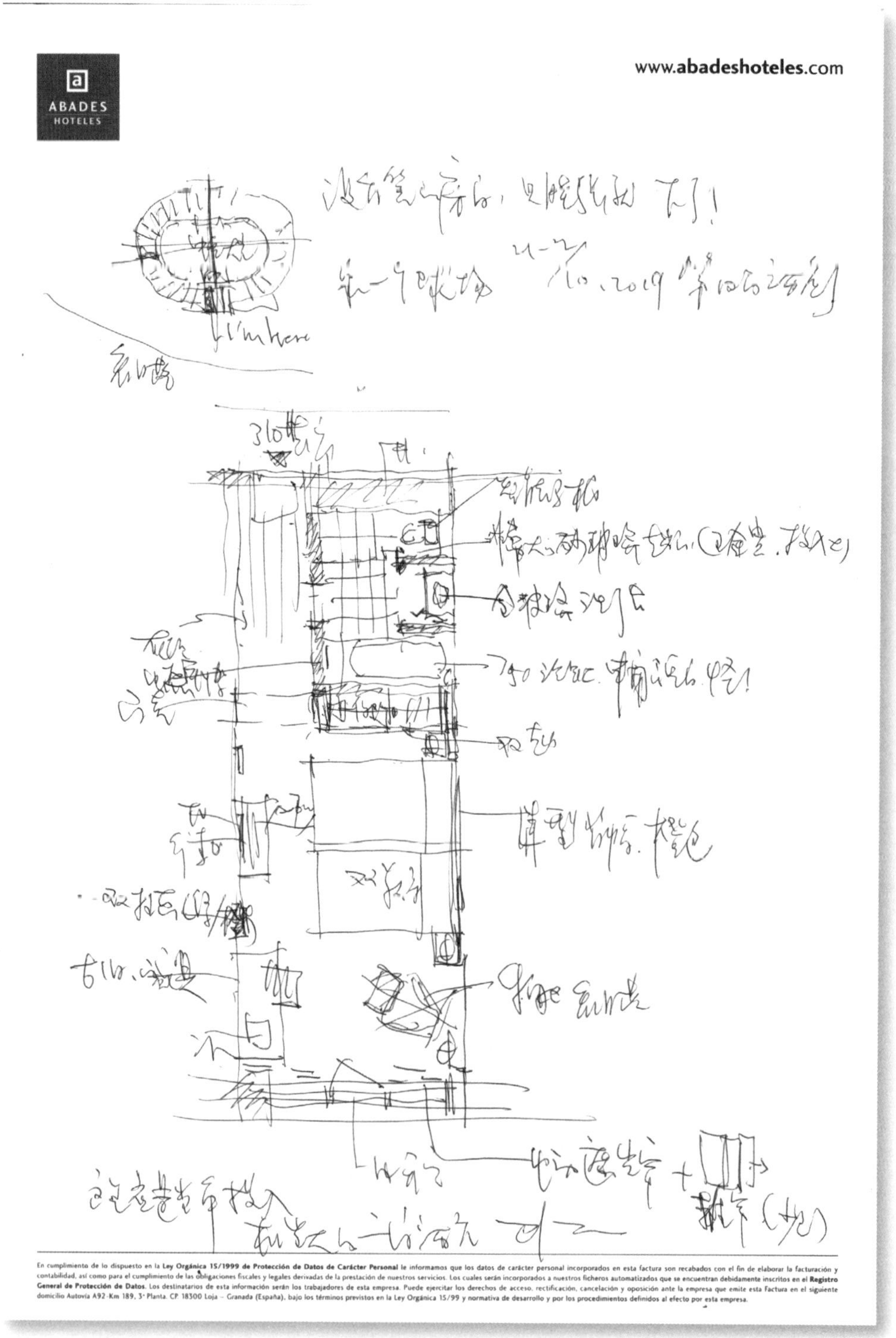

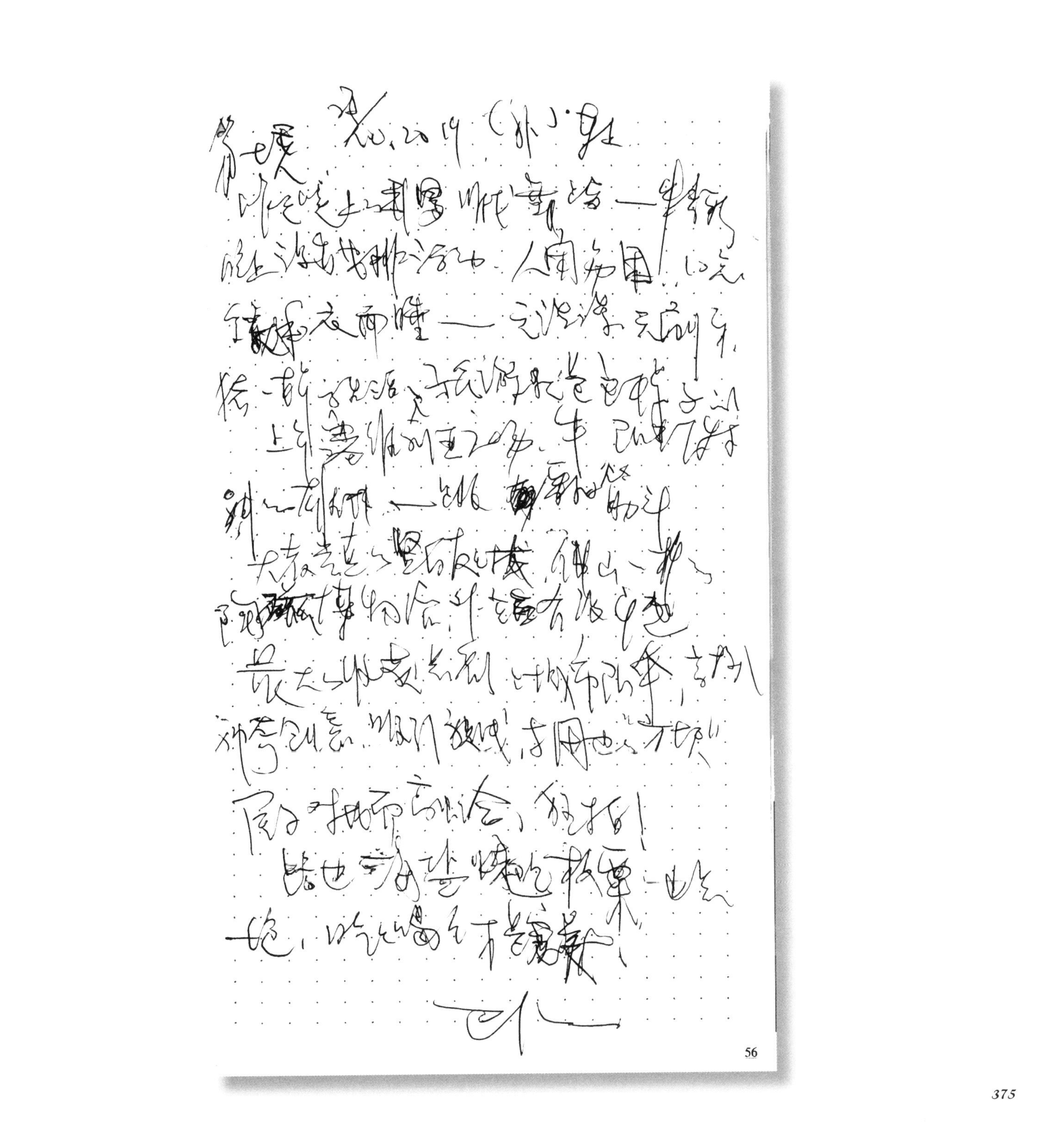

[illegible]

The Seventh Day
October 23, 2019, (Write Back) on the Bus

第七天，2019年10月23日，（补）车上

昨天晚上的弗拉门戈舞蹈——精彩。

How wonderful the Flamenco Show we saw last night!

晚上没有安排活动，人闲易困，10点多就和衣而睡——无洗澡，无刷牙，猪一样的生活，旅游就是这样子的！

There weren't any activities in the evening. People felt sleepy easily if nothing to do. I felt asleep in my clothes without taking a shower or brushing my teeth. What we did during the travel was to live like a pig.

上午的塞维利亚广场，牛！阳光下特别的雄伟——跳、翻筋斗。

大教堂真正的竖向挺拔，佛山一样的陶瓷博物馆，外立面有设计感。

最大的收获是看了城市阳伞，高投入的神奇创意，吸引视线，与周边的“方块”房子对比而融洽，狂拍！

In the morning, we visited Square of Seville. I'm pressive! It seemed imposing in the sun.We could't help jumping and turning somersaults.

Catedral de Sevilla was really tall and straight. It was a Ceramic Museum like Foshan(in Guangdong Province), with a designed facade.

The most rewarding was Metropol Parasol. High cost made creative magic. It really caught our eyes and also integrated with the square houses around it. We took a lot of pictures crazily.

路边吃海盐烤板栗，也是一绝，吃吃喝喝才是完美。

Eating sea-salt roasted chestnuts by the road was also great fun. Eating and drinking were perfect.

每一次、每一景的跃起，每一个人的姿态都是一道风景，帅酷。塞维利亚广场的雄伟建筑、大海沙滩等背景，瞬间凝固。

Every jump in different scene, every pose was a landscape. How cool! The great buildings in Square of Seville, the sea, the beach and other backgrounds all stopped at this moment.

塞维利亚广场上的跳跃
Jumping on the Caixaforum-Sevilla Centre

城市阳伞的爬高爬低，这样子的创意和投入确实让我感受到中外思维的不同，给我们以后的设计项目以启发。

We climbed up and down along the Metropol Parasol. The creativity and cost showed me the ideology difference between home and abroad. We were inspired a lot for our future design projects.

Hesperia
hotels & resorts

91 HESPERIA酒店

★★★★

HESPERIA HOTEL & RESORTS SEVILLA SPAIN

Address : Avda. Eduardo Dato 49, Sevilla, Spain 41018
Telephone : (+34)954548300

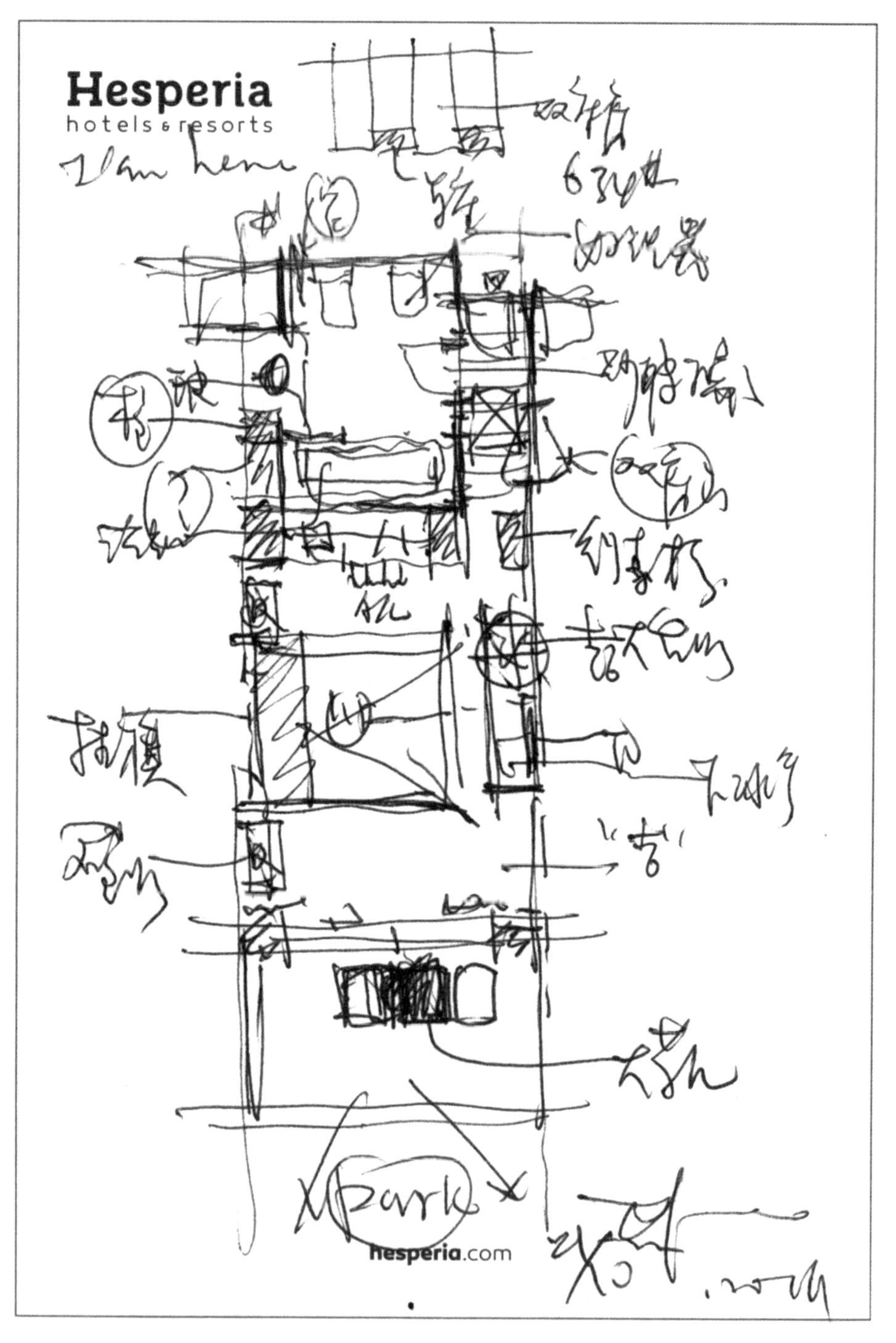

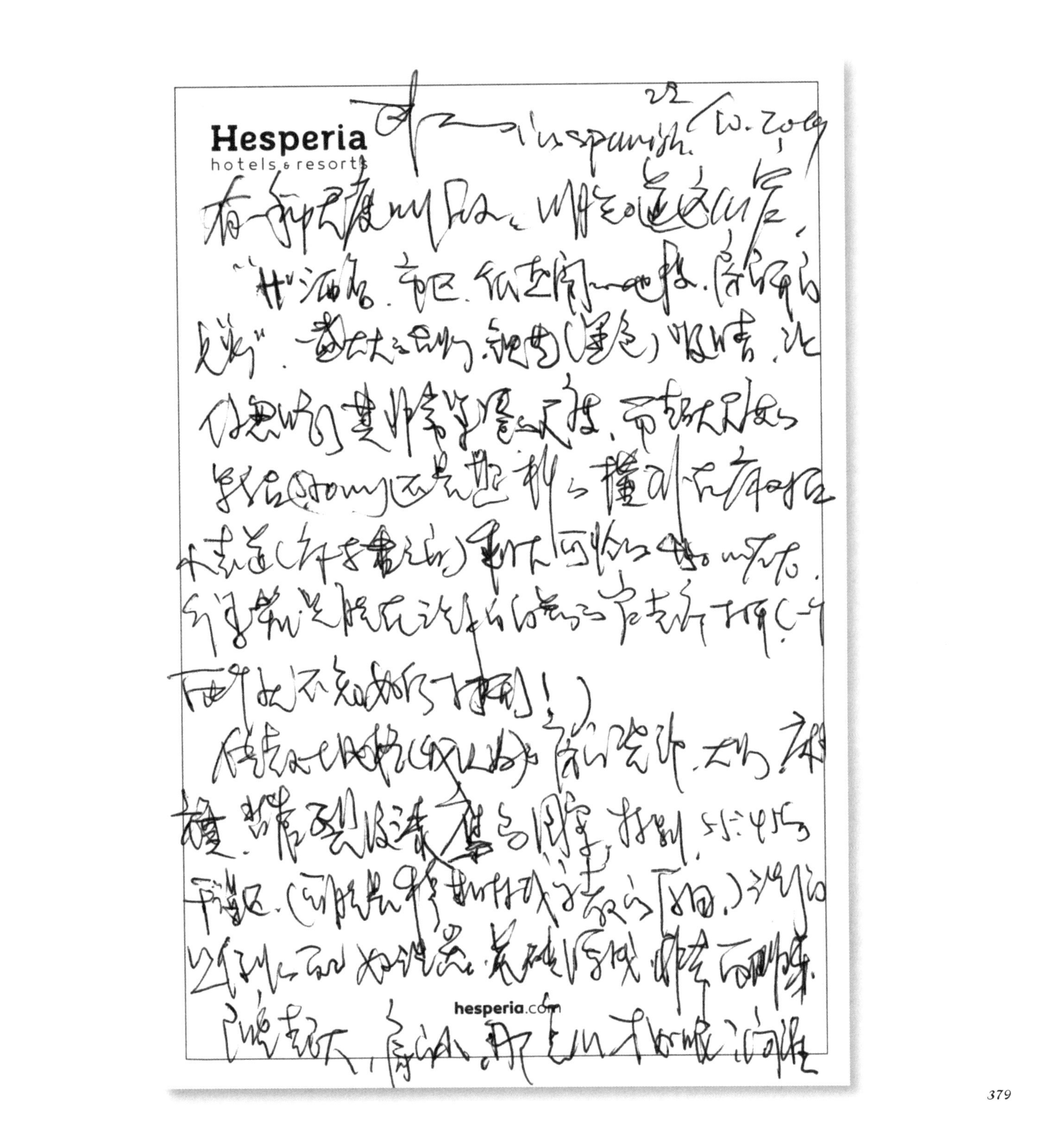

Hesperia
hotels & resorts

spanish. 22 10. 2019

hesperia.com

A New Rule of Narrowness

"窄"出新尺度

2019年10月22日于西班牙

October 22, 2019, in Spain

"H"酒店，市区，很热闹的地段，房间开门见"灯"，一盏大大的台灯，铁艺（黑色）吸睛，让你忽略了其非常紧凑的尺度，而超大尺度的写字台(550mm)还是照样的，横卧在床对面的小走道（床与台之间），剩下可怜的450mm 左右。行李箱只能在洗手间前面的窄走廊打开（一个，两个就不知如何打开了！）。

"H" Hotel lay in a busy urban area. A lamp was seen when opening the door. The black iron table lamp drew your eyes so you might completely ignore the narrow measure. The supersized table(550mm) still lay on the hallway across the bed (between the bed and the table), only 450mm was left. The suitcase had to be opened on the narrow corridor in front of the bathroom (just one, I didn't know how to open two suitcases at the same time).

传教士风格（我认为）的房间装饰，大灯，床背旗，背靠裂纹漆鹰鸟图案，特别。55：45的干湿区（可能是穆斯林或宗教的原因），洗手间照例地配妇洗器，花砖腰线，非常西班牙。

I thought the decor of the guestroom was Missionary style, there was a big lamp.the back of the bed was a flag with a special crackle-paint eagle pattern. The wet and dry areas were 55:45. Because of Muslim or the religion, there was also a women wash device in the bathroom. The tiles waist line was quite Spanish style.

阳台超大，房间小，那怎么才好呢？问谁？

The balcony was very big while the guestroom was so small. So what to do? Who knows?

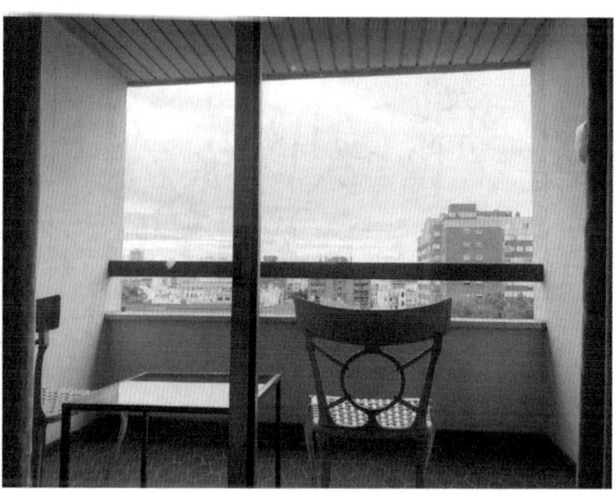

超大的阳台，可以感受城市的声音与气息

You can feel the sounds and smells of the city on the huge balcony.

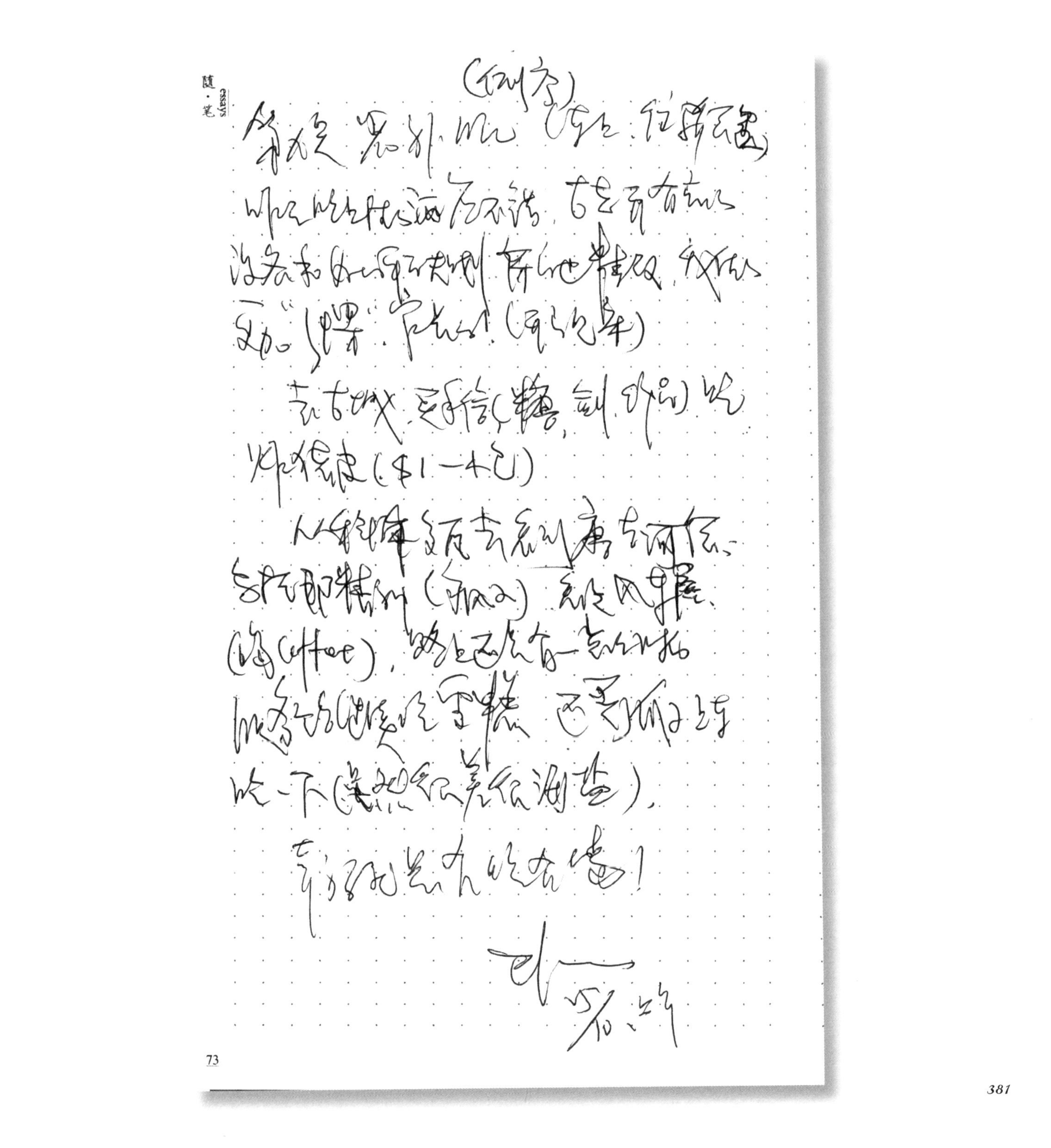

[illegible]

The Ninth Day,October 25
Write Back the Diary (Reverse order) (In a car, in Madrid)
第九天，10月25日，补日记（倒序）（车上，住马德里）

10月25日上午

On the morning of October 25

昨天晚上住的酒店不错，古老而有新意，好的设备和平面规划，房间也精致，我住的更加“了椤”，窄长的！（开门见床）。

The hotel which we stayed in last night was great. Old met new. The facilities and the plane were good, the guestrooms were also quaint, especially mine, particularly quaint—narrow and long! (I saw the bed while opening the door)

走古城，买手信（糖、剑、饰品），吃炸猪皮（1欧元/小包）。

We walked around the old city and bought some gifts (candies, a sword and some decorations), also tried fried pig rinds (a small bag cost € 1.)

从科尔多瓦去看唐吉坷德：乌托邦精神（“疯子”），看看风车屋，喝coffee，路上还是有一点点时间。服务站继续吃雪糕，还买了瓜子上车吃一下（虽然很差，海盐放太多）。

We went to visit Don Quixote from Catedral de Cordoba, the Utopian Spirit (the “mad” man). It was great to see the windmills and enjoyed coffee. We still had enough time to eat ice cream at the service station on the way,I bought some seeds to eat on the bus. (Although it tasted terrible and salty.)

幸福就是有吃有喝！

Eating and drinking means happiness!

O.'s File **(区生词典)**

了椤是粤语的叫法，特别的、刁钻的、独树一帜 。（百度知道）

It is Cantonese.It means special quaint and unique.

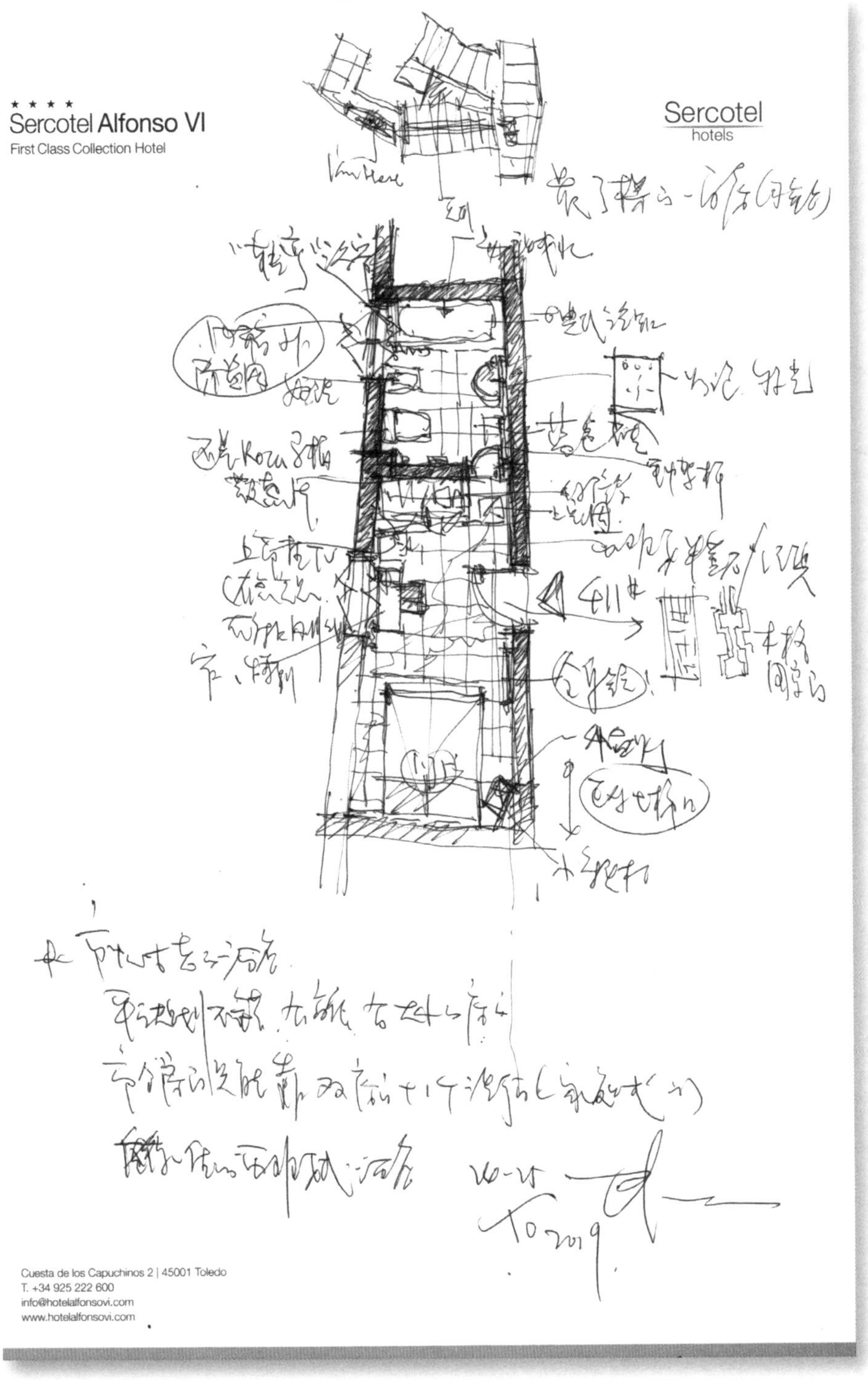

★ ★ ★ ★
Sercotel Alfonso VI
First Class Collection Hotel

Sercotel
hotels

Cuesta de los Capuchinos 2 | 45001 Toledo
T. +34 925 222 600
info@hotelalfonsovi.com
www.hotelalfonsovi.com

Sercotel
hotels
SERCOTEL酒店

92

★★★★

SERCOTEL ALFONSO VI TOLEDO SPAIN

Address : Cuesta de los Capuchinos 2, Toledo, Spain 45001
Telephone : (+34)925222600

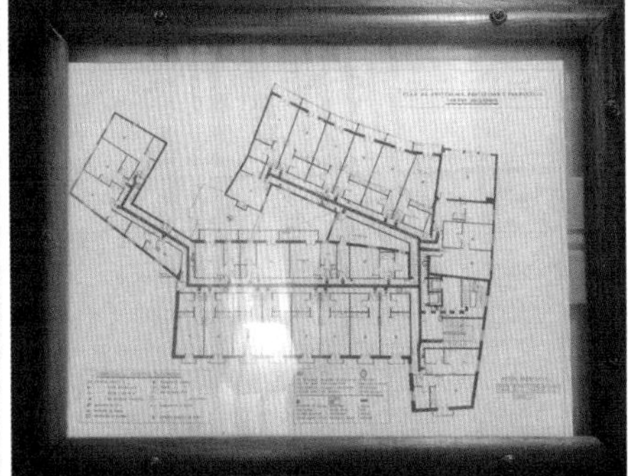

SERCOTEL酒店
Sercotel Alfonso VI, Toledo, Spain

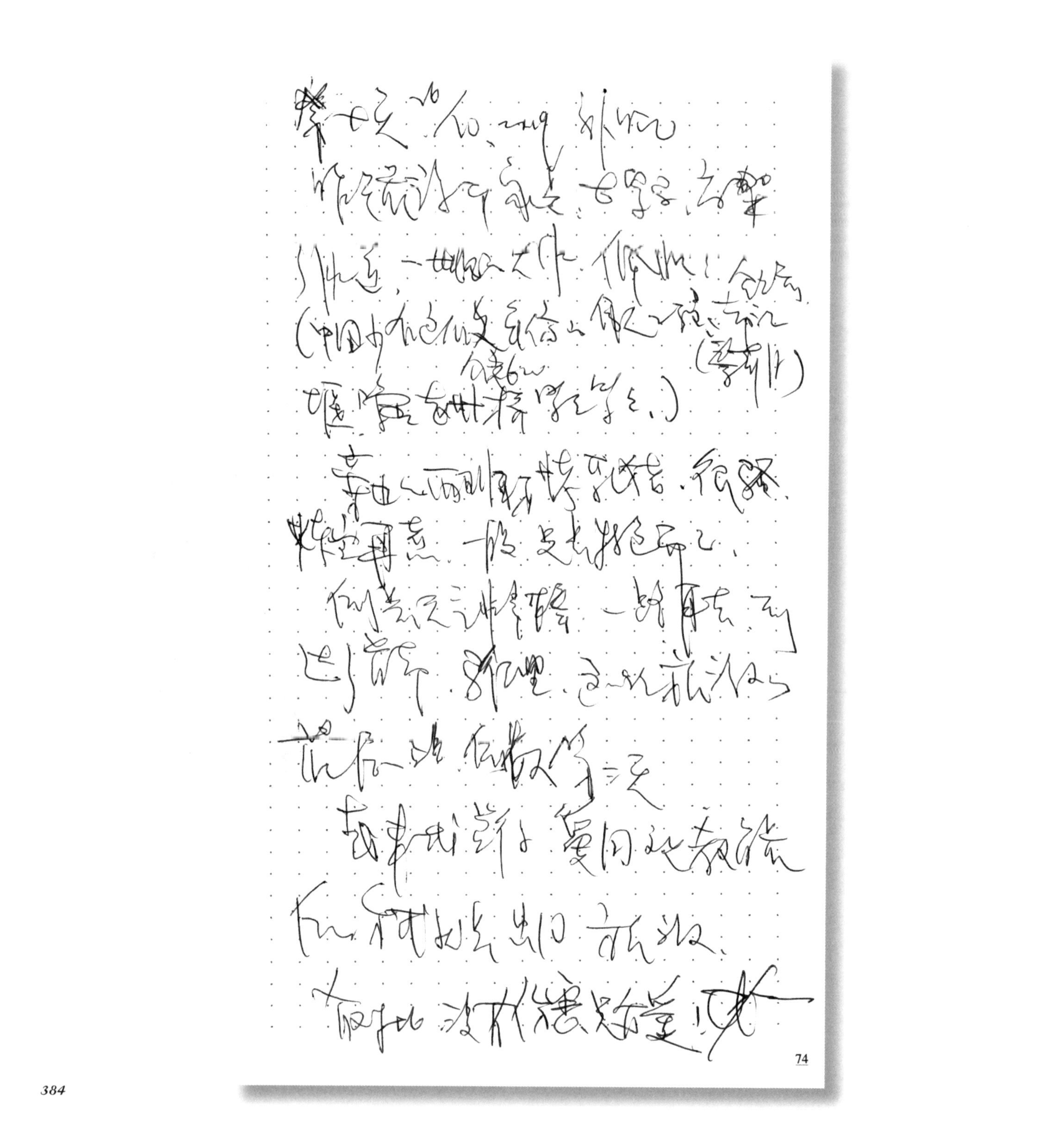

74

艳阳下的高架引水道，更显“高人一等”

In the bright sunlight, the Ancient Rome elevated aqueduct seemed super high.

The Tenth Day, October 26, Write Back the Diary
第十天，2019年10月26日，补日记

昨天印象最深的是：古罗马高架引水道（桥）1世纪的大作，佩服！公元前，中国少有这么久年份的伟大工程（秦朝都江堰，赵州桥建于公元620年，等等）。

Yesterday, what impressed me most was Ancient Rome elevated aqueduct—the masterpiece of a century. I admired it! It was built before century. It was rare to see such great work with a long history in China.(Dujiangyan in Qin Dynasty, the Zhaozhou Bridge was built in 620 A.D., and so on)

O.'s File **(区生词典)**

塞哥维亚古罗马高架水渠建于公元53~117年，迄今保存完好，渡槽用土黄色花岗石干砌（不用灰浆）而成，坚固异常。输水道全长813米，由被128根柱子支撑着的双层拱洞构成，顶端是水渠，到现在还在引导流水，成为塞哥维亚的象征。（百度百科）

Segovia ancient Rome Elevated Aqueduct was built from 53 A.D. to 117 A.D. and is still well preserved. The aqueduct is made of dry earthy- yellow granite (without mortar) and is exceptionally strong. The 813-meter-long waterway consists of double arches supported by 128 columns and topped by a canal. Nowadays, the waterway still guides the flowing water and has become a symbol of Segovia. (From Baidu Encyclopedia)

旁边的西班牙烤乳猪，烤完再煮，一般，只是特色而已。

Beside it, we tried the Spanish roast pig which was cooked after roasting. It had a heavy smell but tasted just so so, though it was the Spanish special dish.

倒是天气非常棒，一路冉去，到达了首都，马德里，这一次旅游的最后一站，倒数第二/三天。

Luckily, the weather was pretty nice. We got to Madrid, the capital of Spain which was the last stop of our trip on the last day but one or two.

越来越觉得爱国主义教育最好的方式就是出国旅游。

I have become increasingly aware that the best way of patriotism education is to travel abroad.

有对比，没有伤害，只有爱！

No harm with comparison, just love!

仪式感十足地上菜，用碟子切烤猪，然后分菜，摔烂碟子，太有意思了，当然餐厅的地点、历史更是吸引人的地方。

Serving dishes was full of ritual sense. The cook cut the roasted pig with a plate, served and then threw the plate. It was really interesting. Of course, the location and the history of the restaurant were more attractive.

爵怡温德姆马德里
阿卡拉611酒店

93

★★★★

SERCOTEL ALCALA 611 HOTEL MADRID SPAIN

Address : Calle de Alcala, 611
Madrid, Spain 28022
Telephone : (+34)917434130

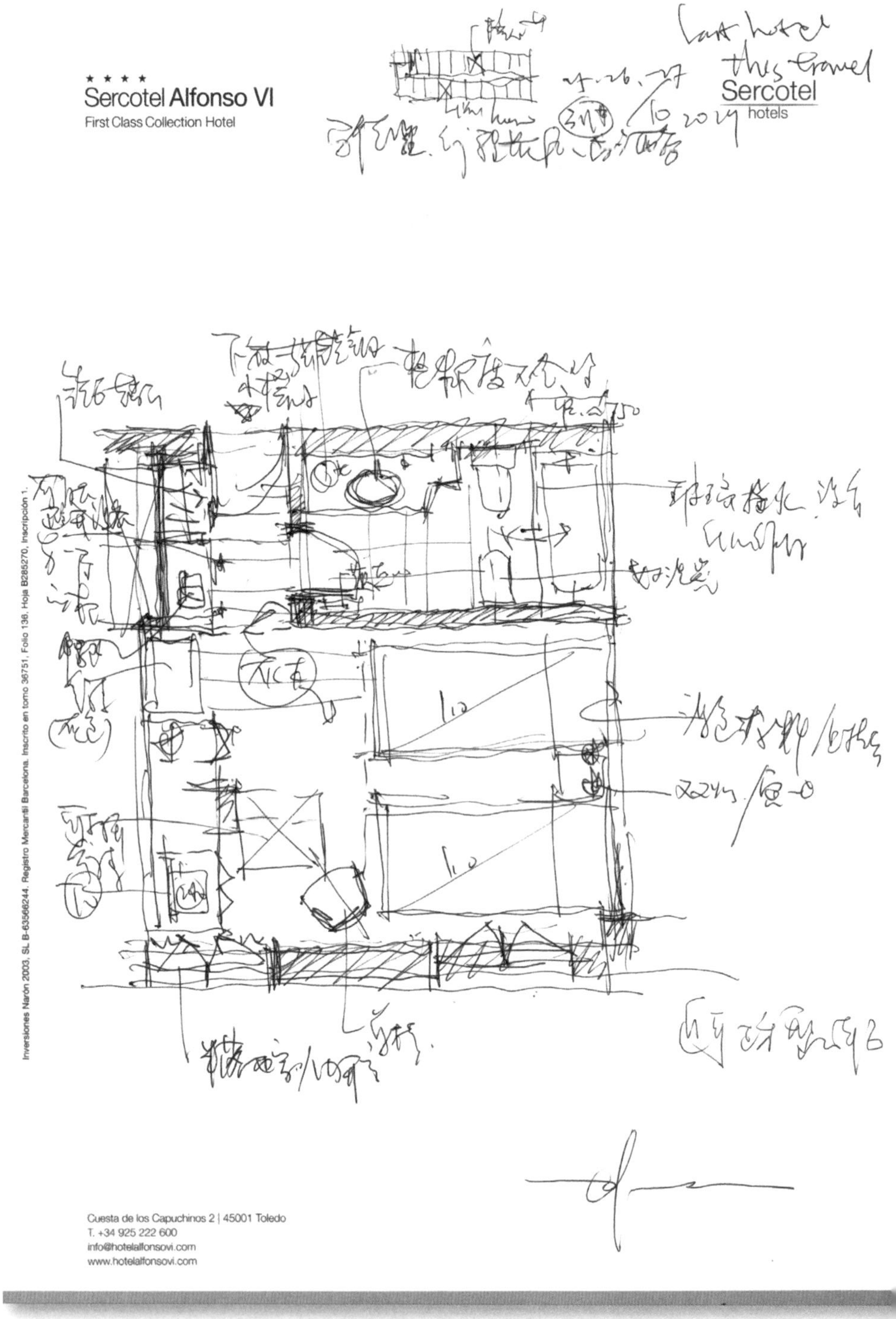

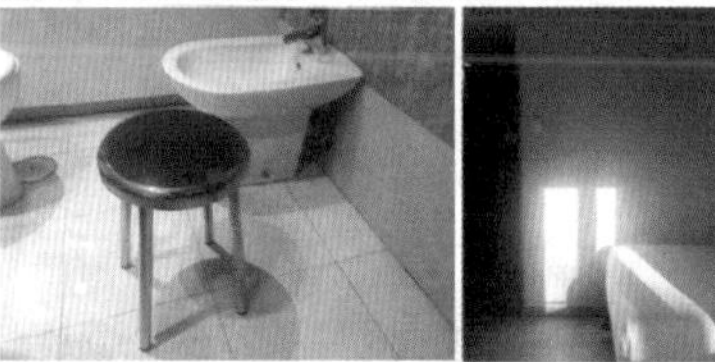

爵怡温德姆马德里阿卡拉611酒店
Sercotel Alcala 611 Hotel, Madrid, Spain

★ ★ ★ ★
Sercotel Alfonso VI
First Class Collection Hotel

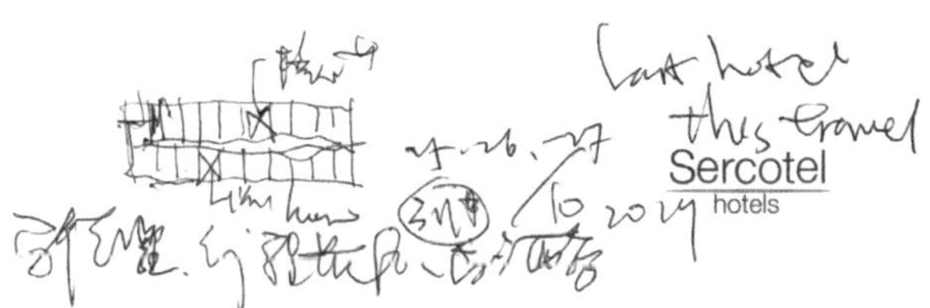

Inversiones Narón 2003, SL B-63568244. Registro Mercantil Barcelona. Inscrito en tomo 36751, Folio 138, Hoja B285270, Inscripción 1.

Cuesta de los Capuchinos 2 | 45001 Toledo
T. +34 925 222 600
info@hotelalfonsovi.com
www.hotelalfonsovi.com

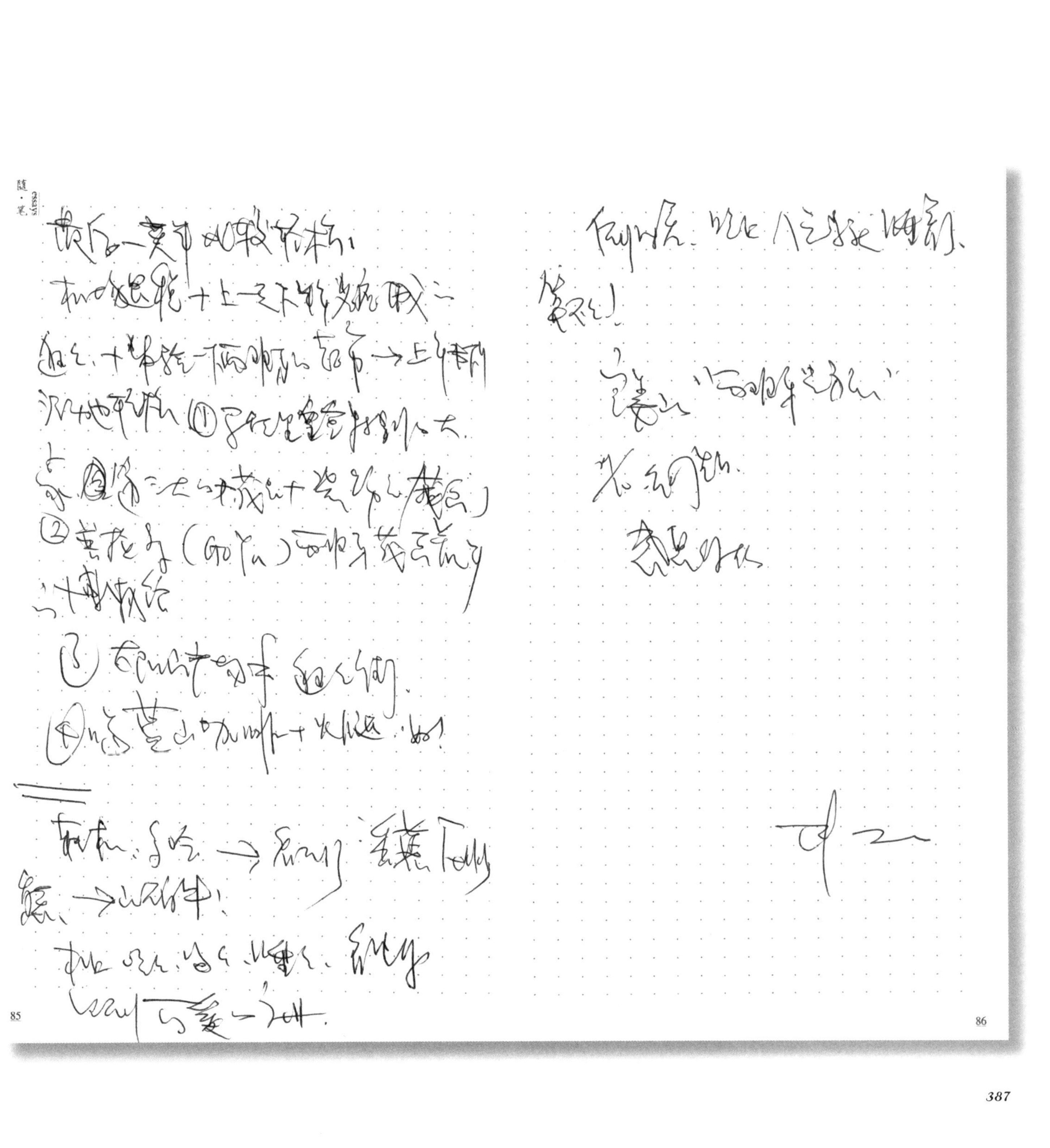

The Last Day is Relaxing

最后一天半比较轻松

机场退税+在名店城逛逛+体验一下西班牙的超市——上午+下午活动也轻松。

1. 马德里皇宫特别的大，豪，欧洲第三大皇宫+藏品+装饰品/藏画。
2. 普拉多（GoYa），西班牙藏画最多的博物馆。
3. 太阳门广场和逛逛街。
4. 喝蓝山咖啡+吃火腿，好！

We got a refund at the airport and spent a half day shopping yesterday. By the way, we experienced the Spanish supermarket. All these activities were relaxing.

1.Palacio Real de Madrid was particularly huge and luxury which it was the third largest palace in Europe with lots of collections decorations and paintings.

2.Prado Museum had the most paintings in Spain.

3.Walked around Sun Gate Square.

4.Enjoyed Blue Mountain coffee and ham.Pretty good!

转机：多哈——看到了香蕉泰迪熊——的确牛！

机上吃吃、喝喝、睡睡、看电影。

回到可爱的广州。

倒时差，晚上八点就睡着了，算不错!

完美的“西班牙之旅”。

10月29日公司生日。

感恩有你们！

Transferred at Doha—we finally saw the yellow Teddy Bear—really cool!

Ate, drank, slept and watched movies on the plane.

Came back to my lovely home—Guangzhou.

I slept at 8 in the evening because of the jet lag, felt good!

The perfect trip to Spain.

October 29 is the birthday of our company.

Thank you for being with you!

THE RITZ-CARLTON

XI'AN

NO. 50, KEJI 2ND ROAD, GAOXIN DISTRICT, XI'AN 710075 SHAANXI, CHINA 86.29.8881.8888 RITZCARLTON.COM/XIAN

陕西省西安市高新区科技二路50号 邮编：710075

THE RITZ-CARLTON®

西安丽思卡尔顿

★★★★★

94

THE RITZ·CARLTON XI'AN CHINA

Address : No.50, Keji 2nd Road, GaoXin District, Xi'an, Shanxi, China
中国陕西省西安市
高新区科技二路50号

Telephone : +86 29 8881 8888

Http : //www.RITZCARLTON.com/xian

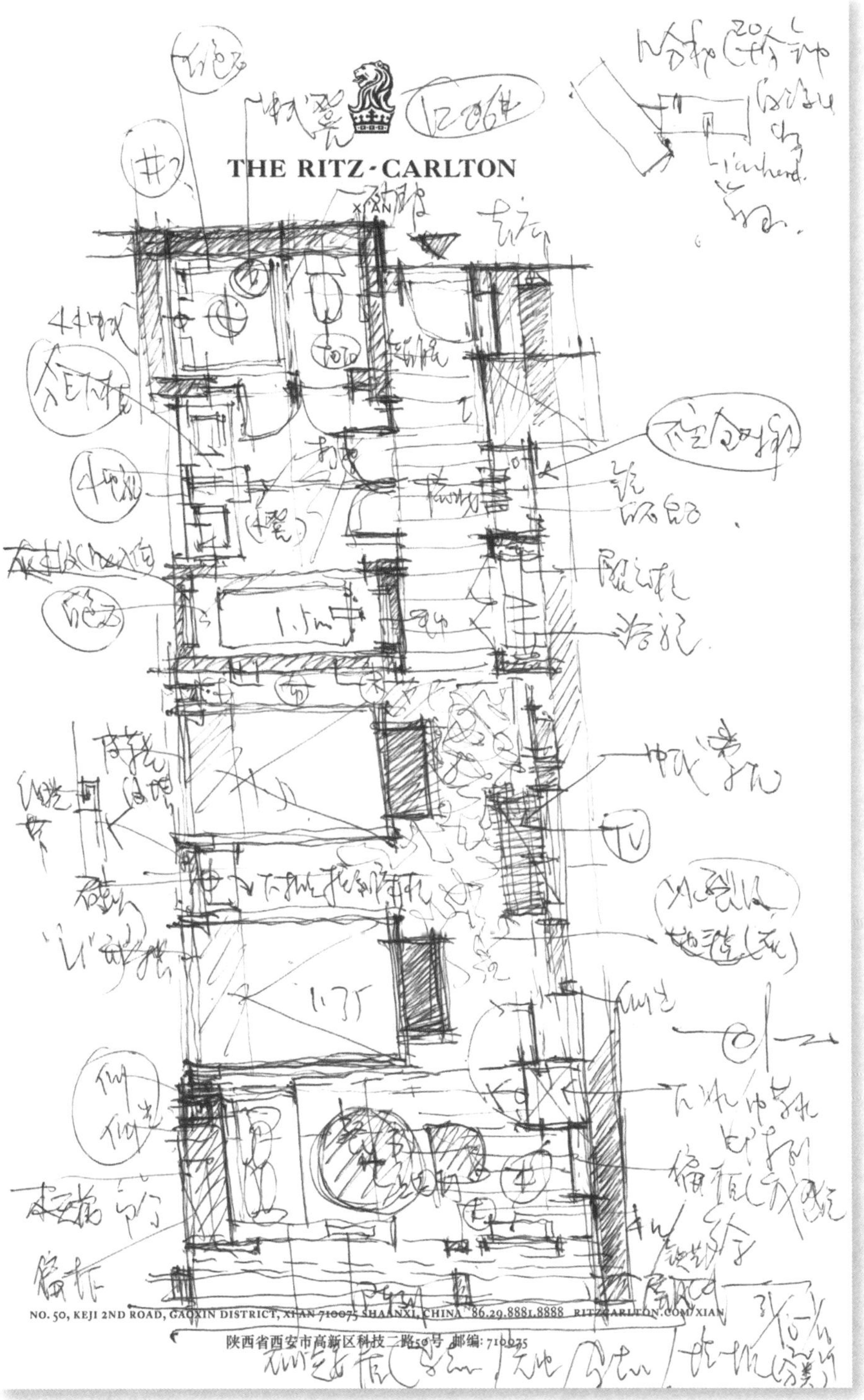

西安丽思卡尔顿
The Ritz·Carlton, Xi'an,China

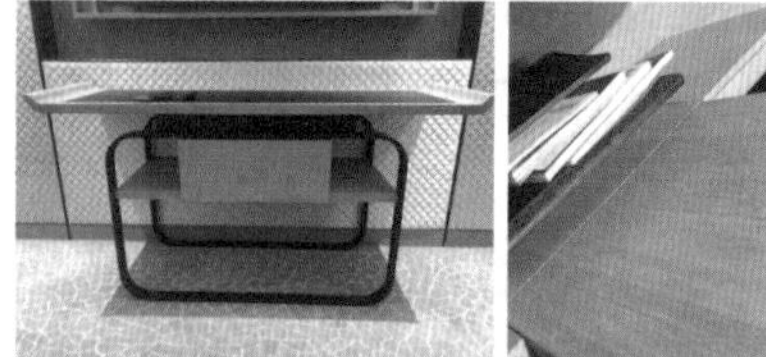

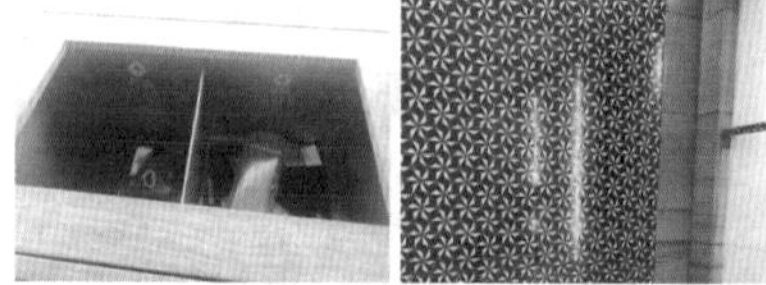

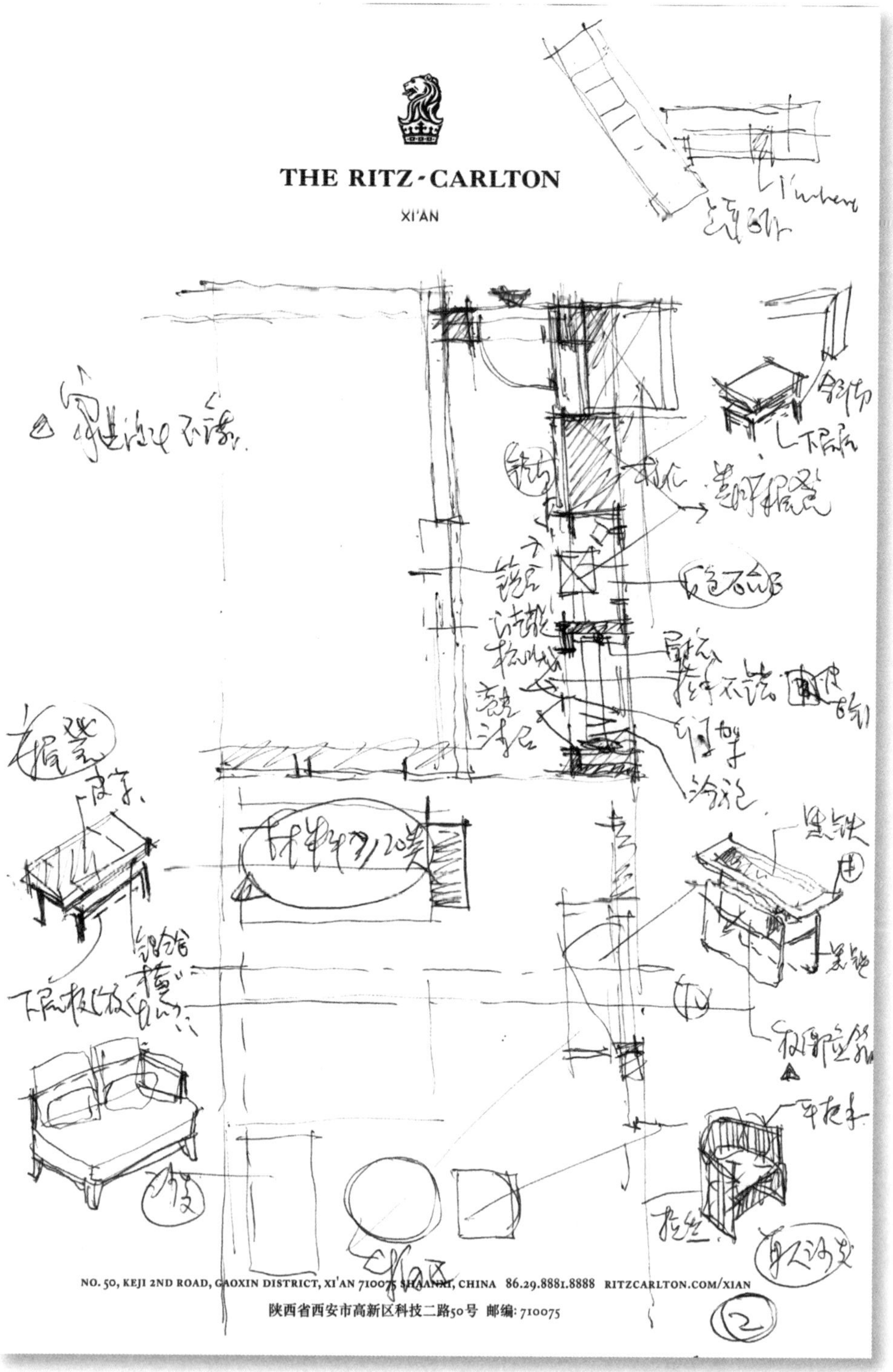

西安丽思卡尔顿

The Ritz·Carlton Xi'an,China

THE RITZ-CARLTON

XI'AN

NO. 50, KEJI 2ND ROAD, GAOXIN DISTRICT, XI'AN 710075 SHAANXI, CHINA 86.29.8881.8888 RITZCARLTON.COM/XIAN

陕西省西安市高新区科技二路50号 邮编: 710075

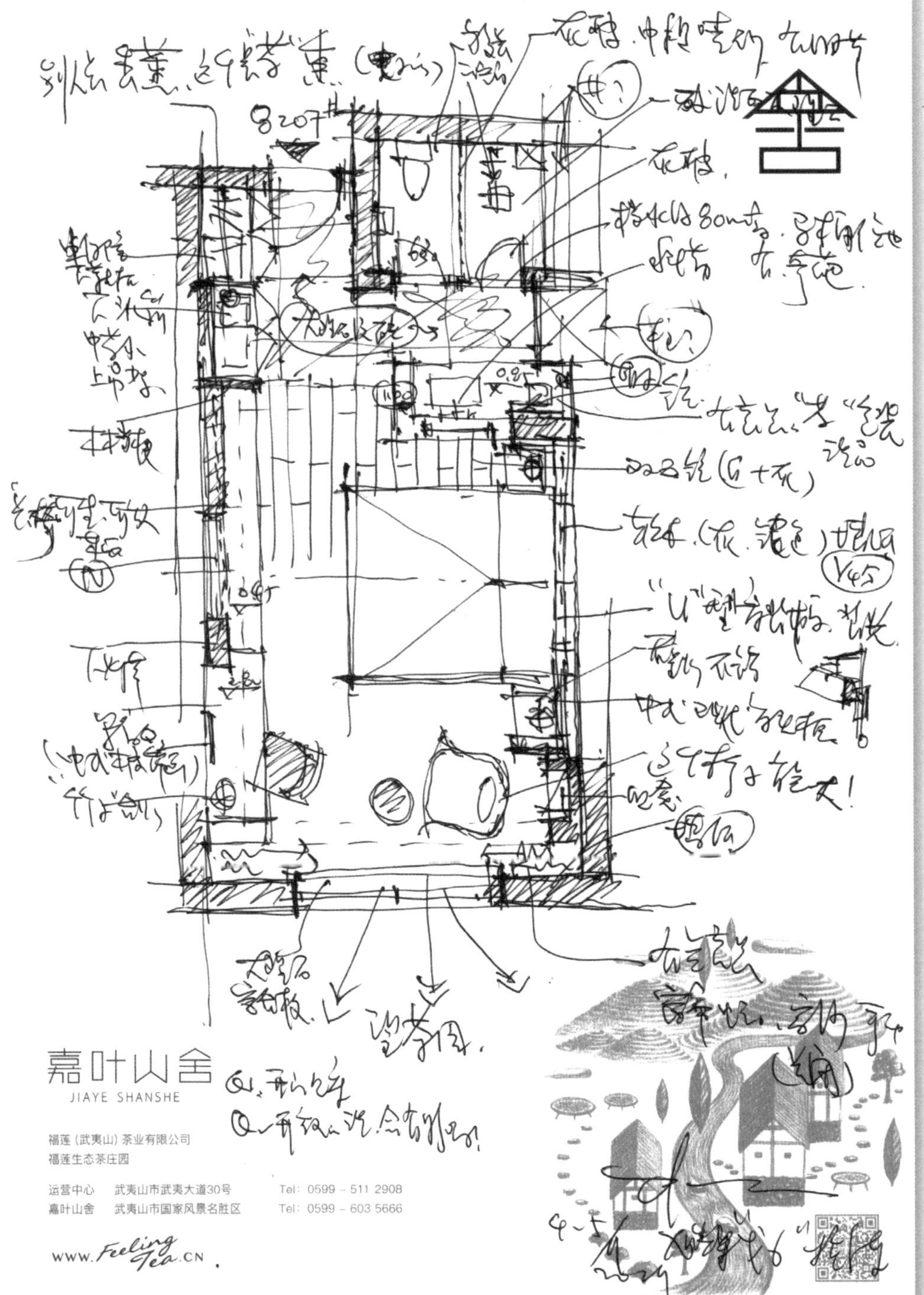

嘉叶山舍
JIAYE SHANSHE

95 嘉叶山舍

★★★★★★

JIAYE SHANSHE FUJIAN CHINA

Address　:　No.30 Wuyi Avenue, Wuyishan,Fujian,China
中国福建省武夷山市
武夷大道30号

Telephone : +86 599 511 2908

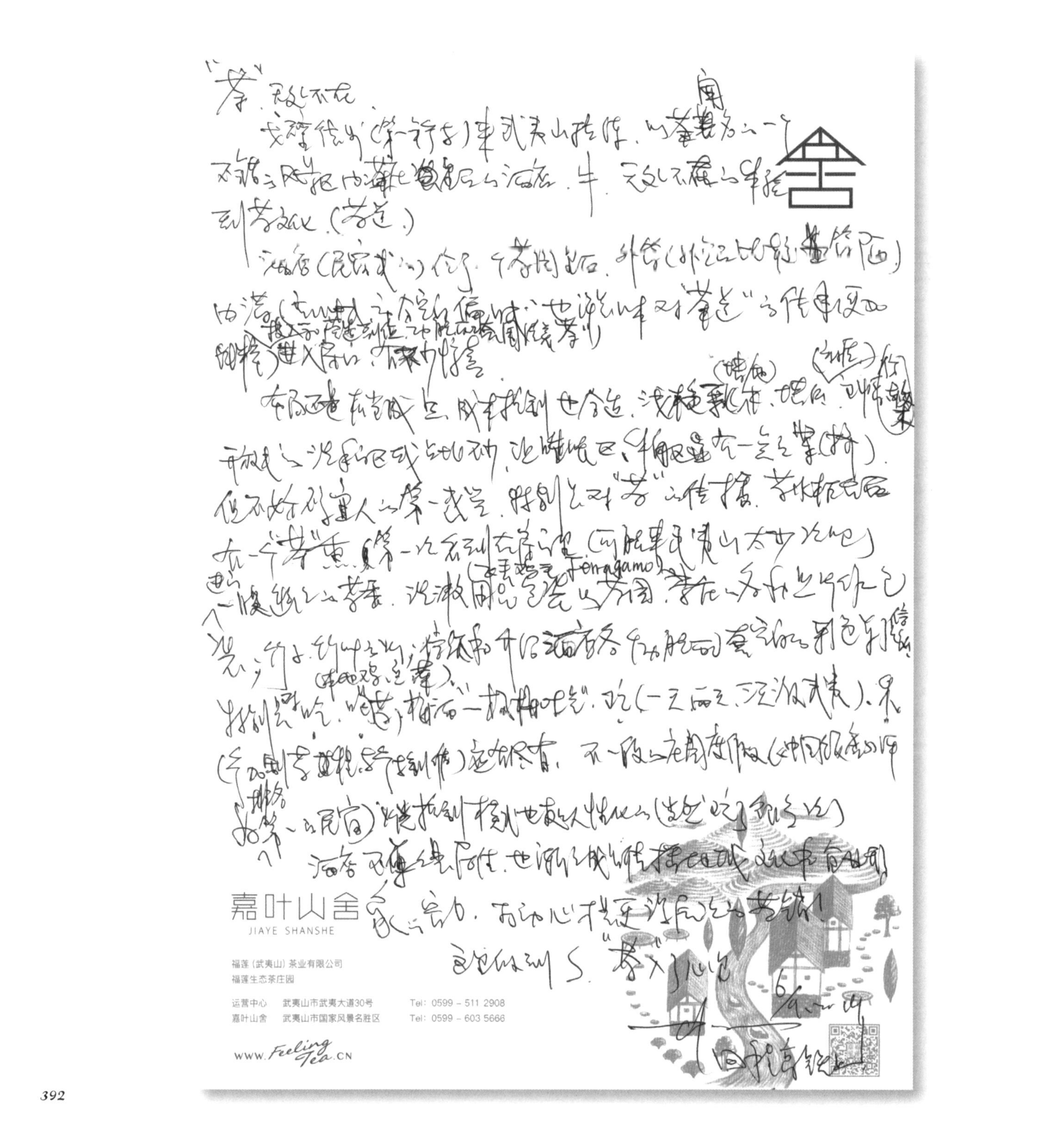

嘉叶山舍
JIAYE SHANSHE

福莲（武夷山）茶业有限公司
福莲生态茶庄园

运营中心　武夷山市武夷大道30号　Tel：0599 – 511 2908
嘉叶山舍　武夷山市国家风景名胜区　Tel：0599 – 603 5666

WWW.FeelingTea.CN

嘉叶山舍

Jiaye Shanshe,Fujian, China

"Tea"is Everywhere

"茶"无处不在

戈壁徒步（第二次参与）来武夷山拉练，入住以茶闻名的一个不错的风景区内茶庄里面的酒店，牛！无处不在地体验到茶文化（茶道）。

I came to Wuyi Mountain to take part in the training of Gebi Hiking (my second time). The hotel was in a beautiful place of interest which was famous for its tea. Cool! I experienced tea culture everywhere.

酒店（民宿式的）位于一个茶园里面，外简(外立面比较显简陋），内满（新中式部分空间偏日式，也许是日本对"茶道"的传承更加纯粹，投入和营运到位，功能配套围绕着"茶"），进入房间，有小小的惊喜。

The hotel like B&B lay in a tea garden. The exterior was simple (the façade seemed a little humble) while the interior was rich (New Chinese Style and some were Japanese, maybe Japanese passed on the tea culture more purely. The investment and operation were in place. The functional facilities were all about "tea") .I felt a little surprised when entering the guestroom.

布局相当成熟，成本控制也合适，浅木色系（墙/地）木，墙纸(麻质)，主幅仿软木，开放式的洗手间区域占比不少，让睡眠区、休闲区显得有一点点紧（挤），但不妨碍官人的第一感觉，特别是对“茶”的传播。茶水柜台面有一个“茶熏”，在房间里第一次看到（可能来武夷山的次数太少吧），进门一股幽幽的茶香。洗漱用品（飞甩鸡毛，Ferragamo）包装以茶园、茶庄的各种照片作素材；竹子、竹叶的灯，信纸和介绍酒店各个功能配套空间的彩色单张明信片，特别是吃，（木地鸡、鱼、菜），喝茶，梅酒——“杨梅吐气”，玩（一天、两天、三天游武夷），和（参加制茶过程与参与制作），应有尽有，不一般的度假庄园（被《中国饭店》评为排名第一的民宿）！灯光控制模式也挺人性化的（当然玩了很多次了）。

The layout was quite mature and the cost was controlled reasonably. The light-wooden-color wall/floor and linen wallpaper was like the cork wood. The open bathroom was much bigger while the sleeping and the leisure area seemed a little crowded. But the first feeling was pleasant, especially for the spread of tea. There was a tea aromatherapy on the tea table. It was my first seeing it. (Maybe I seldom came here.) I smelt the faint scent of tea while entering the room. Toiletries(Ferragamo) were packed by different photos of the tea garden.There was also a lampshade with patterns of bamboo and bamboo leaves. Letter paper and colorful postcards introduced the hotels and every fully-functional area, including delicious food (local chicken, fish and vegetables). Drink tea, taste plum wine, travel in Wuyi Mountain for one or two or three days, take part in the tea making and everything here. It's a wonderful manor hotel.It was chosen to be the No.1 B&B in China Hotel. The control of the lights was quite humanized (of course I tried many times).

酒店不单单是居住，也渐渐成为传播地域文化和企业形象的窗口，打动人心才是深层次的营销。

Hotels are not only for living, but also become the window of spreading regional culture and corporate image. Touching hearts is a high level marketing.

这里做到了“茶入了心里”。

Tea goes into your heart.

嘉叶山舍

Jiaye Shanshe,Fujian, China

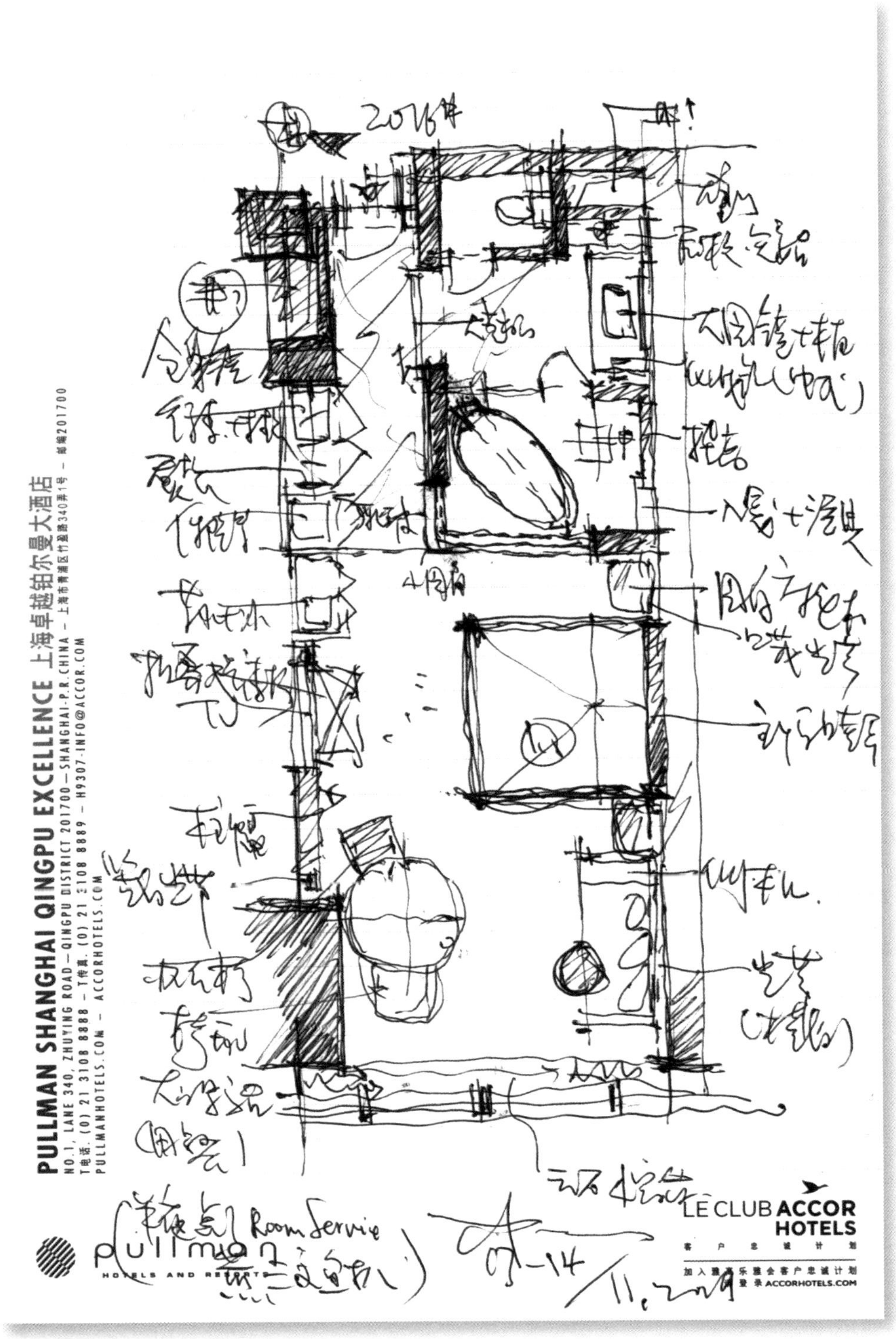

上海卓越铂尔曼大酒店

96

★★★★★

PULLMAN SHANGHAI QINGPU EXCELLENCE CHINA

Address : No.1, Lane 340, Zhuying Road, Qingpu Disrict, Shanghai, China
中国上海市青浦区竹盈路340弄1号

Telephone : +86 21 3108 8888

FAX : +86 21 3108 8889

Http : //www.PULLMANHOTELS.COM-ACCORHOTELS.COM

How to Enjoy Room Banquet

房间大餐，如何吃？

May 5, 2020 the Beginning of Summer

立夏2020年5月5日

上海卓越铂尔曼大酒店
Pullman Shanghai Qingpu Excellence,China

半夜入住远在“天边”的这家酒店，饿“死”了。想想只能在酒店吃饭了。打电话，没有牛扒，只能“冒险”点了煎三文鱼扒。

I checked into the hotel far away at midnight. I was so hungry that I just wanted to have something in the hotel. I called the room service. There was no steak, so I had to order fried salmon filet.

等餐的过程准备一下：想想在哪里吃合适一些。背对或侧对电视的大餐台是第一选择，舒适，随意，可是最致命的是“我想看电视”；沙发有小茶几，勉为其难，好像也可以将就一下；床，哈哈，一个人的床上大餐，似乎不妥……当然，还可以在浴缸上吃大餐，哈哈哈哈！

While waiting for the meal, I tried to find out where to have my big dinner. The best choice was the big table with its back towards the TV or on the sideway. It felt comfortable and relaxing. But I wanted to watch TV! What could I do? Beside the sofa, there was a small tea table. I would just have to take the meal on it. Haha! It was not the proper way to eat on bed… Of course I could also eat in the bathtub.Haha!

忘记了上一次在房间用餐是什么时候，印象深刻的有一些：因为只有拖鞋，露趾头的男性不被允许在大厅用餐，只能房间里用早餐的香港半岛酒店；深夜小饿，一个人不想在大雨天外出吃夜宵，在房间点了水果大餐的海口朗廷酒店。房间用餐也是不可多得的记忆，让我对这间酒店的服务增加了记忆点。当然也是对房间的设施、家具（圆角的）、灯光明亮程度等等的实际用途作了试验。看来，房间用大餐要多来一些。也许，也是“住”的另一种乐趣。

I forgot when I had a meal in the guestroom last time. What impressed me most: I wasn't allowed to take meals in the dining hall because of the slippers (a man with open-toe slippers),so I had to take my breakfast in the guestroom when I stayed in the Peninsula Hong Kong; I felt hungry at midnight and didn't want to go outside for supper on a rainy day, so I ordered a fruit meal in Langham Hotel Haikou. It was a rare memory to take meals in guestrooms and made me remember the room service of the hotel. I also tried the facilities, furniture with (rounded corner), the bright in the room and so on. In my view, eating in the room should be tried more. Maybe it was another kind of pleasure to stay in hotels.

半夜，泡了浴缸，安睡！

After enjoying a tub bath at midnight, I had a good sleep!

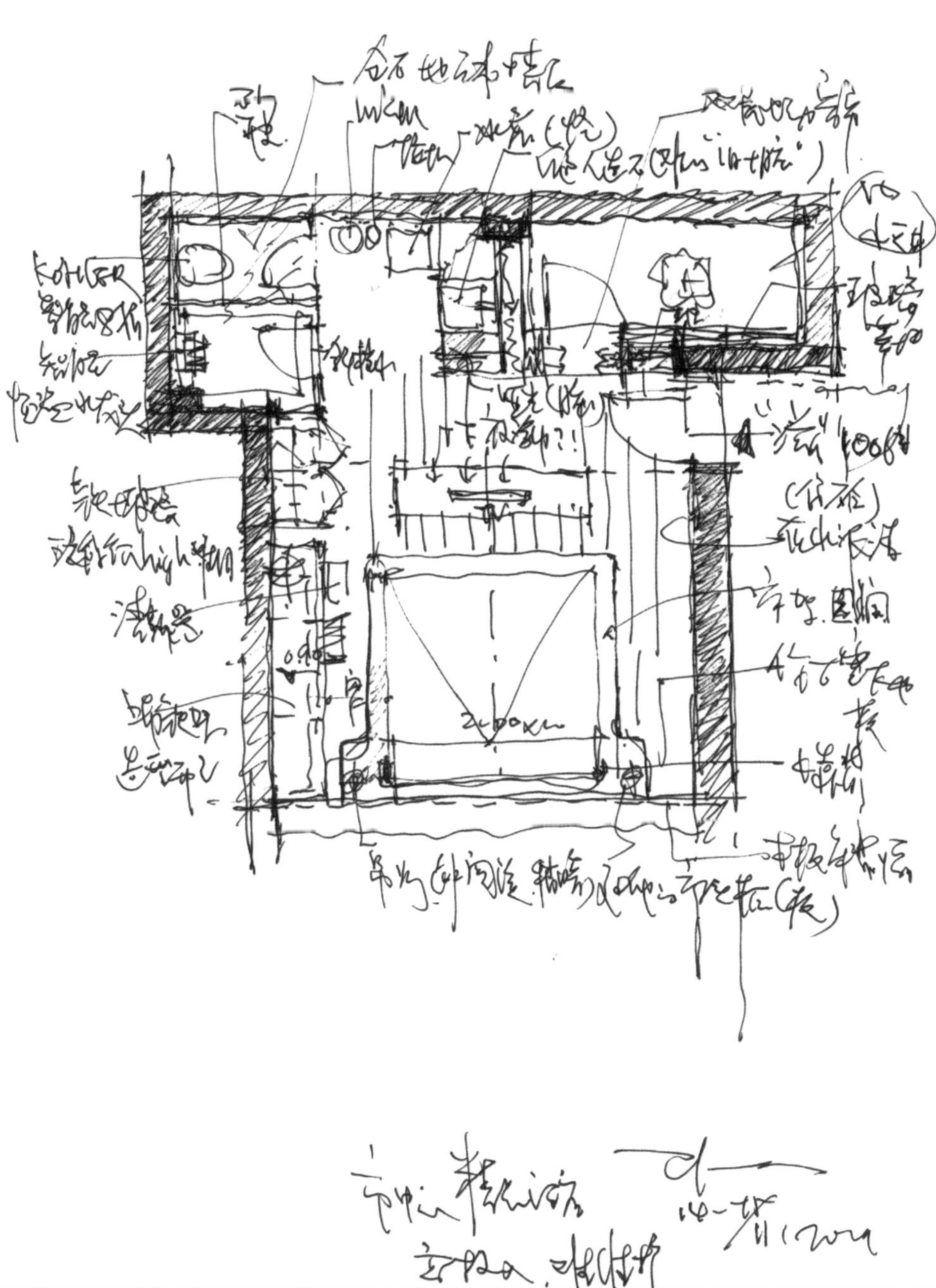

渝舍印象

Yu·Hotel

97 上海渝舍印象

★★★★★

YU • HOTEL SHANGHAI CHINA

Address : No. 439 Fuxing East Road, Huangpu District, Shanghai, China
中国上海市黄浦区
复兴东路439号
Telephone : +86 21 3130 0888

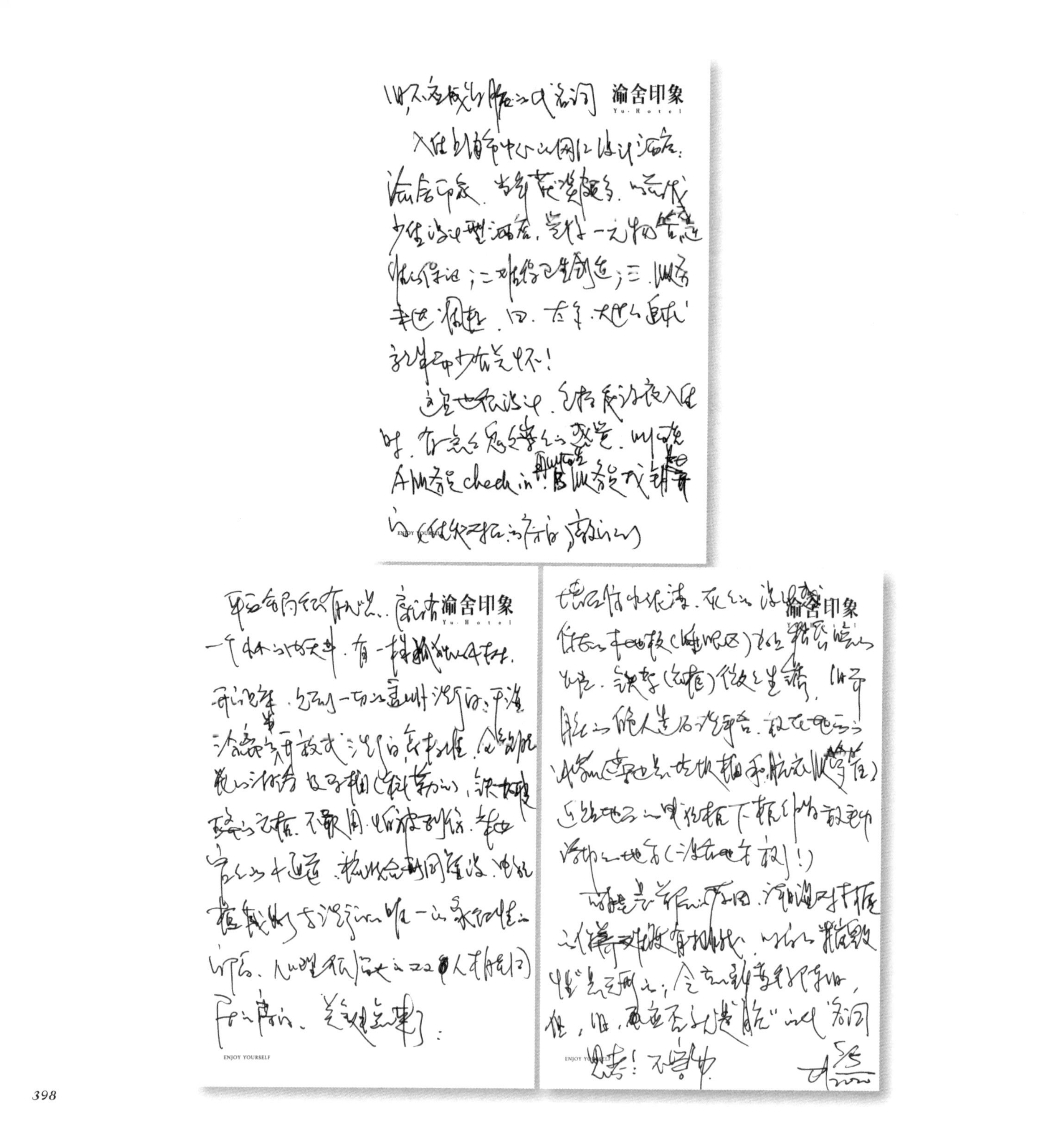

渝舍印象
Yu-Hotel

ENJOY YOURSELF

渝舍印象
Yu-Hotel

ENJOY YOURSELF

渝舍印象
Yu-Hotel

ENJOY YOURSELF

上海渝舍印象

The Impression of YU • Hotel, Shanghai,China

Being Old doesn't Mean to be Dirty

旧，不应成为脏的代名词

May 5,2020

2020年5月5日

入住上海市中心的网红设计酒店：渝舍印象。当年获奖颇多，以前我很少住设计型酒店，觉得一：无物管、无运维（运营和维护）的保证；二：难得卫生舒适；三：服务未达到标准；四：太多，太过地追求效果而少有关怀！

I stayed in an Internet celebrity design hotel in the center of Shanghai: Yu Hotel. It won a lot of awards of the year. I seldom stayed in design hotels because of the following reasons: First of all, there were no guarantees about property management, operation or protection; second, it was hard to keep tidy and comfortable; what's more, the service was not up to the standard; last, it pursued effects too much but little caring!

这里也很"设计"，包括我深夜入住时，有一点点"鬼鬼祟祟"的感觉，叫醒A服务员check in，再叫醒B服务员找钥匙开门（住我对面的房间），敲门的。

The hotel was very well designed including my checking in secretly. I woke up a staff to check in, and then knocked at the next door to wake up another one to open the guestroom for me.

上海渝舍印象

The Impression of YU • Hotel,Shanghai,China

平面布局很有心思，房间有一个小小的内天井，有一株孤独的小树。开门见床，见到一切的透明洗手间：干湿三分离的半开放式洗手间，高标准。全智能化的淋浴及马桶（科勒的），铁加皮革加玻璃的衣柜，不敢用，怕被刮伤。床边窄窄的小通道，梳妆台形同虚设，电视柜成了灯与洗手间之间唯一的象征性的间隔，心理很强大的双人才能同居的房间。关键点来了：墙面仿水泥漆，灰灰的设计感，仿古的木地板（睡眠区），加上稍昏暗的灯光，铁架（衣柜）微微生锈，旧而脏的白色人造石洗手台，放在地面的冰箱（旁边是垃圾桶和脏衣服箩筐），近贴地面的电视柜下柜作为放毛巾浴巾的地方。（没有其他地方放了！）

The layout was quite thoughtful. There was a mini courtyard in the guestroom with a lonely small tree. Seeing the bed while opening the door, as well as a clear washroom which you could see everything in it. A half-opened washroom was divided into three separated dry and wet parts with high standard. Both the shower system and the toilet (Kohler's) were intelligent. The wardrobe was made of iron, leather and glass. I was afraid to hurt myself so I didn't dare to use it. The aisle beside the bed was so narrow that the dresser was but an empty shell. The TV cabinet became the only symbolic interval between the lamps and the washroom. Two tough-minded people would stay in this kind of rooms together. The following were the weaknesses: the walls were grey color with fake cement, the antique wood floor and the dark lights.The iron cabinet was a little rusty. The white artificial stone wash basin was old and dirty, while the fridge was put on the ground beside the trash bin and the laundry hamper. The towels had to be put into the cabinet close to the ground which was under the TV cabinet. (But there was no place for them!)

可能是首层的原因，潮湿对材质的保养难度有挑战。时间的摧毁性是无形的；令新鲜变得陈旧。但，旧，应否就是脏的“代名词”？

It was wet because of the ground.How to maintain the material was a difficult challenge. The destruction of time was invisible. Time could make new become old. However, does being old mean to be dirty?

思考！不容易。

Think it over, it was not easy!

日本游（2019年11月21～29日）

The Trip to Japan

匠心·匠行

有好几年没有去日本了，这一次从头到尾都是"神道"的个性化行程——定位精准，一个风格的寻找"匠人·匠心"之旅，当然也是美食美学之旅、建筑之旅、文化之旅。再一次体会那国家的地道，在各个不同的场所经历我们很难以忘怀的东西，留给[illegible]之处之之。

从长崎出发，一个充满特色和风格的兴趣，时代变迁的答案都表现出来，"更是引发我们对东西方文化的很多好奇。第一站的陆老师讲解当地的书和从的历史、文化、人物（包括设计师），让我们不仅学到，也能得到大开眼界。特别是有一位外国人 Ian Lynam，他到日本十多年，传统、传播和从的"东西"更让我们了解和认识（日本 Yuki 也会给我们各种操作和关键点的东西的美（还是一个下大雨的之后，趣味之作）之后让我们会门我们都有[illegible]各样都[illegible]实加从的人的个性人才等，特别是美术设计的手工作品（布、纺织两件等等）更是让我着迷）

杨楷米的对日本的情感与印象，但也是第一次这么"家们"也接触到其设计（不知道有没有设计的兄弟）以及制作、创造、布料等等的细节。一对可爱的匠人夫妻（橙山夫妇，也是一个日本人和一个外国人的组合，嘿！）让我们在"橙子"里感受到了"时装室"的传承和手艺人的真实一面；一对执着的日本传统文化的布艺人，让我们感受到了"金黑色"中精致的美，在大德寺的[illegible]令人向往的时间[illegible]他们的"手作"。

"神道"的每一次行程中都有惊喜……同行的特别儿在（Drunk Hotel 酒店旁边街的小酒店），大雨，半封闭的雨篷下，挤了几十人，听着雨声享受热闹的"不健康"的串儿（以新鲜的内脏为主材）。大德寺的宁静会[illegible]，红叶[illegible]的环境下，心跟着静下来，美味的[illegible]的享受。茑屋书店的下午和大家，书店的生态模式完全的[illegible]和时代[illegible]了，加入影像和长[illegible]（含咖啡）与设计的结合，书、设计[illegible]的探访。在京都的[illegible]（十二[illegible]）[illegible]我们也会[illegible]一同[illegible]，就是[illegible]了，[illegible]设计，大家的一面。

大德寺，在一间有隐匿的[illegible]里，[illegible]也[illegible]，静静地[illegible]落下，我们的[illegible]在大雨后，[illegible]地

穿过了枫叶的走廊。

去时饭店，隔着嘈杂的人，[illegible]，其次是喜欢的[illegible]隐藏在一条街的一座木木的博物馆：河井宽次郎博物馆展示了各个时代的真迹（[illegible]），各流派各种颜色的精品，目不暇接，眼花缭乱。一个简朴的旧居（[illegible]真正不少）

江之浦测候所：去了同杉本博司的地方，真正了解季节的变化，同时也是对收藏生活：杉本博司先生的各类藏品：石头、木雕佛像、全世界各个古老地方的[illegible]入迷，服了！

第一天入住的[illegible]市中心的某个酒店，入住时[illegible]的金属（[illegible]）[illegible]的[illegible]，[illegible]半露天的温泉池（[illegible]）[illegible]二十[illegible]，细雨下的半夜体验，回味无穷。

银座的[illegible]MiuMiu专卖店，[illegible]大气的[illegible]，像一个[illegible]的宝盒：外衣、内金（[illegible]材），个性、魅力。

每一次的味道都不同：石石之旅、海景（[illegible]）之旅、[illegible]之旅，这一次更加[illegible]不同，[illegible]，[illegible]的“匠心”。

[illegible]
[illegible]

Originality, Handicraft
匠心、匠行

August 05, 2020
2020年8月5日

有好几年没有去日本了，这一次吸引我们的是“朗道”的个性化订制行程——定位精准，一个纯粹的寻找“匠人、匠心”之旅。当然也是美食美景之旅、建筑之旅、文化之旅，再一次体会邻国的“地道”，在多个方面确有值得我们做设计的去走走看看、学习学习之处。

I haven't been to Japan for a few years. We were most attracted by Landao Design's personalized custom tour. It was an absolute trip to look for craftsman and originality. Of course it was also a tour for gourmet, wonderful views, buildings and culture. I could experience the local custom culture of our neighbor again. There were so many things in various ways that we designers should learn.

服装是体现一个民族特色和经济兴衰、时代变迁的简单代表，当然“和服”更是引发我们对大和民族的很多好奇。第一站的专业老师讲解20世纪的和服的历史、故事、人物（包括设计师），让我们印象深刻，也眼界大开，特别是看到一位外国人Ian Lynam，他到日本学习、传承、传播和服的“事迹”更让我们惊讶和佩服（日本Yuki 女士全程示范各种操作和关键点）。短短的半天（还是一个下大雨的天气，在超市买了伞），只让我们学会了或简单拥有了怎样看懂路上穿和服的人的审美的小本事，特别是其复杂又丰富的手工装饰（布、饰绳、配件，等等）更是让我着迷了。

Clothing was a simple sign of a national feature, economic ups and downs and the change of time. Of course, Kimono aroused our more curiosity for Yamato. The first stop of our trip, a professional teacher told us about the history of Kimono of last century, the stories and people, including the designers. It made a good impression on us and let our imagination run wild, especially when we saw a foreigner called Ian Lynam. He went to Japan to study, pass on and spread the culture of Kimono. We felt surprised and admired what he did. (Yuki Sang showed us how to wear kimono and some key points throughout the visit.) It was only half a day. We just learnt to appreciate and understand the people wearing Kimono in the streets. (It was such a rainy day that I had to buy an umbrella in a supermarket.) I became more interested in its complicated and rich hand-made decorations (cloth, thread rope, accessories and so on.)

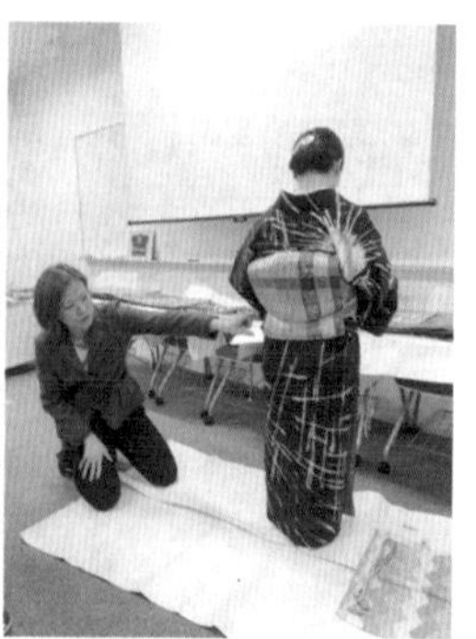

因为好奇加不懂，大家都非常认真地听老师讲课，看Yuki 女士的每一步示范。和服——好神秘。

Because of curiosity and ignorance, all of us listened to Yuki Sang and watched her demonstration carefully. How mysterious Kimono is!

榻榻米，是对日本居屋的最深印象，但也真是第一次这么“亲近”地接触到其设计（居然是有设计的，见笑了！）、礼制、生产制造、市场等环节，一对执着的匠人夫妻（横山充夫妇，也是一个日本人和一个外国人的组合，怪！）让我们在“小巷子”里面感受到了“甘于寂寞”的传承手艺人的真实一面：一对执着的弘扬传统文化的布道人。最有意思的当属看到了“全黑的”榻榻米，在大德寺里的部分功能空间看到了他们的“手作”。

What impressed me most about Japanese houses was Tatami. It was my first close contact with its design(to my surprise, Tatami was actually designed), the ritual system, manufacturing, marketing and so on. An earnest art couple who were willing to be lonely(Mr and Mrs Hengshan Chong, a Japanese and a foreigner.How strange!), showed us the real aspect of craftsman in an alley. This earnest couple was preachers who kept carrying forward the traditional culture. The all black Tatami was the most interesting. I saw some crafts of theirs in some functional space of Da'itokuji.

看到这么多的各式各样的榻榻米，真的好兴奋，也惊叹，制造这些惊世作品的“隐世环境”，纯黑色的最酷！

It was exciting to see so many different kinds of Tatami. I also felt surprised at the secret environment where these amazing works were made. The pure black was the coolest.

“朗道”的每一次行程安排都是满满的，网红的烤串儿店（Drunk Hotel 酒店临街附属小店），大雨，半封闭的雨篷下，挤了几十人，听着雨声享受热热闹闹的“不健康”的串儿（以新鲜的内脏为主材）。大德寺的定食套餐，红叶包裹的环境下，心自然静下来，美味自然而然地浮动。茑屋书店的下午茶和大餐，书店的生态模式完完全全地被市场和时代颠覆了，收入来源都是餐饮（食与咖啡）与文创产品，书沦落为作样子的摆设。百年烤肉店（十二段家）是首相们也会来凑热闹的网红地，就是太甜了，而且店家“拒绝”提供其他酱汁，执着的专一。

Every trip of Landao Design was busy and rich. An internet celebrity barbecue shop on the street belonged to Drunk Hotel. It was raining heavily. Dozens of people crowded under a half-closed awning and enjoyed the delicious but unhealthy barbecue (most were fresh entrails) while listening to the rain. We had a set meal in Da'itokuji. The red leaves all around made me calm down slowly and the smell filled the air naturally. We enjoyed the afternoon tea and super in Daikanyama T-Site. The ecological model of the bookshop was overturned by the market and time. Dining was a major source of income (food and coffee) as well as cultural and creative products. Books fell to pure decoration. Juunidanya was a hundred-year-old barbecue shop. It was also the Internet celebrity which the Prime Minsters liked to come here. The single-minded owner refused to provide other sauces though it was too sweet. How earnest he was!

人气鼎盛的网红点，下着大雨的半露天小店，人们依然欢声笑语，酒店大堂娱乐区更是“人满为患”，手工纸造的海洋生物，点睛之处。

A semi-open popular shop was still full of people and laughter in such a rainy day. The entertainment area in the hotel lobby was even overcrowded. The paper handmade sea animals were the focus.

激动地冲进书店，哈哈，几乎没有看书的人，都是吃东西、饮下午茶、上网的年轻男女，我们就在这里解决了晚餐，只能说这里沦为“有书圈着的餐厅”

I hurried into the bookshop excitedly. Haha! Few people read books. Most young people were eating or having afternoon tea while surfing the internet. We had our dinner here. The bookshop became a dining hall surrounded by books.

大德寺，在一间有暖气的玻璃房里，席地而坐，静静地、静静地等待红叶的落下。午后的阳光在大雨后，恣意地穿透了枫叶的艳红。

Da'itokuji was in a heated glass room. We sat on the ground and waited for the falling of the red leaves quietly. After the heavy rain, the afternoon sunlight pierced the bright redness of the leaves willfully.

要门票的大德寺，走了一段，居然有一间暖房让你可以静心，静默看小雨下，阳光中的红叶，真能想！

I bought a ticket to visit Da'itokuji. After a walk, there was a conservatory. You can sit here and look at the drizzle quietly, or enjoy the red leaves in the sun. How creative!

O.'s File **(区生词典)**

大德寺创建于日本镰仓年间（1325年），位于京都市北区，是洛北最大的寺院，也是禅宗文化中心之一。尤以茶道文化而闻名。其中大仙院的庭园是江户初期枯山水庭园的代表作，景致优雅。（百度百科）

Da'itokuji was founded during the Kamakura Period in Japan (1325 AD). It is located in the north district of Tokyo. Da'tokuji is the largest temple in Luobei and also one of the centers of Zen culture. It is especially famous for its tea culture. The garden in the Daxian Courtyard is the representative of the dry landscape garden in the early Edo Period with elegant scenery. (From Baidu Encyclopedia)

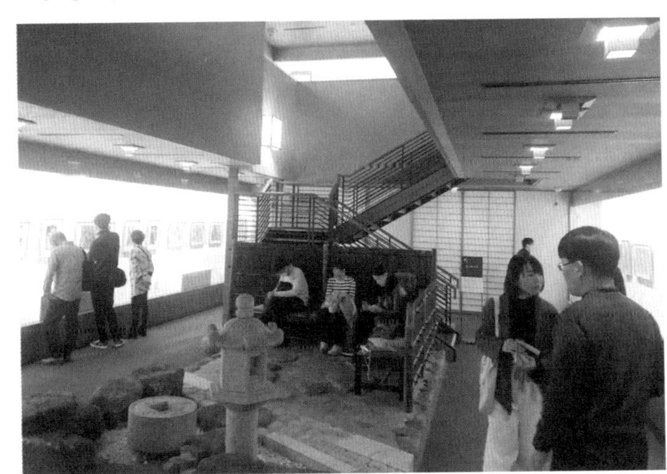

去日本银座，除了看时尚的人、潮的店等外，最为惊喜的收获就是偶遇偏僻一点点的一间小小的博物馆：浮世绘博物馆，其展示的各个时代的真迹（未经核实），各流派、各种主题的精品，让人目不暇接、日瞪口呆，一个闹市的桃源（参观的人真的不少）。

In addition to seeing stylish people and fashion shops in Japan Ginza, the most surprising thing was to run into a remote mini museum: a Ukiyoe Museum. It showed authentic of all times (unverified). They were wonderful art works of various schools and themes. There were too many things to see and made me stunned. It was like a heavenly place in the busy city. (There were lots of people visiting here.)

原来的认知被这么近距离和瞪大眼睛的观看所打破，“浮世绘”对全世界的影响力这么大，真的是有道理的。

My cognitive was broken by a close observation of Ukiyo-e. It has such a great influence on the world, it was quite reasonable.

知道日本的大师们追求极致，但当来到这里，还是有许多让你惊讶的地方：选材、造型，等等，特别是因势而造，更具哲理，受益良多。

I know that the Japanese masters pursue their extreme. However, when I came here I also found a lot of amazing things: materials, modeling and so on. In particular, it was more philosophical to make things based on inner nature. I learnt a lot from it.

江之浦测候所：专门因地而建的地方，真的是为了观测某个时节的日落日出，同时也是为了收藏业主——杉本博司老先生的各类藏品：石头、木屋及其全世界各个特色地方的海平线的摄影作品，服了！

According to the terrain, Enoura Observatory was built specially to observe sunrise and sunset of a certain season, and also for Mr. Hiroshi Sugimoto's all sorts of collections, such as stones, wooden houses and his photos of sea levels from special places around the world. I admired him.

***O.'s File* (区生词典)**

杉本博司是著名的摄影师、艺术家，他将哲学与美学等融入摄影之中，创造了自己的摄影风格。杉本对海很痴迷，最著名的作品便是《海景》。在2009年，身为建筑师的杉本在以海著称的神奈川创建了江之浦测候所。这是一座以自然为灵感，可以观察冬至、夏至、春分、秋分时，光、海与自然不同的景象，沉淀自我的寂静空间。（百度百科）

Hiroshi Sugimoto is a famous photographer and artist, who has created his own photographic style by integrating philosophy and aesthetics into photography. Sugimoto is fascinated by the sea. His most famous work is Seascapes. In 2009, as a designer, Sugimoto created the Enoura Observatory in Kanagawa which it is famous for its sea. It is a tranquil place with the natural inspiration for settling. From here, you can observe winter and summer solstice, spring and autumn equinox. And you can also enjoy different natural scenery such as natural light and the sea. (From Baidu Encyclopedia)

第一天入住的是虹夕诺雅酒店，市中心的奢侈酒店，入住过程就相当棒。而全日式（新派）设计亦是对设计佬（人）的口味，值得一提的是位于顶层的半露天的热汤池（不许拍照），很精细而有空间的特征（近20米高的飘雨方形天井，体验细雨下的半夜，回味无穷）。

I stayed in HOSHINOYA which was the luxury hotel in the center of the city on the first day. The check-in process was wonderful. The new Japanese style was right for the designers. The semi outdoor hot spring on the top was worth mentioning(not allowed to take photos). It was so fine and also the feature space. (The square open patio was nearly 20 meters high.It was unforgettable for me to enjoy the hot spring in the light rain at midnight.)

表参道新开的MIUMIU专卖店，为鸟巢设计师赫尔佐格和德梅隆事务所的大气之作，反传统的少窗外立面，像一个微微打开的宝盒，外灰、内金（紫铜为主材），个性、魅力。

The new MIUMIU Store in Ometesando was a piece of work by Herzog &de Meuron who were Bird's Nest designers. The facade with a few windows was anti-traditional style like a slightly opened treasure box. The outside was grey while the inside was gold, red copper was the main material. The store was full of character and charm.

每一次的日本游都不同：ToTo之旅，温泉（北海道）之旅，安藤忠雄之旅，这一次更有不同收获，可谓别具匠心，乐在其中的“匠行”。

Every visit to Japan was different: ToTo Tour, Hot Spring Tour, Tadao Ando Tour. What I learned this time was quite different. It was a pretty awesome and enjoyable tour of handicraft.

在这么“争妍斗艳”的表参道突围而出真的不容易，室内空间的设计也非常有张力和特色，让人流连忘返！

It is not easy to stand apart from all gorgeous shops in Omotesandou. The interior design is also attractive and characteristic. I didn't even want to leave.

表参道MIUMIU

MIUMIU Store in Ometesando

HOSHINOYA
Kyoto

京都虹夕诺雅酒店

98

★★★★★

HOSHINOYA KYOTO JAPAN

Address：616-0007 11-2 Arashiyama Genrokuzanho, Nishikyo-Ku, Japan
日本京都府京都市西京区岚山元绿山町11-2

Telephone：+81 7587 10001

[illegible]

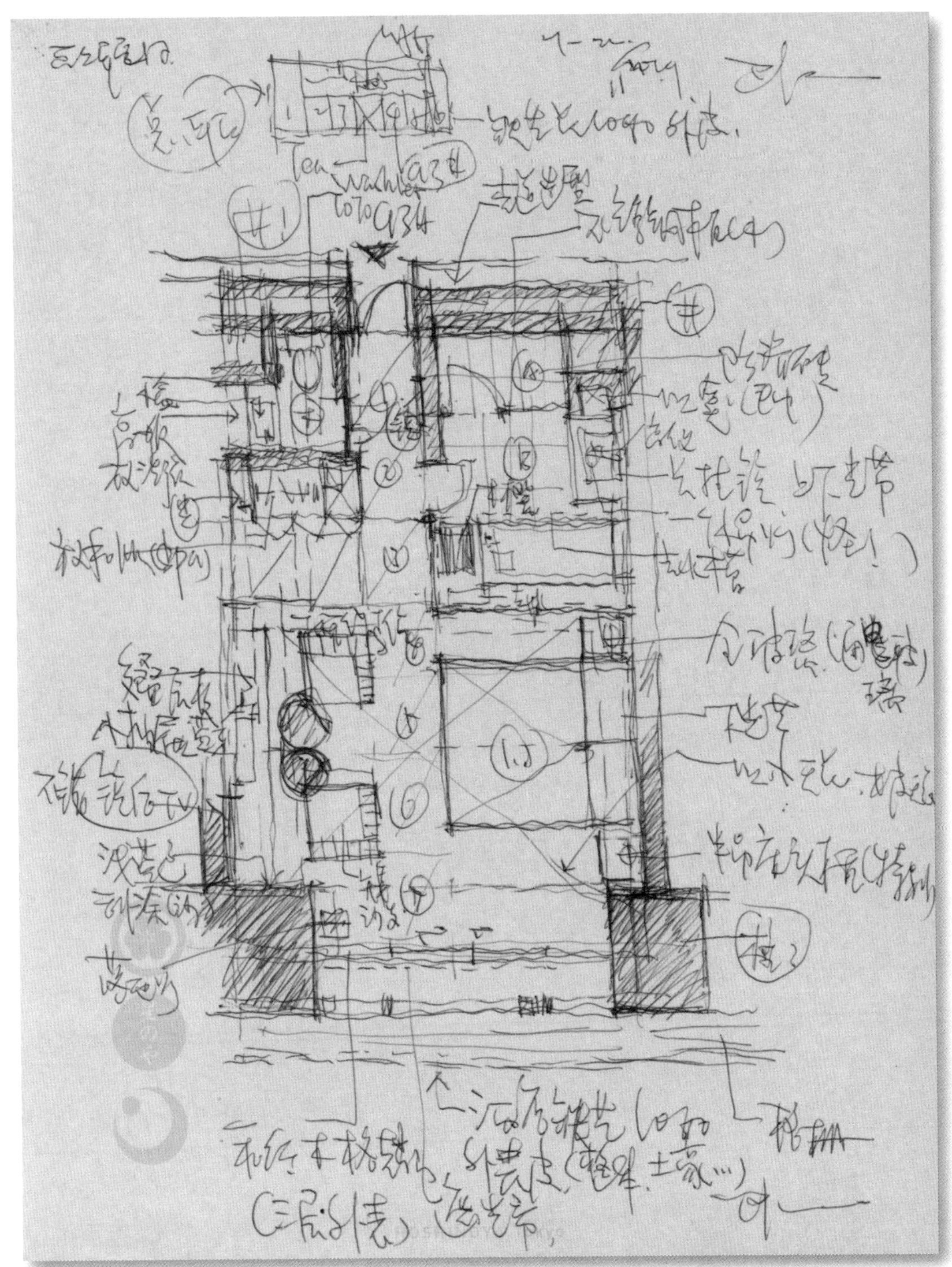

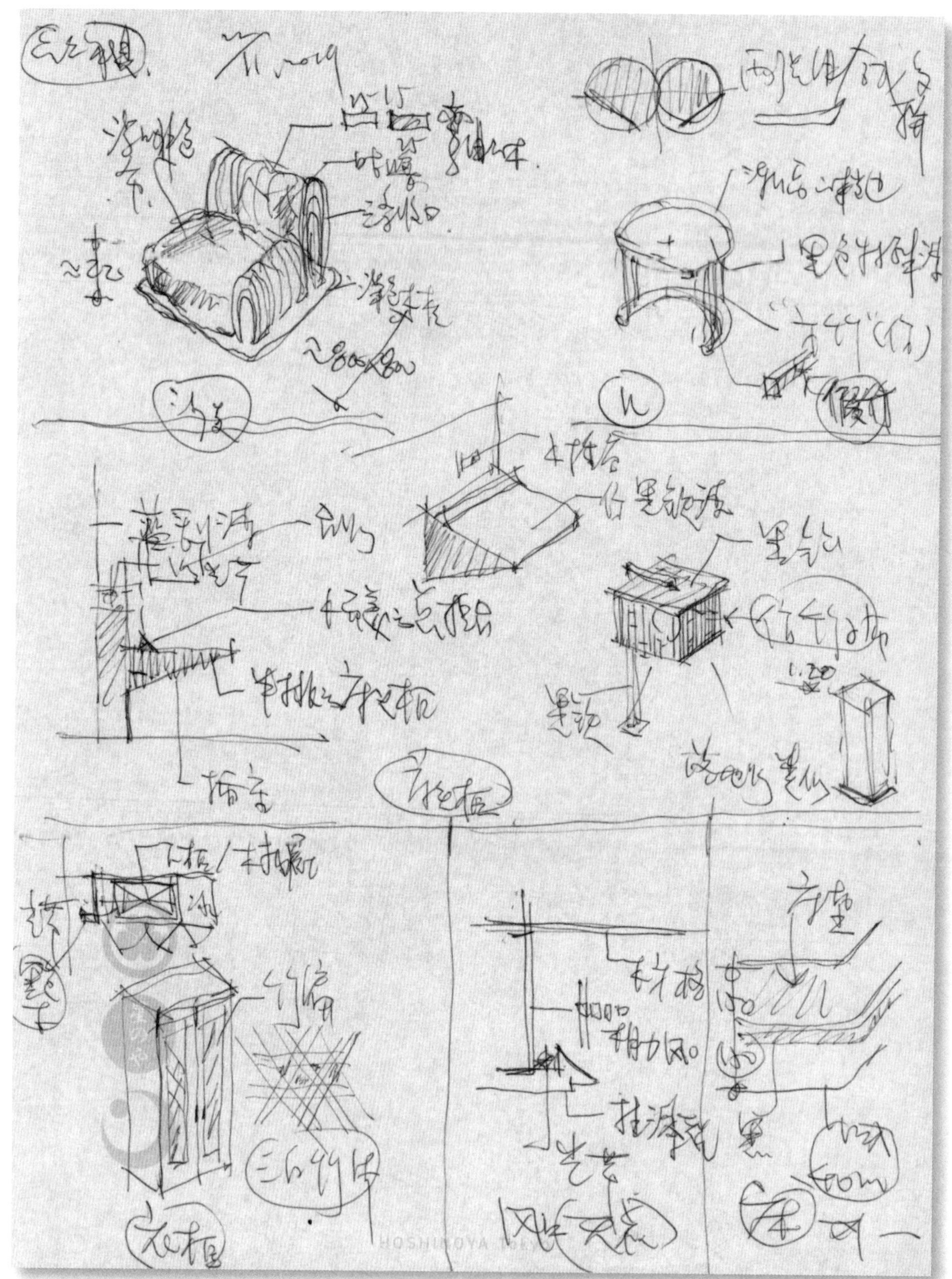

November 22, 2019
Sense of Ritual: Checking in, Arrival and Breakfast (the Experience must be Marked)

2019年11月22日
仪式感：入住，到达与早餐（必须记录下来的体验）

第一次有这样的入住待遇。

It was my first time to be treated well while checking in.

雨天，天黑，落地东京机场，打车去酒店，司机围着大楼找不到入口，原来出租车必须到负二层落客。通过有日式挂布（正名：暖帘）的小电梯门厅，到达首层的窄长的接待大堂，有台阶，原来是让你坐下来，换鞋（脱鞋），每一个人有一个鞋格子柜，这些柜，神奇地组合成巨大的有装饰性的、特别的与城市街道景观交流对话的媒介，真的是形式与功能的完美结合：对我而言是第一次。

It was a rainy day and dark. We took a taxi to the hotel after landing on the Tokyo Airport. The driver circled the hotel and couldn't find the entrance. It came out that taxi had to drop off on the second floor underground. We took the lift which had a quilted door curtain to the long and narrow lobby on the first floor. You could sit on the steps to change your shoes. Everyone would get a lattice to keep their shoes. Magically, the large combination of these lattices became the special decoration and also a media for communication with the street landscape of this city. It was a perfect combination of the form and function. The first time for me as well!

全榻榻米的公共走道（真难以想象怎么能耐用）及房间地面，完成check in 的资料确认，入住都市日式的豪华住房，有大师的手笔和低调的大投资！精美！

Both the public walkway and the floor of the guestrooms were Tatami. (It was hard to imagine how they could last forlong.) After confirming the check-in information,the Japanese luxury guestrooms were masterwork and the investment was high but humble. How stunning!

日式英语与中式（广州）英语对话，了解到一天的服务配套，早餐在房间用，一份传统日式，一份西式早餐，期待。

We talked in Japanese English and Chinglish and learnt about the service, breakfast was serviced in the room, including a Japanese traditional breakfast and a Western breakfast. I was looking forward to it.

果然，不枉此行，摆拍，吃撑了，还是服务最值钱！

Sure enough, there's no regret coming here. I took a lot of pictures and was stuffed. The service was the most valuable!

一组系统的定位（由大至小，由外至内）

[illegible] Benchmark [illegible]

[illegible]

[illegible]

不知不觉中 [illegible]

作品！

[illegible] 2014

HOSHINOYA Tokyo

Quality and Positioning Depend on a Piece of Paper (4 big pieces of paper and four small, white and grass color)

一纸看品质与定位（4大4小，白+草色）

很久没有看到这么会令我兴奋的配品：笔、尺、Bench mark，当然最牛的还是纸：4张双色的大信纸，4张双色的小便笺，可以体味这家酒店的贴心之处，不容易的传统文化的传递，笔也是纸包装的杆子，于是乎：一张大的画平面，一张画家具，一张写入住的到达、仪式和早餐，真的刚刚好。

I hadn't seen the accessories which made me excited for long: pens, rulers and bench mark. Of course, paper was the coolest: four big pieces of double-color letter paper, four small pieces of double-color notepaper. All these made you feel considerate. It was not easy to pass on the traditional culture. Pens were also made of paper. So I drew the plane on a big piece of paper, and drew the furniture on another one, wrote down the arrival, Japanese ritual and breakfast on the others. It was just right.

当然像我这么"认真"地用纸和笔的住客可谓"再世恐龙"，能在都市中心慢慢地花两个小时来写写画画，还是能与这家虹夕诺雅宾馆有一同呼吸的感受。

It was really rare that the guest would use paper and pens carefully like me. I spent two hours drawing and writing slowly in the center of the city, and also breatheg the same air with HOSHINOYA.

文化酒店，用每一份的心情细致的体验，让日式文化在不知不觉中散发出去，那就成功了。

The cultural hotels should be experienced with your heart. It will succeed if it can spread the Japanese culture unconsciously.

一纸品店！

Value the hotel by its paper!

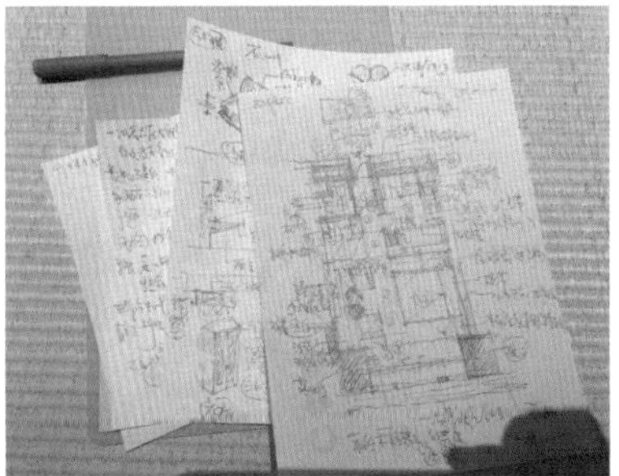

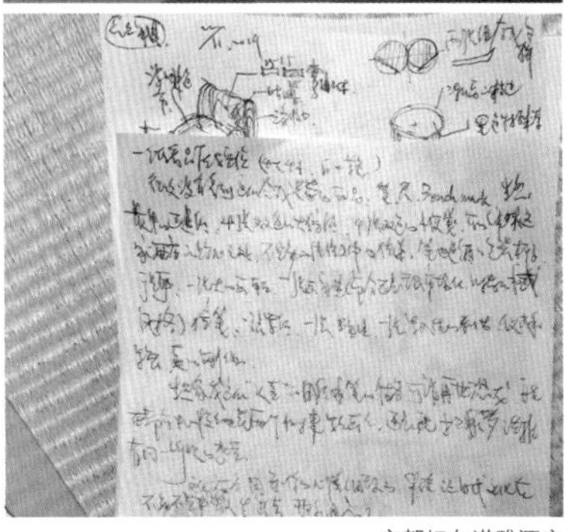

京都虹夕诺雅酒店

Hoshinoya, Kyoto, Japan

竹七半十初提：地台的Spa（西半露天）

新概念的温泉旅馆以庭子、书院半式的空间，第一印象，两浴分离，精巧的两间，室用低调的洗浴区，西边再接[illegible]堂俯瞰区"遥望"，尺度创造，古都分套的空间的意和，大小的层相，的区可打开两个大的分布，半轴式的布局，大半专是室浴区即私人层板檐半边的庭，但是镜面面向黄武的地标似的一也，我们一会检证……城式和时大空间也作第二层窗棂（和当下空间），透光是还在的，更是隔了第三层次，外立面用次西面的从闽宝与统艺覆盖。字是精彩而精准，气氛的测度，对坐的创造，半边新的观的风气参观指出的纪念，第二次的午餐也是在这里吃的。

然而本店自助这个在宵店，也提到和出去楼阁后的半露天浴泉，沉沉的[illegible]，微雨，有气氛!!

Japanese Delicacy & Natural Hot Spring

日式精致与露天SPA

神秘地被服务员引入房子，榻榻米式的客房，第一印象，厕浴分离，精巧的厕所，实用，低调的洗浴区，通电玻璃与睡眠区“遥望”，尺度舒适，有都市奢华客房的卖相，大的衣帽柜，前区可打开两个大的行李箱。横向的中轴式的布局，大床与矮沙发区及陈列层板柜串起来，倒是镜后面隐藏的电视机侧在一边，找了一会才会开启。日式和纸大趟门，也作第二层窗帘（相当于窗纱），遮光帘还是有的，更是围了第三层“皮”。外立面用酒店的图案铁艺覆盖。家具精彩而特设，矮的沙发（竹+木+布艺），难以让人感到舒适，半切角的两张茶几拼起来就是台了。第二天的早餐就是在这里吃的。

The staff led us to the Tatami guestroom secretly. At first glance, the toilet separated from the bathroom. The delicate toilet was functional and the bathroom was humble. The electric glass is opposite the sleeping area in a comfortable size. It was just like urban extravagant hotel with a big wardrobe. The front area is large enough to open two suitcases at the same time. The layout is horizontal and symmetrical. Designers streamlined the bed, the sofas and the showcase. The bed behind the mirror could not be seen first, I turned it on after searching for a while.The big Japanese paper sliding door was used as the second curtain (as window gauze). Also, the window shades were like the third "skin". The facade was covered by the iron patterns of the hotel. The furniture was perfect and special. While the short sofa (made of bamboo, wood and cloth) was uncomfortable. The table was made up of two tea tables with half cut corners. I had my breakfast on it the next morning.

简单的本层自助区小夜宵后，就换了和服去楼顶层的半露天温泉，泡泡汤池，微凉，微雨，有气氛！

After a simple night snack on the same floor, I changed the kimono and went to the semi-outdoor hot spring on the top floor. It was a little cool and rainy while relaxing in the hot spring. What atmosphere!

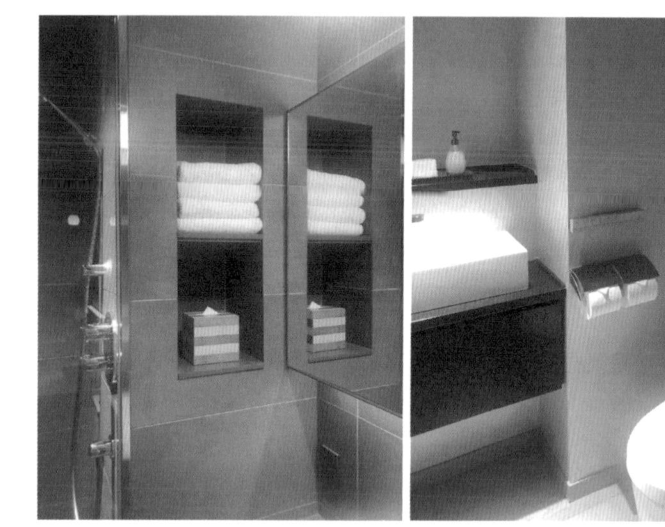

京都虹夕诺雅酒店
Hoshinoya Kyoto,Japan

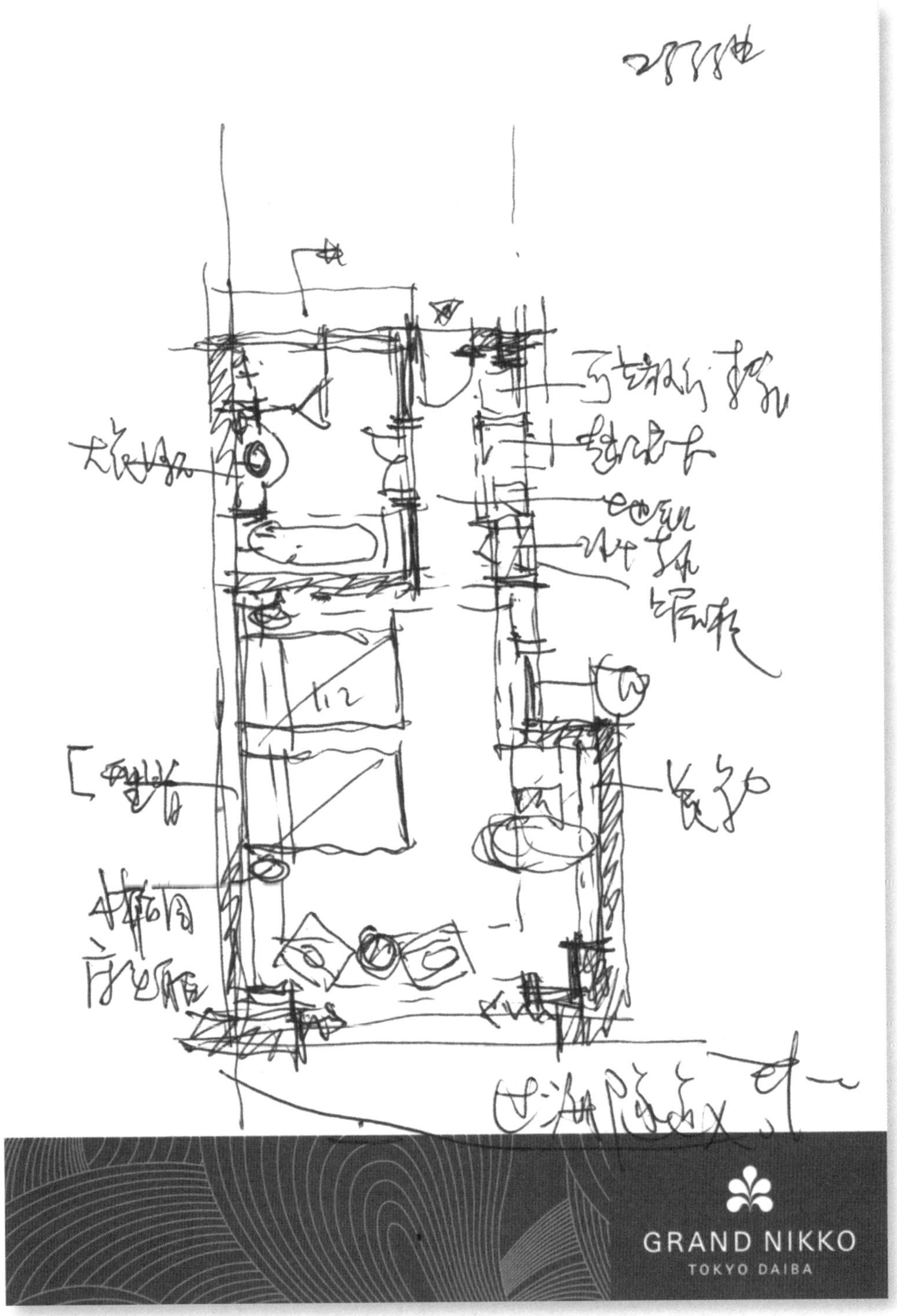

东京台场日航大酒店

99

★★★★

GRAND NIKKO TOKYO DAIBA JAPAN

Address : Tokyo, Minato-ku
Daiba 2-6-1, Japan
Telephone : +03 5500 6711

GRAND NIKKO
TOKYO DAIBA

YOKOHAMA
ROYAL PARK HOTEL

横滨酒店

100

★★★★

ROYAL PARK HOTEL YOKOHAMA JAPAN

Address : 2 2 1 3 Minatomirai, Nishi-ku, Yokohama, 220-8173, Japan
Telephone : +045 221-111
FAX : 045 224-5153
Http : //www.yrph.com

横滨酒店
Royal Park Hotel Yokohama, Japan

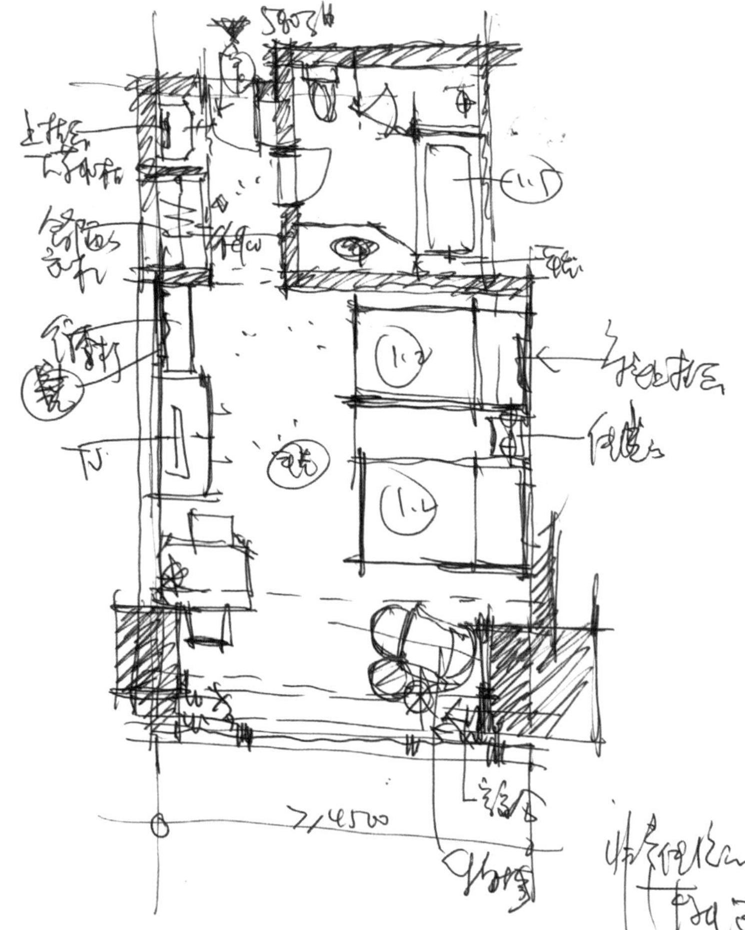

〒220-8173 横浜市西区みなとみらい2-2-1-3
2-2-1-3 Minatomirai, Nishi-ku, Yokohama 220-8173, Japan
Phone : (045) 221-1111 Facsimile : (045) 224-5153
www.yrph.com

京都东急酒店

Kyoto Tokyu Hotel,Japan

京都 東急ホテル

京都东急酒店

101 ★★★★

KYOTO TOKYU HOTEL JAPAN

Address : Kyoto Gojo-sagaru, Horikawa- dori, Shimogyo-ku

Telephone : +8175 341 2411

KYOTO
TOKYU HOTEL
TOKYU HOTELS
京都 東急ホテル

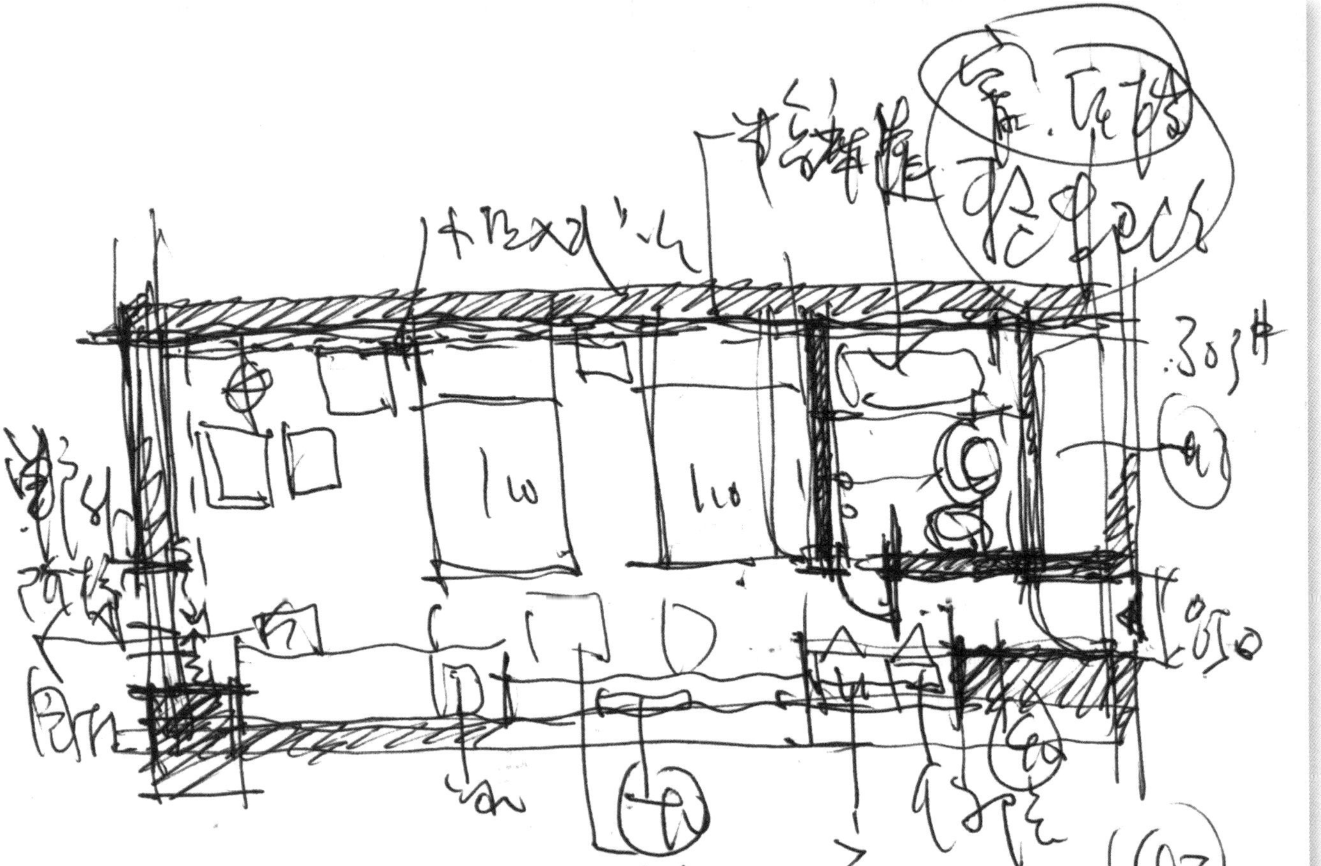

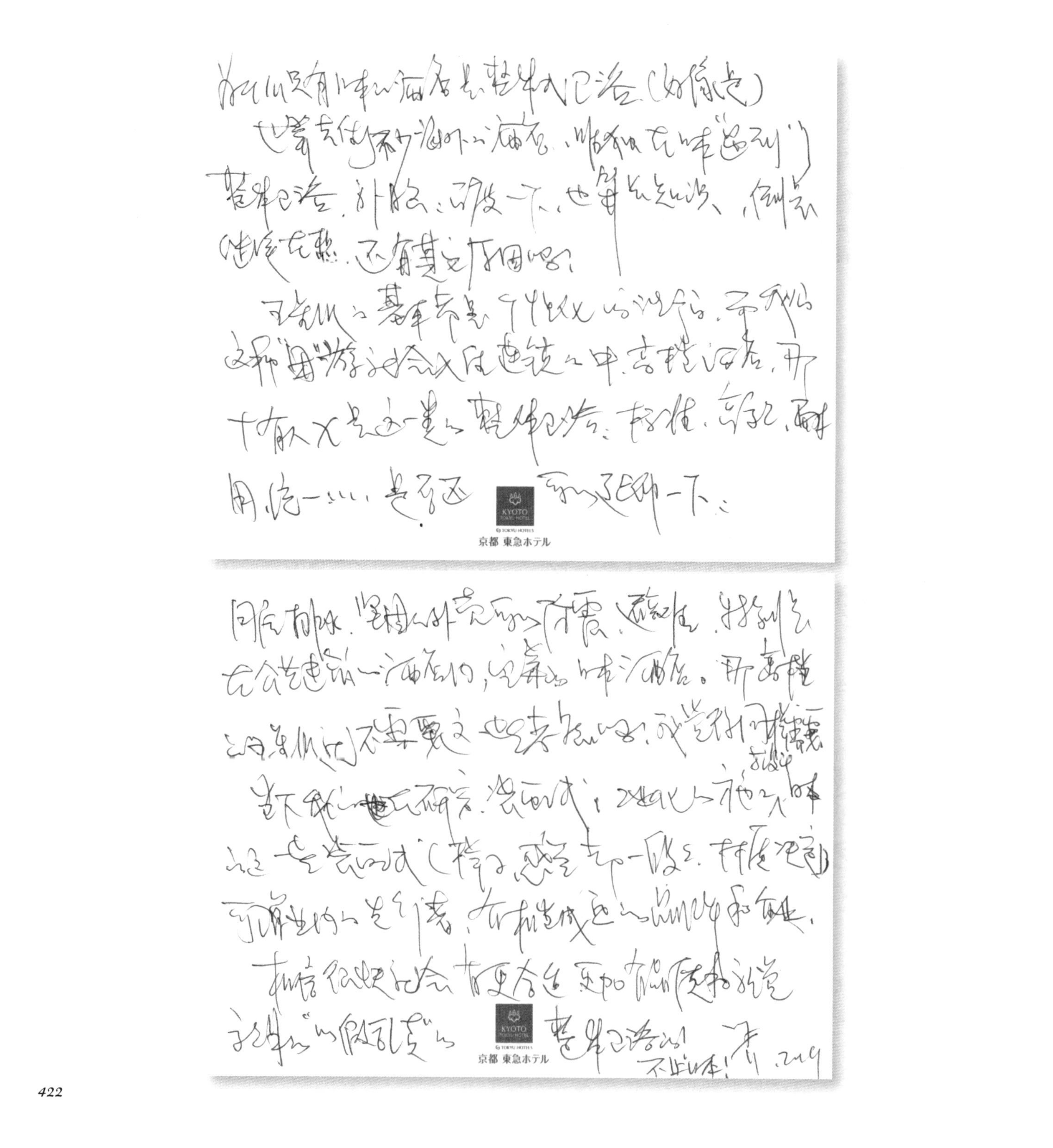

[illegible]

KYOTO
TOKYU HOTEL
TOKYU HOTELS
京都 東急ホテル

[illegible]

KYOTO
TOKYU HOTEL
TOKYU HOTELS
京都 東急ホテル

Why the Integral Bathroom Only in Japan? (It Seems like That)

为什么只有日本的酒店是整体式卫浴（好像是）

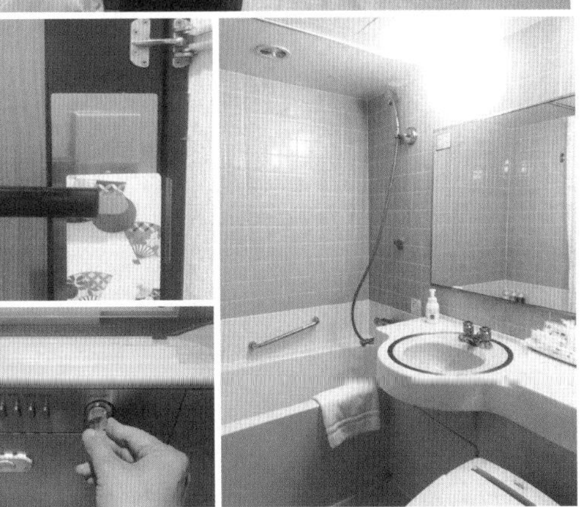

京都东急酒店
Kyoto Tokyu Hotel,Japan

也算去住了不少海外的酒店，唯独在日本“遇到”了整体卫浴。百度一下，也算长知识，继续在想，还有其他原因吗?

I had stayed in lots of hotels abroad, but I met the integral bathroom only in Japan. I searched on Baidu and learned a lot. I was still thinking if there were other reasons.

五星级酒店的基本都是个性化的洗手间，而我们这种“团”游就会入住连锁的中、高档酒店，那十有八九是这一类的整体卫浴：标准，高配，耐用，统一……是否还可以延伸一下?

The five-star hotels have personalized bathroom basically. However, the tour groups like us usually stay in chain hotels which are middle and high grade. Most of them use this type of integral bathroom: standard, high configuration, duration and unification… Whether it can be extended or not?

同层排水，坚固的外壳可以防震，避难，特别是在公共建筑的酒店内。完美的日本酒店！那高档的五星级就不需要这些考虑吗？我觉得同样需要。

The same-floor drain, the hardened housing can be earthquake-proof and a shelter, especially the hotels in public buildings. What perfect hotels in Japan! Five Stars needn't think about it, do you think so? But I don't agree.

当下我们也在研究，装配式、工业化的施工与设计，日本的这些装配式（样子感觉都一般般，由材质决定）可以算业内的先行者，有相当成熟的品牌和企业。

Nowadays, we are studying fabricated and industrialized construction design. The Japanese fabricated design is the pioneer in the field, it has some quite experienced mature brands and enterprises.

相信很快就会有更合适、更加有品质和视觉效果的“以假乱真”的整体卫浴的！不只日本!

I believe there will be more suitable integral bathrooms of high quality and visual effect that look like real soon.Not just in Japan!

MUJI HOTEL

深圳无印良品

102 ★★★★★

MUJI HOTEL SHENZHEN CHINA

Address : 2nd floor, Shenzhen Shangcheng, 5001 Huanggang Road Futian District, Shenzhen,China
中国深圳市福田区皇岗路5001号深业上城二层

Telephone : +86 755 2337 0000

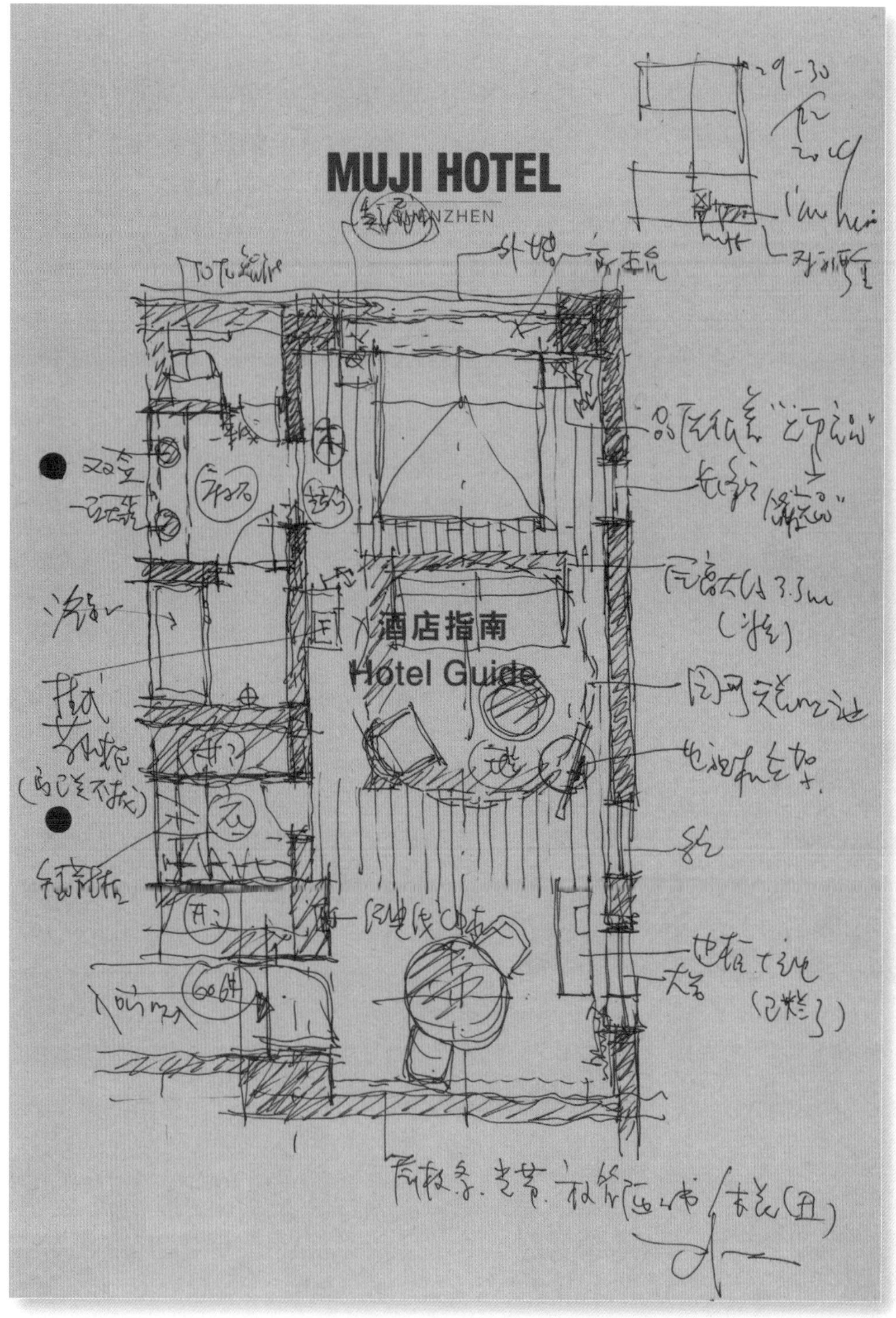

"元印之品"

韩涛同学的[illegible]会。进入位于东区的那个"元印之品"的酒店，也是难得不错的[illegible]之一角。[illegible]建筑的酒店大堂、餐厅，有粗犷的原木板的[illegible]这个酒店的

材质、人文、生态的[illegible]（和北京的那个[illegible]并不同），[illegible]这个[illegible]。

[illegible]所以作为一个[illegible]上海的[illegible]，第一感不错的生态式的设计，也[illegible]"元印之品"（[illegible]）的感觉：本色、[illegible]。

[illegible]

[illegible]"元印之品"[illegible]（一年一月）

[illegible]"元印之品"。

[illegible]

Logo/Mark [illegible]

[illegible] Muji [illegible]

[illegible]

30/11, 2019

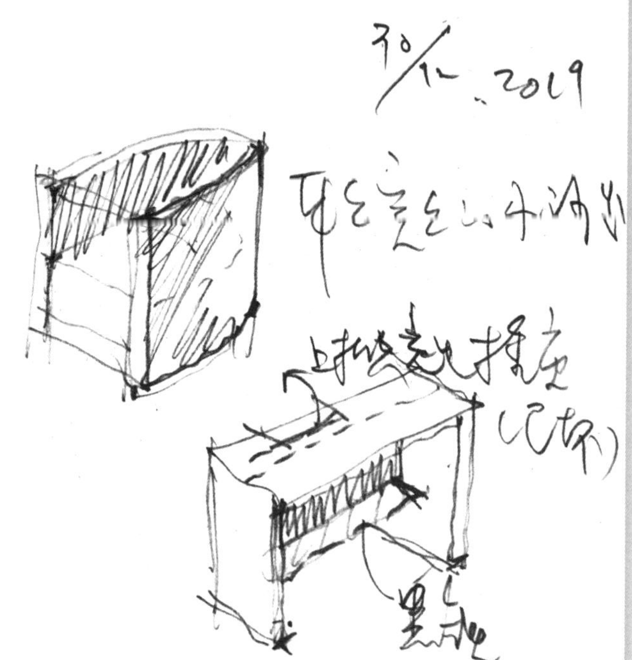

[illegible]

The Taste of Muji

无印之品

December 30, 2019
2019年12月30日

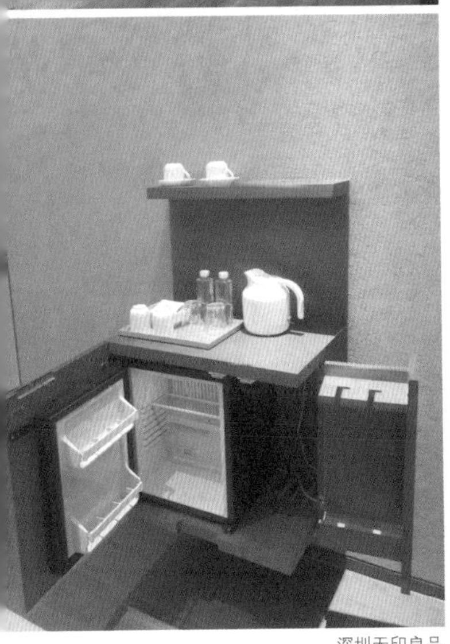

深圳无印良品
Muji Hotel, Shenzhen, China

借清华同学的深圳聚会之机，选择入住华南区唯一的“无印良品”酒店，选址不错。高大上的楼盘的一角，独栋建筑物。

Taking the chance of Qing Hua University reunion, I stayed in the only Muji Hotel in South China. Its location was good, which lay at the corner of a single tall building.

酒店大堂，有意思。粗犷的原木板定义了这个酒店的朴素、人文、生态的主题（和北京的那间有类同与不同），延续这个调子到房间。

The lobby was quite interesting. The hotel was defined as simplicity, humanity and ecology by the rough wood. (It was similar to the one in Beijing but a little different.) The same tune was continued to guestrooms.

尽端的房间为所谓的“套房”。L形的采光房，第一感不错，生态式的设计，也非常之“无印良品”的感觉：木色（橡木家具），墙面是粗面的涂料，洗手间为粗面石材（哑光的，易脏也不怎么防滑，可能是我酒后的原因吧），全部是“无印良品”的感觉，不高档（也许本来就是这样定位的。但，一千多元的房间，价格也不低），可能也不追求高品。使用之后，更觉这里刮花，那里破损，部分还掉下来一些配件。开业才不长的时间!（2018年1月18日，全球首家无印良品酒店）

The guestroom at the corner was the so-called a suite. The day lighting room with the shape of "L" gave me a good impression. The ecological design was quite Muji style: the color (oak furniture) and the rough coating of the wall, the quarry-faced marble in washroom (as well as matte, easy to slip on, maybe after drinking), all were the feeling of Muji, but not fancy. (Perhaps that's Muji. However, more than 1000 yuan for a room was not cheap.) Maybe Muji doesn't go for high quality. After experiencing, I found scratches and breakage everywhere, some accessories even fell down. It opened not long before. (On January 18, 2018, the first MUJI Hotel in the World)

突然也没有了画的兴趣。硬逼着自己找找笔：没有；找找纸，也没有。最后找到了一支无印良品的笔，不错。叫了服务员，半天才告诉我，没有信纸，只有一叠A4复印纸，“无印之出品”！

Suddenly I lost my interest in drawing. I forced myself to look for pens and paper, but nothing could be found. At last I found a Muji pen, not bad. And then I asked the staff for some letter paper. After a long time he told me no letter paper but a pile of A4 copy paper. That was too casual!

逼着我用了（撕下了）酒店的手册指南。这是唯一的可以写东西而又有“酒店”Logo的纸。

I had to tear paper from the Hotel Guide Book. It was the only paper with the hotel logo which could be written on.

细想一下，是否是入住的人（包括我）“一厢情愿”，对无印良品期待过高了？也许他们压根不是追求高品的，但他们一定也不希望看到一间令我“只住一晚”“不想推荐的”酒店。

I thought carefully if the guests expected too much of Muji like wishful thinking, including me. Maybe it didn't go for high quality at all. They surely didn't hope to provide a hotel that I would only stay once and never recommend.

我们是客户（客人），非常想看到一贯的有品的无印良品。（早餐的感觉也一般般，可能是延续了“快”时尚的概念吧！）

We were clients and also guests. We looked forward to seeing the high-quality Muji but not the tasteless one. (The breakfast was so so, maybe it continued the concept of "fast" fashion.)

深圳无印良品
Muji Hotel, Shenzhen, China

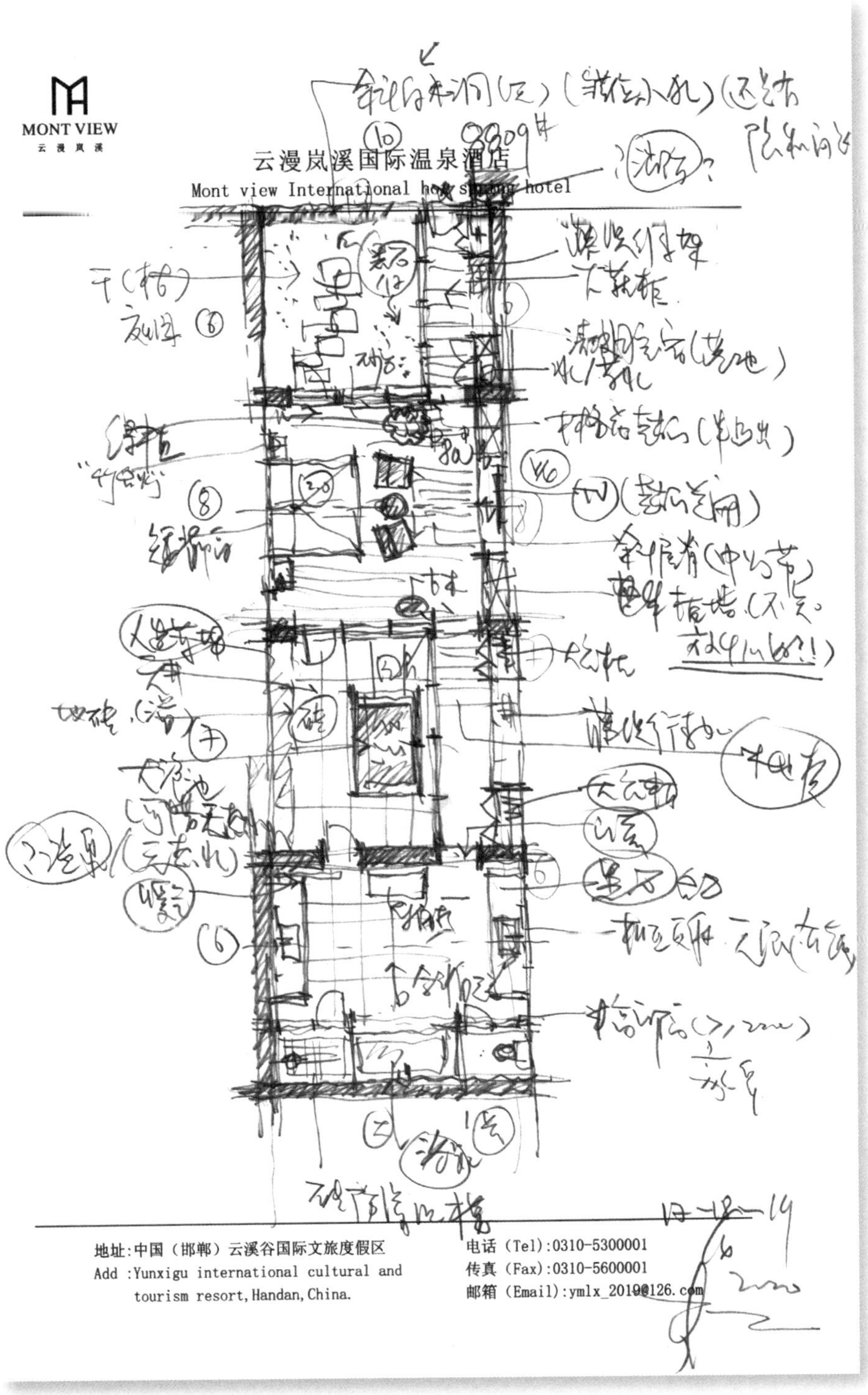

MONT VIEW
云漫岚溪

103

云漫岚溪国际温泉酒店

★★★★★

MONT VIEW INTERNATIONAL HOT SPRING HOTEL HEBEI CHINA

Address : Yunxigu International Cultural and Tourism Resort, Handan, Hebei,China
中国河北省（邯郸）云溪谷国际文旅度假区

Telephone : +86 310 530 0001

Fax : +86 310 560 0001

E-mail : ymlx_2019@126.com

MONT VIEW
云漫岚溪

云漫岚溪国际温泉酒店
Mont view International hot spring hotel

地址：中国（邯郸）云溪谷国际文旅度假区
Add :Yunxigu international cultural and tourism resort, Handan, China.

电话（Tel）:0310-5300001
传真（Fax）:0310-5600001
邮箱（Email）:ymlx_2019@126.com

云漫岚溪国际温泉酒店

Mont view International hot spring hotel

不能不说了！

[illegible]

地址：中国（邯郸）云溪谷国际文旅度假区
Add：Yunxigu international cultural and tourism resort, Handan, China.
电话（Tel）：0310-5300001
传真（Fax）：0310-5600001
邮箱（Email）：ymlx_2019@126.com

Something I Have to Say

不能不说的！

April 17-19, 2020
2020年4月17–19日
Guyu's Birthday is on April 19
2020年4月19日谷雨生日

云漫岚溪国际温泉酒店
Mont View International Hot Spring Hotel,Heibei, China

之前听同事说，这“云漫”酒店与上海养云安缦的设计相似。傍晚来到（一路平原油菜花地、小麦地的村路），一进大堂，啊，这个不是上海吗？（同款不同质）。

I heard from my colleague that this Mont View Hotel was similar to Shanghai Amanyangyun. I arrived late in the afternoon. (The rape flowers and wheat were on both sides of the flat country road.) From the first moment I got on the hotel, I confused if it was Shanghai Amanyangyun(the same design but different quality).

上海安缦刚开时专门去看过，专车停在酒店门口不让进，只有订餐才可以入酒店看（公共区域），当然没有住8000元以上一天的房（我们的某个客户，住了8万多的别墅套房，羡慕）。用（住）不起上海养云安缦酒店，那就住这个也成。这是我们客户自营自建的酒店，也是为客户增加一点收入。

When Shanghai Amanyangyun opened, I went to have a look specially but was stopped at the gate. Only people who made restaurant reservation could enter to visit the public area. Of course I didn't stay in the guestroom which cost over 8000 yuan a day(One client of ours stayed in a villa which cost more than eighty thousand yuan. How envious!) Since Shanghai Aman yangyun was too expensive, I tried the Mont View Hotel. It was built and run by our client so that we could increase incomes for him.

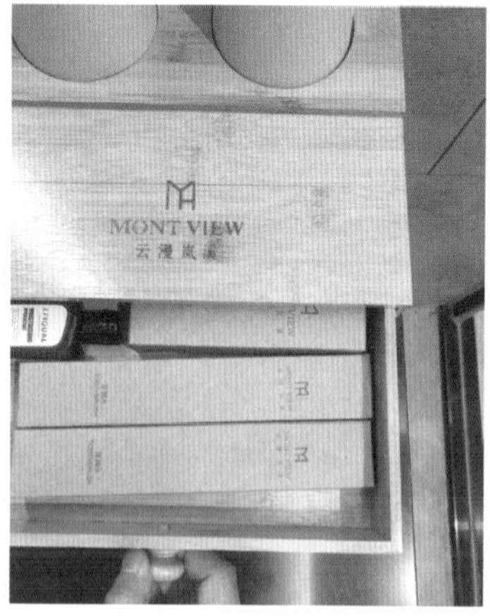

专业“睡客”全面体验房间的一切：1. 开开关关灯，不容易控制；2. 柜子太多，不知道怎么用，也怕忘记拿东西（衣服/其他），就不用了，反而行李架（藤织的）可坐，可放衣服；3. 床太矮，电视柜太复杂，电视里的电影都是收费的（8元看了一部《唐顿庄园》）；4. 洗手间大，不错，热水忽冷忽热，不稳定，常见的毛病；5. 户外小园子的温泉无水；6. 洗手间气味太大了，麻烦……当然，不能用安缦的标准去评价和期待这里。虽是客户，也不能不说。以此为戒，让我们在服务的项目（书院及销售中心等）设计过程中尽量避免！

As a professional sleeper, I looked over the whole guestroom: 1. I turned on and off the lights, it was not easy to control. 2. there were too many cabinets. I didn't know how to use them. I was afraid I would forget something so I didn't put anything in them. I could sit on the cane luggage rack or put clothes on it. 3. The bed was too low and the TV table was too complex, all the films needed a fee(I paid 8 yuan for a film called Downton Abbey). 4. The washroom was big enough. The common problem was the unstable hot water, sometimes hot sometimes cold. 5. There was not any hot spring water in the yard. 6. The washroom was smelly. How terrible! Of course we couldn't expect or make a judgment about it with the standard of Shanghai Amanyangyun. Although it was our client, I had to say that. As a precaution, we would avoid these in our design projects (such as the academy, the sales centers and so on).

云漫岚溪国际温泉酒店
Mont View International Hot Spring Hotel,Heibei, China

WESTIN
HOTELS & RESORTS
威斯汀酒店及度假村

威斯汀酒店及度假村

104

★★★★★

WESTIN HOTELS & RESORTS XI'AN CHINA

Address : No.66 Ci'en Road, Xi'an, Shanxi, China
中国陕西省西安市慈恩路66号

Telephone : +86 29 6568 6568

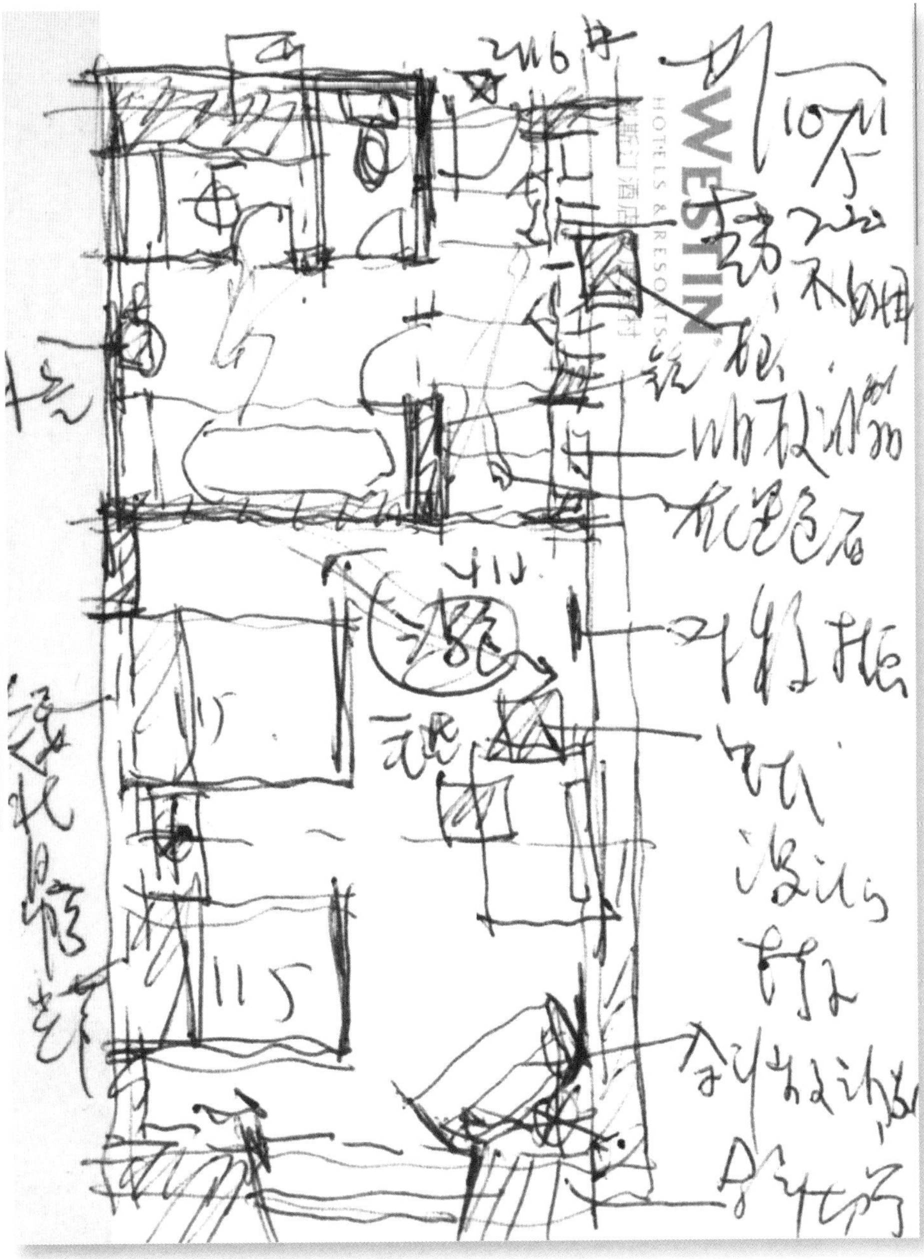

威斯汀酒店及度假村
Westinhotels & Resorts, Xi'an, China

105 希尔顿欢朋酒店

★★★★★

HAMPTON BY HILTON GUANGDONG CHINA

Address : No. 71 Renmin Avenue, Xiashan District, Zhanjiang, Guangdong, China
中国广东省湛江市霞山区人民大道中71号

Telephone :+86 759 8123 888

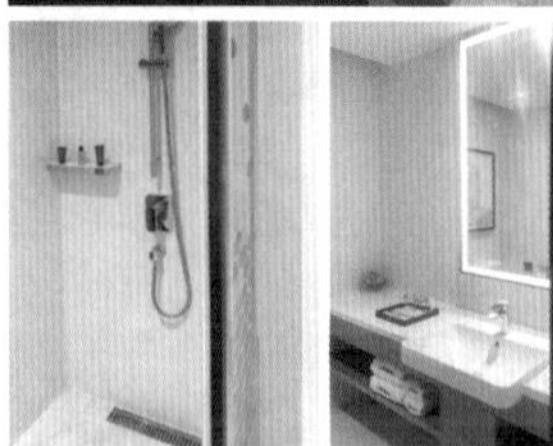

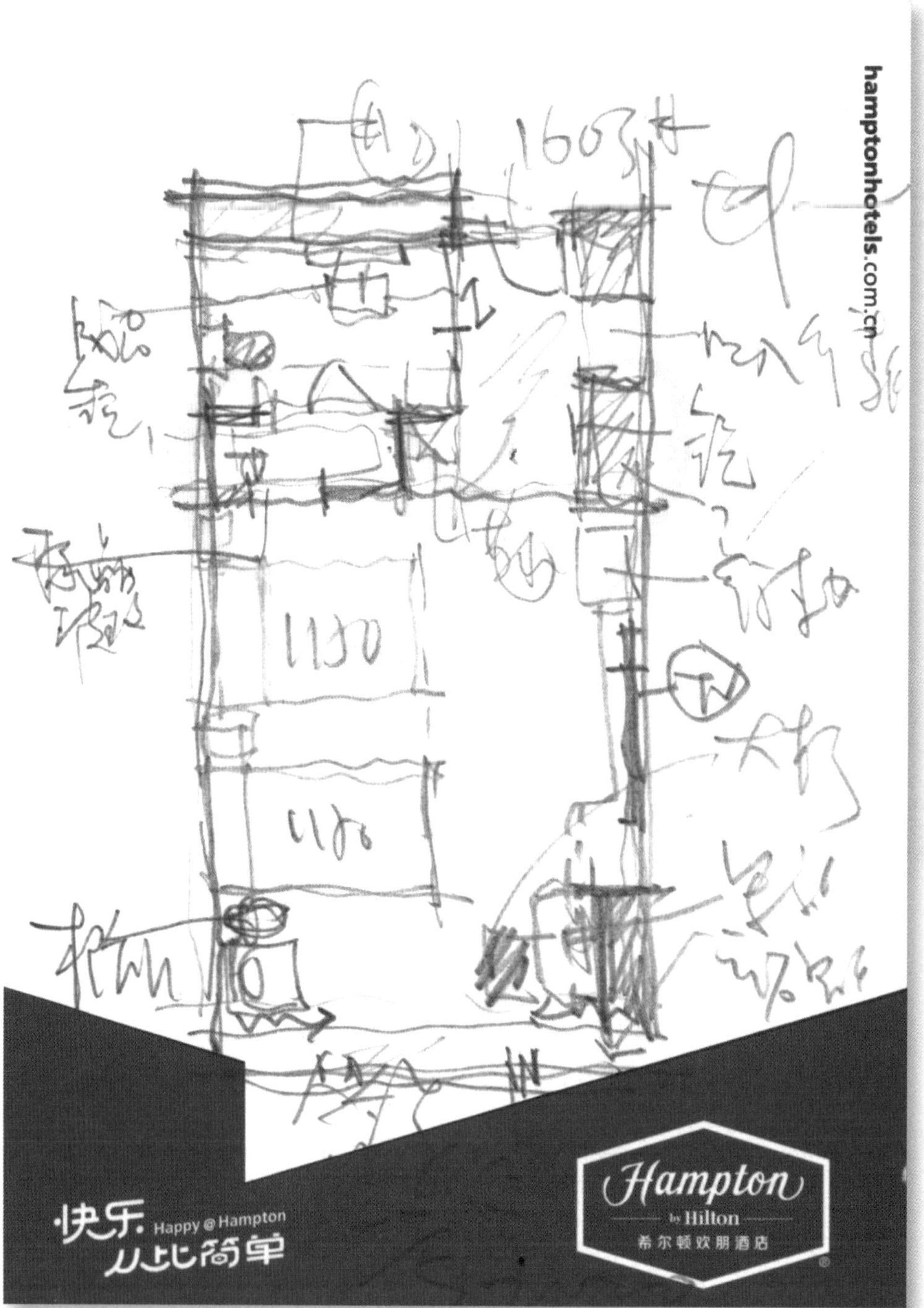

希尔顿欢朋酒店
Hampton by Hiltion, Guangdong,China

hamptonhotels.com.cn

1150

1180

快乐 Happy @ Hampton 从此简单

Hampton
by Hilton
希尔顿欢朋酒店

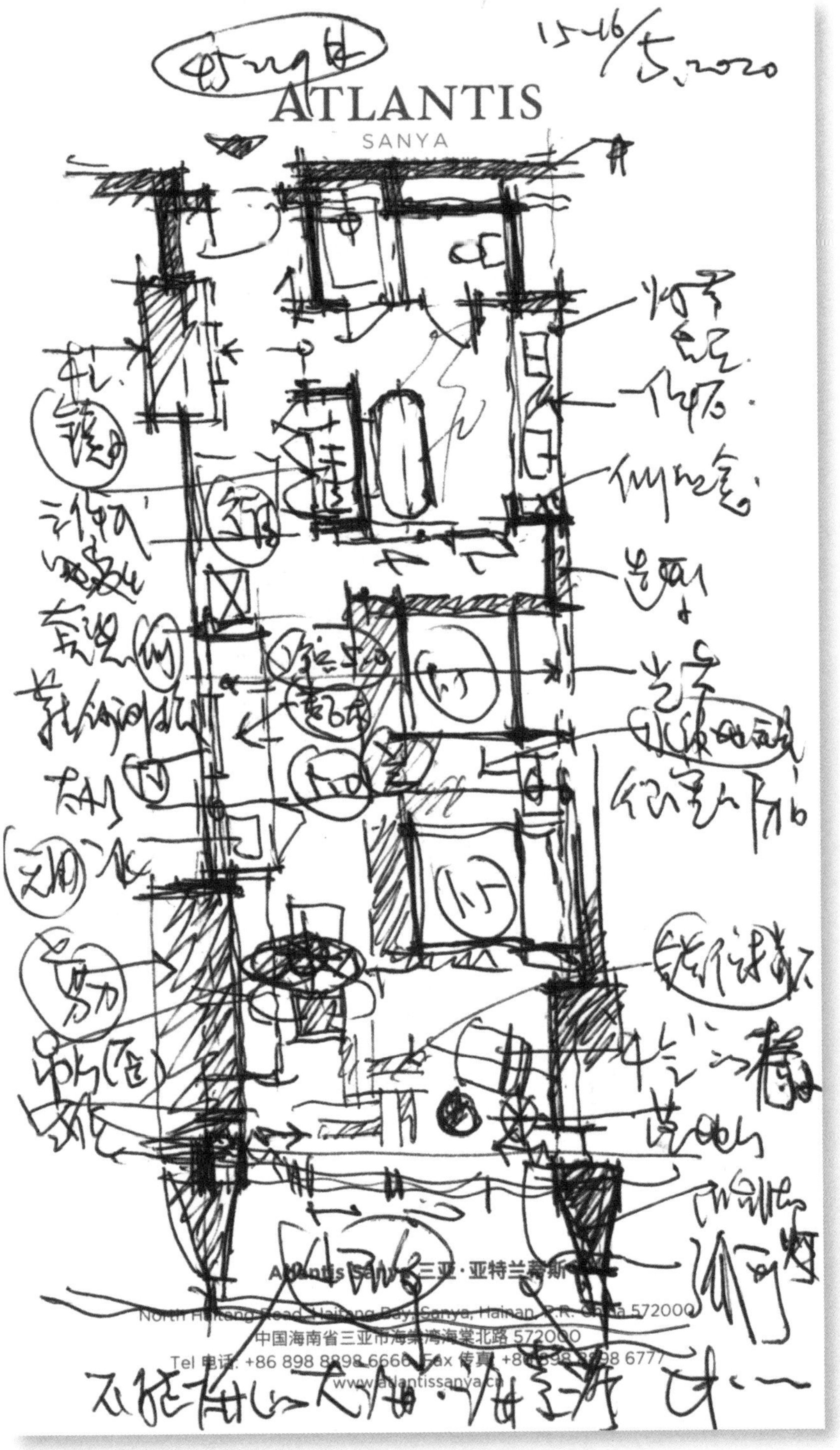

15-16/5, 2020

ATLANTIS
SANYA

Atlantis Sanya 三亚·亚特兰蒂斯
North Haitang Road, Haitang Bay, Sanya, Hainan, P.R. China 572000
中国海南省三亚市海棠湾海棠北路 572000
Tel 电话: +86 898 8898 6666 Fax 传真: +86 898 8898 6777
www.atlantissanya.cn

ATLANTIS
SANYA
三亚·亚特兰蒂斯

三亚·亚特兰蒂斯

106 ★★★★★

ATLANTIS SANYA CHINA

Address	: North Haitang Road, Haitang Bay, Sanya, Hainan, China 中国海南省三亚市 海棠湾海棠北路
Telephone	: +86 898 8898 6666
Fax	: +86 898 8898 6777
http	: //www.atlantissanya.cn

三亚·亚特兰蒂斯
Atlantis, Sanya, China

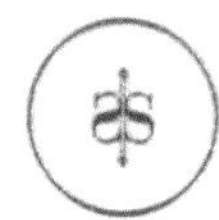

107 三亚保利瑰丽酒店

★★★★★

ROSEWOOD SANYA CHINA

Address : No. 6 North Haitang Road, Haitang Bay, Sanya, Hainan, China
中国海南省三亚市
海棠湾海棠北路6号
Telephone : +86 898 8871 6666
Fax : +86 898 8871 0099

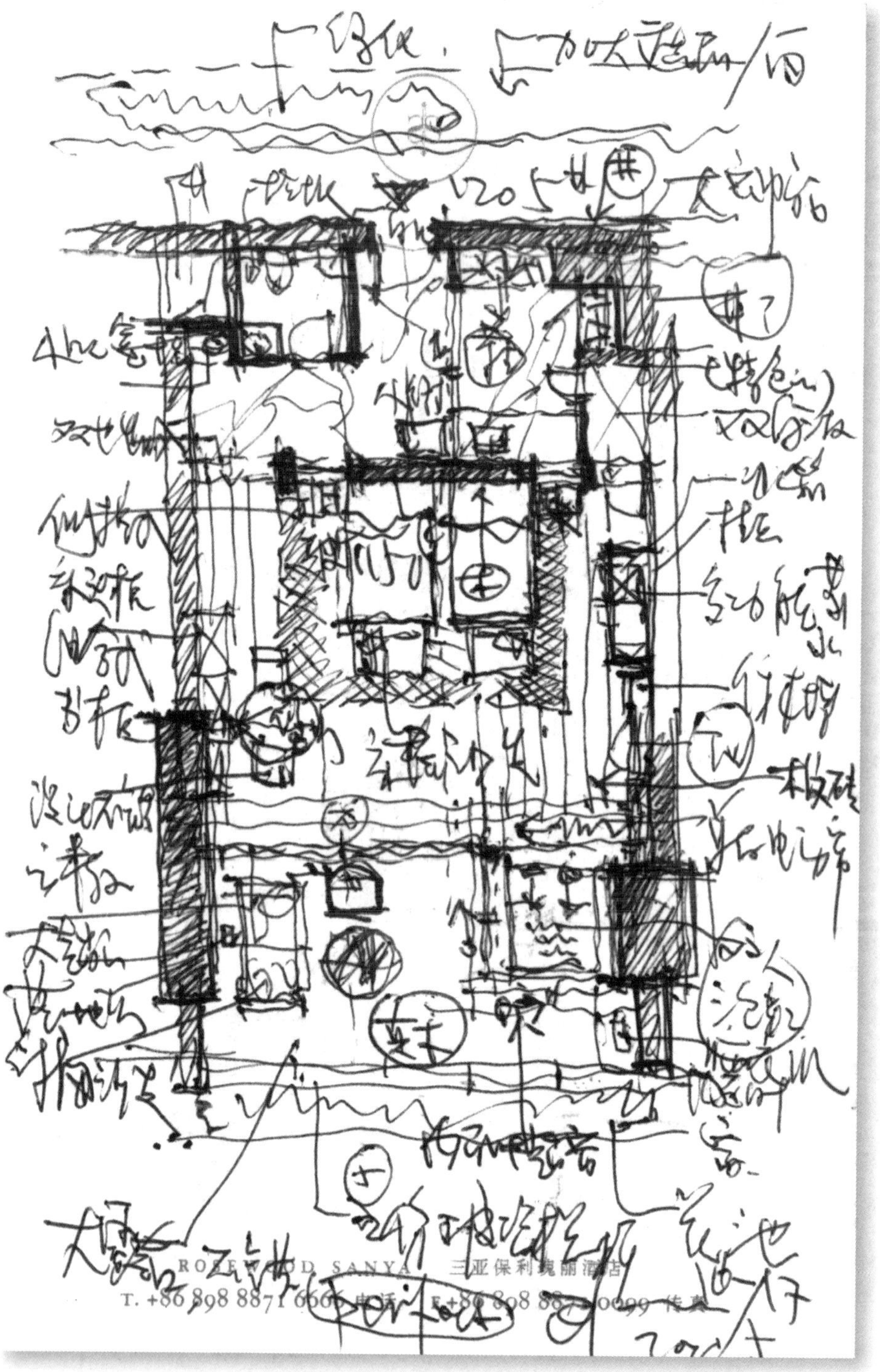

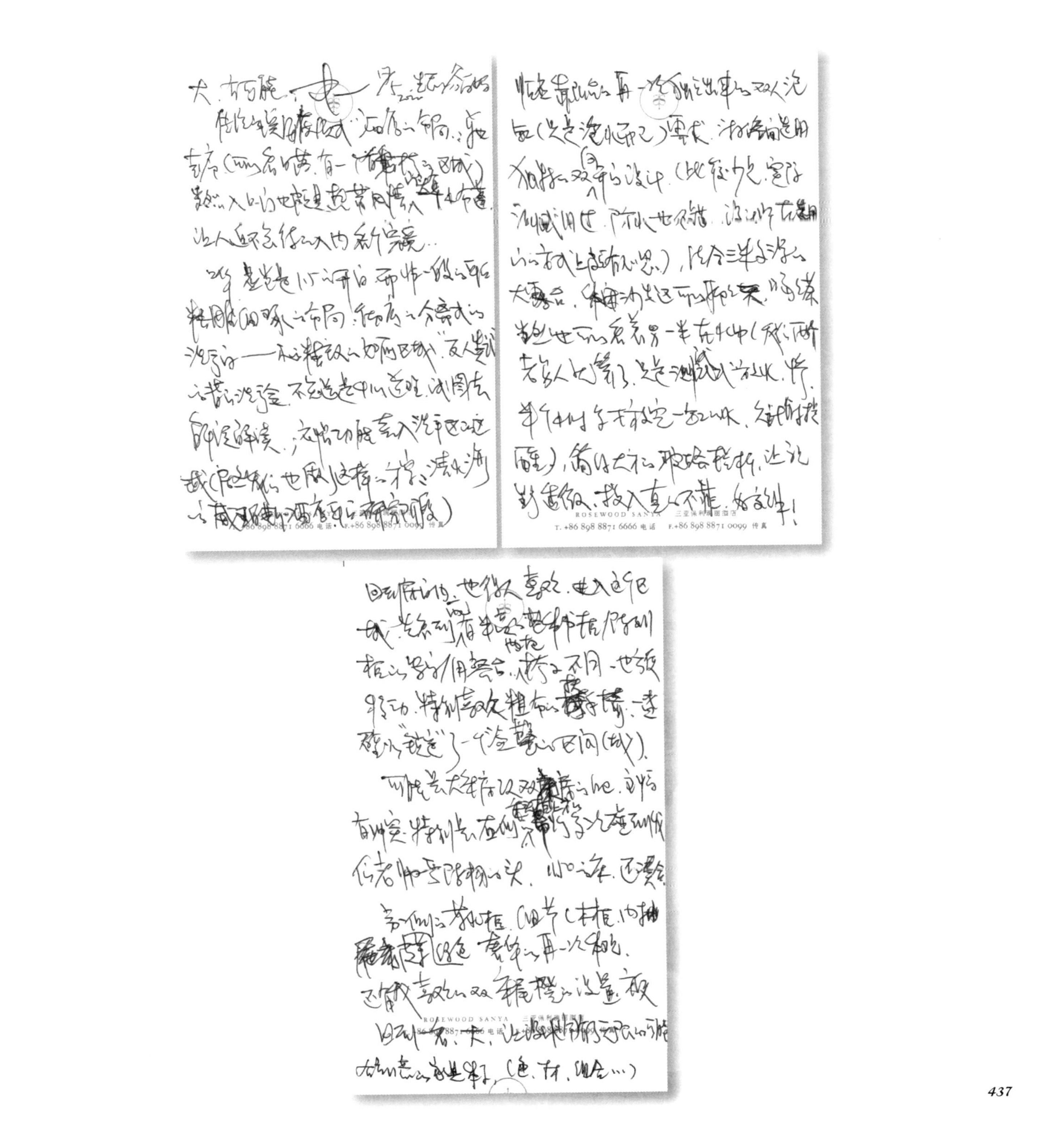

ROSEWOOD SANYA 三亚瑰丽酒店
T. +86 898 8871 6666 电话 F.+86 898 8871 0099 传真

ROSEWOOD SANYA 三亚瑰丽酒店

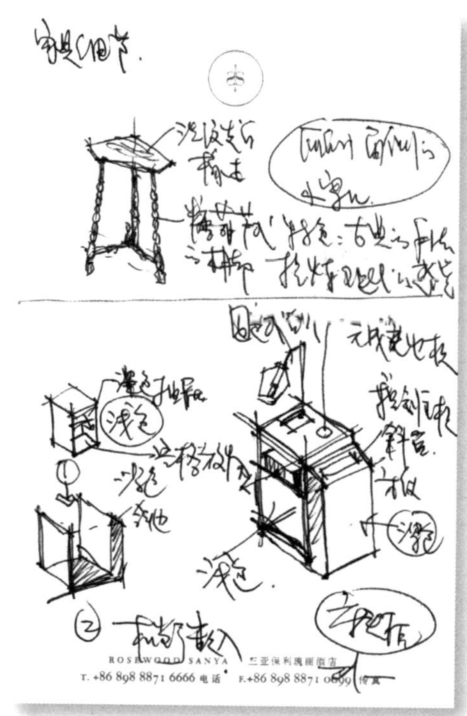
ROSEWOOD SANYA 三亚保利瑰丽酒店
T. +86 898 8871 6666 电话 F.+86 898 8871 0099 传真

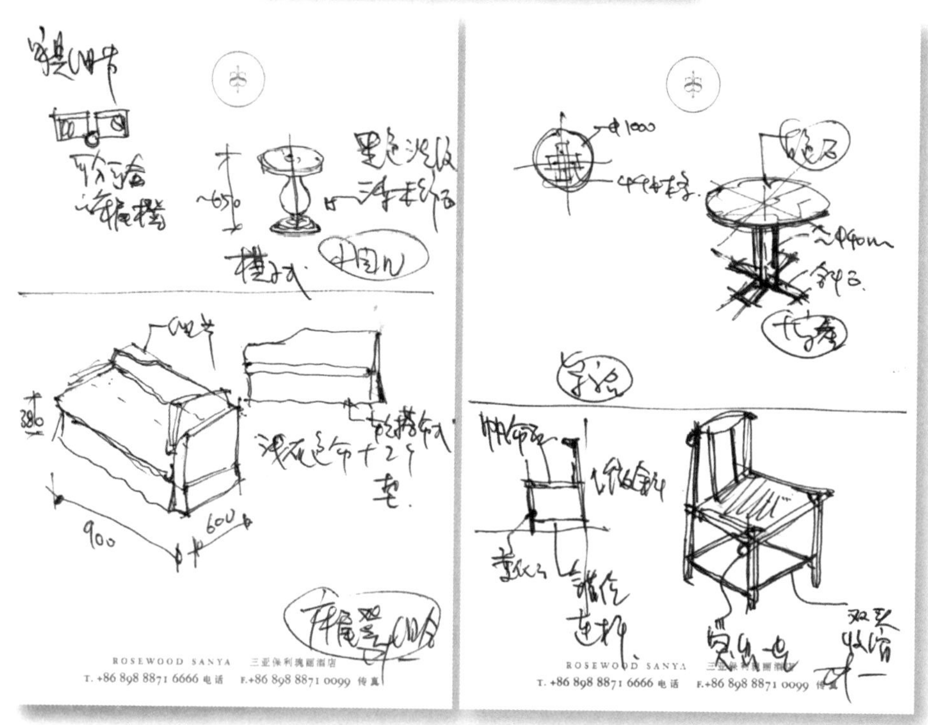
ROSEWOOD SANYA 三亚保利瑰丽酒店
T. +86 898 8871 6666 电话 F.+86 898 8871 0099 传真
900
600
ROSEWOOD SANYA 三亚保利瑰丽酒店
T. +86 898 8871 6666 电话 F.+86 898 8871 0099 传真

Nothing is Impossible Because of Large

大，有可能

May 17, 2020
2020年5月17日

传统的滨海度假式酒店的布局：单边走廊（可以看日落，有一个有躺椅的区域）。当然入口也颇具热带风情,门头有一个小布篷，让人迫不及待地想入内看个究竟。

The traditional resort hotels often had a one-side corridor (you could enjoy the sunset on the other side, and also had an area with deck chairs. Of course the entrance was quite tropical style. A small cloth curtain made people can't wait to see what was inside.

哔！应当是1.5的开间，而非一般的平面，精雕细琢的布局：彻底的分离式的洗手间——小的精致的如厕区域，"反人类式"的背门设置洗手盆，不知道是什么道理，试图去解读解读；衣帽功能套入洗手区的区域（居然我们也做了这样的方案：清水湾的威珀斯酒店平面的研究阶段）。

Wow! The guestroom should be 1.5 bays, not a common plane but an extremely unique arrangement: completely separated washroom— a small exquisite toilet area. I tried to understand why the washbasin was behind the door. It was against humanity.The cloakroom was put in the washroom area. (We made the same project: the study of the plane of Wilpers Hotel in Sanya Clearwater Bay.)

三亚保利瑰丽酒店
Rosewood, Sanya,China

三亚保利瑰丽酒店
Rosewood Sanya,China

顺应靠阳台的再一次独立出来的双人泡/浴缸（只是泡水而已）要求，淋浴间选用独特的双向开门设计（比较少见，实际测试用过，防水也不错，设计师在选用门的方式上颇有心思），结合3米多深的大露台，休闲沙发区可以聊聊天，喝喝茶，当然也可以看着另一半在水中（我们两个老男人就算了，只是测试测试放水。惨，半个多小时才放完一缸水，欠计时提醒），简约大方的玻璃栏杆，让视野透澈，投入真的不菲，好效果!

According to requirements of the independent bath tub near the balcony, the special door that could swing in either direction was used in the shower. (A rare design, I tried and found the waterproof was quite good. It was very thoughtful of the designer to choose the door.) The guests could chat and drink tea on the leisure sofa as well as the three-meter-deep balcony. Of course you could see your roommate take a tub bath. (Forget it, as for us, two old men. We just tested the water. So awful! It took nearly half an hour to fill the tub, no remind timer.) The simple and generous glass balustrades made the view clearer. The investment was so much that the effect was perfect.

回到房间内，也得人喜欢，进入这个区域，先看到配有半高的整体书柜/陈列柜的写字/用餐台，两把椅子不同，也方便移动，特别喜欢粗布的扶手椅。一盏壁灯“锁定”了一个温馨的区间（域）。

We also felt pleased when going back to the room. Entering this area, you would see a half-height bookcase/ a display cabinet/ a writing table / a dining table. Two different chairs were easy to move. I especially liked the coarse armchair. A wall lamp made a cozy area.

可能是大床房改双床房的吧，主幅有冲突，特别是右侧床头柜上的吊灯多次碰到我们老帅哥陈杨的头，1150的床，还凑合。

Maybe a king room was converted into a double room. The main decorations on the wall didn't match; especially the ceiling lamp above the right night table hit our handsome man, Chen Yang's head many times. The 1.15-meter-wide bed was not too bad.

另一侧的茶水柜，有细节（材质，内抽屉带绿色皮革，奢华的再一次体现），还有我喜欢的双床尾凳的设置，方便。

The tea table on the other side was detailed. (The material inside the drawers was green leather which showed luxury again.)I also liked the double bed stools, it was convenient.

回头一看，大，让设计（师）有了无限的可能！

Looked around, it was the big size that made it infinite possibility for the designers.

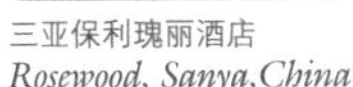

三亚保利瑰丽酒店
Rosewood, Sanya,China

108 三亚嘉佩乐度假酒店

★★★★★

CAPELLA SANYA CHINA

Address : Tufu Resort Area,Sanya, Hainan, China
海南省三亚市海棠湾
镇十福湾度假区

Telephone : +0898 830 999 99

Http : //www.capellahotels.com

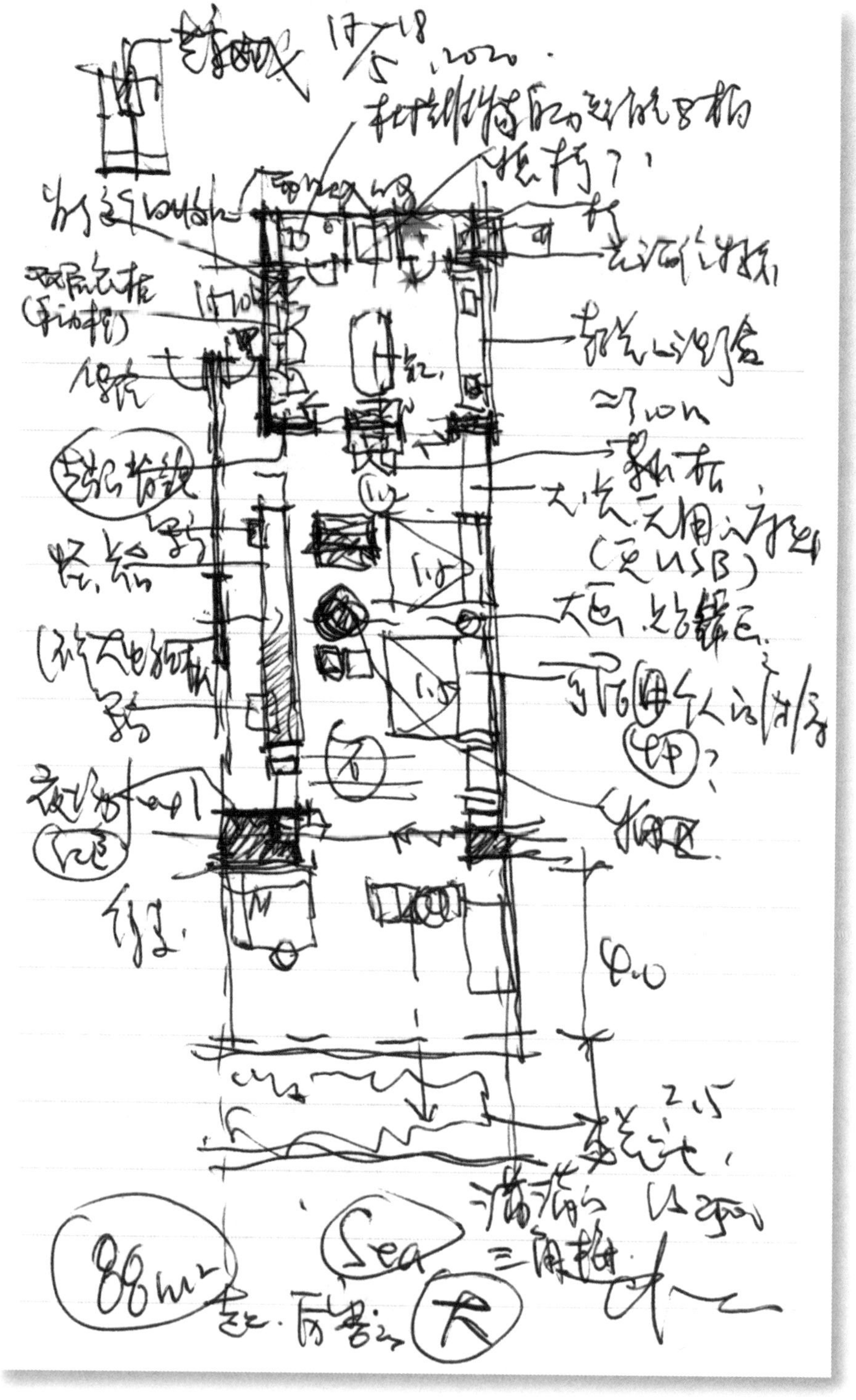

太大了，Villa的浴池。

入住过喜来登酒店，同样感到也[illegible]这些同时的[illegible]，又是浴池。

看过这个酒店的客房是木地板，88m²
心里想，就那几种材料，[illegible]"样
这么多到[illegible]一百板。

入住世纪上一次的[illegible]酒店是不了三个房的一体间的浴缸好，就酒是酒店生意兴隆吧（也可能才开张）

带来好奇，向服务员带路，去过长长、长长的门廊的走廊，然后还[illegible]意，双床套房的[illegible]浴池。（疑问一）五楼、开门，咔，还有到了大大的客厅，第一次住这个[illegible]的双人套房，更加有一个天花大的两边的大阳台。

[illegible]有一个大大的[illegible]区和一个大大的[illegible]区；[illegible]浴池[illegible]用这个开放、[illegible]的；

[illegible]也[illegible]。[illegible]别具一格，"不[illegible]"。但是[illegible]两个[illegible]这个88m²的[illegible]能不到合适[illegible]。（疑问二）

也是一个[illegible]，[illegible]

长长的[illegible]，又[illegible]，[illegible]也是[illegible]浴池，[illegible]也[illegible]，[illegible]一[illegible]一下[illegible]是否[illegible]浴缸[illegible]有问题呢？

[illegible]了。

[illegible]了的也有，[illegible]（像[illegible]酒店）[illegible]大，[illegible]也[illegible]两双房[illegible]一个支撑点（[illegible]）（[illegible]（4P）

回到[illegible]"[illegible]"[illegible]（[illegible]）"[illegible]一样[illegible]，不[illegible]（[illegible]）[illegible]一[illegible]大[illegible]一[illegible]一[illegible]了一个这个客房的浴池（[illegible]了）

浴池也[illegible]。[illegible]但[illegible]一[illegible]，还是[illegible]，[illegible]会[illegible]一样。

① [illegible]，[illegible]一[illegible]。

② [illegible]双[illegible]，[illegible]，甚至[illegible]浴池[illegible]也[illegible]加[illegible]一个"[illegible]"[illegible]。（也[illegible]大）

③ [illegible]还是[illegible]，[illegible]会[illegible]。

④ [illegible]还[illegible]，[illegible]一[illegible]，[illegible]"[illegible]一[illegible]大[illegible]。[illegible]

How to Deal with a Large Room?

房大烧脑

May 17-18, 2020
2020年5月17、18日

入住三亚嘉佩乐酒店，网红打卡地，当然是正圆形的落水中庭，观星池。

Capella Sanya was the Internet celebrity hotel because of its perfect circle water atrium: a star observation pool.

知道这个酒店的客房最小面积：88平方米，心里想，就那几样东西，能怎么“花”掉这么“刺激”的面积。

The smallest guestroom in this hotel was eighty-eight square meters. I thought how to arrange the room in such a big place.

入住也和上一次的瑰丽酒店差不多，三间房间一间一间地清洁好，就当是酒店生意兴隆吧（四五点才能入住。）

It was similar to Rosewood I stayed in last time. Three guestrooms were cleaned one by one because of the good business, so we didn't enter the room until around four or five in the afternoon.

带着好奇，由服务员带路，走过长长的、长长的阳光走廊，感觉还挺惬意。双房共享前厅式设计（疑问一）。五层，开门，哗！终于看到了大大的客房：第一次住这个简单又复杂的双人床房。有一个无比大的面海的大阳台。简单是指只有一个大大的睡眠区和一个大大的洗手间区；复杂是指设计师挖空心思地想怎么用这个大开间、进深、层高，色调颇深，也颇多色彩，有一点像该设计师之前设计的一家夜总会，平面布局也别具一格，“不明觉厉”。但是我们两个老帅哥在这个88平方米的房间里面还是找不到适合放小行李的地方（疑问二）。一体化的洗面梳妆区，3米长，独立浴缸，四周可以跳舞。长长的衣柜，双层，手动杆挂衣，可长期旅居，也是花钱的好设计，厕、淋浴用红色夹丝玻璃，香艳，中间夹有一沙发，可小憩一下，看看电视，是否防止泡浴缸时心脏有问题呢？

With a curiosity, I was led to the guestroom by the staff. It was also a pleasure to go through a long corridor full of sunshine. The antechamber was shared by two guestrooms(Question 1), on the fifth floor. Wow! I saw the large guestroom at last. It was my first time to stay in such a simple and complex double room with a very big balcony facing the sea. Simpleness referred to only a big sleeping area and also a big washroom; while complication meant the designers thought hard to arrange such this room which was so wide, deep and tall. It was not only dark but also colorful, a little similar to the club which was this designer's work before. The plane was quite unique. Although I didn't understand all, I still felt it impressive. We two old handsome guys couldn't find a suitable place to put our luggages in this 88-square-meter guestroom(Question 2).The 3-meter-long wash basin was integrated and there was also a separated bathtub. It was wide enough to dance around it. The long double wardrobe had hand levers to hang clothes. It was suitable to stay for long, surely an expensive design. The red wired glass in the washroom looked gaudy. We could sit on the sofa in the middle to have a rest or watch TV. Did it prevent heart attack while taking a tub bath?

三亚嘉佩乐度假酒店
Capella Sanya,China

想多了。

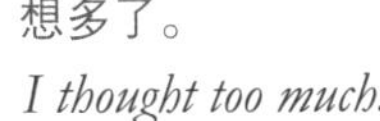

I thought too much.

想多了的也有，衣柜内设计有“脏衣服”回收口（像半岛酒店），我在想用的概率有多大，久住的房客应该会喜欢，也许这个也成为共享双房前厅的一个支持点（有保留）。（我们想到的是共享4个人）

There was something useless. For example, a special design for collecting clothes to be washed was hidden inside the wardrobe (like the Peninsula Hong Kong). I thought how often it was used. The guests who stayed long would like it. Maybe it was the reason why the antechamber was shared by two guestrooms (I disagreed). (What we thought was to share with four people.)

回到主区，最有争议的是长长的“劏猪台”（广州话）式的台，“外父”一样横躺着，不宽（没有什么用），当然，两张一米半的大床舒服，前面有一组沙发，让我们八个同事一起调侃了一会儿这个客房的设计（葡萄酸了）。

Came back to the main area, most contentious was the very long table which was like the one for killing pigs. It was lying in the way like the father in law, not wide and useless. Of course two 1.5-meter beds were comfortable. There was a set of sofa in the front. We six colleagues sat together to discuss the design of this guestroom. (The grapes were sour.)

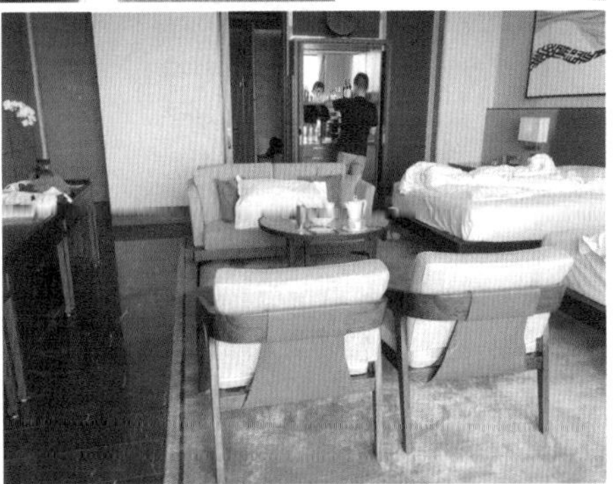

O.'s File **(区生词典)**

劏猪台是杀猪的案板。

It is Cantonese.It is a table for killing pigs.

外父是岳父的意思。（百度知道）

It is Cantonese.It means father-in-law.

三亚嘉佩乐度假酒店
Capella, Sanya, China

设计也很丰富，细节颇多，但施工一般般，还是问自己：我来做，会怎么样?

1.我会控制共享前区的尺度，适合就可以，让入房间的走道长一点点。

2.长台好像会因洗手间的优化而改变，如果针对（兼顾）一家四口的双孩家庭，那厕浴分离，甚至多分区式的洗手间，也许是更加好的一个"耗"面积的方法（也许酒管公司就喜欢这样大）。

3.洗手间还是洗手间，靠床开门（滑门）会严重相互影响。

4.智能控制开关的方式还好，易用。

一间真正的"面朝大海，春暖花开"（阳台），简单、粗暴的一间大房，欠一点点深度。

The design was rich and detailed, but the construction was poor. I asked myself: What if I designed it?

1.I would control the right size of the antechamber to make the aisle a little longer.

2.The long table seemed to change as the washroom was optimized. For a 4-people family with two children, it was a better design to divide the washroom into different zones,such as the separated toilet and bathroom. (Perhaps the management company liked such a big room.)

3.The washroom was still washroom, but the sliding door next to the bed badly affected each other.

4.The intelligent control system was easy to use.

It was a really room towards the sea, with spring blossoms, but too simple and rude, lacked profundity.

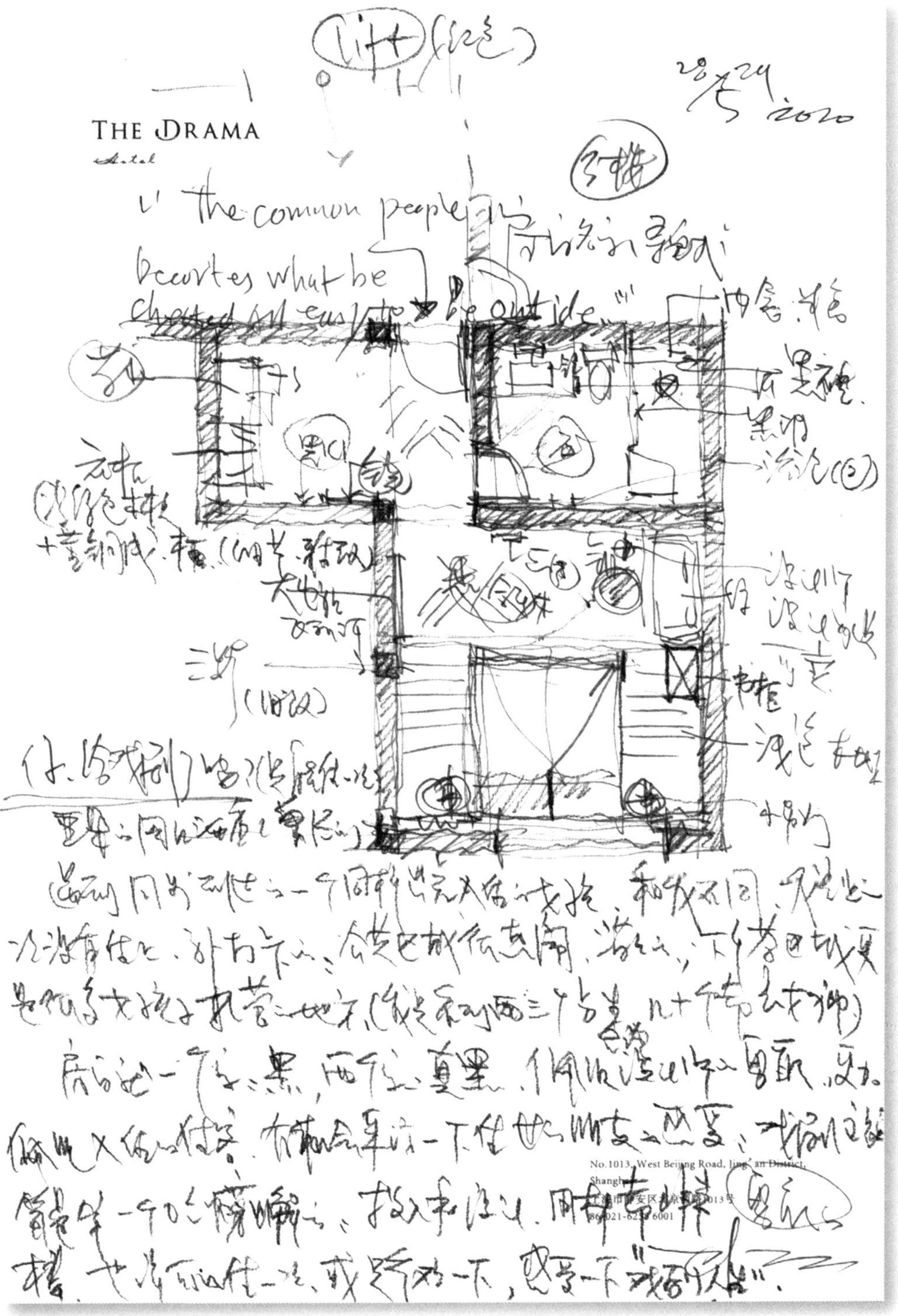

THE DRAMA
Hotel
上海戏剧主题酒店

109

★★★★★

THE DRAMA HOTEL SHANGHAI CHINA

Address : No.1013, West Beijing Road, Jing'an District, Shanghai,China
中国上海市静安区
北京西路1013号尚演谷南楼
Telephone : +86 21 6258 6001
Http : //www.yaduo.com

Do You Feel Like in the Opera? (Only Once)

你，给戏剧了吗？（只能住一次）

May 28-29,2020

2020年5月28、29日

亚朵的网红酒店（曾经的）。

This Internet celebrity hotel belonged to Atour.

遇到同步到达的一个同样出差入住的女孩，和我不同，我是上一次没有住上，补打卡的。公共区域很热闹，满满的；下午茶区域更是很多女孩子“扎营”的地方（我只看到两三个男生，几十个都是女神）。

I met a girl who was on business. We arrived at the same time. She was different from me. I couldn't stay last time so I came again. The public area was a blast and very crowded. Lots of girls came here for afternoon tea. (I saw only a few boys while dozens of pretty ladies.)

房间就一个字：黑，两个字：真黑。佩服室内设计师的勇敢，更加佩服入住的住客。有机会采访一下住过的朋友的感受。戏剧主题算是第一个吃螃蟹的：投入和设计、用材都非常的勇敢。棒，也是可以住一晚或只参观一下，感受一下“戏剧人生”。

One word to describe the guestroom was dark. Two words were very dark. I admired that the interior designer was brave, so were the guests. I hoped to have a chance to ask the people who once stayed here what their feeling was.It was the first Drama Theme Hotel. Its investment, design and material were all brave. Amazing! You could stay for one night or even just visit it to enjoy the drama of life.

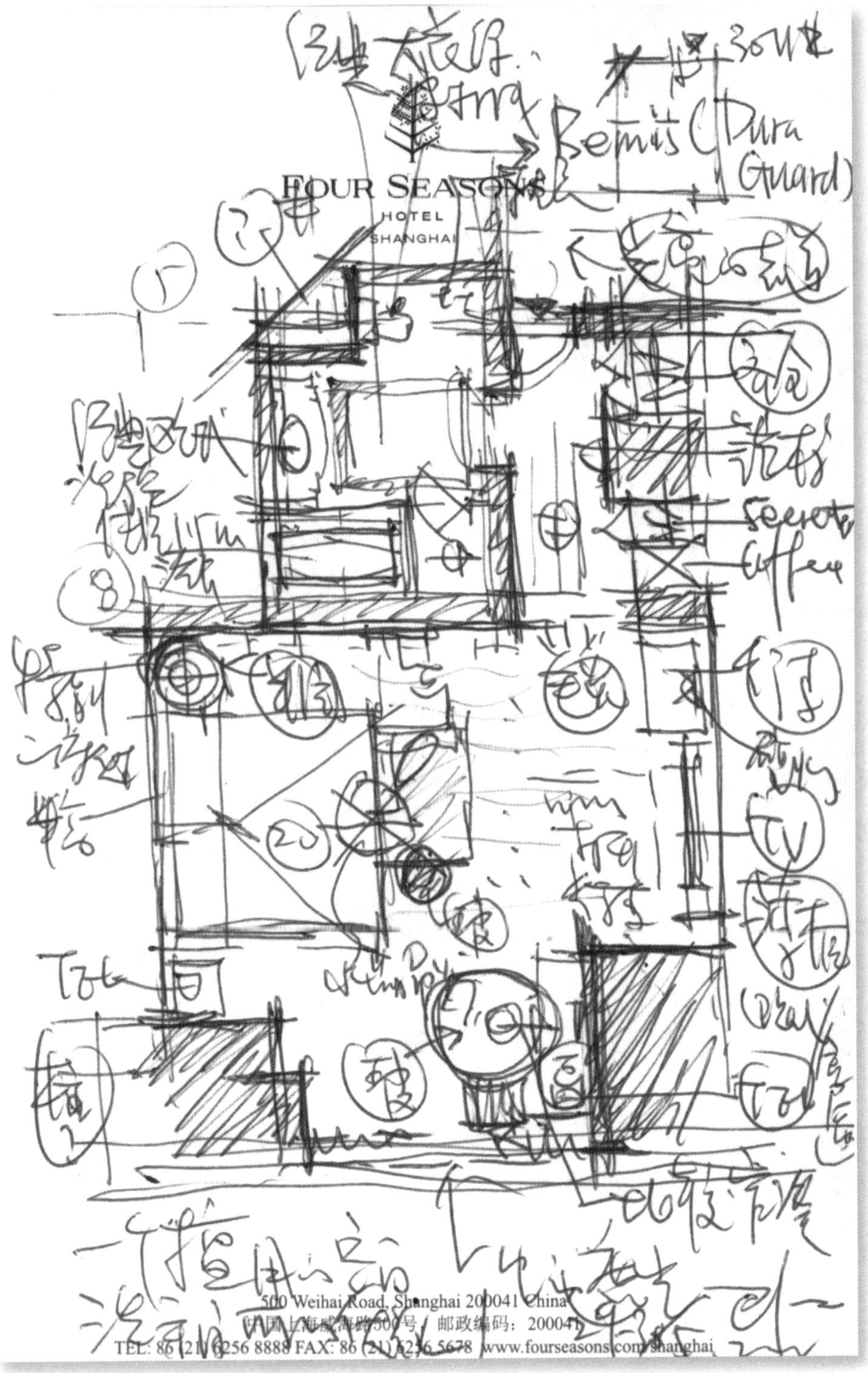

FOUR SEASONS
HOTEL
SHANGHAI

500 Weihai Road, Shanghai 200041 China
中国上海威海路500号，邮政编码：200041
TEL: 86 (21) 6256 8888 FAX: 86 (21) 6256 5678 www.fourseasons.com/shanghai

上海四季酒店

★★★★★★

110

FOUR SEASONS HOTEL SHANGHAI CHINA

Address : 500 Weihai Road, Shanghai, China
中国上海威海路500号

Telephone : +86 21 6256 8888

Http : //www.fourseasons.com/shanghai

FOUR SEASONS
HOTEL
SHANGHAI

500 Weihai Road, Shanghai 200041 China
中国上海威海路500号，邮政编码：200041
TEL: 86 (21) 6256 8888 FAX: 86 (21) 6256 5678 www.fourseasons.com/shanghai

FOUR SEASONS
HOTEL
SHANGHAI

500 Weihai Road, Shanghai 200041 China
中国上海威海路500号，邮政编码：200041
TEL: 86 (21) 6256 8888 FAX: 86 (21) 6256 5678 www.fourseasons.com/shanghai

FOUR SEASONS
HOTEL
SHANGHAI

500 Weihai Road, Shanghai 200041 China
中国上海威海路500号，邮政编码：200041
TEL: 86 (21) 6256 8888 FAX: 86 (21) 6256 5678 www.fourseasons.com/shanghai

FOUR SEASONS
HOTEL
SHANGHAI

500 Weihai Road, Shanghai 200041 China
中国上海威海路500号，邮政编码：200041
TEL: 86 (21) 6256 8888 FAX: 86 (21) 6256 5678 www.fourseasons.com/shanghai

Do Hotels still Need to be cozy?

酒店还要"温馨"吗?

May 30, 2020

2020年5月30日

确实在上海找不到什么值得住的酒店了（哈哈，一家之言），就选了这一家老四季酒店住了，不便宜。当然与新的豪华酒店比，非常合适（嘉佩乐几千，宝格丽好几千……）。同事告诉我可能准备摘牌了（疫情？），也算相当幸运的一次。

I could hardly find any hotels which were worth staying in. (Haha, it was my opinion.) So I chose this expensive old Four Season. However, it was quite suitable, compared with new deluxe hotels(Capella costs several thousand yuan, Bulgari Hotel costs more…).My colleagues told me that it might be delisted because of the CDVID-19. I was so lucky to stay here this time.

老克拉的米黄，温馨的调子，我几乎没有听过身边的人住过，更加别说设计师了。一种让当下设计人"鄙视"的调调。入住，上电梯，走道入房间，一切一切都很老式、老派，倒是让我思考起来：

酒店还需要温馨吗?

Beige color and comfort were like the style of Old carat. I almost never heard that people around me stayed in it, especially designers. The old style was despised by designers today. After checking in, I took a lift and walked to my room. Everything was out of fashion and old. It made me think over:

Do hotels still need to be cozy?

货真价实是我对四季酒店的房间的评价。尽端的房间（面积不小，应是近年翻新），设计合适、合理而尽有细节和关怀，分设的衣柜，变化的色调，包括洗手间外走廊的木地板选搭，洗手间宽敞、明亮（可能是管井的限制）。感觉布局还可以更巧妙些，单盆、马桶使用时感觉一般（特别的品牌Bemis，第一次见到），完整的四件套；灯猛；试着泡泡浴缸，很快放满水，去水更是迅速，热水秒到，佩服。品牌丰富，欧舒丹，适宜，体验感一流。

上海四季酒店

Four Seasons Hotel, Shanghai,China

Bang for the buck was my comment of Four Season. The design of the guestroom at the end was reasonable and detailed as well as considerate. (It should be renovated recent years so it was not small.) The wardrobe was separated, the changeable colors including the wood floor outside the washroom matched perfect. The washroom was both wide and bright .
(Maybe it was restricted by the tub well.)I thought the layout should be more ingenious. Single washbasin and toilet were so so(a special brand Bemis, my first time to see it). There was a complete set of four pieces in the bathroom and the lights were quite bright. I tried to take a tub bath. It was filled with water and emptied quickly. The hot water came fast. I admired. The toiletries were L'Occitane and felt comfortable.

睡眠区有很大的一个打开行李箱的位置，虽然因为建筑的凹凸多（老结构的制约），难以像新酒店的工整，更考验室内设计师的功力（好像没有比这个更好的布局了！）

There was a big place to open the luggage in the sleeping area. Due to the rough surface of the old building, designers needed to work even harder than to work on a brand new hotel, which was neat and order. (It seemed no other layout was better than it.)

倒是睡眠前的“关灯仪式”有一点不顺，是单边床头柜控制，要“爬”到另外一边关掉所有的灯，双边不同的床头柜也许是追求“创新”，但体验感一般。

However, turning off the lights was not easy before sleeping because the control was on one side of the bed. I had to climb over the bed to turn off all the lights. The different night tables on both sides might be an innovation. But it was not practical.

我说过“老（旧）是种嫌弃”，但它在不被你嫌弃之前也许已经“逝去”，也许多住就可以了。

I once said, "Being old is a kind of dislike", but it might "pass away" before you disliked it. Staying more should be OK.

愿酒店给到你一点点温馨。

Hope you feel a little cozy in the hotel.

四季有的，幸运的我。

I was so lucky that I felt cozy in Four Season.

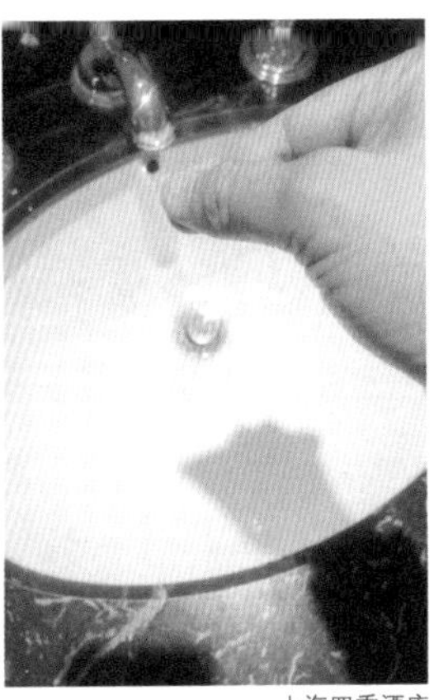

上海四季酒店

Four Seasons Hotel, Shanghai,China

跋

炭笔、油画笔（绘笔）　▲ 81886968　12/2018 最后天

我的绘画从炭笔、油画笔 这一[illegible]

[illegible]、[illegible]，[illegible]一[illegible]笔，[illegible]

[illegible]，笔[illegible]，[illegible]

[illegible]，[illegible]笔[illegible]，[illegible]

[illegible]

也许[illegible]一[illegible]，[illegible]笔[illegible]

[illegible]一[illegible]，[illegible]

[illegible]，[illegible]也[illegible]

[illegible]，这个世[illegible]一种[illegible]"获取"，

[illegible]笔，[illegible]，[illegible]

[illegible]，[illegible]

[illegible]，[illegible]笔们[illegible]

笔，万岁！　//[illegible]　[illegible] 2018

About the Pens in Hotels (Needless Pens)

关于笔，酒店的笔（多余的笔）

写了三次关于酒店的笔。这一次可以怎么再写这支“笔”呢?

I have written about pens in hotels three times. How to write pens this time?

拆、分类；努力地在日常的工作中使用每一支收回来的笔，但还是用不及的。

Take apart and classify the pens.I always try to use every pen in my daily work that I collect in hotels, but they are never used up.

这世界上，笔越来越显得多余，真的吗?

Pens are becoming needless in the world, aren't they?

房间内，偶然遇到笔是全新的，我猜。因为笔尖还是有封头的！原封不动的！

Occasionally I saw some new pens in guestrooms, I guess. Because the tops were unopened. They remain very much intact!

也许酒店可以在信纸上画一点东西，让到访的住客看到笔、画，才能吊起他们的胃口，留下一点点房间的纸、笔的痕迹，微薄的力量和积累，也会有不同趣味，大数据也是要有基础和来龙去脉的，这个也是另一种有趣的“获取”！

Perhaps something should be drawn on paper in hotels. Guests may develop an appetite when they see pens and pictures. And they will use paper and pens to make some marks to accumulate by meager power. Basing on different interests, Big data needs a basis and a context. It is another interesting gain.

房间的笔，不用就显多余，让它有更多的存在意义和实用性，留下印记，那才是最实际的问题。

Pens in guestrooms are needless if nobody uses them. The most practical problem is to make the pens have a meaningful existence and practicability to leave marks.

期待你的使用，让笔有了存在的价值。

Look forward to your using pens to reflect their value of existence.

笔，多余!

Pens, needless!

单手操作的笔是最好的笔。

The best pen can be used by a single hand.

[illegible]：[illegible]

从开始写《你好吗？》，没有想到有这么多的反馈，所以没有《你好吗？》1，是一个特别的也是不可复现的 Bug。还是坚持再计划，决定了：

30年，七本，大约800篇左右的[illegible]，[illegible]记录，记录走过、吃过、[illegible]，所遇到的，所见所闻，[illegible]会[illegible]。

差不多，也就是一直地走走、看看；吃吃、喝喝；玩玩、睡睡……时间久了，也会忘记哪一天，哪一些事情，哪一伙人……所以每次的感动和笔记（手稿）和[illegible]，七本的《你好吗？》好记录[illegible]每一次的脚步、[illegible]、回忆、故事……

[illegible]；每一次的[illegible]，[illegible]；[illegible]，[illegible]，[illegible]；每一次的出版，[illegible]；[illegible]，[illegible]“[illegible]”[illegible]（[illegible]）中国~~[illegible]~~[illegible]出版社[illegible]了[illegible]感谢！（[illegible]！）

我们的《国脚》了，许多的封面也荣誉了吧！

也许路长，路力会阻挡我们，这一些书信伴随我们，我们

愿我们都能一路走来。

真诚

祝老师们，在行在成，大成事。

也许

青年很美。

可！可！可！

11/8 2022

Alzheimer, Perfect!

老年痴呆，正！

从开始写《住哪？》，没有想好有没有后续，所以没有《住哪？1》，是一个特别的，也是不可磨灭的"Bug"，还是坚持并计划，决定了。

From the beginning of writing Where to stay?,I didn't think of any follow-up, so no Where to stay?1at first. It became a special and indelible Bug. And I decide to insist and make a plan for it.

30年，十本，大约800间左右的五星级或相当级别的酒店的记录。记录足迹，吐槽或肯定，向酒店业学习，向同行学习，相信也会收集到越来越多的韦格斯·杨设计的酒店。

Lasting for 30 years, ten books will record about 800 five-star hotels or the same level, as well as my footprints, complaints or appreciations. Learn from the hotel industry and learn from other designers. I believe I will collect more and more hotels which are designed by GGC.

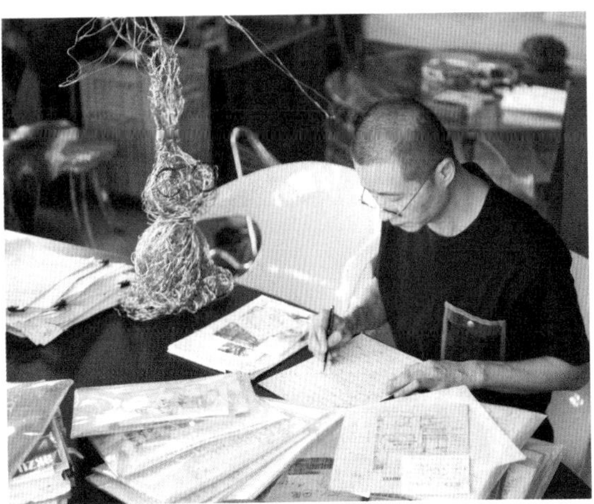

差旅，也就是一直走走，看看，吃吃，喝喝，玩玩，睡睡……时间久了，就会忘记是哪一天，哪一些事情，哪一伙人……以惊人的毅力和坚持（自诩），相信，十本的《住哪？》会让自己记住每一次的脚步，朋友，地点，故事……

Travel is just to walk and watch, to eat and drink, to play and sleep…As time goes by, I may forget when and where I have even been, what I have done, who I have travelled with…To make me remember my every step, friends, places and stories, I believe I will insist on writing 10 books with my surprising willpower and persistence.

有你就好，点点滴滴。每一次的整理工作，自己来；最痛苦的莫过于筛选照片了：住店的、行走的、主题的，等等等等；每一次的出版，整个周期约9个月；同事们、朋友们“听到就惊”（听到就害怕），中国建筑工业出版社大约3个月反反复复的校对。感谢！（据说都是一些长者把关，招牌啊，值得信赖！）《住哪？》干干净净的封面也获奖了啊！

It's wonderful to be with you, little by little.I must collect and collate the materials myself every time. The most painful thing is to choose the photos: the photos of hotels, travels, different subjects and so on. It takes about nine months to publish a book. My friends and colleges feel nervous while hearing it. It also takes three months to proof read the books repeatedly by the editors of China Architecture & Building Press. (It is said that some senior editors make checks strictly, they are trustworthy!)The pure simple cover of Where to stay?won the prize.

也许再过十年，记忆力会渐渐减弱，这些书能让我、我们看到青春一路走来。
真的是，
渐老的，有你，有我，有《住哪？》。
也许，
老年痴呆，
正！正！正！

Perhaps in another ten years, my memory will get weak gradually. These books can remind us, you and me of our youth.

It's true that those who get older and older are you and me, as well as Where to stay?

Maybe Alzheimer is so perfect!